Springer Series on Atomic, Optical, and Plasma Physics

Volume 117

The Springer Series on Atomic, Optical, and Plasma Physics covers in a comprehensive manner theory and experiment in the entire field of atoms and molecules and their interaction with electromagnetic radiation. Books in the series provide a rich source of new ideas and techniques with wide applications in fields such as chemistry, materials science, astrophysics, surface science, plasma technology, advanced optics, aeronomy, and engineering. Laser physics is a particular connecting theme that has provided much of the continuing impetus for new developments in the field, such as quantum computation and Bose-Einstein condensation. The purpose of the series is to cover the gap between standard undergraduate textbooks and the research literature with emphasis on the fundamental ideas, methods, techniques, and results in the field.

More information about this series at http://www.springer.com/series/411

Miron Ya. Amusia · Larissa V. Chernysheva

Computation of Atomic and Molecular Processes

Introducing the ATOM-M Software Suite

Springer

Miron Ya. Amusia
Racah Institute of Physics
Hebrew University
Jerusalem, Israel

Larissa V. Chernysheva
Ioffe Institute
St. Petersburg, Russia

ISSN 1615-5653 ISSN 2197-6791 (electronic)
Springer Series on Atomic, Optical, and Plasma Physics
ISBN 978-3-030-85142-2 ISBN 978-3-030-85143-9 (eBook)
https://doi.org/10.1007/978-3-030-85143-9

Translation from the Russian language edition: *ATOM_M, Алгоритмы и программы для исследования атомных и молекулярных процессов* by Miron Ya. Amusia, and Larissa V. Chernysheva, © Санкт-Петербург «Наука» 2016. Published by Санкт-Петербург «Наука». All Rights Reserved.

This Springer imprint is published by the registered company Springer Nature Switzerland AG
The registered company address is: Gewerbestrasse 11, 6330 Cham, Switzerland

Preface

This book is devoted to numerical methods for studying the structure of the electron shells of atoms and diatomic molecules and their interaction with various particles—electrons, positrons, photons, mesons.

There are two reasons why the problems considered here are of significant interest to scientists and engineers. First, atoms are the main element of various natural formations, from huge distant stars to the smallest biological and chemical objects—molecules. The photons emitted by excited atoms and molecules are a source of very important, if not unique data that carry information about the structure of various objects, as well as the state and structure of the substance that forms them. The spectrum of atomic radiation makes it possible to determine, for example, whether a given atom included in the substance under study is neutral or ionized and how many electrons are removed from it or added to it during the formation of a chemical compound. The spectra and probabilities of photoionization (i.e., the removal of an electron from an atom by a photon) are not even a business card, but fingerprints that allow us to unambiguously identify one or another type of atom or ion in a polyatomic molecule or in a more complex object under study.

Second, the interaction of fluxes of different particles of light, i.e., photons with atoms and simplest molecules, is a widespread natural process. Our entire planet, its atmosphere, is constantly bombarded by streams of cosmic rays, irradiated by sunlight. And, it is very important from the point of view of life itself to know how, and with what probability, this radiation interacts with atoms and molecules of the atmosphere, to what extent the radiation passes through almost unchanged, and to what extent it leads to excitation or ionization of atoms or molecules. Sometimes, knowing these features is crucial. For example, the formation of a heat balance on the Earth's surface, the ability to establish the reasons for changes in its temperature, if any, is impossible without detailed knowledge of the cross sections for the absorption of photons by nitrogen, oxygen, and carbon atoms and molecules, primarily carbon dioxide, of which they are constructed.

There is also great interest in the study of electron interaction processes for studying the structure of atoms (ions), as well as molecules and solids. Recall that the flow of fast electrons interacting with a specially selected solid is the physical basis of an X-ray tube, a kinescope, and a number of other modern electronic devices. Laser technology also significantly demands the knowledge of the structure of atoms and their interaction with electromagnetic radiation. Thoroughly investigating the electronic structure of many atoms and simple molecules is also impelled by the fact that often the electron shells are a difficult to remove and by no means always a useful factor that interferes with our study of the atomic very core—its nucleus. So, in the study of electron scattering, wishing to obtain information about the nucleus with a high degree of accuracy, it is important to know what is introduced by atomic electrons and what is introduced by the nucleus itself.

Relatively recently, a new reason has arisen for a detailed study of the atomic structure. It turned out that the forces of interaction between electrons, as well as the forces of interaction of electrons with the nucleus, have, along with an electromagnetic, an additive of different nature. Study of this small parity-violating additive that violates the equivalence between "left" and "right" in our world requires a very accurate account of much stronger effects of an electromagnetic nature.

The complexity of the object of research—the electron shell of atoms—makes it necessary to use in their study the capabilities of modern computers, which naturally implies the creation of appropriate numerical models. The simplest molecules are of interest, since their study, in comparison with their constituent atoms, shows the mutual influence of electrons, mainly under the influence of different centers of attraction.

This book presents numerical methods for solving a wide range of problems in atomic physics associated with both the structure of the atom and the simplest molecules, and with their interaction with the electromagnetic field, electrons, and other particles. The ATOM-M software package discussed in detail here is a software system for automating atomic and molecular calculations based on the achievements of the well-known ATOM system and developing it. The ATOM system was the first to consistently implement numerical methods in the physics of atomic processes and has been widely used for more than 50 years for theoretical research in atomic physics and related fields of science. The programs in the system were written in the ALGOL language, then in FORTRAN, which are widely used in many countries. Therefore, the programs included in the ATOM-M system can be easily transferred from one computer system to another.

Since the publication of the book Miron Ya. Amusia and Larissa V. Chernysheva, *Computation of atomic processes* (IOP Publishing Ltd, Bristol and Philadelphia, 1997), a considerable time has passed. As a result of the development of computer technology in general and the widespread use of personal computers in particular, it became necessary to revise and radically modernize a number of existing programs and create new ones. Methods for calculating diatomic molecules have been developed, which required the creation of a number of programs based largely on the ATOM system, and in many respects being its development. All this led to the need to write this book. Its continuity in relation to its predecessor will be reflected in references to it, as [AC].

As the title of this book suggests, we will focus in it not on obtaining physical results, but on developing algorithms and programs that allow us to study a very wide range of processes of interaction of atoms and diatomic molecules with photons, electrons, positrons, and mesons. Therefore, although physics plays a fundamentally important role in understanding the structure and properties of atoms and molecules, in this book, it is subordinate and acts as a source of various approximations used and the corresponding equations. The justification for these approximations is outside the scope of this book.

Our goal as authors is to teach the reader to calculate what he wants from the variety of processes that are discussed in the book. We hope to help also in the case when the process is not directly discussed, but can be considered in the same way as we have done in other cases.

We offer the user of our programs a number of approximations, describing qualitatively their foundations, advantages, and difficulties, but we leave it to the reader to choose what and how to do. In this sense, this book is a user's guide and not a detailed description of various problems of atomic and molecular physics. It is not our task to discuss and show by what and why one approach is better than another in describing many processes: the Hartree–Fock method, perturbation theory, the random phases approximation with exchange, or something else. However, we do not hide our preferences and give reasons in their favor.

All necessary references for understanding the physical approaches used are presented in the book. However, this book is by no means a textbook on the physics of atoms and molecules and processes with their participation. This book aims to show how to apply numerical methods in order to find with good accuracy many different characteristics of atoms, molecules, and processes with their participation. To a certain extent, we believe that the reader himself knows what he really needs: from textbooks, personal experience, recommendations of colleagues, scientific advisor, recommendation from a manager, etc.

As in our previous book [AC], the goal is to create a unified set of computational programs that allow the researcher himself, even on a personal computer, to calculate a large number of characteristics of atomic and molecular processes. A distinctive feature of our approach is the creation of a single complex of programs that allows for sequential expansion and development as tasks become more complex and computational capabilities increase. For example, it is already easy to foresee that not only spherically symmetric and axially symmetric atoms and molecules, but also arbitrary three-dimensional molecules will soon become objects of study using computational programs similar to those described in this book.

Interest in computations is great, and creating a separate program for one specific, even relatively simple task is not just a difficult and time-consuming job for a non-specialist in programming. Such a "handicraft" program, inevitably using the results obtained with the help of programs made by others, can and often lead to a large computational error due to two or more accounting for the same effects. That is why it is advisable to use the ATOM-M complex, which contains the entire set of programs necessary up to a certain level. Further complication and expansion should be carried out on the basis of this complex.

There is often a belief among physicists, which we consider a delusion, that every theoretician can and should create the programs he needs. At the same time, they proceed from the assumption that the programming technique is simple and quite accessible to every theoretician. As a result, apart from our book [AC], which is already outdated, there are no more books devoted specifically to calculations. And the need for them, as the fate of [AC] has shown, is great.

Let us note a significant difference in the methods of computational and theoretical physics. The latter are based, as a rule, on the presence of a *small parameter* that permits to present the characteristics of interest in the form of a series in the powers of this parameter. However, in most cases, in both atomic and molecular physics, there is no clearly visible parameter. Therefore, it is impossible to justify the application of this or that physical approximation *a priori*. The use of machine computation on a large scale makes it possible to test the applicability of the physical method *a posteriori* demonstrating good agreement, if any, between the calculated results and experimental data, and sometimes even predicting them.

For the experimentator, as a rule, it is desirable to have not too crude calculated results already before the experiment in order to simplify and refine its planning. So, the need for the calculation of a given characteristic or process is constantly increasing, as well as the requirement for the accuracy of such a calculation. As a rule, this scientist is interested in one or more specific processes. That is why we tried to make the different parts of the book as independent as possible. It is understood that the reader of the book will not necessarily study it all, from A to Z, but will immediately choose what he needs.

We believe, relying, among other things, on our own vast experience, that there should be a clear division of duties between the theorist and the programmer. The task of a theoretical physicist is to select an object of research, a theoretical model of its description, and an assessment of the likelihood of the calculated result. The task of the programmer is to find the available and to develop new algorithms and programs necessary for solving the problem. Of course, the programmer, who often leads several theoretical problems at the same time, closely interacts with the physicists who set these problems. This approach has made it possible to conduct many hundreds of studies that have a good reputation among specialists all over the world. In this book, some of the results obtained will be presented as an illustration of the developed methods.

The interest in the algorithms and programs discussed in this book is determined by the fact that the considered characteristics of atomic and molecular processes can in many cases be calculated with very high accuracy and reliability. At the same time, this book is not a textbook either on the physics of the processes under consideration or on the programming of mathematical formulas for calculating the characteristics of these processes. Therefore, the theoretical description of each process considered in the book is brief and includes only the formulation of the main ideas underlying the methods used, as well as key formulas, which then become the object for the development of computational programs. As for the programs, only the main ideas of their creation and the formation of the input database are given. The Appendix to this book contains proper reference to Springer database that direct the reader to working versions of all programs and test cases, as well as gives explanations for entering the initial data for the calculation.

We assume that this book will be useful to very many people, from qualified specialists to students, theorists, and experimentators, to all those who are interested in atomic and molecular calculations, as well as in other areas, for example nuclear physics, where the methods developed in this book can be applied.

Entirely, we will present in this book a lot of computational programs for various physical processes, almost all associated with the interaction of atoms and ions with photons, electrons, and positrons and also consider the probabilities of decay of atomic vacancies. For most of these processes, we present the necessary computational programs that make it possible to carry out calculations in the one-electron Hartree–Fock (HF) approximation and taking into account many-electron correlations in the framework of the random-phase approximation with exchange (RPAE) or many-body perturbation theory (MBPT).

As reference books on quantum mechanics and electrodynamics in describing the structure of atoms and processes with their participation, we will use the following two fundamental ageless guides—books: L. D. Landau and E. M. Lifshits, *Quantum Mechanics. Nonrelativistic Theory* (Pergamon Press, Oxford, 1977) and V. B. Berestetskii, E. M. Lifshits, and L. P. Pitaevskii, *Quantum Electrodynamics* (Butterworth-Heinemann, Oxford, 1982). As a reference book on computing, we will use S. Berezin and N. P. Zhidkov, *Computing Methods. Volume* I*and* II. (Pergamon Press, Oxford—London—Edinburgh—New York—Paris—Frankfurt, 1965).

When calculating atomic characteristics and scattering cross sections, the radial, angular, and spin variables are usually separated and the integration over the angular variables and the summation over the spin projections are performed analytically. As a guide for these actions, one can even now use the ageless book by I. I. Sobel'man, *Introduction to the theory of atomic spectra* (Pergamon Press, Oxford, 1972). Further references to all these sources will be given as [LL], [BLP], [BZ], and [IS], respectively.

There is significant overlap, especially in notations, diagrams that represent physical processes, and in the presentation of the mathematical formulas used in this book, with previous publications: M. Ya. Amusia, *Atomic Photoeffect* (Plenum Press, New York and London, 1990), M. Ya. Amusia and L. V. Chernysheva, *Computation of atomic processes* (IOP Publishing Ltd., Bristol and Philadelphia, 1997), and M. Ya. Amusia, L. V. Chernysheva, and V. G. Yarzhemsky, *Handbook of theoretical Atomic Physics, Data for photon absorption, electron scattering, and vacancies decay* (Springer, Berlin, 2012). References to these books will be given as [AM], [AC], and [ACY]. A much shorter initial version of [ACY] was published in Russian as M. Ya. Amusia, L. V. Chernysheva, and V. G. Yarzhemsky, *Photon absorption, electron scattering, vacancies decay* (Nauka, Saint-Petersburg branch, St. Petersburg, 2010) [ACYr].

Throughout the book, we use the so-called atomic system of units, assuming Planck's constant $\hbar$, electron charge e, and its mass m to be equal to 1, since otherwise formulas, calculations, and expressions will be overloaded with different powers of $\hbar$, m, и e. When passing to the SI system of units, the following numerical values of the charge and mass of the electron are used: $e = 1.6{\cdot}10^{-19}C$, $m = 9108{\cdot}10^{-31}$ kg, as well as the relationship of the basic SI units with atomic units: length $a_0 =$

$\hbar^2/me^2 \simeq 0.529 \cdot 10^{-10}\ m = 5.29 \cdot 10^{-1}\ nm$; energy $\varepsilon_0 = 2\ Ryd = me^4/\hbar^2 \approx 27.21\ eV = 43.59 \cdot 10^{-10}\ nJ$; cross section $a_0{}^2 \approx 27.98 \cdot 10^{-18}\ \mathrm{cm}^2 = 27.98 \cdot 10^{-4}\ \mathrm{nm}^2 = 27.98$ Mb (1 Mb is equal to $10^{-18}\ \mathrm{cm}^2$); speed $v_0 = (\varepsilon_0/m)^{1/2} = e^2/\hbar \approx 2.188 \cdot 10^6$ m/s; time $t_0 = a_0/v_0 = \hbar^3/me^4 \simeq 0.242 \cdot 10^{-16}$ s. The units of angular momentum and momentum in the atomic system of units are $\hbar$ and $\hbar/a_0 = me^2/\hbar$, respectively. The fine structure constant is equal to $\alpha = e^2/\hbar c \simeq 1/137.04$, and in atomic system $\alpha = 1/c$ and the speed of light is $c \simeq 137.04$.

The texts of the mathematical programs included in the ATOM-M system are stored separately under the name "ATOM-M system programs for calculation of atomic and molecular structure and processes and examples for calculating various physical characteristics" (Springer database). The proper link to them one can find in the Appendix to this book.

We considered it appropriate, in the same Appendix, to provide a reference to the List of publications in which the ATOM-M system was used. An important contribution to this book has been made by Dr. S. K. Semenov, who was an expert in calculations of molecular processes and creation respective computing codes. His untimely death at a relatively young age of 62 stopped his continuous and effective efforts in creating and refining the two-atomic molecules codes that are presented in this book. We are grateful to him for fruitful collaboration on those parts of this book that deal with two-atomic molecules. We had huge plans to study molecular endohedral, but were forced to stop with only one publication with him "Photoionization of Molecular Endohedrals," JETP Letters, **110**, 2, 102–106 (2019) that considered endohedral $H_2@C_{60}$ only.

The programs described in this book, at the first stages, were constructed with participation of the listed below our collaborators. In this regard, we consider it our duty to express our gratitude for the contribution to this work, and, thus, indirectly for this book, first of all, to the late N. A. Cherepkov and to V. K. Ivanov. We also thank N. B. Avdonina, A. S. Baltenkov, G. F. Gribakin, B. V. Gul'tsev, late D. Davidovich, V. A. Kilin, A. N. Kolesnikova, A. V. Korol', M. Yu. Kuchiev, I. S. Li, V. F. Orlov, V. Radojevic, S. G. Shapiro, S. A. Sheinerman, S. I. Sheftel, A. S. Kheifets, K. L. Tsemekhman, V. G. Yarzhemsky, and V. L. Yakhontov.

We are grateful also to readers of the previous book [AC] for their comments, which have been accounted in this book.

Jerusalem, Israel — Miron Ya. Amusia
St. Petersburg, Russia — Larissa V. Chernysheva

Dedication

In memory of

Miron Ya. Amusia

(18.11.1934–15.09 2021)

Professor Miron Amusia was one of the world's outstanding theoretical physicists and a long-time leader in the field of atomic physics. He made a significant contribution to the physics of the nucleus, solids, condensed states, authored and co-authored 17 books, over 500 scientific articles in physics journals and presented over 100 reviews and 450 reports at national and international conferences. In Russia, Prof. Miron Ya. Amusia created a scientific school comprised of talented students. Nowadays, many of his students have become professors, authors of fruitful ideas, scientists with world renown. He and his research group have been pioneers in the field of atomic physics for decades. The state-of-art software package "ATOM" developed together with his colleagues for studying collective effects in atoms and the results thus obtained made a decisive contribution to the development of AMO science. Professor Miron

Ya. Amusia has received many scientific awards. Thus, he was a laureate of the A. von Humboldt Research Prize (1990), was elected a Fellow of the American Physical Society (since 1995), an Honored Associate Member of the Center for Theoretical Research of Physical Systems at Clark-Atlanta University (from 1998), was elected an Academician of the Russian Academy of Natural Sciences (since 2002), honorary professor of many universities, laureate of many scientific Prizes and Medals: N. N. Semenov (2018), P. L. Kapitsa (2004) Medals and A. F. Ioffe (2017, RAS) Prize, as well as A. I. Frenkel (2013) and B. P. Konstantinov (2001) Medals and Prizes from Ioffe Institute.

In addition to the great scientific activity of Prof. Amusia, his social activity was striking as well: over the past twenty years, he has written a large number of notes and essays on many resonant public events taking place in the world.

Professor Miron Ya. was not only a very talented scientist, but also an extremely decent person. In his person, world science has suffered a huge and irreplaceable loss.

We, friends, colleagues and associates, will forever preserve in our hearts the memory of Prof. Miron Ya. Amusia.

Contents

General References and Abbreviations

General References

[AM] M. Ya. Amusia, *Atomic Photoeffect* (Plenum Press, New York and London, 1990).

[AC] M. Ya. Amusia and L. V. Chernysheva, *Computation of atomic processes* (IOP Publishing Ltd, Bristol and Philadelphia, 1997).

[ACY] M. Ya. Amusia, L. V. Chernysheva, and V. G. Yarzhemsky, *Handbook of theoretical Atomic Physics, Data for photon absorption, electron scattering, and vacancies decay* (Springer, Berlin, 2012).

[ACYr] M. Ya. Amusia, L. V. Chernysheva, and V. G. Yarzhemsky, *Photon absorption, electron scattering, vacancies decay* (Nauka, Saint-Petersburg branch, St.-Petersburg, 2010) (In Russian).

[AICC] M. Ya. Amusia, V. K. Ivanov, L. V. Chernysheva, N. A. Cherepkov, *Processes in multielectron atoms* (Nauka, Saint-Petersburg branch, St.-Petersburg, 2006) (In Russian).

[BLP] V. B. Berestetskii, E. M. Lifshits, and L. P. Pitaevskii, *Quantum Electrodynamics* (Butterworth-Heinemann, Oxford, 1982).

[BZ] I. S. Berezin and N. P. Zhidkov, *Computing Methods. Volume I and II* (Pergamon Press, Oxford—London—Edinburgh—New York—Paris—Frankfurt, 1965).

[IS] I. I. Sobel'man, *Introduction to the theory of atomic spectra* (Elsevier, New York, 1972).

[LL] L. D. Landau and E. M. Lifshits, *Quantum Mechanics. Nonrelativistic Theory* (Pergamon Press, Oxford, 1977).

Abbreviation for Basic Methods and Processes

DWBA	Distorted waves Born approximation
FRPAE	RPAE with accounting of the reflection from the fullerene shell with finite thickness potential
GFRPAE	RPAE with accounting of reflection from a fullerene shell with finite potential thickness and polarization of fullerene
GRPAE	Generalized RPAE
GRPAE-II	GRPAE with accounting of an inner vacancy decay
GZRPAE	RPAE with accounting of reflection from a fullerene shell with zero-thickness potential and polarization of fullerene
HF	Hartree–Fock approximation
MBPT	Many-body perturbation theory
ORPAE	RPAE for atoms with open-shells
PWBA	Plane wave Born approximation
RPAE	Random phase approximation with exchange
SP RPAE	Spin-polarized version of RPAE
SRPAE	Simplified RPAE
ZRPAE	RPAE with accounting of the reflection from the fullerene shell with zero-thickness potential

Chapter 1
Short Review

This chapter contains all the information you need to read and understand this book. The main physical features of the structure of atoms, simple molecules, fullerenes, and endohedrals are discussed, definitions of concepts that are used throughout the book are given, and a diagrammatic description of all the processes under consideration is presented.

The main theoretical approaches used are described, such as the Hartree–Fock (HF) self-consistent method, the random phases approximation with exchange (RPAE), generalized RPAE (GRPAE), simplified RPAE (SRPAE), spin-polarized approximation HF and RPAE, and a number of others. Thus, all the approaches used in describing the properties of the objects mentioned above and the cross sections of the processes, total and differential, involving these objects—their photoionization, elastic and inelastic scattering of electrons, positrons, mesons, and probabilities of the decay of emerging vacancies are mentioned and briefly explained.

We discuss not only how to take into account the self-consistent one-electron field, but also the electron–electron interactions or electron correlations. The capabilities of the ATOM-M system are also described, which make it possible to study various atomic–molecular processes and characteristics, in addition to those directly discussed in this book.

1.1 Some Qualitative Notes

The purpose of this book is to develop a set of mathematical programs that allow for fairly accurate calculations of the probabilities of the processes that describe the interaction of atoms, diatomic and some polyatomic molecules with photons,

---

Supplementary Information The online version contains supplementary material available at (https://doi.org/10.1007/978-3-030-85143-9_1).

M. Ya. Amusia and L. V. Chernysheva, *Computation of Atomic and Molecular Processes*,
Springer Series on Atomic, Optical, and Plasma Physics 117,
https://doi.org/10.1007/978-3-030-85143-9_1

electrons, and mesons. Attention will also be paid to the compilation of programs that allow calculating the probability of decay of vacancies formed in the above-mentioned objects in the course of these processes.

From a computational point of view, the most important issue is the spatial symmetry of the system under consideration. The simplest calculations are in spherically symmetric objects. The subject of this book is systems with two types of symmetry—central (spherical) or axisymmetric. Most of the results presented by us refer to the case of the central symmetry of the main potential that determines the structure of the system. As applied to an atom, this is the Coulomb potential of the nucleus.

For the case of an axisymmetric potential, programs for calculating diatomic molecules are presented. Nevertheless, the methods presented below and the corresponding programs can be rather easily extended to solve other problems of the same symmetry. As will be seen from the following description, the features of the theoretical consideration presented here and the corresponding programs relate primarily to the symmetry of the main potential, which is often referred to as the "core" potential.

In the central (spherical) field, which determines the main features of the structure of the atom, electrons occupy certain states. These states are characterized by energy E, angular momentum of the electron ℓ and its projection m, as well as the projection of the spin σ (intrinsic internal angular momentum) of the electron. In a spherically symmetric field, each value E, called an energy level, corresponds to a number of states that differ in values ℓ, m, and σ. In the case of axial symmetry of the ground potential, corresponding, in particular, to the case of diatomic molecules, the states of electrons are described by the energy E, the projection of the angular momentum onto the axis of symmetry m and also the projection of the spin σ.

Electrons with the same energies form a *shell*. The electronic energy of the system under consideration, i.e., the total energy of the electron shells is minimal if as many lower states are filled as necessary to accommodate a given number of electrons in the system under consideration—an atom or a molecule. This electronic state of the system is called *electronic ground state.*.

The Coulomb repulsion of electrons can be effectively taken into account by introducing some field that includes the averaged action of all electrons—generally called *the core field.* The movement of electrons in this field is considered as independent. In it, as before, similar to a purely Coulomb field, shells are the main feature of the electronic structure. However, states with different angular momentum already have different energies in the core field. States with the same energy and different projections of spin and angular momentum form a *subshell.* A shell (subshell), all states of which are occupied by electrons, is called a filled, or closed shell (subshell).

The crudest description of the atom assumes that atomic electrons move independently. This approach is called the *one-particle approximation*. A more accurate account of the Coulomb interelectron repulsion indicates a certain deviation from the independent motion of atomic electrons, or their *correlation.*

In addition to the ground state, an atom can be in *an excited state*, where one or more of the higher levels are occupied, while one or more of the lower levels are free. One or more electrons can be removed from the atom, i.e., to be in *the continuous spectrum*. This process is called *ionization*. When an electron is removed from the highest energy subshell, *a residue ion* is formed. When one or more electrons from an atom or molecule are removed from the inner shell, *an excited state of the ion* is formed. The smallest amount of energy required to remove an electron from an atom or molecule is called the *ionization potential*.

Excitation and/or ionization of an atom or molecule can occur due to application of a time-dependent field. As an example, we will consider an electromagnetic field, which is represented as a flux of quanta called photons. An electron, when it goes from level to level, can absorb or emit a photon. The energy of incident particles (electrons, positrons, mesons) can be conserved in course of the collision process, as well as scattered partially or completely due to interaction with atomic or molecular electrons. The scattered energy provides excitation and/or ionization of an atom or molecule. These two cases correspond to *elastic* and *inelastic scattering* of the incident particle, respectively.

In the one-electron approach, it is assumed that a photon causes a change in the state of only one electron that absorbed it. In the case of inelastic scattering, the incident particle excites all electrons to which its energy is transferred. In the case of elastic scattering, the interaction between an incident particle and atomic and molecular electrons changes their states, thus distorting the electron shells. Therefore, an additional *polarization potential* arises, acting on the scattered particle.

In one-particle approximation, vacant internal levels formed as a result of excitation or ionization can be filled with electrons from external levels, which is accompanied by emission of a photon. The filling of *a vacancy* by an external electron is called *vacancy decay*. Decay accompanied by the emission of a photon is called *radiative decay*.

The interaction between atomic electrons makes describing the absorption of a photon and the scattering of an incident particle much more complicated. First of all, the transfer of energy from one electron to another can occur in such a way that the electron that originally absorbed the photon returns to its initial state. In this case, the number of electrons at the excited levels does not change. Secondly, it is possible to transfer the energy of one excited electron to several other electrons. As a result, the number of electrons at excited levels increases. The third possibility is that vacancies do not decay by radiation, but transfer their energy to electrons with a lower binding energy, which leave the system under consideration.

1.2 Quantum–Mechanical Description of the Electronic Structure of Atomic–Molecular Systems

An electronic system with a number of electrons N_e in terms of quantum mechanics is described by the Schrödinger equation:

$$\hat{H}\Psi \equiv \sum_{i=1}^{N_e}\left[-\frac{\hbar^2\Delta_i}{2m} - V(\mathbf{r}_i) + \frac{1}{2}\sum_{k\neq i=1}^{N_e}\frac{e^2}{|\mathbf{r}_i - \mathbf{r}_k|} + \frac{1}{2}\sum_{i,k\neq 1}^{N_e} f(x_i \ldots x_k)\right]$$
$$= E\Psi(x_1 \ldots x_{N_e}), \qquad (1.1)$$

where Δ_i is the Laplace operator acting on the coordinate of the ith electron; m, e, and $\hbar$ are the mass, electron charge, and Planck's constant, respectively, which further, using the atomic system of units throughout this book, we will set equal to 1, $e = m = \hbar = 1$; E is the total energy of the system, Ψ is its *total wave function*, depending on N_e coordinates of all atomic electrons r and the values σ of their spin projections $x \equiv (\mathbf{r}, \sigma)$, f is the spin-dependent part of the interaction, $\hat{H}$ is the Hamilton operator. The potential $V(r_i)$ describes the interaction of electrons with immobile nuclei. For an atom or ion, it is simply a Coulomb spherically symmetric potential $V(r_i) = -Z/r_i$. For a diatomic molecule, this is the axisymmetric potential $V(r_i) = -(Z_1/r_i + Z_2/|\mathbf{r}_i - \mathbf{R}|)$, where Z_1 и Z_2 are the nuclear charges, and R is the internuclear distance.

Solutions (1.1) must satisfy the Pauli principle: Two electrons with equal spin projections cannot have the same coordinates, that is $\Psi(x_1 \ldots x_{N_e}) = 0$, if at least a pair of electron spins and coordinates coincide. This follows from Pauli's principle, a direct consequence of which is that the wave functions are antisymmetric with respect to the permutation of the coordinates of any two electrons:

$$\Psi(x_1 \ldots x_i \ldots x_k \ldots x_{N_e}) = -\Psi(x_1 \ldots x_k \ldots x_i \ldots x_{N_e}). \qquad (1.2)$$

An atom can be in various states. *The ground state* as a function of time does not change, while the higher-energy *excited states* are non-stationary in the sense that ultimately the excited atom decays to the ground state, emitting photons and/or electrons.

The state of an atom is determined by its energy, the *total angular momentum* J, which is the sum of the *orbital angular momentum* L and the electron spin S, and the projection of J onto an arbitrary, albeit definite, axis. The interaction, which depends on electron spins $f(x_1 \ldots x_k)$, is usually small compared to the Coulomb potential of nuclei. Therefore, it is often neglected. In this case, the orbital angular momentum $\mathbf{L} = \sum_{i=1}^{N_e} \mathbf{l}_i$ and atomic spin $\mathbf{S} = \sum_{i=1}^{N_e} \boldsymbol{\sigma}_i$ are kept separately. The absolute values L and S determine the *term* of the atomic system in the so-called *LS*-coupling approximation, which we, as a rule, will use.

The large number of variables makes exact numerical solution of the Schrödinger wave equation, even with modern computers, impossible. That is why approximations are usually made when describing a many-electron atom or molecule. The simplest case from this point of view is the case of atoms. When considering them, at the first step, the problem is reduced to solving a one-dimensional differential or integradifferential equation, or a set of a certain number of such equations. In this approximation, the physical process with the participation of an atom, whether it be the interaction between an atom and an external field, electrons or other particles, is considered the most roughly. Then the description of the structure of the atom and the interaction between the atom and projectile particles is consistently improved. Each stage requires numerical calculations, and therefore appropriate software must be developed. This will be done in this book. The problem of studying a diatomic molecule is more complicated, and already at the first step it is not reduced to the numerical solution of a one-dimensional equation, but requires the use of a two-dimensional equation. This task requires solving a system of equations from the very beginning.

The goal of the ATOM-M system is to develop methods for the numerical simulation of atomic and molecular properties and structure and interactions between atoms and molecules with photons, electrons, and other particles. These methods should allow one to obtain the appropriate characteristics of atomic and molecular processes with the desired accuracy.

1.3 Hartree–Fock Approximation

In this book, the structure of the electron shells of the system under consideration is described on the basis of the Hartree–Fock approximation [1, 2]. In the Hartree–Fock model, it is assumed that all electrons move in a potential field, which is called a self-consistent field, so, instead of Eq. (1.1), the following equation should be solved:

$$\begin{aligned}\hat{H}_{\mathrm{HF}}\Psi_E(x_1\ldots x_{N_e}) &\equiv \sum_{i=1}^{N_e}\left[-\frac{\Delta_j}{2}-V(r_j)+\hat{U}_{\mathrm{HF}}(x_j)\right]\Psi_E(x_1\ldots x_{N_e})\\ &= E\Psi_E(x_1\ldots x_{N_e}),\end{aligned} \tag{1.3}$$

where $\hat{U}_{\mathrm{HF}}(x_j)$ is a rather complicated operator that represents the self-consistent Hartree–Fock potential and depends only on the coordinates of the electron and its spin. The solution to Eq. (1.3) can be represented as an antisymmetrized product of one-particle functions $\phi_j(x_j)$:

$$\Psi_E(x_1\ldots x_{N_e}) \equiv \Psi_E^{\mathrm{HF}}(x_1\ldots x_{N_e}) = \frac{1}{\sqrt{N_e!}}\hat{A}\prod_{j=1}^{N_e}\phi_j(x_j) \tag{1.4}$$

where ϕ_j are functions of one variable, $\hat{A}$ is the antisymmetrization operator. In the simplest case $\Psi_E^{\mathrm{HF}}(x_1\ldots x_{N_e})$, as is known, it is a determinant formed from one-electron functions ϕ_j, known in theory as Slater's determinant.

The potential $\hat{U}_{\mathrm{HF}}(x_j)$ is determined from the condition for the minimum of the electron energy E with varying one-electron functions ϕ_j. The field $\hat{U}_{\mathrm{HF}}(x_j)$ is called *a self-consistent field*, since $\hat{U}_{\mathrm{HF}}(x_j)$ is determined by the wave functions of all electrons. Solution (1.3), as well as solution (1.1), can be determined up to an arbitrary normalization constant.

Here, as usual, it is assumed that the total electron wave function (1.4) is normalized to unity, that is, the condition

$$\int \left|\Psi_E^{\mathrm{HF}}(x_1 \ldots x_{N_e})\right|^2 \mathrm{d}x_1 ... \mathrm{d}x_{N_e} = 1, \tag{1.5}$$

where the one-particle wave functions in $\Psi_E^{\mathrm{HF}}(x_1 \ldots x_{N_e})$ are orthogonal and normalized:

$$\int \phi_j^*(x)\phi_k(x)\mathrm{d}x = \delta_{ik} \equiv \begin{cases} 1 \; j = k \\ 0 \; j \neq k \end{cases}. \tag{1.6}$$

For the ground state, the total electron energy is, taking into account (1.5), given by relation

$$E_0^{\mathrm{HF}} = (\Psi_0^{\mathrm{HF}}|\hat{H}|\Psi_0^{\mathrm{HF}}), \tag{1.7}$$

where $\hat{H}$ is the Hamilton operator. From the condition of minimum energy with respect to small variations $\phi_j(x_j)$, we obtain the HF equation for an atom:

$$-\frac{\Delta}{2}\phi_j(x) - \frac{Z}{r}\phi_j(x) + \sum_{k=1}^{N_e} \int \phi_k^*(x') \frac{\mathrm{d}x'}{|\mathbf{r} - \mathbf{r}'|} \left[\phi_k(x')\phi_j(x) - \phi_j(x')\phi_k(x)\right]$$
$$= E_j \phi_j(x) \tag{1.8}$$

where E_j is the one-electron energy.

Diatomic molecules, neglecting the motion of their nuclei, are described by the HF equation

$$-\frac{\Delta}{2}\phi_j(x) - \left(\frac{Z_1}{|\mathbf{r} - \mathbf{R}_1|} + \frac{Z_2}{|\mathbf{r} - \mathbf{R}_2|}\right)\phi_j(x)$$
$$+ \sum_{k=1}^{N_e} \int \phi_k^*(x') \frac{\mathrm{d}x'}{|\mathbf{r}' - \mathbf{r}|} \left[\phi_k(x')\phi_j(x) - \phi_j(x')\phi_k(x)\right] = E_j \phi_j(x) \tag{1.9}$$

where Z_1, Z_2 are the charges of the nuclei of the molecule, and the vectors $\mathbf{R}_1$, $\mathbf{R}_2$ set the coordinates of the nuclei in space. The generalization to the case of polyatomic molecules is obvious.

Each electron in an atom or molecule is described by its wave function $\phi_j(x)$. In the ground state, all the lowest electronic levels are sequentially filled, and the index sum k includes as many levels as there are electrons in the atom. Let us introduce the concept of the *Fermi level*, meaning the filled level with the highest energy.

This limitation in the amount of k can also be expressed symbolically as a condition $k \leq F$.

Solutions of Eqs. (1.8) and (1.9) for $j > F$ are the wave functions of an additional electron, since they describe the state of its motion in the field of the nucleus (or nuclei—for the case of molecules) and all electrons of the atom. Solutions to Eqs. (1.8) and (1.9) can correspond to both discrete ($Ej < 0$) and continuous ($Ej > 0$) spectra. These two types of solutions have significantly different behavior at large r, and, hence, normalization. The wave functions corresponding to the discrete spectrum are normalized according to (1.6), and their energy is strictly defined. For $Ej > 0$, solutions (1.8) and (1.9) exist for any value of E. At large distances from the atom, they oscillate and are normalized as follows

$$\int \phi_E^*(x)\phi_{E'}(x)\mathrm{d}x = \delta(E - E'), \tag{1.10}$$

where $\delta(E - E')$ is the Dirac delta-function.

The wave function of the simplest, called one-electron, energy excitation $E > E_0$ is obtained by replacing one of the functions $\phi_j(x)$ with $j < F$ in (1.8) or (1.9) with the wave function of a discrete $\phi_k(x)$ or continuous $\phi_E(x)$ spectrum. This is how *electron–hole excitation* is formed. This term implies that the excitation is associated with an electron moving to a free level, while its original location remains free, forming *a hole* or *vacancy*. The lowest energy at which excitation or ionization takes place is called the threshold energy, or simply *the threshold*.

More complex excitations arise due to transitions of two, three, or more electrons from the levels they originally occupied to free ones, with the formation of two, three, or more electron–hole pairs.

The atomic core is considered *frozen* when the states of the electrons do not change, except for those that are directly involved in the transition.

The interaction between the electron of the system and the external field is described by the operator $\hat{B}(r)$. Under its influence, the electron passes from level $j \leq F$ to level $k > F$, that is, an electron–hole pair is created. The transition probability is determined by the matrix element [3]:

$$\langle i|M|f\rangle = \int \Psi_i^*(x_1 \ldots x_{N_e}) \sum_{p=1}^{N_e} B(r_p)\Psi_f(x_1 \ldots x_{N_e}) \prod_{q=1}^{N_e} \mathrm{d}x_q, \tag{1.11}$$

where Ψ_i and Ψ_f are the wave functions of the initial and final states, respectively. When calculating functions Ψ in Hartree–Fock approximation with a frozen core, the matrix element (1.11) reduces to the sum of integrals over a single variable:

$$\langle j|M|k\rangle = \int \varphi_j^*(x)\hat{B}(x)\varphi_k(x)\mathrm{d}x. \tag{1.12}$$

Let us look at atoms first. We will return to diatomic molecules at the end of this chapter. The spherical symmetry of the Coulomb field of an atom makes it possible to significantly simplify Eq. (1.8). The wave functions of the ground and excited one-electron states in an atom $\phi_j(x)$ can be represented as the product of the radial part $R(r) \equiv P_{nl}(r)/r$ where $P_{nl}(r)$ is the so-called radial function, the angular function $Y_{lm}(\mathbf{r}/r)$, and the spin function X_σ. The first depends only on the magnitude r, $Y_{lm}(\mathbf{r}/r)$ is determined only by the direction $\mathbf{r}$ and X_σ depends on the spin projections. After the angular and spin variables are separated, which can be done analytically, a relatively simple set of one-dimensional functions arises, which can be obtained by numerically solving the HF equations.

The electronic energy E_i in an atom is defined by the fundamental quantum number n_j and angular momentum l_j, which together describe the *radial term* of the wave function. Electronic levels of the same energy form a subshell, while all states with the same fundamental quantum number form a shell. The number of such filled levels in each subshell describes the electronic configuration of the atom.

After separating the angular and spin parts in $\phi_j(x)$ and $\phi_k(x)$, the matrix elements of the external field are expressed through one-dimensional radial integrals:

$$\langle j|M|k\rangle = \int P_{n_j l_j} B(r) P_{n_k l_k}(r) \mathrm{d}r, \tag{1.13}$$

which are calculated using the ATOM-M system.

An incident electron can conserve its energy after interacting with a target atom, or lose this energy partially or completely. In the first case, *elastic scattering* takes place, in the latter one speaks of *inelastic scattering*.

Coulomb interaction between incident and atomic electrons,

$$V(|\mathbf{r}_1 - \mathbf{r}_2|) = \frac{1}{|\mathbf{r}_1 - \mathbf{r}_2|}, \tag{1.14}$$

causes excitation or ionization of the atom.

An incident particle loses energy not only for the excitation of an atom or its ionization, but also for *bremsstrahlung*—the emission of a photon that accompanies the interaction between electrons and atoms. The probability of an atomic electronic transition caused by an incident electron is described by the interaction matrix elements, which include the initial and final functions of the incident particle $\phi_{\nu_1}(x)$, $\phi_{\nu_3}(x)$, and atomic electrons $\phi_{\nu_2}(x)\phi_{\nu_4}(x)$:

$$\langle \nu_1\nu_3|V|\nu_2\nu_4\rangle = \iint \phi^*_{\nu_1}(x_1)\phi^*_{\nu_2}(x_2)V(|\mathbf{r}_1 - \mathbf{r}_2|)\phi_{\nu_3}(x_1)\phi_{\nu_4}(x_2)\mathrm{d}x_1\mathrm{d}x_2, \tag{1.15}$$

where ν denotes a set of quantum numbers $nlm\sigma$ ($Elm\sigma$ in the case of a continuous spectrum). Like (1.13), after the angular and spin parts in $\phi(x)$ are separated and integrated over the directions $\mathbf{r}_1/|r_1|$ and $\mathbf{r}_2/|r_2|$, only integrals over the variables r_1 and r_2 remain:

$$\langle \nu_1\nu_3|V^l|\nu_2\nu_4\rangle_r \equiv \int\limits_0^\infty\int\limits_0^\infty P_{\nu_1}(r_1)P_{\nu_3}(r_2)V^l(r_1,r_2)P_{\nu_2}(r_1)P_{\nu_4}(r_2)\,\mathrm{d}r_1\,\mathrm{d}r_2 \quad (1.16)$$

where V^l is the element of the potential V that depends only on the radii $|\mathbf{r}_1|$ and $|\mathbf{r}_2|$.

Just like an electron, any other charged particle of mass m_1 and charge e_1 (positron, meson, proton) can either preserve its energy after interacting with an atom or molecule, i.e., elastically scatter, or transfer it to atomic or molecular electrons due to inelastic scattering. To study this kind of processes, one must first find the wave function of this particle in the field of an atom or molecule by solving equations like (1.8) or (1.9), respectively:

$$-\frac{\Delta}{2m_1}\chi_j(x) - \frac{Ze_1}{r}\chi_j(x) - \sum_{k=1}^{N_e}\int \phi_k^*(x')\frac{1}{|\mathbf{r}-\mathbf{r}'|}\phi_k(x')\chi_j(x) = E_j\chi_j(x), \quad (1.17)$$

$$-\frac{\Delta}{2m_1}\chi_j(x) - \left(\frac{Z_1e_1}{r} - \frac{Z_2e_1}{|\mathbf{r}-\mathbf{R}|}\right)\frac{Ze_1}{r}\chi_j(x) - \sum_{k=1}^{N_e}\int \phi_k^*(x')\frac{1}{|\mathbf{r}-\mathbf{r}'|}\phi_k(x')\chi_j(x) = E_j\chi_j(x). \quad (1.18)$$

with different mass, charge, and, naturally, without taking into account the exchange with atomic electrons.

The interaction between the incident particle and the atom is determined by a matrix element similar to (1.15) with the functions $\phi_{\nu_1}^*(x)$ and $\phi_{\nu_3}(x)$, replaced by $\chi_\varepsilon^*(x)$ and $\chi_{\varepsilon'}(x)$. For an atomic target, just as in $\phi(x)$, in $\chi_\varepsilon(x)$ one can separate the radial and angular parts, carry out the integration over the angular variables analytically, and reduce the matrix element of the interaction of the incident particles with an atomic electron to a computer-calculated two-dimensional integral over r_1 and r_2.

1.4 Residual Interaction

The description of the electronic structure of an atom or molecule, given by Eq. (1.3) with $\Psi_E(x_1 \ldots x_{N_e})$, does not fully take into account the electron–electron interaction. This is obvious since the Hamilton operator $\widehat{H}$ is not identical to $\widehat{H}^{\mathrm{HF}}$, i.e., $\widehat{H} \neq \widehat{H}^{\mathrm{HF}}$ due to the fact that

$$\sum_j \widehat{U}_{\mathrm{HF}}(x_j) \neq \frac{1}{2}\sum_{j,k}\frac{1}{|r_j - r_k|}. \quad (1.19)$$

If necessary, the description of the ground states, as well as the excited states, can be improved by taking into account the *residual interaction* $V_{\text{res}}(r_j, r_k)$, defined as follows

$$\frac{1}{2}\sum_{j\neq k} V_{\text{res}}(r_j, r_k) = \frac{1}{2}\sum_{j\neq k}\frac{1}{|r_j - r_k|} - \sum_j U_{\text{HF}}(r_j). \tag{1.20}$$

Under the action of V_{res} electrons in the ground or in excited states interact with one another. As a result, they can move to other levels: from ν_1 and ν_2 to ν_3 and ν_4 $(E_1 \neq E_3, E_2 \neq E_4)$ and remain at these levels for a period limited τ by the Heisenberg uncertainty relation:

$$\tau \sim \frac{\hbar}{E_1 + E_2 - E_3 - E_4}, \tag{1.21}$$

thus, modifying their energy so that $E_1 + E_2 \neq E_3 + E_4$. Such states temporarily occupied by electrons are called *virtual* (intermediate) excited states.

Excitation or ionization of an atom can only take place under the influence of time-dependent fields or incident particles. In nature, weak fields (compared to those that operate inside atoms) are common. It is sufficient to take such fields into account in the lowest order of the perturbation theory in the action of this field. This book deals mainly with weak electromagnetic fields. The interaction between atomic or molecular electrons can also be written as a matrix element similar to (1.15), but with different states ν_1, ν_2, ν_3, ν_4 of the electrons involved, i.e., for the process of scattering between holes and electrons and the creation (or destruction) of electron–hole pairs.

The response of an atom or molecule to an external field or an incident charged particle is represented as their interaction with atomic electrons. As a result, the atomic electrons become excited. Then, atomic electrons, both in the initial state and in the excited state, are scattered on each other or generate electron–hole pairs. Considering the interaction between atomic or molecular electrons, or the interaction of these electrons with an incident particle or external field, we can get the probability amplitude of any atomic or molecular process. For this purpose, as described above, it is necessary to be able to calculate the wave functions of the ground and excited states, matrix elements of the form (1.12) and (1.15), and then sum up or, where required, integrate the products of various matrix elements over all intermediate, real, and virtual states.

1.5 Diagrammatical Technique

An important element of the theoretical description of atomic and molecular characteristics is a schematic sketch of the process under study—a diagram that uses the notation presented below (see, e.g., [4]). Exactly, dotted line

 (1.22)

denotes a non-stationary external field; right arrow line

 (1.23)

represents an electron in a state ν_1 described by a wave function in the Hartree–Fock approximation (1.8); double arrow line

 (1.24)

denotes an electron in a discrete excited state ν_1 described by the wave function in the Hartree–Fock approximation (1.8), and the line with an arrow to the left

 (1.25)

represents a vacancy (hole) in the state ν_2 described by the wave function in the Hartree–Fock approximation (1.8); double line with right arrow

 (1.26)

is used to denote an incident particle (other than an electron) described by the wave function in the Hartree–Fock approximation (1.17) or (1.18), and the wavy line

 (1.27)

represents the Coulomb interaction between atomic electrons, or between atomic electrons and an incident particle (1.14).

Filled circle or bullet

$$\bullet \tag{1.28}$$

represents the act of interaction between an electron and an applied external field.

Using this notation, we can construct elementary diagrams describing the interaction between atomic or molecular electrons, or between atomic and molecular electrons and an external field.

The following diagram describes the interaction of an external field with the electrons of an atom or molecule, which leads to the creation of an electron–hole pair and the absorption of a field quantum (photon in the case of an electromagnetic field) with energy ω

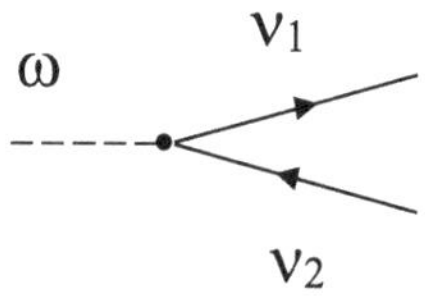

(1.29)

Annihilation (*self-destruction*) of an electron—hole pair with the emission of released energy in the form of a quantum is represented by the following diagram:

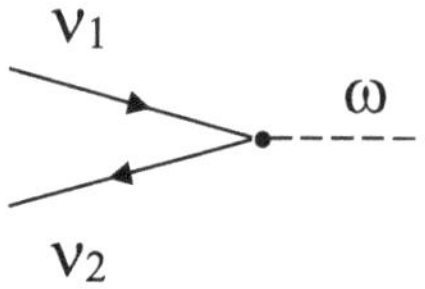

(1.30)

Virtual elementary processes: "creation" of an electron–hole pair and a quantum is represented by the following simplest diagram

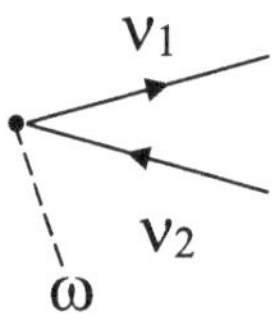

(1.31)

and the "annihilation" of an electron—hole pair and a quantum

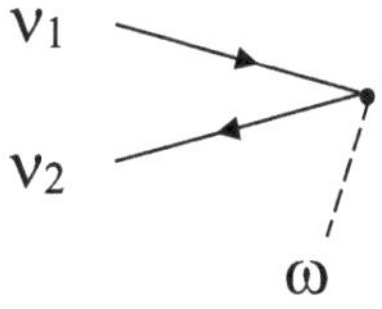

(1.32)

The interaction between electrons is depicted as an electron scattered by another electron

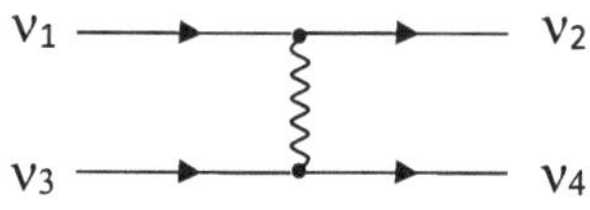

(1.33)

hole scattered by another hole

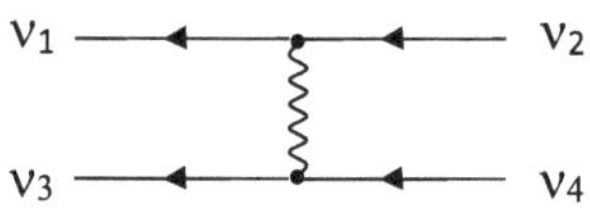

(1.34)

and a hole scattered by an electron or vice versa:

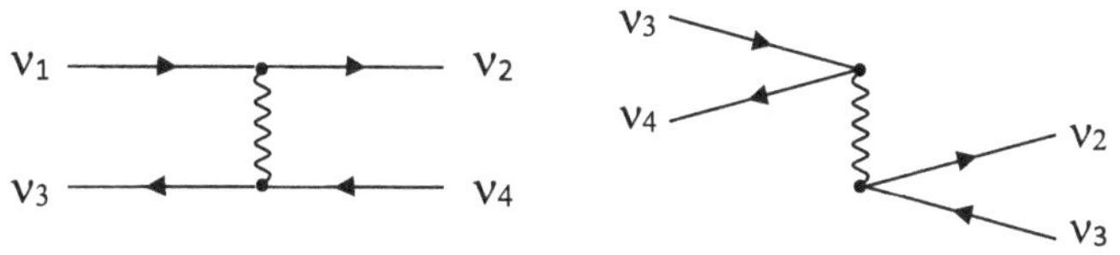

(1.35)

The second diagram in (1.35) represents the destruction of a pair in one state, followed by the creation of a pair in another state.

The propagation of an electron (hole) can be accompanied by the creation of a pair $\nu_3\nu_4$. It can be described as follows:

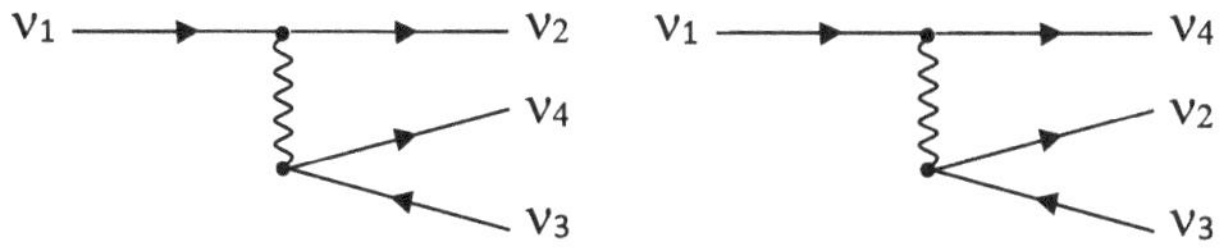

(1.36)

The propagation of an electron or a hole can also be accompanied by the destruction of a pair:

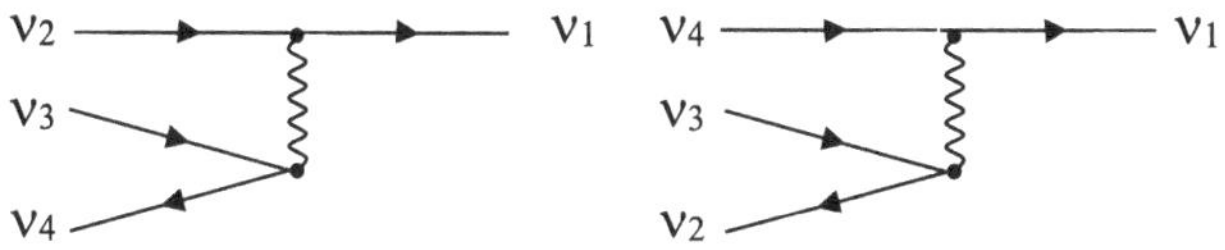

(1.37)

There are the following virtual elementary processes: "creation" of two electron–hole pairs

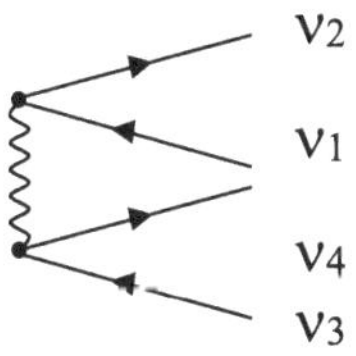

(1.38)

and "annihilation" of two electron–hole pairs

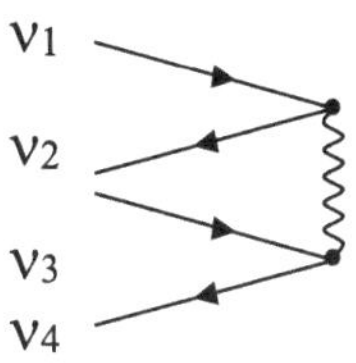

(1.39)

The interaction between atomic electrons and an incident particle other than an electron (e.g., a positron, nucleon, or meson) are represented by diagrams depicting the scattering of this particle by an electron

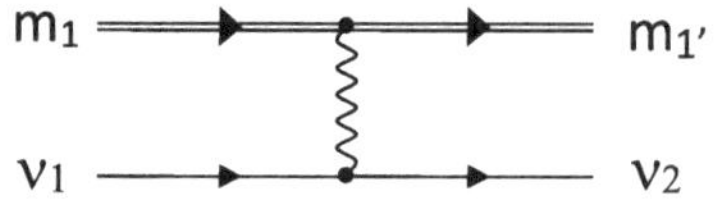

(1.40)

hole

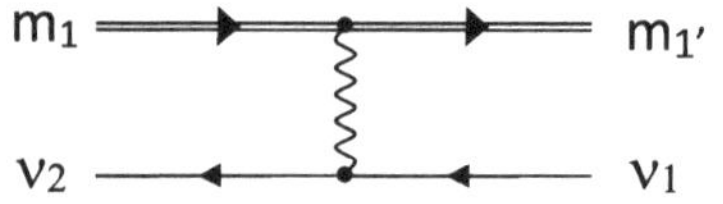

(1.41)

creation of an electron–hole pair

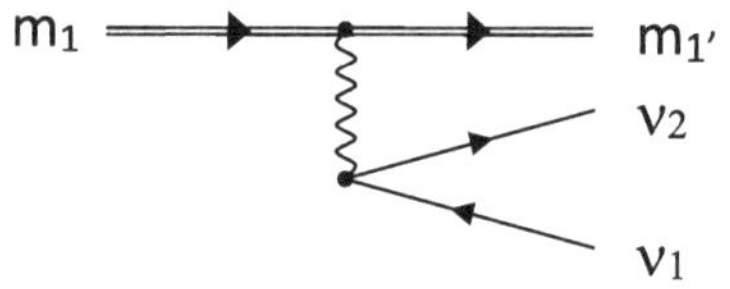

(1.42)

or its destruction

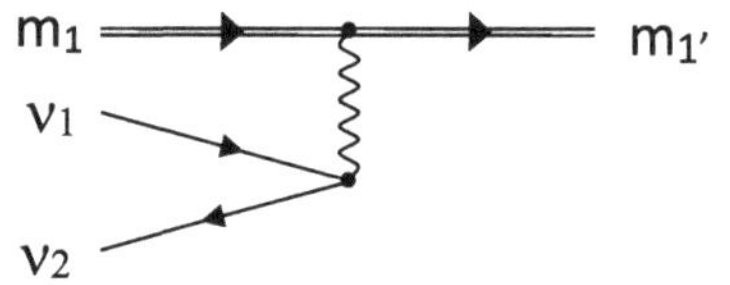

(1.43)

Here sign in 'm_1' emphasizes that energy and angular momentum of the projectile can change in the considered process.

Any investigated physical process can be represented as a combination of the above elementary diagrams.

In order to draw a diagram of an atomic process, first of all, one should represent the initial and final states by means of the elementary diagrams given above. For example, to describe one-electron photoionization, the photon (1.22) and the electron–hole pair are given in the initial and final states, respectively. For elastic scattering of an electron, both the initial and final states are represented by a line with an arrow directed to the right (1.23).

For the process of vacancy decay, in which a vacancy passes from one state to another, and the transition energy is released to create an electron–hole pair, the initial state is a single vacancy (1.25), while the final state is a vacancy and an electron–hole pair. Elastic and inelastic scattering of a particle other than an electron is represented by a double line (1.26) in both the initial and final states.

The creation of the diagram begins with the lowest order in the electron–electron interaction. The goal is to present the process under study as a transition from its initial to final state in the form of the shortest series of elements of interparticle interaction. For example, the simplest diagram of photoionization is (1.29), and (1.36) is the simplest diagram of inelastic scattering of an electron with the creation of one electron–hole pair.

The initial and final states can also be interconnected by a number of elementary events. For example, the following diagram represents the contribution to the amplitude of the one-electron photoionization process with inclusion of the electron–hole interaction in the lowest order:

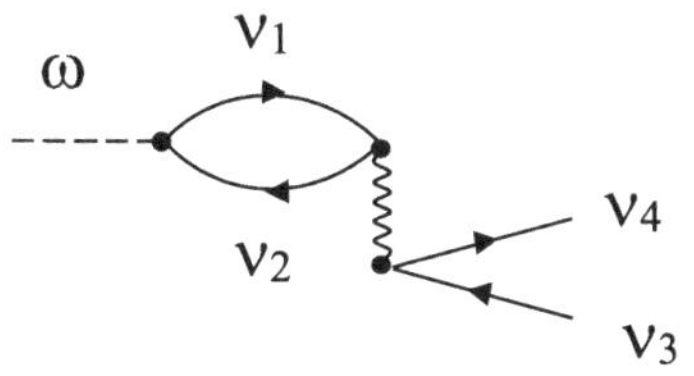

(1.44)

The diagrams represent the process as developing in time, which is considered to be increasing from left to right.

The following *correspondence rules* allow to match an analytical expression to each diagram:

- introduce designations $\nu_1, \nu_2 ... \nu_n$ for all electronic and hole states;
- introduce the concept of "dissection" of the diagram, meaning by this actually the vertical lines for separating each act of interaction between atomic electrons or an atomic electron with an external field or other particle;
- relate to each "cut" the energy denominator

$$\left(\sum E_h - \sum E_e + E_{\text{in}} + i\eta\right)^{-1} \tag{1.45}$$

  where $\sum E_h$ and $\sum E_e$, accordingly, are the total energy of all hole and electronic states included in the "dissection", and E_{in} is the energy of the initial state; the parameter $\eta \to +0$ shows the direction of bypassing the singularity when integrating expressions containing (1.45);
- match matrix element

$$\left\langle \nu_1 \middle| \widehat{B} \middle| \nu_2 \right\rangle \equiv \int \phi^*_{\nu_1}(x)\widehat{B}(r)\phi_{\nu_2}(x)\mathrm{d}x \tag{1.46}$$

  to each act of interaction with the external field (1.29)–(1.32), and matrix element

$$\langle \nu_1\nu_3|V|\nu_2\nu_4\rangle \equiv \int \phi^*_{\nu_1}(x)\phi^*_{\nu_3}\frac{e^2}{|r-r'|}\phi_{\nu_2}(x)\phi_{\nu_4}(x')\mathrm{d}x\mathrm{d}x' \tag{1.47}$$

  to interelectron interaction (1.33)–(1.39),
- sum up and integrate over all states of an electron and a hole;
- multiply the resulting expression by $(-1)^Q$ where Q is the sum of the numbers of hole lines and closed electron–hole loops in the diagram. The result obtained in this way is the contribution of the diagram to the amplitude of the process under consideration.

For example, the contribution of the diagram (1.44) is

$$\Delta B_{\nu_3\nu_4} = \sum_{\nu_1 > F, \nu_2 \leq F} \frac{\left\langle \nu_1 \middle| \widehat{B}(r) \middle| \nu_2 \right\rangle \langle \nu_2\nu_3 | V | \nu_1\nu_4 \rangle}{E_2 - E_1 + \hbar\omega + i\eta} (-1)^3 \tag{1.48}$$

Here $Q = 3$, because the diagram has one electron–hole loop and two hole lines ν_1, ν_2 in the diagram. More complex diagrams are constructed in the same way, and analytical expressions are related to them in the same way.

An infinite variety of diagrams corresponds to any physical process with a given initial and final state. Forced to limit the consideration to only some diagrams, we can connect them with the corresponding analytical expressions and thus obtain the amplitude of the process under study, the probability of which is proportional to the square of the modulus of the amplitude. The total probability of a process per unit time is the sum of the squares of its amplitudes over all possible final states. This sum must then be averaged over the initial states, taking into account the fulfillment of the energy conservation law and the total angular momentum of the system in its initial and final states.

The total probability per unit time is proportional to the flux density of the incident particles causing the reaction. The probability divided by the flux density is the dimension of the area, which is called the *cross section* of the process. The cross section is a characteristic of the effective size or diameter of a particle in this particular reaction. The total cross section of the process takes into account any changes in the state of the incident particle or absorption of the incident particle, regardless of the final state of the process.

The differential cross section of the process corresponds to a certain energy and direction of particles and quanta in the final state or other features of the final state.

1.6 Accounting for Residual Interaction

After introducing the residual interaction V_{res} (1.20), any physical characteristic of the ground or excited state of an atom or molecule, or the amplitude of the process with their participation, which determines the amplitude of the transition probability under the action of an external field $B(r)$ or incident particles (electrons, positrons, nucleons, mesons) can be represented as an infinite series in powers of the matrix element containing this interaction. It is not possible to sum this entire series, which would be equivalent to an exact solution of the corresponding many-body problem. Therefore, it is very important to choose an approximation that is able to provide the desired accuracy of the calculation results.

Usually, the external field that we are considering is weak, so that it is sufficient to take it into account by terms of the first order in $B(r)$. The interaction between the incident particle and atomic electrons, if it is small, is also sufficient to take into account the terms of the first or, as a maximum, also of the second order in this interaction. This approach is called, respectively, the first and second *Born approximations*, respectively. They are sufficient mainly for *fast* incident particles, i.e., those whose speed is much higher than the speed of atomic electrons. At low (*slow* particles) and intermediate speeds, some simplifications are also inevitable, because the exact mathematical description of the interaction between the incident particles and the atomic electron in the field of the atom is a very difficult task.

There are two different approaches to accounting V_{res}. The first is known perturbation theory in the residual interaction. In this approach, only a few first terms in power expansion in matrix elements V_{res} are preserved. The advantage of this method is that the rough approximation is simple and the method for considering it is clear. For example, a process in which a single photon removes two electrons from an atom (i.e., creates two electron–hole pairs) is represented as follows:

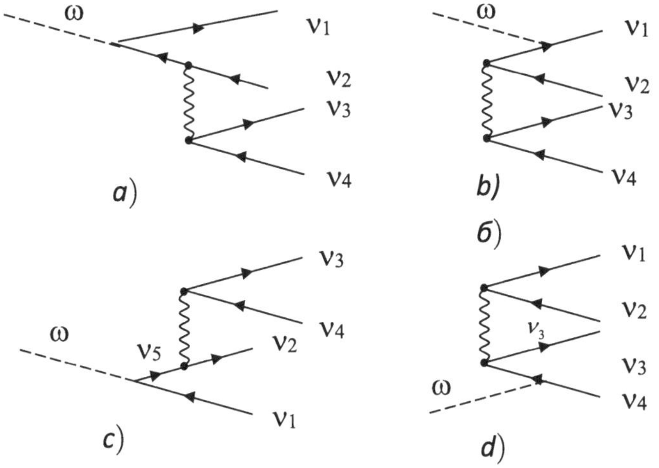

(1.49)

where diagrams (1.49)—(a) and (1.49)—(b) describe the emission of an electron and the transition from state ν_3 to ν_4 due to scattering of an electron (or hole) ν_5 accompanied by the transition of the latter to a state ν_2. Diagrams (1.49)—(c) and (1.49)—(d) correspond to the interaction between electrons in the ground state. As a result, the electrons are removed from the atom or occupy excited states.

However, when the number of interelectronic interactions exceeds three, the corresponding diagrams can in principle be shown, but it is very difficult to calculate the matrix elements for the following reasons. First of all, with the number of acts of interaction being N_{int}, the number of diagrams is $N_{int}!$, i.e., grows rapidly with increasing N_{int}. Second, calculations become more and more complex with growth of N_{int}, since the number of intermediate states increases, and with it the number of integrations and summations.

A different approach is often used. Its idea is to selectively sum infinite sequences of diagrams. In each order q containing the product of q matrix elements V_{res}, not all matrix elements V_{res} are taken into account, but only a limited number of them.

The choice is dictated by physical considerations or by some rough approximate calculations, which allow us to assume that some terms exceed the total contribution of all others in a given expansion order in V_{res}. It is obvious that even a part of the matrix elements of all degrees of V_{res} cannot be summed sequentially, one order after another. However, in some cases, such a summation can be performed by finding and solving an equation (or a set of equations) called a closed equation, in which all the functions that will be found are expressed in terms of already known functions. One of the variants of this approach is called the random phase approximation with exchange (RPAE).

1.7 Random Phase Approximation with Exchange

The random phase approximation with exchange (RPAE) [4] is a generalization of the Hartree–Fock approximation for an atom or a molecule in a weak non-stationary external field or in an excited state arising, for example, in the collision of an atom or molecule with an incident particle, etc. The method assumes that both in the presence and in the absence of an unsteady applied field (see Sect. 1.3), the wave function of an atom or molecule at each moment is a Slater determinant composed of one-electron wave functions $\phi_j(x,t)$. They can be obtained by solving equations that differ from (1.8) or (1.9) by the term $B(r,t)\phi_j(x,t)$ on the right-hand side, and the operator $i\hbar\partial/\partial t$ instead of the energy E_i.

These equations are greatly simplified on the assumption that the variation of one-particle functions under the action of an external field $B(r,t)$ or atomic (molecular) electronic excitation is small, that is $\left|\phi_i(x,t)-\phi_j(x)\right| \ll \phi_j(x)$. Under this assumption, the RPAE equations are obtained from the non-stationary Hartree–Fock equations. The RPAE equations take into account not only the action of an external field on an individual atomic or molecular electron, but also the interaction of the external field with all other electrons. The virtual transitions caused by this interaction change the average atomic (molecular) field and therefore lead to the removal or excitation of an atomic (molecular) electron.

In terms of the diagrams, the elements of which are given above, RPAE corresponds to taking into account the sequential interaction between the hole and the electron. Several excited electron–hole pairs can coexist, but in each case only one pair can make a real, physically observable transition. Below are a few diagrams from their infinite collection, which describe the response of an atom to an external field that creates an electron–hole pair in the final state:

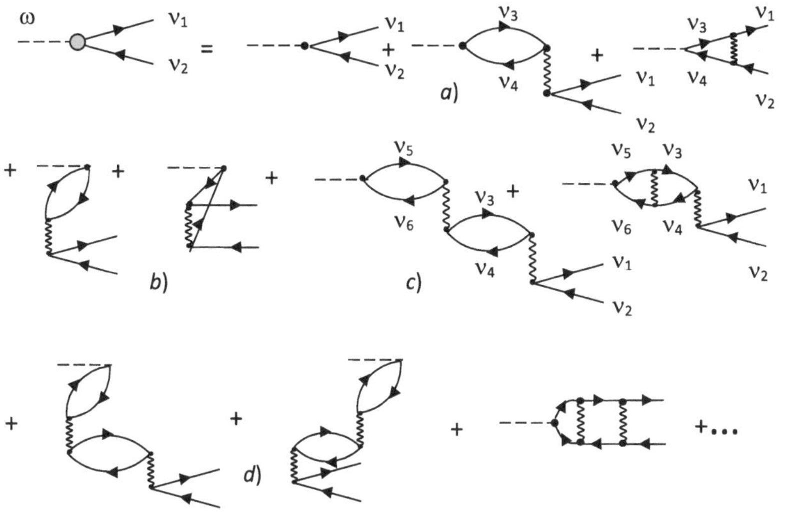

(1.50)

The infinite set (1.50) can be presented also in a different way, expressing the diagrams in the right-hand side via themselves:

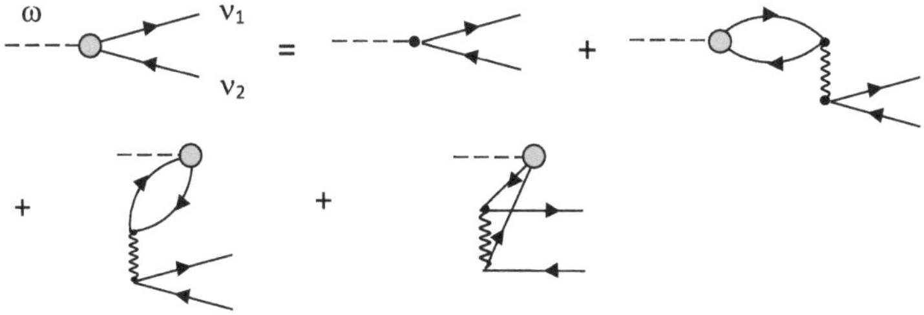

(1.51)

This set of diagrams can be analytically represented as an integral or differential equation. This book uses the integral representation for matrix elements (1.51):

$$\langle \nu_1|D(\omega)|\nu_2\rangle = \langle \nu_1|d|\nu_2\rangle + \left(\sum_{\nu_3>F,\nu_4\le F} - \sum_{\nu_4>F,\nu_3\le F} \right) \frac{\langle \nu_3|D(\omega)|\nu_4\rangle\langle \nu_4\nu_1|U|\nu_3\nu_2\rangle}{\omega - E_3 + E_4 + i\eta(1-2n_3)}, \quad (1.52)$$

where $i\eta$ determines, as noted above, the direction around the pole when integrating over energy, n_3 is the *Fermi step*:

$$n_3 = 1\ \nu_3 \le F,\ n_3 = 0\ \nu_3 > F, \quad (1.53)$$

d is the operator of interaction of an atomic or molecular electron with an external field $\widehat{B}(r,t)$. In the case of an electromagnetic field, d is equal to the vector $\mathbf{r}$ of the position of the electron (in the so-called length form), or $\vec{\nabla} \equiv \partial/\partial\vec{r}$ (in the "velocity form"). Matrix elements $\langle\nu_1|d|\nu_2\rangle$ and $\langle\nu_4\nu_1|U|\nu_3\nu_2\rangle$ are determined by Hartree–Fock one-electron functions $\varphi_\nu(x)$ (1.8) or (1.9), for an atom or a molecule, respectively:

$$\langle\nu_1|d|\nu_2\rangle = \int \phi^*_{\nu_1}(x)\hat{d}\phi_{\nu_2}(x)\mathrm{d}x$$
$$\langle\nu_4\nu_1|U|\nu_3\nu_2\rangle = \iint \phi^*_{\nu_4}(x)\phi^*_{\nu_1}(x')\frac{1}{|\mathbf{r}-\mathbf{r}'|}\left[\phi_{\nu_3}(x)\phi_{\nu_2}(x') - \phi_{\nu_2}(x)\phi_{\nu_3}(x')\right]\mathrm{d}x\,\mathrm{d}x' \tag{1.54}$$

Summation over ν_3 or $\nu_4 > F$ includes also integration over the states of the continuous spectrum, so that, in fact, (1.52) is an integral equation.

An important element of the RPAE is the effective interaction $\Gamma(\omega)$. It differs from the pure Coulomb interaction V by corrections caused by the virtual excitation of other atomic or molecular electrons during the interaction of the electron–hole pair under consideration. Now we can express, in terms of the effective potential $\Gamma(\omega)$, the corrections of the RPAE to the physical characteristics of atoms, molecules, and processes with their participation, obtained in the Hartree–Fock approximation. They are called *correlation corrections.*

The following considerations make it possible to write an equation for $\Gamma(\omega)$. The impact of one atomic (molecular) electron on another can be considered as an external field

$$B_{\nu_3\nu_4}(\mathbf{r}) = \int \phi^*_{\nu_3}(x')\frac{1}{|\mathbf{r}-\mathbf{r}'|}\phi_{\nu_4}(x')\mathrm{d}x'. \tag{1.55}$$

Replacing in (1.52) $\langle\nu_1|D|\nu_2\rangle$ by (1.55), we obtain the expression for the "*effective field*" or, in other words, the effective electron–electron interaction in the RPAE $\Gamma(\omega)$:

$$\langle\nu_2\nu_3|\Gamma(\omega)|\nu_1\nu_4\rangle = \langle\nu_2\nu_3|U|\nu_1\nu_4\rangle + \left(\sum_{\nu_5>F,\nu_6\le F} - \sum_{\nu_6>F,\nu_5\le F}\right)\frac{\langle\nu_2\nu_5|\Gamma(\omega)|\nu_1\nu_6\rangle\langle\nu_6\nu_3|U|\nu_5\nu_4\rangle}{\omega - E_5 + E_6 + i\eta(1-2n_5)} \tag{1.56}$$

The amplitude of interaction with the external field (1.52) is expressed in terms of $\Gamma(\omega)$ in the following way:

$$\langle\nu_1|D(\omega)|\nu_2\rangle = \langle\nu_1|d|\nu_2\rangle + \left(\sum_{\nu_3>F,\nu_4\le F} - \sum_{\nu_4>F,\nu_3\le F}\right)\frac{\langle\nu_3|d|\nu_4\rangle\langle\nu_4\nu_1|\Gamma(\omega)|\nu_3\nu_2\rangle}{\omega - E_3 + E_4 + i\eta(1-2n_3)}. \tag{1.57}$$

In the language of diagrams, effective interaction can be represented as an infinite series:

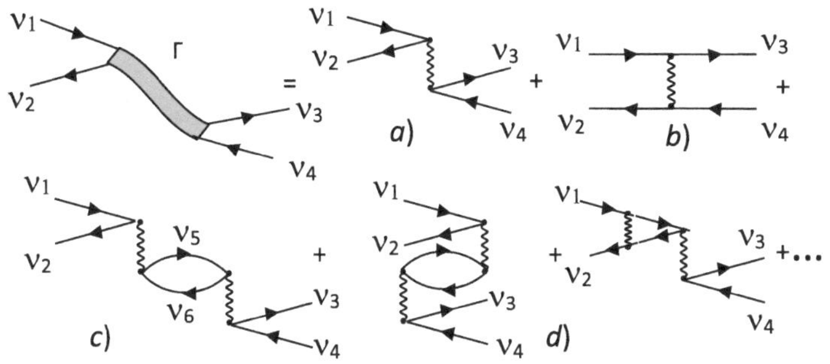

(1.58)

The RPAE takes into account all possible permutations of interactions in time in the loop diagrams (1.50)—(c) and the corresponding exchange diagrams. In them, the interacting partial and hole lines are interchanged. For example, the permutation between the states ν_2 and ν_3 in the *direct* diagram (1.58)—(a) leads to *exchange* diagrams (1.58)—(b).

The point of the external field (1.28) or the lines of interelectron interaction (1.27) determine the sequence in time of the process under study. Loop diagrams (1.50)—(a), (c) and corresponding exchange diagrams in higher orders with respect to the electron–electron interaction determine the time order. The first sum in Eq. (1.56)) represents the so-called time-forward diagrams. Diagrams (1.50)—(b), (d) and (1.58)—(d), and the corresponding exchange diagrams, are called "time-backward diagrams".

The sum of the "time-forward diagrams" in the RPAE, which corresponds to the first sum with $\nu_6 \leq F$ in (1.56), where only one term with $\nu_6 = \nu_4 = \nu_2$ is included, forms the so-called simplified random phases approximation with exchange (SRPAE). On the other hand, SRPAE diagrams, that is, a part of RPAE diagrams, can be taken into account by the appropriate choice of one-electron wave functions in the Hartree–Fock approximation [4]. This partial summation technique will be widely used in the future.

The RPAE equations have a feature that is very useful in solving them. Suppose we represent the equation for $\Gamma(\omega)$ (1.56) in symbolic form:

$$\hat{\Gamma} = \hat{U} + \hat{U}\hat{\chi}\hat{\Gamma}. \tag{1.59}$$

According to (1.56), the energy denominator and summation are included in the definition of the operator $\widehat{\chi}$. Equation (1.59) can be written as a system of two equations:

$$\begin{aligned} \widehat{\Gamma}_1 &= \widehat{U} + \widehat{U}\widehat{\chi}_1\widehat{\Gamma}_1, \\ \widehat{\Gamma} &= \widehat{\Gamma}_1 + \widehat{\Gamma}_1\widehat{\chi}_2\widehat{\Gamma}. \end{aligned} \tag{1.60}$$

provided that the operator $\hat{\chi}$ is presented as a sum:

$$\hat{\chi} = \hat{\chi}_1 + \hat{\chi}_2 \tag{1.61}$$

This "two-step" approach is especially useful if $\hat{\chi}_1$, for example, is a singular operator that has a pole and the considered value of ω is close to the pole. The probabilities W of the considered process for the exact wave functions and calculated in the RPAE have some common features. For example, in the case of an electromagnetic field, W^r in the length form and W^∇ in the velocity form for exact wave functions coincide. The same is true within the RPAE: $W^r_{\text{RPAE}} = W^\nabla_{\text{RPAE}}$. Some integral relations for probabilities or quantities that differ from them by a numerical factor—the sum rules—also lead to similar results, both with exact wave functions and within the framework of the RPAE. These features are very convenient to use to control the accuracy of calculations.

The RPAE equation can be represented in differential and integral form. Equation (1.52) in integral form does not always provide very good accuracy, since the error arises from summation over unbound intermediate states. This disadvantage, however, is not very important due to the rapid decrease in matrix elements with increasing energy of intermediate states. On the other hand, such equations permit immediately derive expressions for matrix elements that are amplitudes determining the probability of the considered atomic or molecular process. This advantage makes the integral representation of the RPAE equations more convenient than the differential form, from which some radial functions are obtained, which are used later in the integrations necessary to calculate the matrix elements.

Note that in any form of presentation, differential or integral, there is an error, characteristic to the RPAE technique. It is due to violation of the Pauli principle in time-reversal diagrams. Indeed, in the RPAE, the states of two hole lines can coincide, while, according to the Pauli principle, they should always be different. In such diagrams, the error of the RPAE is of the order of $1/N^e$, where N^e is the number of electrons in an atom. This error is actually much larger when the interaction of electrons in taken into account within only one or two subshells, with the number of electrons $N_q < N_e$. For this reason, there is no point in overly exact numerical solution of the RPAE equations.

The RPAE approach permits to describe atoms and molecules with a non-degenerate ground state, which implies the existence of a single total electron wave function corresponding to each value of the total energy of an atom or molecule. In the case of a degenerate state, several wave functions (i.e., several levels) correspond to each energy value. This makes it difficult to apply the perturbation theory, since one should take into account virtual transitions between degenerate states, which correspond to the zero value of the energy denominator of the form (1.45). In this case, in order to obtain the final result, one should also take into account all divergent matrix elements that lie outside the scope of the RPAE, which is technically very difficult.

1.8 Rearrangement Method

In addition to, along with, and also in conjunction with the RPAE, the so-called *rearrangement* method [5] is used, which allows one to take into account consistently the role of the residual interaction V_{res} in calculating the probabilities of electronic transitions during ionization. This technique assumes that the initial state of the system, which existed before the external field acted on it, and the final state, can be described in the Hartree–Fock approximation. It is assumed that not only the electron, which is under the direct influence of the external field $B(r)$ changes its state, but all the others as well. These changes occur because the removal (or excitation) of one or several atomic or molecular electrons under the action of the applied field changes the self-consistent field of the residual ion, which leads to "unfreezing" of the levels of atomic and molecular electrons under the influence of an external field.

As a result, the wave function of the final state of an ionized atom or molecule is represented as a determinant similar to (1.4), i.e., constructed from one-electron wave functions. They are found by solving equations of the form (1.8) or (1.9), in the absence of terms corresponding to the removed electrons in the sum over k. Thus, the rearrangement of the atomic or molecular core due to the removal (excitation) of one or several atomic (molecular) electrons is taken into account. The solutions $\tilde{\varphi}_\kappa(x)$ of Eqs. (1.8) or (1.9) modified in this way are generally not orthogonal to the one-electron wave functions of the initial state $\varphi_j(x)$, that is,

$$\int \varphi_j^*(x)\tilde{\varphi}_k(x)\mathrm{d}x \neq \delta_{jk} \tag{1.62}$$

The total wave function of the final state is the antisymmetrized product of the wave functions of the residual ion and remote (excited) electrons in the field of the rearranged core.

The matrix element of the probability amplitude for the transition of an atom under the action of an external field is written in the form (1.11). This matrix element in the Hartree–Fock approximation with a frozen core can be expressed in terms of one-electron transition wave functions (1.12). In terms of the rearrangement method, this is the sum of three groups of terms that include the matrix elements of the operator $\widehat{B}(r)$ between all one-electron states.

The first group includes matrix elements similar to (1.12) between the states that were occupied in the atom (molecule) and became free in the residual ion:

$$\left\langle \nu_1|\hat{B}|\tilde{\nu}_2\right\rangle = \int \phi_{\nu_1}^*(x)\hat{B}(r)\tilde{\phi}_{\nu_2}(x)\mathrm{d}x,\ \nu_1 \in \Theta. \tag{1.63}$$

Matrix elements are multiplied by the products of the overlap integrals of the atomic (molecular) one-electron wave functions of the atom (molecule) $\phi_\nu(x)$ and the residual ion $\tilde{\phi}_\nu(x)$ with the same quantum numbers:

$$\prod_{\Theta}\langle\nu|\tilde{\nu}\rangle = \prod_{\nu}\int\phi_{\nu}^{*}(x)\tilde{\phi}_{\nu}(x)\mathrm{d}x, \tag{1.64}$$

where Θ is the set of levels from which electrons not removed; the product Π_{ν} includes the overlap integrals of all atomic or molecular states other than those from which electrons are removed. Due to rearrangement and violation of the orthogonality condition (1.62) $\prod_{\Theta}\langle\nu|\tilde{\nu}\rangle < 1$.

The second group includes matrix elements of the operator $B(r)$ between the wave functions of excited electronic $\phi_{\nu_3}(x)$ and atomic (molecular) electrons, which are not directly removed by the external field:

$$\left\langle\nu_3|\hat{B}|\tilde{\nu}_2\right\rangle = \int\phi_{\nu_3}^{*}(x)\hat{B}(r)\tilde{\phi}_{\nu_2}(x)\mathrm{d}x,\ \nu_3 \notin \Theta \tag{1.65}$$

Each matrix element (1.65) must be multiplied by the overlap integrals corresponding to the occupied states with the same quantum numbers in the original atom and the final residue ion $\langle\nu|\tilde{\nu}\rangle$ and to the states with different quantum numbers, $\langle\nu_1|\tilde{\nu}_3\rangle$, where $\nu_1 \in \Theta$. In the so-called frozen core approximation, integrals like $\langle\nu_1|\tilde{\nu}_3\rangle$ are equal to zero.

The third group of terms includes the matrix elements of the operator $B(r)$ between the states occupied in the initial atom and the remainder ion of the type:

$$\left\langle\nu_1|\hat{B}|\nu_2\right\rangle = \int\phi_{\nu_1}^{*}(x)\hat{B}(r)\tilde{\phi}_{\nu_2}(x)\mathrm{d}x,\ \nu_3 \notin \Theta \tag{1.66}$$

These matrix elements are multiplied by the overlap integrals associated with the filled states of the initial atom with the same quantum numbers and the filled states of the residual ion with different quantum numbers. Among them, the overlap integrals between the occupied initial states and the excited final states must be included.

The rearrangement technique significantly improves the calculated value of the atomic or molecular ionization potentials. It is not assumed to be equal the ionization potential energy I_{ν}^{HF} and the external Hartree–Fock energy level $|E_{\nu}|$, $I_{\nu}^{\mathrm{HF}} \neq |E_{\nu}|$. In this approximation, the ionization potential is determined by the difference between the total energies of an atom (molecule) and a residual ion with a vacancy ν_1 in the Hartree–Fock approximation:

$$I^{\mathrm{rear}} = -(\varepsilon_{A}^{\mathrm{HF}} - \varepsilon_{I(\nu)}^{\mathrm{HF}}). \tag{1.67}$$

The ionization potential I_{ν}^{rear} differs from I_{ν}^{HF} by the value of the rearrangement energy, which is a consequence of the change in the state of the ion as a result of the creation of a vacancy. For a slow electron removed from an atom, the rearrangement technique takes into account the fact that atomic and molecular electrons, due to the creation of a vacancy, find themselves in a stronger field and therefore are more strongly attracted to the nucleus, thereby screening it. Therefore, the self-consistent field of the ion, in which the removed electron moves, becomes weaker than in the "frozen" core. Rearrangement can also be important in virtual states, if their lifetime is long compared to the time of rearrangement.

Rearrangement is the result of the presence of residual interaction and in this sense is a manifestation of *electronic correlations* in the atom. It is advisable to take it into account when the lifetime of an excited discrete level or the time of removal of an electron from an atom exceeds the transformation time, which can be estimated using the uncertainty relation:

$$\tau_{\text{rear}} \sim \frac{\hbar}{I_\nu^{\text{exp}} - |E_\nu|}, \tag{1.68}$$

where $I_\nu^{\text{эксп}}$ is the measured value of the ionization potential of the level ν_1, E_ν is the electron energy at this level in the Hartree–Fock approximation.

The rearrangement method can also be applied in the case of a rapid sudden change in the atomic field, for example, when one or more fast electrons are removed from the atom, or as a result of nuclear decays when its electric charge changes. When the field changes abruptly, the electronic state does not immediately follow the field variation, and it is very likely that the electrons at the old levels become excited relative to the new atomic field. The amplitude of the probability of an electronic transition from a level in the initial field to a level in a new field is determined by the overlap integral $\langle \nu_1 | \tilde{\nu}_2 \rangle$, where $\phi_{\nu_1}(x)$ is the solution (1.8) or (1.9), and $\tilde{\phi}_{\nu_2}(x)$ is the solution (1.8) or (1.9) with the modified field. The field can be taken into account regardless of the nature of the change. For example, this approach works in the case of photoionization with a rapidly removed electron, as well as α- or β- nuclear decay.

It is rather difficult to establish exactly which infinite series of diagrams correspond to and are taken into account in the framework of the rearrangement method. However, it can be definitely indicated that these sequences include a refinement of the wave function of the initial electron, which changes due to the interaction of an atomic or molecular vacancy with more complex excited states "two holes—one clectron":

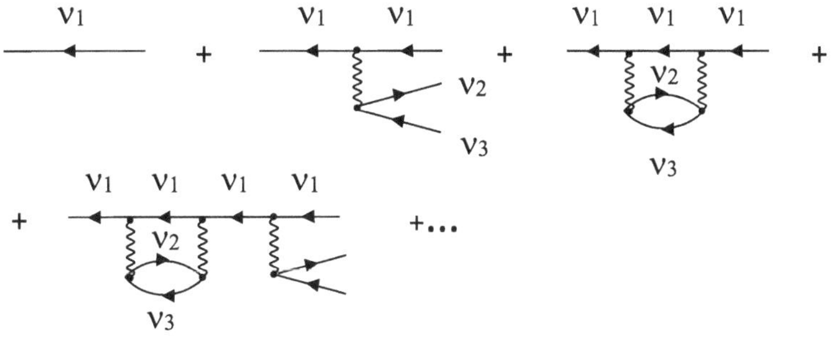

(1.69)

In contrast to perturbation theory, where the summation is performed over all possible virtual states, the rearrangement method takes into account only those matrix elements, moreover, of any degree in V_{res}, which include the initial vacancy ν_1.

In addition to the indicated series (1.69), the rearrangement method also allows one to include the interaction of a hole ν_1 with a newly created hole ν_3 and an electron ν_2. Only those processes that do not affect the vacancy ν_1 are taken into account, as well as the interaction of an electron–hole pair, in which it passes into another electron–hole pair, or its state does not change.

This method is used to study the processes of two or more electronic photoionization, atomic and molecular ionization accompanying α- and β- nuclear decay, etc.

When the wave functions of excited states in discrete or continuous spectra are used in the RPAE equations, taking into account rearrangement, then this technique of taking into account the correlation corrections is called the *generalized random phase approximation with exchange* (GRPAE).

Both experimental and theoretical studies of diatomic molecules, not to mention more complex ones, in their accuracy reached the level achieved in studies of atomic processes with a delay of several decades. Therefore, in theoretical calculations on the photoionization of the K-shells of diatomic molecules, up to the present time, the rearrangement method has been used in a simplified manner. Namely, as the wave function of the final ionized state of molecules, a certain intermediate function between the "frozen" core and the core rebuilt after ionization is chosen. This wave function is also found by solving the system of HF equations, in which the values of the coefficients are determined as some intermediate between the coefficients of the initial and final (ionized) states. One numerical parameter that determines these intermediate values of the coefficients (see the formulas in Chap. 4) is selected based on the requirement of the best match of the calculation results with experimental data on the position of the maximum $\sigma *$- the so-called shape resonance. Calculations show that in such a simplified model, simply the average values of the coefficients between the ionized and the initial state lead to good agreement with experiment.

Approaches based on the Hartree–Fock approximation, RPAE, SRPAE, methods of perturbation and rearrangement theory, GRPAE are implemented in the ATOM-M system. That is why this system can be widely used to study the structure of atomic and molecular structures and the interaction of such structures with external fields.

1.9 Objects of Research and Considered Processes

The proposed system ATOM-M and its earlier versions [6] were used to investigate processes involving atoms, diatomic molecules, fullerenes, and endohedrals. The main attention was paid to photoionization, one- and two-electron, calculation of its total and differential cross sections, scattering cross sections of fast electrons and angular distributions of secondary electrons knocked out in this process. The scattering cross section for fast electrons is expressed in terms of the so-called generalized oscillator strengths (GOS). Radiative and non-radiative, i.e., Auger decays, were considered. The systems ATOM and ATOM-M have been successfully used to calculate the cross sections for elastic and inelastic electron scattering of slow and intermediate velocities.

Photoionization of noble gas atoms, corresponding to the angular distributions, has been studied most fully. The programs presented below were applied without any modifications to the same calculations for other atoms with filled shells, for example, alkaline earth atoms. Considerable attention was paid to atoms with semi-filled shells, as well as to alkalis. Several atoms with open shells, in particular, chlorine and iodine, were considered as an example.

Calculations of the inelastic scattering of fast electrons and the angular distribution of secondary electrons knocked out of the target atom are more laborious than photoionization, are less studied experimentally, and therefore were limited only to atoms of noble gases.

The situation is similar with vacancy decays, which were considered on the example of just a few transitions in a pair of noble gas atoms.

The calculation of the scattering cross sections for slow electrons and, especially, for electrons of intermediate velocities requires very laborious calculations, and therefore was carried out only for neon, argon and xenon, as well as manganese and chromium.

When using a number of simplifying assumptions about the structure of the electron shell of fullerenes, it is possible to use the proposed software package for calculating the photoionization cross sections of endohedral atoms, i.e., systems consisting of fullerene, inside which an ordinary atom is placed.

The most complete available calculation results are presented in the monograph [6].

Calculations for molecules are limited so far by a few examples of H_2 and N_2 photoionization cross sections. The restriction to symmetric diatomic molecules is dictated by their simplicity, since, compared to atoms with filled shells that are spherically symmetric, these molecules are axisymmetric, thus losing only one degree of symmetry. There exists here an enormous scope for considering a range of processes and their characteristics.

1.10 Description of the ATOM-M System of Programs

The ATOM-M system is the latest, modern version of the ATOM system, the main capabilities of which were previously described in [7] with additional procedures that allow performing a number of similar calculations for systems with axisymmetric, and not only centrally symmetric, potential.

The ATOM-M system provides the calculation of the following characteristics. First of all, in the Hartree–Fock approximation, the electronic wave functions of the ground states of atoms or molecules are calculated. Also, the wave functions of excited states of an electron, meson, and positron in the field of an atomic or molecular ground state or in the field of an ion formed as a result of the removal of one electron without changing the state of the rest are calculated. The ATOM-M system also allows the calculation of wave functions and energy levels in the Hartree–Fock–Dirac approximation.

The use of the wave functions of the ground and excited states in the Hartree–Fock approximation makes it possible to calculate the matrix elements of the interaction between the electron and the applied electromagnetic field and the photoionization cross section in the one-electron approximation, in the frame of which the reaction of other electrons to the removal of the electron under consideration is neglected. The wave function of the removed electron is considered in the field of an ion created as a result of the interaction between an atom or molecule and a photon. The response of other electrons is described in the framework of the RPAE, the main element of which is the effective interaction Γ determined by Eq. (1.56), which differs from the pure Coulomb interaction.

The ATOM-M system enables the user to calculate the cross section taking into account the influence of electrons in one, two or more shells. To obtain the angular distribution of photoelectrons, the elastic scattering phases in the ion field are also found. The reaction of an atom to a fast electron (i.e., an electron with an energy much exceeding the atomic ionization potential I) leads to the same result as the action of some electric external field of arbitrary multipolarity. The characteristic of this reaction, the generalized oscillator strength (GOS), which determine the scattering cross section of all fast charged particles, is considered in the ATOM-M system both in the framework of the one-electron approximation, which ignores electronic correlations, and the RPAE, which takes into account these correlations.

RPAE is applied most simply to atoms with closed subshells. However, since in the ground state of atoms with semi-filled subshells the spins of all electrons are directed in the same way (Hund's rule), these atoms can be considered as atoms with filled (closed) subshells containing, however, two types of electrons: with spin "up" and with spin "down". Accordingly, modified accounting for exchange and the calculation of the effective interaction in the RPAE makes it possible to find the wave functions and photoionization cross sections for atoms with semi-filled and open shells.

Polarizability is an important characteristic of an atom (or ion) that determines the cross section for the scattering of light. It is also calculated within the ATOM-M system.

Inelastic scattering of electrons of intermediate energy leads mainly to the removal of an atomic electron, i.e., ionization of an atom or molecule by electron impact. The ATOM-M system calculates the electron ionization cross section for a given scattering angle of the incident particle, the scattering angle of the ejected electron, and the energy of both. Other electrons, both in the ionizable shell and in other subshells, can take part in such processes. The ATOM-M system can also take into account the corresponding corrections within the framework of the RPAE and perturbation theory.

In the case of photoionization of a deep shell or inelastic scattering near its threshold, the vacancy decays by the Auger mechanism takes place before the photoelectron leaves the atom. As a result, this electron is not in the field of the initial vacancy (the field of the primary ion), but in a much stronger field (at least as strong as the field of the secondary ion). This noticeably affects the photoionization cross section. This effect, known as *post-collision interaction* (PCI), can also be accounted for in the ATOM-M system.

An electron removed from an atom by photoionization can knock out or excite another atomic (molecular) electron. The ATOM-M system allows calculating the probability of this process. The ATOM-M system makes it possible to find the ionization potential in the Hartree–Fock approximation, to correct it within the framework of the SRPAE, or to calculate it using the rearrangement method.

As a rule, in the study of the scattering of slow electrons by an atom or molecule, one should take into account the contribution to the scattering cross section of the polarization interaction caused by the virtual excitation of the atom by the scattering electron. Considering the creation of only one electron–hole pair and limiting the allowance for the polarization interaction to the lowest order, the ATOM-M system makes it possible to find the elastic scattering phases of slow electrons in the framework of the SRPAE, and with their help, the total elastic scattering cross section. The cross section of the inelastic scattering process, in which the energy of a slow incident electron is transferred to an atomic electron, leading to its excitation or ionization, is determined by the imaginary part of the scattering phase.

Since the polarization interaction plays an important role for a number of atoms, the ATOM-M system provides for the possibility of its accurate accounting.

An important characteristic of a vacancy formed as a result of the ionization process is the so-called level width γ. It is related to the lifetime τ of the level by the ratio $\tau = 1/\gamma$. The level width is proportional to the probability of its decay. What we call decay is a process in which one of the outer electrons makes a transition to a lower vacant level, while excess energy is transferred to another atomic electron, leading to its ejection from or excitation in an atom or molecule. This decay process is known as the Auger effect, and its probability is calculated within the ATOM-M system. In calculating the lifetime of a hole (vacancy), a certain role is played by the interaction with other electrons. This interaction using the ATOM-M system can be taken into account in the framework of the RPAE or perturbation theory. The ATOM-M system allows calculating the probabilities of radiative decay, in which the energy of the transition of an electron from one level to another is emitted in the form of a photon. The calculation of the probability can be carried out both within the framework of the Hartree–Fock approximation, and in RPAE.

Due to interaction between electrons and vacancies, two or even more vacancies can be created even as a result of single-photon ionization or inelastic scattering of fast electrons by an atom or molecule. These vacancies can then decay independently or correlatively. In the case of independent decay, its total probability for two or more vacant states is equal to the product of the decay probability of each vacancy calculated using the ATOM-M system. The presence of several vacancies, however, should be taken into account when determining the transition energies.

Correlated decays are also possible, for the realization of which it is essential to take into account the interaction between the initially formed vacancies. Of particular interest is the decay of a state with two vacancies, which proceeds with the emission of a single electron or photon. This type of process is responsible for the appearance of characteristic lines in the spectrum of emitted photons or electrons with energies often much higher than the ionization potential of the inner atomic shell. Such decays are called the double Auger effect and the one-photon decay of the two-vacancy state, respectively. To find the probability of a process, we need to know the transition energy and decay matrix elements. The corresponding calculation is performed in the ATOM-M system. The system also allows one to calculate the probabilities of different modes of correlated decay, in which the energy of the initial two-vacancy state is acquired by two or more emitted electrons or photons, or both.

Bremsstrahlung is an important mechanism of inelastic scattering of fast electrons and electrons of intermediate energy. This process is investigated within the ATOM-M system directly in the Coulomb and Hartree–Fock fields of an atom or molecule (ignoring their dynamic reaction), as well as taking into account the influence of other electrons of the atom or target molecule within the framework of the RPAE and GRPAE.

The calculation of the probability of positron scattering by an atom or a molecule begins with the numerical solution of the radial equation for χ_E obtained from (1.17) and is carried out in the ATOM-M system.

Inelastic scattering of any other charged particle, for example, mu-mesons, as well as the transition from level to level in a meso-atom, is accompanied by the emission of atomic or molecular electrons. The ATOM-M system allows one to obtain the probabilities of these processes, as well as the phases of mu-mesons elastic scattering by atoms, taking into account the polarization interaction between them.

Almost every program in the ATOM-M system provides more characteristics than those for which a specific program is aimed. For example, atomic form factor $F(q) = \int \exp(i\mathbf{qr})\rho(r)\mathrm{d}r$ (where q is the momentum acquired by an atom from an incident particle, and $\rho(r)$ is atomic or molecular electron density) or phases of elastic scattering of an electron by an atom, ion, or molecule and matrix elements of interaction between atomic electrons, etc. Another example is the calculation of the autoionization profile shape of a state taking into account the excitation "two electrons—two holes", which is achieved by combining the matrix elements calculated in different programs of the ATOM-M system.

The structure of the ATOM-M system allows it to be easily expanded and combined with other programs, allows one to study a larger number of processes and a larger number of properties of interaction of atoms with external fields and incident particles.

Making some reasonable approximations about the structure and behavior of a fullerenes shell, ATOM-M permits to study a whole variety of processes in so-called endohedral atoms.

Of interest are not only the amplitudes of atomic and molecular processes that determine their cross sections, but also the energy derivatives of the phases of the amplitudes that determine the duration in time of respective physical processes.

Our experience of using the ATOM-M system has shown that, although it is implemented in the FORTRAN language, it is relatively easy to translate it into another computer language.

References

1. D. Hartree, *The Calculation of Atomic Structure* (Wiley, New York, 1957)
2. V.A. Fock, Z. Phys. **61**, 126; **62**, 795 (1930)
3. L.D. Landau, E.M. Lifshits, *Quantum Mechanics: Non-Relativistic Theory*, 3rd edn. (Pergamon Press, Oxford, 1977)
4. M.Ya. Amusia, *Atomic Photoeffect* (Plenum Press, New York, 1990)
5. M.Ya. Amus'ya, V.K. Ivanov, S.A. Sheinerman, S.I. Sheftel', JETP **51**, 3, 458 (1980)
6. M.Ya. Amusia, L.V. Chernysheva, V.G. Yarzhemsky, *Handbook of Theoretical Atomic Physics, Data for Photon Absorption, Electron Scattering, Vacancies Decay* (Springer, Heidelberg, 2012)
7. M.Ya. Amusia, L.V. Chernysheva, *Computation of Atomic Processes* (Institute of physics Publishing, Bristol and Philadelphia, 1997)

Chapter 2
Physical Content of the ATOM-M System Programs

This chapter describes the main programs of the ATOM-M system. Each program corresponds to an independent physical problem related to the study of the structure of atoms and the calculation of the probabilities of atomic processes. The choice of approach is described for each investigated problem.

2.1 Wave Functions of the Ground State of an Atom in the Hartree–Fock Approximation

The one-particle approximation underlies almost every modern approach to the study of atomic or molecular properties. In this approximation, the atomic or molecular electron wave function is represented as a set of one-electron wave functions (1.4), which can be conveniently written in the form of Slater's determinant. We find the wave functions in the Hartree–Fock approximation, which is the best one-particle approximation. The equations defining these functions are obtained on the basis of the assumption that the total electron energy of an atom or molecule is minimal when Hartree–Fock one-electron wave functions are chosen as the one-electron ones.

These wave functions for the ground state of the atom are obtained using the program called ***scfhf***[1] [1–3]. One-particle wave functions of excited states of an atom in a self-consistent field or in the field of the so-called frozen core are obtained using the program called ***fchf***[2] [2, 4]. Both groups of functions are used as zero approximations when calculating various physical properties of atoms and atomic processes with inclusion of many-electron correlations. These two programs, ***scfhf*** and ***fchf***, are essential parts of a package designed to accommodate residual interactions V_{res}, i.e., investigate the role of correlations in an atom.

[1] ***scfhf***—Self-consistent functions in Hartree–Fock.

[2] ***fchf*** = Frozen core Hartree–Fock.

M. Ya. Amusia and L. V. Chernysheva, *Computation of Atomic and Molecular Processes*, Springer Series on Atomic, Optical, and Plasma Physics 117,
https://doi.org/10.1007/978-3-030-85143-9_2

The ***scfhf*** program calculates one-electron radial wave functions in the non-relativistic one-configuration Hartree–Fock (HF) approximation for an arbitrary atom or ion in the absence of a given total term or in the LS-coupling scheme, which corresponds to one Slater determinant for atoms with closed shells/subshells or a linear combination of several Slater determinants for atoms with open shells/subshells.

The ***scfhf*** program also calculates the wave functions and ground state energies for some negative ions, where the affinity of an additional electron to an atom is calculated using the Hartree–Fock potential.

Although a number of computer programs for calculating wave functions in the Hartree–Fock approximation have already been developed earlier [5, 6], we present here their new version, which has several advantages. One of them is that it can be used to calculate the wave functions of the ground and excited states of electrons, positrons, mesons, protons in discrete and continuous spectra and various physical characteristics of atomic processes.

Similar programs for the molecule will be discussed in Sect. 2.5.

2.2 Wave Functions of the Ground State of an Atom in the Hartree–Fock-Dirac Approximation

Atomic calculations, which constitute an important part of the study of many processes occurring in heavy atoms, should be based on the one-electron relativistic self-consistent approximation known as the Hartree–Fock–Dirac (HFD) approximation. This makes it possible to take into account the fact that in heavy atoms the average velocities of the electrons of the inner and some intermediate shells are close to the speed of light. Within the framework of the HFD approximation, the wave function of an atom is written in the form of Slater's determinant, composed of one-electron relativistic wave functions, which are obtained by solving a system of integral–differential equations of the HFD. This system is obtained from the same as HF requirement for a minimum of the total relativistic energy of an atom on the basis of relativistic one-electron wave functions [7, 8].

In the ***rscfhf***[3] program [9], one-electron radial wave functions of the ground state are calculated in the Hartree–Fock–Dirac approximation for an arbitrary atom or ion in the absence of a term or in the *jj*-coupling approximation. However, in contrast to [7, 8], all subshells of an atom are divided into two groups. For the inner and part of the intermediate subshells, a relativistic calculation should be carried out, and for the rest, a non-relativistic approach is sufficient. In general, the wave function of an atom is sought in the form of Slater's determinant, which includes functions of both types. For an atom with open shells, the wave function is a linear combination of several Slater determinants.

It is rather difficult to correctly calculate the exchange interaction operator (see, for example, [10]). It is especially difficult to do this when numerically solving a set of

[3] ***rscfhf***—relativistic self-consistent functions in Hartree–Fock approximation.

HFD equations. Our algorithm accurately takes into account the exchange, including that between the relativistic functions of the inner and non-relativistic functions of the outer shells. The program we are presenting has certain advantages over similar, already known, programs [7, 8, 11–14]. One of them is that the main set of equations is obtained neglecting the higher orders in the small parameter in the equations of the exact set of HFD.

The small parameter is understood here as the fine structure constant $\alpha = 1/c \approx 1/137$. The HPD equations also include terms of the order $Z\alpha = Z/137$, which for heavy atoms is of the order of unity. The approximation we have adopted preserves all terms of any degree of the parameter $Z\alpha$, but is limited to no more than the second degree of corrections in powers of α. The calculation of this simplified system [9] is much easier than in [7, 8, 11–14], and the results coincide up to two or three decimal places, which guarantees a sufficiently high accuracy of calculations.

2.3 Wave Functions of the Excited State in the Hartree–Fock Approximation

The wave functions of excited states in the one-particle Hartree–Fock approximation can be found in two ways.

First of all, Eq. (1.8) can be derived for a term and a configuration corresponding to a certain excited state of the atom. The ***scfhf*** program is used for such calculations. In it, however, the condition of orthogonality of the wave functions of the excited and ground states is ignored, so that the one-electron wave functions of various excited states turn out to be eigenfunctions of various Hamilton operators.

In another version of this program, one-electron wave functions of excited states in the field of a "frozen" core are calculated. In this case, the wave functions of the core are assumed to be the same as the wave functions of the ground state. As a result, the one-electron wave functions of excited states, both in the discrete and in the continuous spectrum, are found as eigenfunctions of the same Hamilton operator (1.3); that is, they are solutions of the same Eq. (1.8). Such a set of wave functions is required as a basis for the expansion of the corrections taken into account the matrix elements V_{res}, which is necessary when including the electron correlations, in the framework of perturbation theory. Thus, obtained wave functions are used to calculate a number of physical characteristics of an atom and to describe the elastic and inelastic scattering of electrons by atoms, their photoionization, and other processes in the Hartree–Fock approximation and taking into account multielectron correlations.

The most commonly used method is the calculation of one-electron wave functions in the Hartree–Fock approximation with a "frozen" core. The ***fchf*** program allows one to find the total atomic wave functions in the form of a single Slater determinant or to reduce them to a given type of coupling between the angular momenta and spins of individual electrons into the total atomic spin S and angular momentum L

(e.g., the LS-coupling approximation), which corresponds to linear combination of several Slater determinants. A number of options are possible here.

(A) In the program ***fchf*** (variant "exchange before orthogonalization"), the wave functions of the electron can be calculated. In this program, the convergence of the iterative sequence of solving the radial Hartree–Fock integral–differential equation essentially depends on the so-called exchange parameters, which determine the magnitude and sign of the non-local exchange potential. The exchange potentials are calculated taking into account the term and configuration of the excited state. As a rule, they are small and this ensures the convergence of the iterative process.

(B) However, in an entire class of excited states (monopole excitations), one of the exchange parameters is a negative number with a large absolute value. For example, in Ar ($3p^5np\ ^1S$) and Pd ($4d^9nd\ ^1S$), the exchange parameters are equal $\gamma(3p^5, np\ ^1S) = -5$ and $\gamma(4d^9, nd\ ^1S) = -9$, respectively. In such cases, the wave functions of an excited electron cannot be calculated as described in the previous version. On the other hand, monopole excitation spectra are of significant interest. It is necessary to know them when solving a number of problems in which it is important to have the monopole polarizability of the outer subshells. This polarizability determines the corrections to the Hartree–Fock energy for the electrons in the outer shells. The ***fchf*** program (orthogonalization before exchange) then treats the Hartree–Fock equation as a linear differential equation with the exchange potential placed on the right. The essence of the corresponding algorithm is that to solve this equation iteratively, it is necessary to orthogonalize the wave functions that are taken as initial values. As a result, the exchange potential decreases rapidly, reaching its normal value, and the convergence of the iterative procedure turns out to be fast. Using this program, the wave functions of the excited states of the aforementioned term ^1S were calculated. In addition, the described procedure reduces the computation time for the wave functions of a low-energy electron in other cases at large off-diagonal energy parameters.

(C) The main difference of this version of the program is that the off-diagonal energy parameters are expressed through the matrix elements of the Coulomb interaction. With each new iteration, the parameters are calculated using the wave function found in the previous iteration. As a result, this version of the program allows calculating the wave function, including sometimes those cases for which the calculation procedure based on algorithms A and B does not converge.

So, the program ***fchf*** [2, 4] permits to choose the appropriate option for determining the wave function, saving the computer time of computation in cases where the off-diagonal parameters are small. In addition, this program can be used to calculate wave functions that are not orthogonal to the main wave functions, with arbitrary off-diagonal parameters. This choice is required, in particular, for calculating the photoionization cross section for atoms with semi-filled shells in a modified spin-polarized random phase approximation with exchange—SP RPAE. In this approach, wave functions are employed that are not orthogonal to the wave functions of the occupied states, while the off-diagonal energy parameters are determined from the condition of equality of the photoionization cross sections in the "length" and "velocity" forms.

The ***fchf*** program is also used to calculate the wave functions of mesons, positrons, and protons. This solves the Hartree equations, since there is no exchange term in the equation. Of course, the difference between the charge and mass of the incident particle from their values for an electron is also taken into account.

2.4 Wave Functions of an Excited State of an Atom in the Hartree–Fock–Dirac Approximation

When calculating the cross section for ionization of an atom by photons of high energies, of the order of the electron rest mass, it is necessary to take into account both the difference between the interaction operator of a photon with an electron from a purely dipole one and the relativistic nature of the wave function of an electron in its initial and final states. In addition, there are atomic transitions that are completely forbidden by the condition of conservation of the angular momentum in the non-relativistic approximation and are allowed when relativism is taken into account.

Considering the inclusion of electron correlations in the inner and intermediate shells of heavy atoms also requires knowledge of the relativistic one-particle wave functions. It is also advisable to use them when calculating the decay of vacancies in the inner shells of atoms.

The wave function of the excited state of an atom in the Hartree–Fock–Dirac approximation [9], as well as in the HF approximation, is determined in the field of the ground state of the atom found in the HFD approximation, by analogy with calculating the ground state, i.e., with the same restrictions.

The ***rfchf***[4] program allows one to find the wave functions of the excited state in the Hartree–Fock–Dirac approximation.

Based on the same approximation as the program for calculating the ground state function ***rscfhf,*** the ***rfchf*** program avoids taking into account negligibly small relativistic corrections for those states and transitions whose velocities and energies are far from relativistic.

2.5 Wave Functions in the Hartree–Fock Approximation for Diatomic Molecules

Molecules as an object of investigating the role of electronic correlations are of interest not less, but more than multielectron atoms. The reason for this is in the large

[4] ***rfchf***—relativistic function continuous in Hartree–Fock.

number of correlating electrons and in the distribution of their confining attractive charge among several nuclei. As a result, the relative role of the residual electron–electron interaction in them should be greater than in even heavy atoms. In addition, in a molecule, due to the presence of several attraction centers, specific, close in energy electronic states arise, the interaction between which significantly affects the processes in molecules in comparison with similar atomic processes.

Photoionization of a molecule is an interesting experiment, since an electron can arrive at an observation point in different ways, starting from different centers. There is a close analogy between the interference pattern that occurs when an opaque screen with a number of narrow, equidistantly spaced slits is installed between a point light source and an observation screen. As a result, a complex interference pattern is observed due to the wave nature of light. The photoelectron wave behaves like a light wave and the angular distribution of photoelectrons is determined both by which atom they are removed from and by which atoms they collide with when leaving the molecule.

However, the study of the electronic structure of molecules and processes with their participation is complicated by the fact that the field of molecules is in general not spherically symmetric that makes it impossible to significantly simplify the problem by reducing the numerically solved equations to one-dimensional radial equations. As is known, in a spherically symmetric field, the angular parts are separated and the associated calculations are performed analytically. The molecular problem is multicenter and the corresponding calculations in the HF approximation for one-electron functions require the numerical solution of three-dimensional equations, and even with a multicenter potential. It should be kept in mind that the movement of the atomic nuclei included in its composition is also important in the molecule, which can both vibrate and rotate as a whole.

As a result, poorly controlled simplifications are inevitable, which sharply reduce the value of calculating molecules from the point of view of studying and separating the role of multielectron correlations in them.

A useful exception is diatomic molecules, which have axial symmetry that simplifies calculations. Examples of very interesting objects are polar molecules of the type KCl or CsI, the photoionization of the outer shells of which would be very interesting to compare with the data for $3p^6Ar$ and $5p^6Xe$, respectively. From the point of view of calculation, the situation is especially simple with diatomic molecules consisting of identical atoms. We will pay special attention to them. A decisive simplification is achieved here by going over to spheroidal coordinates.

The modern approach used here to study the properties of diatomic molecules is based on the Hartree–Fock approximation [15–17]. The wave functions are presented in spheroidal coordinates as expansions in spherical harmonics (see Chaps. 4 and 6).

The ***scfhfm***[5] program solves the Hartree–Fock equation for the ground state of a diatomic molecule in spheroidal coordinates, i.e., calculates the sets of smooth functions $P_{j\lambda}(\xi)$ included in expansion (4.15). Thereby, a set of energy eigenvalues

[5] ***scfhfm***—self-consistent field Hartree–Fock for molecules.

ε_j from (4.23) and the corresponding wave functions $\varphi_t(\mathbf{r})$, $t = 1 \div N_s$ for all electron shells t present in the ground state of the molecule are found.

The ***fchfmd***[6] program solves the Hartree–Fock equation in a frozen core for discrete excited states of an electron in a diatomic molecule. Those for a given number of excited states, the negative energy eigenvalues ε_f and the corresponding wave functions $\varphi_f(\mathbf{r})$ are found (see (6.17) and (6.18)).

The ***fchfmc***[7] program solves the Hartree–Fock equation in a frozen core for a continuous spectrum of excited states of an electron in a diatomic molecule, i.e., solves Eq. (6.18). As a rule, this is done for a set of positive energies ε_f corresponding to equidistant values k_f of the electron momentum. For each value $\varepsilon_f(k_f)$, the solutions of the system of Eq. (6.18)—functions $X_{f\lambda}(\xi)$ are found. Thus, the wave functions $\varphi_f(\mathbf{r})$ of the continuous spectrum (6.16) are obtained.

The position of the nuclei, i.e., the internuclear distance R_Z is considered fixed in each calculation. When performing a series of calculations for various R_Z, it is possible to study the photoionization parameters as a function of the internuclear distance.

2.6 Wave Functions in the Hartree–Fock Approximation for Endohedrals

Endohedrals are relatively recently discovered objects that represent an ordinary atom or a simple molecule placed inside a closed fullerene shell. Potentially important practical applications of endohedrals for creating new materials and their complex structure stimulate the research of them as, first of all, many-electron objects with a nucleus distributed in space. In general, it is appropriate to consider endohedrals as quasi-atoms with electron shells much larger than those of ordinary atoms. In the endohedral, the fullerene polarizes and deforms the atom placed in it and in turn is affected by this atom.

The interaction between an atom placed in a fullerene and the fullerene itself is clearly manifested in the processes of photoionization of a "trapped" atom, and the decay of vacancies in a "trapped" atom with the participation of the fullerene shell. We will focus on the development of computational programs for considering these processes. We take into consideration two types of effects—the deformation or polarization of the fullerene electron shell by the photon field and the modification of the photoelectron wave function due to its scattering by the static field of the fullerene. The latter is described either by a potential of zero thickness, and, accordingly, an infinite depth well (so-called bubble potential), or by a potential of finite width and depth. Considerations that allow calculating the parameters of the fullerene potential and the modification factor of the external electromagnetic field under the influence of a fullerene will be discussed in Sect. 2.29.

[6] ***fchfmd***—frozen core Hartree–Fock for molecules, discrete.

[7] ***fchfmc***—frozen core Hartree–Fock for molecules, continuous.

We assume that the bubble potential does not affect the wave functions of the initial state of atomic electrons. The situation is different for a potential of finite width and depth [18]. To account for potential influence, two programs have been developed.

The ***scfhfw***[8] program (a version of the ***scfhf*** program) allows solving the Hartree–Fock equation with a potential of finite depth and width for the ground state of atoms placed inside the fullerene shell.

The ***fchfw***[9] program (version of the ***fchf*** program) solves the Hartree–Fock equation with a potential of finite depth and width for the excited state of electrons of atoms enclosed in a fullerene shell.

2.7 Photoionization Cross Section and Oscillator Strength

Multielectron correlations are included in the random phase approximation with exchange (RPAE), which allowed us for the first time to obtain good agreement between the calculated and measured photoionization cross sections for noble gas atoms in a wide range of photon energies. The great advantage of this approach is that the cross sections found with the coordinate and momentum matrix elements (i.e., in the form of "length" and "velocity") coincide. In addition to the total photoionization cross section, the angular distribution of photoelectrons was obtained, which, as a rule, is also in good agreement with the measured data. The spin polarization and the probabilities of vacancy decay were also calculated with account ща RPAE correlations.

Within RPAE, it is natural to distinguish two types of correlation corrections defined by the second term in the equation for $\langle \nu_1 | D(\omega) | \nu_2 \rangle$ (1.52): intra- and inter-transitional or intra- and intershell corrections.

Intra-transitional correlations are associated with the same transition. They correspond to terms with different energies and equal orbital angular momentum $l_3(l_4)$ summed over intermediate states $\nu_3 > F(\nu_4 > F)$ and terms with equal energies $E_4 = E_2$ and angular momentum l_2 summed over $\nu_4 \leq F(\nu_3 \leq F)$, respectively. After separating the angular terms in the one-electron wave functions and integrating over the angular variables, intra-transition correlations are taken into account if the sum on $\nu_3 \leq F(\nu_4 \leq F)$ includes the term with $E_4 = E_2$, and the sum on $\nu_3 > F(\nu_4 > F)$ includes terms corresponding to different energies of the discrete and continuous spectra, but with $l_2 = l_1 \pm 1$.

Inter-transitional correlations correspond to terms with different vacancy energies, but with $l_2 = l_1 \pm 1$ summed over $\nu_3 > F(\nu_4 > F)$.

The difference between intrashell and intershell correlations is that in the first case, the sum on $\nu_4 \leq F(\nu_3 \leq F)$ includes only one term, with $E_4 = E_2$, for one

[8] ***scfhfw***—self-consistent field Hartree–Fock, finite width well.

[9] ***fchfw***—frozen continuous Hartree–Fock, finite width well.

subshell, while in the second case, there may be two or more shells, depending on the number of shells considered.

Usually, the strong transition, that is, the one that increases the orbital angular momentum $l_2 \rightarrow l_1+1$, makes the most important contribution to the photoionization cross section, while the weak transition $l_2 \rightarrow l_1 - 1$ mainly manifests itself in the angular distribution of photoelectrons.

Accounting of the interaction between different transitions showed the influence of a strong transition, for example, $np - \varepsilon d$ on a weak one $ns - \varepsilon p$ near its ionization potential, changing the cross section of the latter in some cases by several times in comparison with the single-particle calculation. On the contrary, at very high energies, the opposite effect is very noticeable—a weak transition upon a strong one. This begins to manifest itself in the region of photon energies, where the relative role of their contributions is reversed. The reason is that, with increasing energy, the contribution of the weak transition becomes larger than that of the strong one. The experiment is usually in good agreement with these calculations.

Taking multielectron correlations into account in RPAE provides reliable data for the total photoionization cross section of atoms, their partial cross sections for individual transitions, and the angular distribution of photoelectrons. The ***phoabs***[10] program [2, 19] calculates the photoionization cross section and oscillator strengths for an atom or ion in the form of "length" and "velocity" in the one-configuration Hartree–Fock approximation. It also takes into account multielectron correlations in RPAE. This program can only be used for atoms or ions with all filled or semi-filled shells.

The ***ospp***[11] program determines intrashell and intershell correlations in atoms with open shells [2, 20]. It also calculates the photoionization cross sections and the oscillator strength for atoms (ions) in the form of "length" and "velocity," taking into account many-electron correlations in an arbitrary number of transitions. It is possible to calculate the photoionization cross sections in a narrow energy range with small fixed intervals of the photon energy, and thus numerically find the position of autoionization resonances if they make a significant contribution to the cross section.

Finally, the ***polar***[12] [21] program calculates the dipole dynamic polarizability of the atom for the given energy values, and the ***level***[13] program allows you to find the so-called photoionization quasi-cross sections. Quasi-cross sections are obtained from the discrete oscillator strengths multiplying them by a numerical factor and by the energy interval to the adjacent level. This concept is convenient in that region of the spectrum of discrete excitations where they are located so close to each other that they cannot be distinguish experimentally.

In the ***phoabs*** program, you can also get the matrix elements of the operator of interaction of a photon with an electron on the mass shell. By "mass shell" we mean the fulfillment of the energy conservation law—the photon energy ω—is equal to

---

[10] ***phoabs***—photoabsorption.

[11] ***ospp***—open shell photo process.

[12] ***polar***—polarizability.

[13] ***level***—discrete oscillator strength into smooth cross section.

the sum of the energies of the photoelectron ε_p and the ionization potential of the hole I_i. Some problems require knowledge of the matrix elements of the operator of interaction between an electron with a photon and the Coulomb interaction between electron off the mass shell. These matrix elements are found using ***pabsout***[14] that is a version of ***phoabs***.

2.8 Angular Distribution of Photoelectrons in the Dipole Approximation

The angular distribution of photoelectrons in the dipole approximation is determined, in contrast to the cross sections, not only by the moduli of the photoionization dipole matrix elements, but also by their phases. Investigation of the angular distribution provides information on the scattering phases of photoelectrons and on the phases of the matrix elements of the RPAE. In addition, the angular anisotropy parameter of photoelectrons contains, along with the squares of the moduli of the matrix elements of $l_2 \to l_1 + 1$ (strong) and $l_2 \to l_1 - 1$ (weak) transitions, also their product, enhanced by a numerical factor 6. As a result, the angular anisotropy parameter permits to reveal the contribution of a weak transition against the background of a strong one much more accurately, than the study of the partial cross sections of photoionization.

In the ***phoabs*** and ***ospp*** programs, there is a reference to the ***beta***[15] procedure, which allows obtaining the angular distribution of photoelectrons in the dipole approximation [2, 19].

2.9 Scattering Cross Section and Generalized Oscillator Strength for Fast Electrons

It is known that a number of characteristics of inelastic scattering of fast charged particles (electrons, mesons, protons, etc.) by atoms can be expressed in terms of generalized oscillator strengths (GOS) [22]. In contrast to the oscillator strength in photoionization, the generalized oscillator strength gives information not only about dipole transitions with $L = 1$, but also about transitions of various multipoles (with $L = 0, 1, 2, 3$, etc.). GOS describes the response of an atom to the transfer of any momentum, including a large one, q, in contrast to photoionization, where the momentum transferred to the target atom is small, since it is equal to $\kappa = \omega/c$. GOS can be used to compute differential cross sections for inelastic scattering of any charged particle over a wide range of energies. The only, albeit strict, limitation is that the speed of the incident particle must be much higher than the speed of atomic

[14] ***pabsout***—photoabsorption out of the energy shell.

[15] ***beta***—angular anisotropy parameter β.

electrons, so that we can restrict ourselves to the first Born approximation in the interaction of the incident particle with an atomic electron.

The calculated data for GOS in the one-particle approximation for intermediate and heavy atoms, as a rule, are in poor agreement with experiment. Even for the best one-particle approximation, the Hartree–Fock one, the calculated results differ from the measured data quite often by 1.5–2 times. In addition, in the HF approximation, the generalized oscillator strength obtained with the matrix elements of the operators $exp(i\mathbf{qr})$ and $[exp(i\mathbf{qr})]\vec{\nabla}$, on average, differs by a factor of two, while for exact wave functions they must coincide, both for operators $\mathbf{r}$ and $\vec{\nabla}$ in photoionization.

We have developed a method for accounting for multielectron correlations for the GOS in the RPAE. This approach ensures that the GOS obtained with the matrix elements of the operators $exp(i\mathbf{qr})$ and $[exp(i\mathbf{qr})]\vec{\nabla}$ are matched.

The method applied to electrons inside a single subshell provides good agreement between the calculated and measured GOS and GOS density for noble gas atoms, for example, Ar, over a wide range of energy ω and momentum transfers q.

Taking into account, the interaction between transitions of different subshells demonstrates the mutual influence of one transition on another. In some cases, the transitions essentially mix with each other, increasing their GOS by a factor 1.5–2. Examples are the effect of a dipole transition $np - \varepsilon d$ on $ns - \varepsilon p$ or a quadrupole transition $ns - \varepsilon d$ on $np - \varepsilon f$ in atoms of noble gases such as Ar or Xe. Usually, it is sufficient to include correlations within one transition to obtain the GOS of discrete excitations or the GOS density. This is not the case, however, when one considers transitions from few-electron subshells, for example, $ns - \varepsilon p$ or the GOS density of electrons nl in the outer shells near the ionization threshold of the inner subshells (e.g., $3p - \varepsilon f$ near the threshold for $3s^2$-electrons in Ar). In these cases, it is required to take into account the correlations within these two transitions.

The ***gos***[16] program [2, 23] determines the GOS of discrete excitation and the GOS density corresponding to one-electron ionization in the Hartree–Fock approximation and random phases with exchange within one or more transitions.

2.10 Scattering Cross Section for Fast Electrons in the Logarithmic Approximation

Knowledge of the GOS makes it possible to calculate not only the probability of an individual discrete transition, but also to find their sum, including excitation and ionization in scattering with large linear momentum transfer. There is a closed expression for the total cross section for inelastic scattering of fast particles on a certain subshell of an atom—representing this cross section as an integral of the photoionization cross section of the same subshell [LL]:

[16] ***gos***—generalized oscillator strength.

$$\sigma^B_{nl,In}(E) = \frac{c}{2\pi E}\int\limits_{I_{nl}}^{E} \frac{\sigma_{nl}(\omega)}{\omega} \ln\left(\varsigma_n \frac{4E}{\omega}\right) d\omega \tag{2.1}$$

Here c is the speed of light, E is the energy of the incident electron, I_{nl} is the ionization potential of the nl subshell; $\ln \varsigma_n \sim 1$ cannot be found without detailed calculations. The main contribution to the integral is made by the values $\omega \simeq I_{nl}$. Therefore, $\ln(4E/\omega) \gg 1$, all discarded terms are of the order of unity. This approximation is called logarithmic.

The ***scatlog***[17] program [24] calculates the scattering cross section for fast electrons in this approximation.

2.11 Angular Distribution of Photoelectrons Beyond the Dipole Approximation

In the total and partial cross sections for the photoionization of atoms and simple molecules, relativistic corrections become significant at very high photon energies, and at photoionization of only heavy atoms. The reason for this is that the relativistic corrections in the cross sections enter with a factor $\alpha^2 \approx 5.3 \times 10^{-5}$, and besides, they are also suppressed by the presence of a small number for the corresponding contributions. The situation is different in the angular distribution, which includes the product of non-relativistic amplitudes and relativistic corrections to them. As a result, the anisotropy parameters include relativistic effects as corrections $\sim \alpha$, and quite often, the radial matrix elements corresponding to relativistic corrections are much larger than non-relativistic ones.

Especially large are the corrections caused by taking into account the nonzero linear moment of the photon. They are often distinguished from the relativistic corrections to the electron wave function in the initial and final states. We will focus on the role of nonzero photon momentum $\kappa = \omega/c$. When these corrections are taken into account, the corresponding dimensionless parameter is the ratio of the radius of the ionizable subshell r_{nl} to the wavelength of the ionizing photon $\lambda \sim c/\omega$, i.e., $\kappa r_{nl} \ll 1$ for photons with energies up to several MeV. Taking into account the nonzero momentum of a photon leads to the fact that, instead of the dipole operator of the interaction of a photon with an electron in the form of, for example, velocity $(\mathbf{e}\vec{\nabla})$, a more complex expression $\exp(i\kappa\mathbf{r})(\mathbf{e}\vec{\nabla})$ appears, where $\mathbf{e}$ is the polarization vector of the photon, and $\mathbf{p} = i\nabla$ is the linear momentum of the photoelectron.

Due to action of the operator $(\kappa\mathbf{r})(\vec{e}\vec{\nabla})$ additional to dipole $l \rightarrow l \pm 1$, also quadrupole $l \rightarrow l \pm 2, l$ transitions contribute to the photoionization cross section. We restrict ourselves to the photoionization amplitude by the first correction to $\kappa r_{nl} \ll 1$, i.e., along with $(\mathbf{e}\vec{\nabla})$ we take into account and $(\kappa\mathbf{r})(\mathbf{e}\vec{\nabla})$.

[17] ***scatlog***—scattering of fast electrons in logarithmic approximation.

In the photoionization amplitude, the dipole and quadrupole matrix elements are summed. As a result, the photoelectron angular anisotropy parameter contains first-order terms in the photon momentum $\kappa = \omega/c$, while the terms with the momentum of the photon make a much smaller $\sim \kappa^2$ contribution to the total and partial cross sections. For this reason, the study of the so-called non-dipole corrections to the angular anisotropy parameter of the photoelectron attracts considerable attention. To calculate them, the following set of programs is used.

The ***anglaniz0***[18] program [24] calculates the photoelectrons non-dipole angular anisotropy parameters γ and η, γ^C, δ^C and $(\gamma^C + 3\delta^C)$ (11.6–11.8), as well as of secondary electrons in the process of inelastic scattering of fast electrons γ^{in}, η^{in} (11.25) for *s*-subshells in the HF and RPAE approximations for the transferred momentum in the limit $q = 0$.

The ***anglaniz1***[19] program [24] calculates the photoelectron non-dipole angular anisotropy parameters γ and η, γ^C, δ^C and $(\gamma^C + 3\delta^C)$ (11.6, 11.36 and 11.37), as well as secondary electrons in the inelastic scattering process of fast electrons, (11.27 and 11.28) for *p*-subshells in the HF and RPAE approximations for the momentum q = 0.

The ***anglaniz2***[20] program [24] calculates the photoelectron non-dipole angular anisotropy parameters γ and η, γ^C, δ^C and $(\gamma^C + 3\delta^C)$ (11.6, 11.39 and 11.40), as well as secondary electrons in the process of inelastic scattering of fast electrons, (11.30 and 11.31) for *d*—subshells in the HF and RPAE approximations for the transferred momentum $q = 0$.

Difgos0[21] [ACY] program calculates differential GOS for the magic angle $P_2(\cos\theta_m) = 0$ and *s*-subshells of the atom at different q (13.7–13.9).

Difgos1[22] [ACY] program calculates differential GOS for magic angle $P_2(\cos\theta_m) = 0$ and *p*-subshells of an atom at different q (13.10–13.12).

Difgos2[23] [ACY] program calculates differential GOS for the magic angle $P_2(\cos\theta_m) = 0$ and *d*-subshells of the atom at different q (13.13–13.15).

2.12 Spin Polarization Parameters of Photoelectrons

As N. A. Cherepkov showed several years ago (see references in [AICC]), the spins of photoelectrons emitted even from atoms with completely filled shells are not directed arbitrarily but are oriented relative to the photon polarization vector both for linear and for circular polarization of ionizing photons. The polarization of a photoelectron depends on its energy. It can be observed experimentally by studying

---

[18] ***anglaniz0***—angular anisotropy for s-subshells.

[19] ***anglaniz1***—angular anisotropy for p-subshells.

[20] ***anglaniz2***—angular anisotropy for d-subshells.

[21] ***difgos0***—differential GOS for s-subshells.

[22] ***difgos1***—differential GOS for p-subshells.

[23] ***difgos2***—differential GOS for d-subshells.

the photoionization of individual sublevels, with full momentum $j = l \pm 1/2$, fixing photoelectrons at a certain angle of emission.

In general, the angular distribution of photoelectrons from the level nlj in the dipole approximation is determined by five independent parameters, namely the photoionization cross section, the dipole angular anisotropy parameter, and three spin parameters—A, α^p, ξ. On the other hand, the photoionization of the nl subshell under consideration is determined in the dipole approximation by two matrix elements $l \rightarrow l \pm 1$, their moduli and phases, i.e., four parameters. This means, first, that the measurement of the spin polarization of electrons makes it possible to carry out the so-called complete experiment, i.e., determine all matrix elements, their modules and phases, directly from experiment. At the same time, the excess of the number of experimentally determined parameters over the number of matrix elements and their phases makes it possible to establish a connection between them, the consequence of which is a relationship between the parameters A, α^p, ξ, which is valid to the extent that in the considered photon energy range it is possible to restrict ourselves to only the dipole approximation.

The spin anisotropy program in the dipole approximation is determined using the ***spinpol***[24] program [25], which allows obtaining the dipole parameters of the electron spin polarization.

Taking into account the non-dipole parameters in the angular distribution of photoelectrons increases the number of experimentally determined parameters by two, which can be seen from the corresponding formula for the angular parameter of anisotropy (see [AICC], (1.77)). Non-dipole corrections are introduced by five additional parameters—three modules of transition $nl \rightarrow \varepsilon(l \pm 2, 0)$ matrix elements and two relative phases.[25] Naturally, the number of parameters that is used to represent the spin polarization of photoelectrons, taking into account non-dipole corrections, also increases. To calculate them, there exists the ***spinpolq***[26] program [25], which allows to find non-dipole corrections for the quadrupole terms of d-subshells.

2.13 Angular Distribution of Secondary Electrons in the Dipole Approximation

The interaction of a fast charged particle with an atomic electron has much in common with the interaction of a photon with an atomic electron. The difference is that the interaction operator in this case $\exp(i\mathbf{qr})$ corresponds to the exchange not of a transverse photon, but a virtual, longitudinal one. As an analog of the parameter $\kappa r_{nl} \ll 1$ in photoionization, serves the parameter qr_{nl} that can be much larger than $\kappa r_{nl} \ll 1$. It is significant that while the momentum of a photon is rigidly related to its energy $\kappa = \omega/c$, the relationship between the energy transferred in elastic

[24] ***spinpol***—spin polarization dipole.

[25] The common phase is not included in the physical measured characteristics.

[26] ***spinpolq***—spin polarization quadrupole.

scattering and the transferred momentum is more complicated. Roughly speaking, in this process, $\omega \approx \mathbf{pq}$. This means that a given transmitted energy ω corresponds to a whole range of values q and vice versa.

Thus, it is natural to regard inelastic scattering of fast charged particles as a kind of photoeffect with a controlled momentum transfer. This means that in the process of inelastic scattering of fast particles, it is possible, at the will of the experimenter, to control the contribution of the non-dipole components, increasing their influence in comparison with the dominant dipole component. Another feature of inelastic scattering of fast particles in comparison with photoionization is the presence of a contribution from the monopole transition.

The analogy between inelastic scattering of fast particles and photoionization is especially complete if we study the angular distribution of electrons knocked out of an atom with respect to the momentum $\mathbf{q}$ transferred by a fast particle to the target atom. For this research, an appropriate theoretical approach has been developed [24] and a number of programs have been created.

The ***anglaniz0***[27] program [25] calculates the angular anisotropy dipole parameters of secondary electrons β^{in} for s—subshells in the HF and RPAE approximations.

The ***anglaniz1***[28] program [25] calculates the angular anisotropy dipole parameters of secondary electrons β^{in} for p—subshells in the HF and RPAE approximations.

The ***anglaniz2***[29] program [25] calculates the angular anisotropy dipole parameters of secondary electrons β^{in} for d—subshells in the HF and RPAE approximations.

2.14 Angular Distribution of Secondary Electrons Beyond the Dipole Approximation

In the scattering of fast charged particles, the role of non-dipole corrections in the angular distribution of electrons knocked out of the atom with respect to the direction of the moment $\mathbf{q}$ transferred to the atom during scattering can be much greater than in photoionization. Therefore, when deriving formulas, not only the terms of the first but also of higher orders in q were retained. The corresponding formulas were obtained in [26]. Formulas are given that include all Legendre polynomials $P_L(\cos\theta)$ as functions of the scattering angle θ up to $L = 4$ terms, while non-dipole corrections to the angular distribution of photoelectrons include $1 \leq L \leq 3$ terms. A number of computer programs have been developed to calculate the non-dipole parameters of secondary electrons.

The ***anglaniz0***[30] program [27] calculates the angular anisotropy non-dipole parameters of the secondary electrons γ^{in}, η^{in} for s—subshells in the HF and RPAE

[27] ***anglaniz0***—angular anisotropy for s-subshells.

[28] ***anglaniz1***—angular anisotropy for p-subshells.

[29] ***anglaniz2***—angular anisotropy for d-subshells.

[30] ***anglaniz0***—angular anisotropy non-dipole for s-subshell.

approximations for momentum $q = 0$, and, separately, for an arbitrary but small momentum $q \neq 0$.

The ***anglaniz1***[31] program [27] calculates the angular anisotropy non-dipole parameters of secondary electrons γ^{in}, η^{in} for p—subshells in the HF and RPAE approximations for momentum $q = 0$, and, separately, for an arbitrary but small momentum $q \neq 0$.

The ***anglaniz2***[32] program [27] calculates the angular anisotropy non-dipole parameters of the secondary electrons γ^{in}, η^{in} for d—subshells in the HF and RPAE approximations.

2.15 Compton Effect

With increase of photon energy ω, the photoionization cross section of an atom or molecule decreases rather rapidly. Starting from about ten keV, the Compton effect [BLP] becomes dominant. This is a process in which an atom or molecule is ionized or excited by a photon, while also emitting a secondary photon. While the photoionization process is presented schematically as $\gamma + A \rightarrow A^+ + e$, the Compton effect is presented as $\gamma + A \rightarrow A^+ + e + \gamma'$.

An important feature of the Compton effect is that, in a very wide range of photon energies ω, the operator determining the amplitude of the probability of this process and the operator determining the amplitude of inelastic scattering of fast charged particles $\exp(i\mathbf{qr})$ coincide up to factors that are insignificant in the calculation. The only difference is that much larger momenta q can play and play an important role in the Compton effect than in GOS.

Thus, the ***gos*** program [23] allows calculating the cross section for ionization and excitation of an atom in the Compton effect. Examples of calculation results can be found in [27].

2.16 Inelastic Scattering of Low and Intermediate Energies Electrons

The triple-differential scattering cross section, studied by the reaction denoted $(e, 2e)$, allows one to obtain very detailed information about both the target atom or molecule and the ionization process itself. In this reaction, an electron inelastically scattered and knocked out of a target—an atom or molecule—are simultaneously fixed. The reaction cross section determines the likelihood of simultaneous detection of two electrons of a given energy emitted at fixed angles.

---

[31] ***anglaniz1***—angular anisotropy non-dipole for p-subshell.

[32] ***anglaniz2***—angular anisotropy non-dipole for d-subshell.

To calculate the triple-differential cross section even in the simplest one-electron approximation, it is necessary to solve the three-body problem, since the interaction between the incident particle, the emitted electron, and the residual ion must be taken into account.

There are several methods of approximate calculation, in which it is assumed that certain components of the electron–electron or electron–nuclear interaction make an insignificant contribution, and therefore, they can be neglected. For example, if the incident particle and the ejected electrons are fast, i.e., their velocities are high in comparison with the velocities of electrons in the target, the interaction with it and the residual ion can be neglected, then the wave function is taken in the form of a plane wave $\phi(\mathbf{r}) = exp(i\mathbf{pr})$, where the energy of the electrons is $E = p^2/2$ and $\mathbf{p}$ is the momentum of the incident or knocked out electron. Scattering in this case can be represented as a collision of two free electrons, one of which (atomic electron) is described by its velocity distribution in a bound atomic or molecular state. This distribution is equal to the Fourier transform of the wave function $\phi_\nu(\mathbf{r})$ of the knocked-out electron:

$$\int \phi_\nu(\mathbf{r})exp(i\mathbf{qr})d\mathbf{r} \tag{2.2}$$

where $\mathbf{q}$ is the recoil momentum, i.e., the difference between the momentum of the projectile and the total linear momentum of the two emitted electrons. One of the emitted ones, which has undergone inelastic scattering, has a momentum $\mathbf{p}_1$, and the second, knocked out of the target, has a momentum $\mathbf{p}_2$, so that the following relation holds:

$$\mathbf{q} = \mathbf{p}_1 - \mathbf{p}_2 - \mathbf{p}_3 \tag{2.3}$$

The following equation takes into account the law of conservation of energy:

$$E_1 + E_4 = E_2 + E_3 \tag{2.4}$$

where $E_{1,2,3} = p_{1,2,3}^2/2$ is the energy of the incident electron, $E_4 = -I_{nl}$ and I_{nl} is the ionization potential of the electron removed from the target. The problem of taking into account the interaction of two free electrons is reduced to the problem of electron scattering by a Coulomb field. This problem has an exact analytical solution.

In the case of atomic ionization by electron impact, the approach based on the description of the wave functions of the incident and both outgoing electrons by plane waves is called the impulse approximation with plane waves. Its accuracy can be improved by introducing an averaged effective potential, which allows one to take into account the fields of the atom and ion. In this approach, ionization is still considered as the scattering of two free electrons, but their momenta and energies are related by the following relations:

$$E_{2,3,1} + W_{2,3,1} = \frac{p_{2,3,1}^2}{2} \tag{2.5}$$

As noted by subscripts 2, 3, 1, different effective potentials W_i can be used for the atomic field and the residual ion field. This approach is called impulse approximation with distorted waves. The amplitude of the process probability is determined by expressions (2.2)–(2.4), while the momentum and energy of each electron are related by (2.5).

For intermediate and low energies of the incident particle and the ejected electron, the role played by the atomic field is essential. On the other hand, at significantly different velocities of two emitted electrons, their interaction is unimportant and it is sufficient to take it into account in the first Born approximation, while the interaction of electrons and the atom itself must be described, at least, within the framework of the Hartree–Fock self-consistent field method. The electron wave function is then derived from Eq. (1.8). The atomic field effect can be limited only by taking into account its interaction with the knocked-out electron (plane wave Born approximation, or PWBA[33]), or both electrons (distorted wave Born approximation, or DWBA[34]).

The ***impact*** and ***impact1*** programs [29] calculate the triple differential atomic ionization cross sections in PWBA and DWBA, respectively. The ***impact2*** program [29] determines the differential cross section for atomic excitation by electron impact in DWBA, taking into account the atomic field in the Hartree–Fock approximation.

2.17 Near-Threshold Ionization with Account for Post-collision Interaction

The emission of a slow electron is typical for ionization processes near its threshold. If the state formed as a result of ionization is unstable relative to decay with the emission of electrons, then this decay can strongly affect the ionization process itself. If the excited state resulting from ionization decays with the emission of a fast electron, then the self-consistent field acting on the slow escaping electron changes significantly and rapidly. If the decay occurs faster than the time it takes a slow electron to leave the atom, the cross section of the ionization process acquires some special features associated with the decay process. Along with the cross section, the distribution of the energy of electrons emitted during decay also changes.

A change in the atomic field due to Auger decay of an internal vacancy arising during ionization or autoionization decay of an excited state leads to significant changes in the wave function of the slowly emitted electron. Due to the fact that it finds itself in the field, for example, of a double ion formed during Auger decay, it is affected from the side of the core by a stronger attraction than before the vacancy decay. Therefore, its energy drops, so that, ultimately, it may even find itself at one

[33] PWBA—Plane Wave Born Approximation.

[34] DWBA—Distorted Wave Born Approximation.

of the discrete levels in the field of a double ion formed as a result of the decay of an internal vacancy. The phenomena occurring as a result of a change in the field due to the decay of a non-stationary excited state are called *post-collision interaction* (PCI) [30].

The ***gos*** program calculates the GOS density near the threshold of the inner shell, giving the cross section for photoionization and inelastic electron scattering taking into account the interaction after collision, and the ***autpci***[35] program calculates the cross section for inelastic electron scattering near the autoionization threshold of the triplet state, i.e., states with total spin $S = 1$, taking into account the effect of the post-collision interaction (PCI).

2.18 Self-energy Part of the One-Particle Green's Function

In theoretical consideration of the scattering of a slow electron by an atom or a molecule, serious complications arise due to the need to take into account the exchange of the incident and atomic electrons and the deformation of the target's electron cloud by the Coulomb field of the incident particle. Among the methods describing the elastic scattering of electrons or other particles by an atom or molecule, the following approach is very widespread: the field acting on the incident particle is replaced by a phenomenological potential with a real and imaginary parts. It is usually called *optical potential* [31].

There are a number of approximate methods for solving the problem of collisions between an electron and an atom without free parameters, which can be used to compare the degree of agreement between calculated and measured data [32]. Unfortunately, all of these methods involve more and more complex calculations, the volume of which increases rapidly with the number of electrons.

The diagrammatic descriptions and the related general many-body theory can be very effectively applied to solve this problem, especially when studying the scattering of slow electrons. It can be shown that the wave function of an electron Ψ_E with energy E elastically scattered by an atom, taking into account the polarization of the target particle, is described by an exact equation similar to (1.8), which, in addition to the Coulomb potential of the nucleus and the self-consistent Hartree–Fock field, also includes a non-local additional term $\hat{\Sigma}$, a kind of potential that depends on the energy of the electron. This potential is determined by the residual interaction and takes into account the interaction of the incident particle and target electrons, as well as these electrons with each other. Symbolic equation for Ψ_E is:

$$(H^{\mathrm{HF}} + \hat{\Sigma}_E - E)\Psi_E = 0. \tag{2.6}$$

[35] ***autpci***—Autoionization in Post Collision Interaction.

It is more convenient to consider this problem in terms of the one-electron Green function G rather than $\hat{\Sigma}$. These two functions are related in operator form by the following equation

$$(H^{\mathrm{HF}} + \hat{\Sigma}_E - E)\hat{G} = \hat{1} \tag{2.7}$$

whose solution is

$$G = (H^{\mathrm{HF}} + \hat{\Sigma}_E - E)^{-1}. \tag{2.8}$$

The equation in terms $\hat{\Sigma}$ can be represented in integral form

$$\Psi_E = \phi_E - \hat{G}_0\hat{\Sigma}_E\Psi_E, \tag{2.9}$$

or

$$\Psi_E = \phi_E - \hat{G}\hat{\Sigma}_E\phi_E. \tag{2.10}$$

Here, ϕ_E is the solution of the Hartree–Fock Eqs. (1.8, 1.9) for energies in the range $E > 0$, $\hat{G}_0 = \left(H^{\mathrm{HF}} - E\right)^{-1}$, $\hat{G}$ и $\hat{G}_0$ are related by the following integral equation:

$$\hat{G} = \hat{G}_0 - \hat{G}_0\hat{\Sigma}_E\hat{G} \tag{2.11}$$

If we multiply (2.9) by $\hat{\Sigma}_E$ and introduce a new operator $\hat{\overline{\overline{\sum}}}_E$ using the relation

$$\hat{\Sigma}_E\Psi_E \equiv \hat{\overline{\overline{\sum}}}_E\phi_E, \tag{2.12}$$

we get the following equation for the function $\hat{\overline{\overline{\sum}}}_E$:

$$\hat{\overline{\overline{\sum}}}_E = \hat{\sum}_E - \hat{\sum}_E\hat{G}_0\hat{\overline{\overline{\sum}}}_E \tag{2.13}$$

In many-body theory, $\hat{\sum}_E$ and $\hat{\overline{\overline{\sum}}}_E$ are called, respectively, irreducible and reducible self-energy part of the one-particle Green function.

Operator Eqs. (2.9), (2.11) and (2.13) are in fact integral equations. They include summation and integration over intermediate states, since the functions Ψ_E, $\hat{G}_0$, $\hat{\overline{\overline{\sum}}}_E$ on the right side of the equations are under the integral.

The approach described by relations (2.9) and (2.10) makes it possible to express the most important characteristic of scattering, the differences in the phases of electronic wave functions due to the interaction, through matrix elements $\hat{\overline{\overline{\sum}}}_E$. Thus, the physical meaning of the operator $\hat{\Sigma}_E$ is the same as that of the phenomenological potential of the optical approach in scattering theory (see [31, 32]). Using in calculations $\hat{\overline{\sum}}_E$, rather than $\hat{\Sigma}_E$, provides a more accurate account of the self-energy part $\hat{G}$.

The calculated value of the phase at low energies of the scattered electron gives an answer to the question of whether there really is a negative ion in the field of a given atom or molecule in a certain partial wave. Equation (2.6) can have a solution corresponding to negative energy, that is, describe a negative ion, and in this case, Ψ_E is a wave function of a negative ion [33].

The ***scat***[36] [34] program computes the diagonal matrix element of $\hat{\Sigma}_E$ in the SRPAE. The advantage of this approach over other non-phenomenological methods for calculating the scattering phases is that no assumptions are made about a change in the energy of an incident particle during its deformation of the electron shells of a target n atom or a molecule, even in virtual states. This statement is called the adiabatic approximation.

Another advantage is that the calculations are relatively simple, and therefore, the scattering phases of a slow electron can be calculated for almost every atom with closed or semi-closed subshells.

Many processes and characteristics can be described using function $\hat{\sum}_E$. In itself, $\hat{\sum}_E$ is a complex and unknown function. In order to find it, you have to make various additional assumptions. The imaginary part of the matrix element $\hat{\sum}_E$ between the electronic wave functions in the continuous spectrum determines the difference in the scattering phase due to inelastic processes: excitation or ionization of an atom or a target molecule by an incident particle. Corrections to the Hartree–Fock ionization potential, taking into account the interaction of atomic electrons, can be expressed in terms of matrix elements $\hat{\sum}_E$ between the wave functions of the occupied states. The imaginary part of this matrix element takes into account the total width of the vacancy level relative to Auger decay.

In addition, the so-called *spectroscopic factor* of the electron (hole) level can be expressed in terms of the energy E derivative $\partial\hat{\Sigma}_E/\partial E$ if the matrix element is calculated between vacant (occupied) states. The spectroscopic factor shows what fraction of time an electron (vacancy) is in a pure single-particle state and how long this state is a mixture with more complex configurations, of which the simplest example is an electron–hole pair.

At electron energies E above the atomic ionization potential, the imaginary part $\hat{\sum}_E$ determines the cross section of inelastic scattering of electrons by atoms or molecules. Using $\hat{\sum}_E$, we can calculate the wave function of the scattered electron and use it then in further calculations.

If an atom or molecule has a negative ion state, its photoionization amplitude can be calculated in the crudest approximation as the matrix element of the "length" or "velocity" operators between the negative ion wave function (solution of Eq. (2.6) at $E < 0$) and the continuous spectrum wave function. This function describes an

---

[36] ***scat***—scattering of electrons.

electron moving in the field of a neutral atom or molecule (the solution of Eq. (2.6) at $E > 0$).

The ***scat*** program calculates the *reducible self-energy part* $\hat{\tilde{\Sigma}}_E$, which is the solution to the equation symbolically represented by (2.13). The kernel of this equation includes an *irreducible self-energy part* $\hat{\sum}_E$. The latter can be calculated in the SRPAE [34] or in other more accurate approaches.

To describe the scattering of an electron or other particle by an arbitrary potential A, matrix $\langle|A|\rangle$ must be introduced instead of $\langle|\sum_E|\rangle$. The wave functions in these matrix elements are expressed in terms of the matrix elements $\hat{\tilde{\Sigma}}_E$. The scattering phases are expressed as matrix elements of c. This implies a more accurate accounting of the self-energy part than in the ***scat*** program.

The electron scattering phases obtained with the ***scat*** program make it possible to find the total cross sections for elastic and inelastic scattering of an incident electron. The ***anglscat***[37] program is used to obtain the cross section differential in the electron scattering angle.

2.19 Photoabsorption and One-Electron Ionization with Account for the Self-energy Part of an Electron

A photoelectron, escaping from an atom or molecule, as a result of absorption of a photon by them, especially from the inner and intermediate shells, on its way from the atom with a noticeable probability collides with the outer electrons of the target. As a result, both excitation and ionization of these electrons become possible. Measurements have shown that the probability of this process, called *internal inelastic scattering*, is quite high. Therefore, this process significantly affects the properties of photoionization. First of all, this manifests itself in a decrease in the energy of some of the photoelectrons. The loss is spent on the excitation or ionization of the *residual ion*.

As a result, the one-electron photoionization cross section of the inner or intermediate shell becomes smaller than the value that it would have if the internal inelastic scattering was neglected. An important manifestation of inelastic scattering of a photoelectron in the field of a residual ion is not only its energy, but also the angular momentum, which manifests itself in a change in the angular distribution of photoelectrons.

The ***phosep***[38] program [35] allows calculating the photoionization cross section accounting for the interaction of photoelectrons and atomic electrons of the outer shell. This is achieved by using the self-energy part of the one-particle Green's function found in the ***scat*** program. The ***phosep*** program calculates the wave function of a photoelectron that undergoes elastic scattering in the ion field, with accounting

[37] ***anglscat***—angular distribution of scatted electrons.

[38] ***phosep***—photoabsorption with the Self-Energy Part.

of the dynamic polarization of the residual ion during the scattering process. The effect of the self-energy part of the one-particle Green's function is included both below the threshold of electron inelastic scattering and above, that is in the range of photoelectron energies at which there exists a possibility of inelastic processes with photoelectron's participation.

2.20 Scattering of Positrons

At first glance, the cross section for the scattering of a positron by an atom or molecule is a simpler process than the scattering of an electron, since there is no exchange between the projectile and the target electron. However, a different process becomes important in the scattering of positrons. Interacting with the target electrons, the positron transforms them into real and/or virtual excited states, the size of which is much larger than atomic or molecular. Excited electrons, being at a considerable distance from the target ion, interact very strong with the incident positron, together forming a state called virtual positronium. It is this interaction that has a decisive influence in the scattering of slow positrons by atoms or molecules.

Accurate accounting for the formation of virtual positronium is a very difficult task, since it requires at least taking into account the strong interaction of a positron and an electron moving in the static field of the target atom. We simplify the problem greatly, taking into account the formation of virtual positronium, by the fact that in matrix elements of the second order in the interaction of the positron with the target electrons, the binding energy of the positron in the intermediate state is shifted by the binding energy of positronium I_{Ps}.

The aforementioned is implemented by the ***scat*** program [33], which thus also makes it possible to study, along with electrons, scattering of positrons.

2.21 Non-radiative or Auger Vacancy Decay

In a large number of processes, such as the interaction of electrons, positrons, photons and heavy particles (ions) with atoms and molecules, vacancies are created in various electron shells. These vacancies can be filled by the outer electrons of an atom or molecule, and the energy released in this process can be emitted in the form of photons or transferred to one or more atomic or molecular electrons, knocking them out of the atom or molecule. The decay of a vacancy with the emission of electrons is called the Auger effect. Vacancies decay mainly with the emission of one electron, the energy of which is determined by the energy of holes in the initial and final states. That is why the study of Auger spectra provides important, sometimes unique, information about the electron levels energy, if one takes into account the recoil of the target in the process of decay and rearrangement that occurs in the target as a result of the creation of a vacancy in it.

In addition to the energy of the emitted electron, the lifetime of a vacancy is an important characteristic of Auger decay. The lifetime is simply related to the width of the level in its energy spectrum or the probability of its decay. The lifetime of a vacancy is determined by the value of the residual interaction of the target electrons with each other V_{res}.

The ***auger***[39] program [36] is designed to calculate the widths of atomic levels due to Auger decay with the emission of one electron. The effective interaction, the matrix element of which determines the probability of Auger decay, is found by accounting for the multielectron correlations in two or more transitions within the framework of the RPAE. The effect of two transitions is included, because in several cases our calculations have shown the presence of strong intershell correlations, which noticeably affect the probability of Auger decay, both increasing and decreasing them. It is also important to include the alteration in the energy of, say, a two-vacant state in comparison with the sum of the energies of individual two vacancies. This can be achieved in the ***auger*** program if, for the energies of states, we choose not the Hartree–Fock values, but the experimental ones.

In photoionization or inelastic scattering of an electron by an atom, two vacancies can be formed instead of one. This possibility significantly enriches the radiation spectra of atoms and molecules. In addition to the characteristic line, groups of new lines, called satellites, appear in the spectra. Of particular interest are satellite lines arising from the so-called correlative decay of states with two holes [37, 38]: The decay of two vacancies occurs in such a way that the energy of their decay is transferred to one photon or electron. The first is called one-photon decay, and the second is called the double Auger effect [39, 40].

The study of the double Auger effect is interesting because it helps to identify a number of lines in photoelectron spectra, makes it possible to predict new lines in Auger spectra, and provides important information about multielectron correlations in atoms. In the program ***dauger***[40] [41], the probability of double Auger decay is calculated in the lowest order of the perturbation theory.

The ***phodec***[41] program [42] investigates single-photon decay between states with two vacancies. This decay channel also leads to the appearance of a rather wide class of new lines, not only in the range of hard X-rays, as originally thought [39, 40], but also in the range of soft X-rays, and sometimes even in the range of far ultraviolet radiation [43–45]. Therefore, their position and intensity are important for the correct identification of the spectra observed in experiments. The program calculates both matrix elements and transition energies. The ***dauger*** and ***phodec*** programs are intended to be used in conjunction with the ***auger*** program to calculate the probability of a number of other processes, where the initial state decays with the emission of two electrons, an electron and a photon, or two photons. The ***dauger*** program also calculates the amplitudes of other atomic processes, such as the Auger effect, when a state with one hole ν_1 decays by emitting of two electrons $\nu_2\nu_3$ (2.14)

[39] ***auger*** = Auger-decay.

[40] ***dauger***—Double Auger-decay.

[41] ***phodec***—one Photon Decay of two-vacancy states.

that leads to the formation of three vacancies $\nu_4\nu_5\nu_6$, i.e., to a triply charged ion. It also allows us to describe the process in which the autoionization state $\nu_1\nu_2$ can decay emitting two electrons $\nu_3\nu_4$ (2.15), and the two-hole state $\nu_1\nu_2$ that decays with the excitation of the state $\nu_3\nu_6$—an electron at a discrete level (2.16). This also includes the processes of decay of a one-vacant state ν_1 with excitation and emission of electrons ν_3 and ν_2 (2.17):

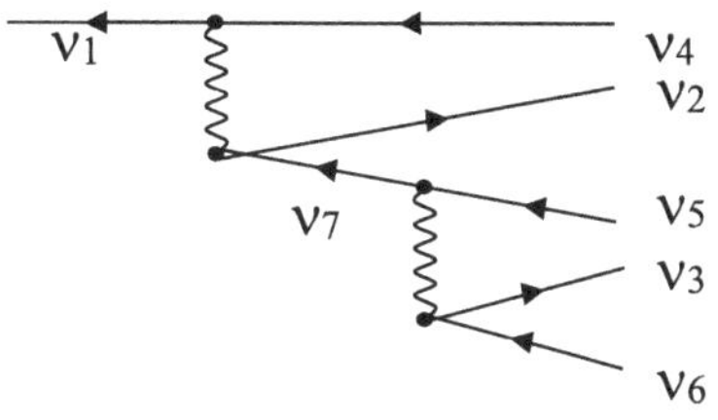

(2.14)

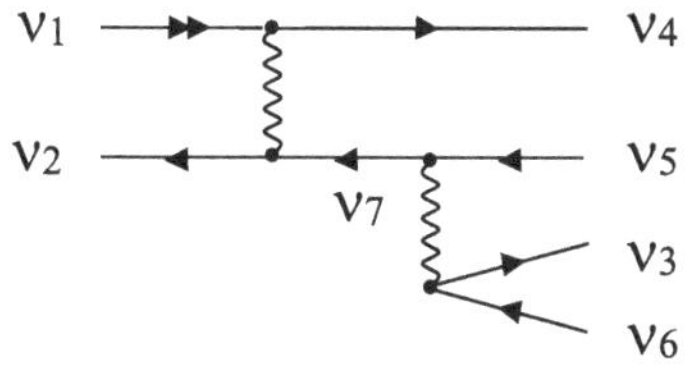

(2.15)

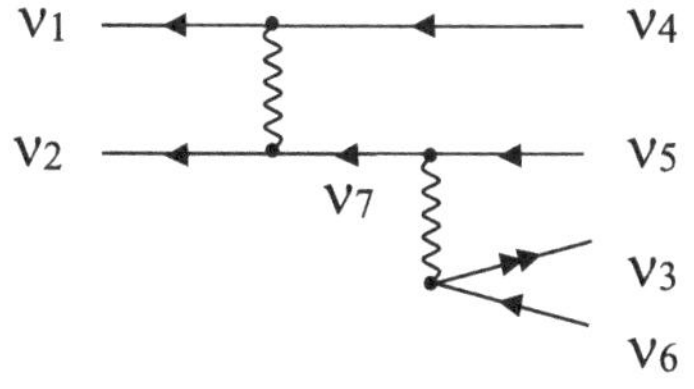

(2.16)

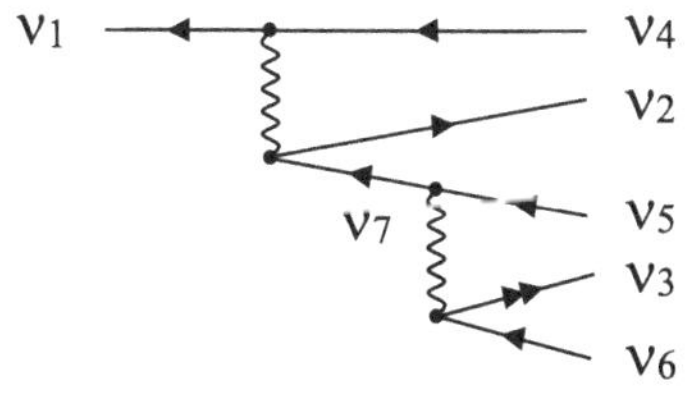

(2.17)

The double arrow implies a discrete excited electron state.

In ***phodec*** program, the angular and spin factors can be calculated or included in the input data. This greatly expands the possibilities of the program application. In particular, using the ***phodec*** program, the amplitudes and probabilities of some single-photon processes can be calculated, such as:

1. So-called semi-Auger radiative effect, in which the decay energy of an internal vacancy ν_1 is emitted in the form of a photon and simultaneously causes a discrete excitation of an atomic electron. Diagram (2.18) gives an example for determining the amplitude of the semi-Auger radiation effect.

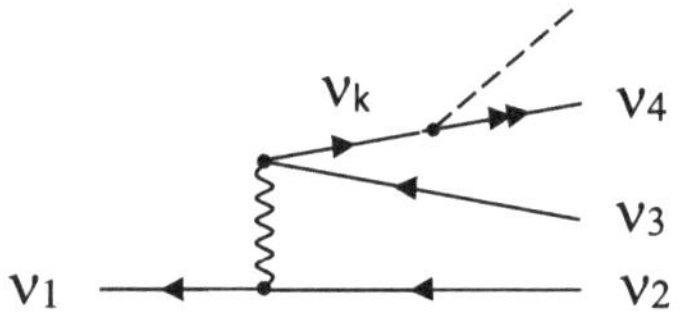

(2.18)

2. Photoionization of the excited state $\nu_2\nu_1$, by accounting for the corrections of the first order of the perturbation theory in the electron–electron interaction. The following is an example of a corresponding diagram:

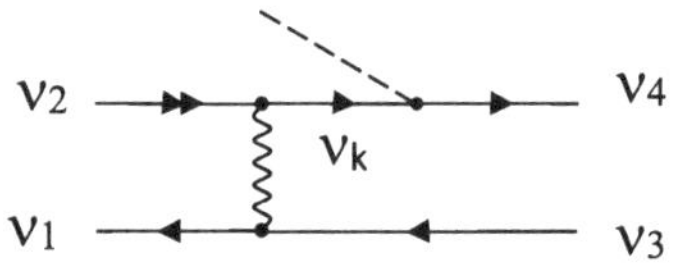

(2.19)

3. Discrete excitation of an electron ν_4 with simultaneous ionization ν_1 as a result of absorption of a photon. For example:

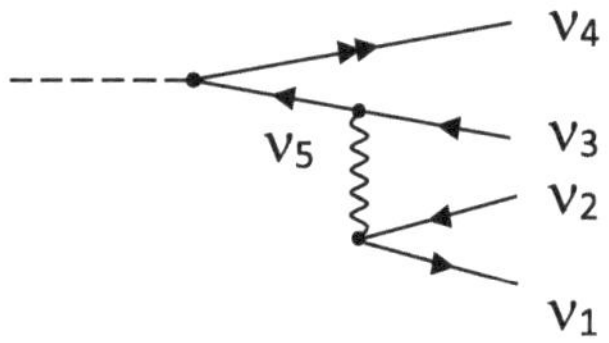

(2.20)

4. Corrections of the first order of the perturbation theory in the electron–electron interaction to the amplitude of the radiative decay of a vacancy ν_1, for example:

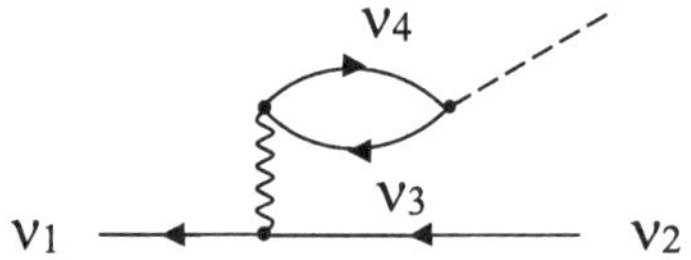

(2.21)

5. Corrections of the first order of the perturbation theory to the amplitude of the so-called hyper-satellite transitions, where one of the vacancies ν_1 decays in the presence of the other ν_2, for example:

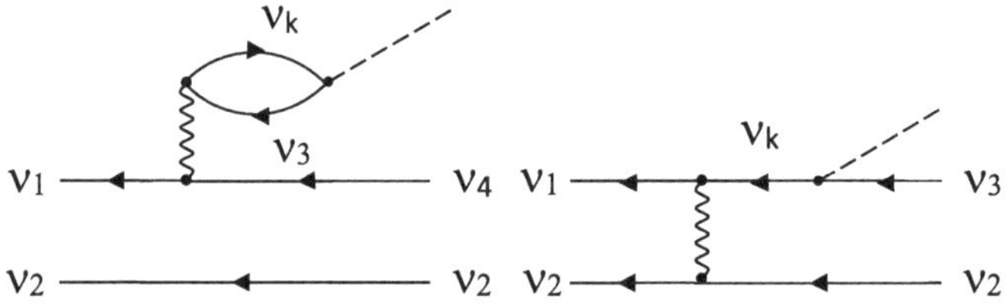

(2.22)

To determine the total amplitude of the above processes, for each partial amplitude, the individual angular and spin factors must be found and used as input parameters.

In the case of simultaneous decay of two vacancies, the angular factors are calculated in the program itself. The energy of an emitted or absorbed photon is an input parameter. This also makes it possible to calculate the probability (width) of the processes under consideration.

2.22 Radiative Decay of Vacancies in HF and RPAE

An important characteristic of the decay of a vacancy is its radiative width, which is determined by the probability of emission of one photon upon the transition of one of the outer electrons to the vacant level located below. In some cases, for example, for the outer subvalent shells of noble gas atoms ns^2, the radiative transition $ns^{-1} \to np^{-1} + \gamma$ is the only possible one, since the vacancy in the subvalent shell simply does not have enough energy for Auger decay [27].

In the HF approximation, the amplitude of radiative decay is simply the matrix element of the length or velocity operators between HF one-electron functions. The radiative transition is influenced by multielectron correlations. By analogy with photoionization and electron scattering, they are described in this book within the framework of the RPAE. Electronic correlations can increase the probability of radiative decay in comparison with its Hartree–Fock value or decrease it, in principle, up to the complete disappearance of the corresponding line in the radiation spectrum. The role of intershell correlations is especially important.

2.23 Electron-Induced Triplet Excitations in the First and Second Orders of the Distorted Wave Method

In the theoretical consideration of the excitation of atoms caused by an incident electron at low and intermediate energies, the Born approximation that accounts for the interaction of the incident particle and an excited atomic electron in the first order in the residual interaction is very inaccurate. The error is primarily determined by the strong influence of the atomic field on the incident particle and the scattered electron. This defect can be accounted within the framework of various approximate approaches, of which the so-called DWBA is one of the most effective.

Many calculations have been performed of the differential in angle and total excitation cross sections for atoms caused by a fast electron in the first Born approximation, taking into account the distortion of the incident and outgoing waves by various (mainly phenomenological) potentials [43]. However, by including second-order processes within the framework of the distorted wave method (i.e., the second-order approximation in the interaction of an incident particle and an atomic electron), it is possible to significantly improve the agreement between the results and experimental data. This is especially true when one considers the excitation of triplet atomic levels (i.e., levels with the total spin $S = 1$). By taking into account the second-order corrections, it is possible to describe the measured cross sections quite accurately at not too high values of the incident electron energy. In addition, accounting for the second order in the interaction of the incident and atomic electrons would make it possible to apply the distorted wave method in a wider range of scattering angles.

The ***intriple***[42] program [47] calculates the probability amplitudes for electron-induced excitation of triplet atomic states at low energies in the first and second order of the distorted wave method. The ***twosec***[43] program [47] calculates the differential in the scattering angle cross section of excitation of triplet monopole levels by low energies incident electron.

2.24 Bremsstrahlung Spectrum

Bremsstrahlung is the process of emission of electromagnetic radiation that accompanies the scattering of a charged particle by an atom or molecule. It is assumed that the target, atom or molecule, as a result of scattering, remains in the ground state. The emission of a photon, which accompanies the deceleration of a fast electron in the field of an atom or molecule, occurs with the help of two different physical mechanisms. First, it is the direct emission of a photon by an incident particle, decelerating in the static field of the atom. This process is very well studied [48]. Secondly, this is the emission of a photon by atomic or molecular electrons, virtually excited by the incident particle. The importance of this mechanism was noted in [49].

In programs ***brem***[44] and ***brem1***[45] [50], the bremsstrahlung spectrum is calculated for energies of the emitted quantum above (***brem***) and below (***brem1***) the ionization potential of the shell in question. The interaction of an incident electron and an atom is treated in the lowest order of perturbation theory in the interaction between the incident particle and an atomic electron. The direct or ordinary radiation amplitude is expressed in terms of the so-called form-factor of the target atom, i.e., in terms of the Fourier transform of the atomic electron density calculated with the Hartree–Fock single-particle wave functions for the atomic ground state. The radiation amplitude of atomic electrons is presented as the dynamic response of an atom to the impact of an incident electron. This amplitude is also calculated in the lowest order of the perturbation theory in the interaction of a fast incident particle and an atomic electron. Excited atomic states are considered only within one transition.

The ***brem*** and ***brem1*** programs also calculate the angular anisotropy coefficient of the emitted photon. This characteristic describes the probability of a photon escaping at different angles. The ***brem*** and ***brem1*** programs allow calculating the

---

[42] ***intriple***—inelastic triplet states excitation.

[43] ***twosec***—two-step differential cross section.

[44] ***brem***—bremsstrahlung.

[45] ***brem1***—bremsstrahlung.

bremsstrahlung spectra of various fast charged particles: electrons, positrons, protons, ions.

The ***bremco***[46] program [50] allows one to find the cross section for bremsstrahlung of fast electrons, taking into account correlations in the RPAE within one transition for an incident electron energy exceeding the ionization threshold of the subshell under study. The generalization to several transitions in the RPAE frame is straightforward.

In contrast to the ***brem*** program, where the amplitude of ordinary bremsstrahlung is calculated in the Born approximation, the ***bsiea***[47] program [51] performs a more accurate calculation using the "length" or "velocity" matrix elements calculated between the wave functions of the continuous spectrum in the Hartree–Fock approximation. They describe an electron moving in the field of an atom (ion). The amplitude of ordinary bremsstrahlung is presented as a series of partial waves. The amplitude of the atomic bremsstrahlung arising from the polarization of the target by the incident particle in the process of scattering is also calculated with the Hartree–Fock wave functions.

2.25 Capturing Negative Mesons

There is no exact solution for the problem of the interaction of a meson with an atom, as well as for a slow electron, since too many particles take part in the process. However, some features of this process are of considerable interest, for example the determination of the energy of a meson, for which the probability that it will be captured to a discrete atomic level reaches a maximum. The mass of a relatively light meson is approximately 200 times that of an electron. The absolute values of negative energies of muon and electronic states and radii of electronic and meson wave functions with equal quantum numbers are determined by the same relations. However, one should pay attention to the fact that, due to the difference in masses, the wave functions with the principal quantum number $n \leq (m_1/m_e)^{1/2} \approx 14$ (m_1 and m_e are the mass of a muon and an electron, respectively) are located within the atomic orbit.

The energy lost by a meson is acquired by atomic electrons or emitted in the form of photons. The probability of the latter process is relatively small. Due to loss of energy, an electron can be on one of the discrete atomic levels (Auger capture) or in a continuous spectrum (inelastic scattering). In order to calculate the probability of any process, it is desirable to have as accurate as possible wave functions for the meson, atomic electrons, and outgoing electrons. For electrons, the Hartree–Fock wave functions are quite accurate, and for the meson, the Hartree functions, since there is no exchange between the electron and the meson.

[46] ***bremco***—bremsstrahlung with correlations.

[47] ***bsiea***—bremsstrahlung.

The ***meson***[48] program [52] calculates the Auger meson capture cross section by an atom ($E > 0$), or the Auger transition probability ($E < 0$) [53]. The peculiarities of elastic and inelastic scattering of negative mesons by an atom are calculated using the ***scat*** program. The specificity of the problem is that, due to the large mass of the meson, even at its low energy, many partial waves have to be included in calculations. But, since the phases of the wave functions of a meson are smooth functions of its angular momentum, it is enough to actually calculate one phase out of five, so that the determination of phase shifts at five points corresponds to inclusion of all partial waves in the range $0 \leq l \leq 25$.

The interaction of the meson and atomic electrons, leading to the scattering of the meson, is taken into account in the lowest order of the perturbation theory, i.e., in the first Born approximation. In elastic scattering, in addition to the Hartree–Fock potential, the polarization potential determined in the SRPAE is taken into account, as in the case of electron scattering.

The ***meson*** program also calculates the capture of negatively charged mesons by the target atom. It can also be used in calculating scattering of positively charged heavy particles, for example positive mesons.

2.26 Photoionization of Diatomic Molecules

Photoionization is calculated, in principle, in the same way as for isolated atoms. A significant difference lies in the two-center potential of nuclei, which dictates the use of spheroidal coordinates in calculations. With their help, the axisymmetric problem of calculating a diatomic molecule with identical nuclei is simplified to the maximum extent.

The ***pabsm***[49] program calculates the photoionization cross section and oscillator strengths for diatomic molecules in the Hartree–Fock approximation and, if necessary, permits inclusion of multielectron correlations in the RPAE [17]. Namely, these quantities are calculated by the ***pabsm*** program using the wave functions found in advance of the ground and excited states of the molecule (see Sect. 2.5). Calculations in the Hartree–Fock and RPAE approximations are performed on the basis of general formulas (1.52)–(1.54).

Chapter 9 presents the details of these formulas for the case of axial symmetry, accounting for the selection rules corresponding to this case. Inclusion of electron correlations in RPAE makes sense for shells close in energy. As a rule, these can be either the outer shells of a molecule [17] or inner shells of different symmetry, close in energy—for example $1\sigma_g$ and $1\sigma_u$- shells of a homonuclear molecule [54].

---

[48] ***meson***—Auger meson capture.

[49] ***pabsm***—photoabsorption by molecules.

2.27 Photoionization of Endohedrals $A@C_N$ and $A@C_{N1}@C_{N2}$

As noted in Sect. 2.6, endohedrals are atoms enclosed in a fullerene shell [27]. The same section presents also programs used to calculate needed electron wave functions. The study of endohedrals photoionization is on the agenda of a number of laboratories. The existing methods of electron spectroscopy make it possible to reliably separate the contribution of the trapped atom against the background of the much larger photoionization cross section of the fullerene shell itself. Electronic waves emitted by an atom placed in a fullerene "illuminate" it from the inside, allowing one to obtain additional information about the structure of the fullerene shell and to judge changes in it under the influence of an atom or molecule placed inside.

The ***pabfulrf1***[50] program calculates the photoionization cross section and oscillator strength for an atom or ion placed in a fullerene shell, $A@C_N$, in the length and velocity forms and in one-configuration Hartree–Fock approximation. It also accounts for the multielectron correlations in the RPAE. This choice is only available for atoms or ions with closed or semi-closed subshells. A fullerene shell of zero thickness is taken into account both according to perturbation theory in the form of a factor in the scattering amplitude and phase or outside the framework of perturbation theory, adding factors and phases to intermediate states, over which the summation is performed in the RPAE equations.

The ***pabfulrf2***[51] program calculates the photoionization cross section and oscillator strength for an atom or ion enclosed in a fullerene shell, $A@C_{N2}$, in the length and velocity forms and in one-configuration Hartree–Fock approximation. The potential of the fullerene shell has a finite thickness Δ, chosen on the basis of indirect experimental data, and a depth, chosen so that, in agreement with experiment, describe the affinity energy of fullerene and an electron that is known from experiment.

The ***pabful2***[52] program calculates the photoionization cross section and oscillator strength for an atom or ion placed in two fullerene shells, $A@C_{N1}@C_{N2}$, in the length and velocity forms and in the one-configuration Hartree–Fock approximation. The calculation assumes that the radius of the outer fullerene R_2 is much larger than the radius of the inner fullerene R_1, which, in turn, is larger than the atomic radius R_a. The potential of the fullerene shell is infinitely thin.

The indicated programs, in addition to partial and total cross sections, allow calculating both dipole and non-dipole angular anisotropy parameters. As specific examples, we consider fullerenes with $N = N_1 = 60$ and $N_2 = 240$.

---

[50] ***pabfulrf1***—photoabsorption by an atom in a fullerene, $A@C_{N1}$ in random phase (i.e. RPAE).

[51] ***pabfulrf2***—photoabsorption by an atom in a fullerene, $A@C_{N2}$ in random phase (i.e. RPAE).

[52] ***pabful2***—photoabsorption by an atom in two fullerenes, $A@_{N1}@C\ C_{N2}$ (in HF and RPAE).

2.28 Inelastic Scattering of Fast Electrons and GOS in Endohedrals $A@C_N$ and $A@C_{N1}@C_{N2}$

Inelastic scattering of fast charged particles, primarily electrons, makes it possible to study the changes that an atom undergoes when it is placed inside a fullerene shell. The scattering cross section, total and differential with respect to the energy transferred to the atom and the scattering angles of both the fast electron and the secondary one knocked out of an endohedral atom under the influence of the energy transferred to it in the process of scattering, is expressed in terms of generalized oscillator strengths (GOS) [27] (see Sect. 2.9). So far, GOS for endohedrals have not been measured. But such measurements are possible, and their results will be informative.

The ***gosfulrf1***[53] program calculates the GOS for an atom or ion enclosed in a fullerene shell, $A@C_N$ in the length and velocity forms and in one-configuration Hartree–Fock approximation. It also takes into account multielectron correlations in the RPAE. This choice is only available for atoms or ions with closed or semi-closed shells.

The ***gosfulrf2***[54] program calculates the GOS for an atom or ion enclosed in a fullerene shell, $A@C_{N2}$ in the length and velocity forms and in one-configuration Hartree–Fock approximation and RPAE.

The ***gosful2***[55] program calculates the GOS for an atom or ion enclosed in two fullerene shells, $A@C_{N1}@C_{N2}$, in the length and velocity forms and in the one-configuration Hartree–Fock approximation and RPAE.

2.29 Decay of Vacancies in Endohedrals $A@C_N$

The fullerene shell significantly affects the decay of vacancies of an atom or molecule placed inside its shell (see [27] and references therein). It not only changes the contribution of decay channels already existing in the atom, but also adds new ones to them due to the participation of electrons of the endohedral shell. The change in probability here is akin to what happens in an isolated atom when intershell correlations are taken into account (see Sects. 2.21 and 2.22) and is described in Sects. 20.1.6 and 20.1.7.

Of great interest, however, is the opening of new channels. So, if vacancies in, e.g., isolated noble gas atoms subvalent ns^2 subshells can decay only due to the emission of a photon in the transition $ns^{-1} \rightarrow np^{-1} + \gamma$, the presence of a fullerene shell adds a non-radiative decay channel arising due to the filling of ns^{-1} vacancy by an

[53] ***gosfulrf1***—generalized oscillator strength of an atom in a fullerene, $A@C_N$ (in Hartree–Fock, RPAE).

[54] ***gosfulrf2***—generalized oscillator strength of an atom in a fullerene, $A@C_{N2}$, (in HF, RPAE).

[55] ***gosful2***—generalized oscillator strength of an atom in two fullerenes, $A@C_{N1}@C_{N2}$, (in HF, RPAE).

electron from the fullerene shell. The released energy is sufficient to emit another fullerene electron. As a result, the probability of ns^{-1} vacancy decay increases by 4–5 orders of magnitude.

2.30 Angular Distribution of Photoelectrons from Endohedrals $A@C_N$ Beyond the Dipole Approximation

The fullerene shell significantly affects the angular anisotropy parameters of the A atom in the $A@C_N$ endohedral. They are calculated in a number of programs.

The ***anglanizful0***[56] program calculates the photoelectrons non-dipole angular anisotropy parameters γ and η, γ^C, δ^C and $(\gamma^C + 3\delta^C)$ [see (20.68 and 20.69)], as well as secondary electrons in the scattering of fast electrons on the endohedral γ^{in}, η^{in} and λ^{in} (21.29–21.31) for the *s*-subshells $A@C_N$ in the HF and RPAE approximations for the transferred momentum $q = 0$.

The ***anglanizful1***[57] program calculates the photoelectrons non-dipole angular anisotropy parameters γ and η, γ^C, δ^C and $(\gamma^C + 3\delta^C)$ [see (20.68 and 20.69)], as well as in the processes of scattering of fast electrons on the endohedral of secondary electrons γ^{in}, η^{in} and λ^{in} (21.34–21.36) for *p*-subshells $A@C_N$ in the HF and RPAE approximations for the transferred momentum $q = 0$.

The ***anglanizful2***[58] program calculates the photoelectrons non-dipole angular anisotropy parameters γ and η, γ^C, δ^C and $(\gamma^C + 3\delta^C)$ [see (20.68), (20.69)], as well as secondary electrons in the processes of scattering of fast electrons on endohedrals γ^{in}, η^{in} and λ^{in} (21.39–21.41) for *d*—subshells $A@C_N$ in the HF and RPAE approximation for the momentum $q = 0$.

Difgosful0[59] [ACY] program calculates differential GOS for the magic angle and *s*-subshells $A@C_N$ for different q (21.16).

Difgosful1[60] [ACY] program calculates differential GOS for magic angle and *p*-subshells $A@C_N$ for different q (21.18).

Difgosful2[61] [ACY] program calculates differential GOS for magic angle and *d*-subshells $A@C_N$ for different q (21.21).

The complex of programs is designed as a single system, the elements of which are easily combined with each other. Therefore, the combination of different programs of the complex makes it possible to calculate a number of characteristics of atoms and simple molecules not mentioned above, and processes with their participation.

---

[56] ***anglanizful0***—angular anisotropy parameters for s-subshells of endohedrals.

[57] ***anglanizful1***—angular anisotropy parameters for p-subshells of endohedrals.

[58] ***anglanizful2***—angular anisotropy parameters for d-subshells of endohedrals.

[59] ***difgosful0***—differential GOS for s-subshells of endohedrals.

[60] ***difgosful1***—differential GOS for p-subshells of endohedrals.

[61] ***difgosful2***—differential GOS for d-subshells of endohedrals.

References

1. M.Ya. Amusia, L.V. Chernysheva, *Automated System for Studying the Structure of Atoms* (Leningrad: Science. 1983) (In Russian)
2. M.Ya. Amusia, L.V. Chernysheva, *Computation of Atomic Processes* (Institute of Physics Publishing, Bristol and Philadelphia, 1997)
3. L.V. Chernysheva, N.A. Cherepkov, V. Radojevič, Self-consistent field Hartree-Fock program for atoms. Comp. Phys. Comm. **11**, 57 (1976)
4. L.V. Chernysheva, N.A. Cherepkov, V. Radojevič, Frozen-core Hartree-Fock program for atomic discrete and continuum states. Comp. Phys. Comm **18**, 87 (1979)
5. Ch. Froese Fischer, *The Hartree-Fock Method for Atoms* (Wiley–Interscience, New York, 1977)
6. V.F. Bratsev, *Tables of Atomic Wave Functions* (Nauka, Moscow, 1966). ((In Russian))
7. P. Grant, Relativistic calculations of atomic structures. Adv. Phys. **19**, 747 (1970)
8. M.A. Goulthard, Proc. Phys. Soc. **9**, 144 (1967)
9. L.V. Chernysheva, V.L. Yakhontov, Comp. Phys. Comm **119**(2–3), 232 (1999)
10. D. Hartree, *The Calculation of Atomic Structure* (Wiley, New York, 1957)
11. F.C. Smith, W.R. Johnson, Phys. Rev. **160**, 136 (1967)
12. J.P. Desclaux, D.F. Mayers, O'Brien, J. Phys. B: Atom. Mol. Phys. **4**, 631 (1971)
13. I.M. Band, V.I. Fomichev, *System of Computer Programs RAINE* (The Institute of Nuclear Physics, No. 498, Leningrad, 1979) (In Russian)
14. V.A. Dzyuba, O.P. Sushkov, V.V. Flambaum, *A Software Package for Calculating Atomic Wave Functions and Energy Levels* (The Institute of Nuclear Physics, No. 82–89, Novosibirsk, 1982) (In Russian)
15. E.A. McCullough Jr., J. Chem. Phys **62**, 3991 (1977)
16. L. Laaksonen, P. Pyykko, D. Sundholm, Comput. Phys. Rep., **4**, 313 (1986)
17. S.K. Semenov, N.A. Cherepkov, G.H. Fecher, G. Schönhense, Phys. Rev. A**61**, 032704 (2000)
18. V.K. Dolmatov, *Photoionization of atoms encaged in spherical fullerenes*, in *Theory of Confined Quantum Systems, Part 2*, ed. by J. R. Sabin, E.J. Brändas, *Advances in Quantum Chemistry*, vol. 58, 13 (Academic Press, USA, 2009)
19. L.V. Chernysheva, M.Ya. Amusia, N.A. Cherepkov, *Numerical Calculation of the Photoionization Cross Sections of Atoms with Allowance for Correlations in One and Two Transitions.* Ioffe Institute No. 459 (Leningrad, 1974) (In Russian)
20. L.V. Chernysheva, N.A. Cherepkov, *Numerical Calculation of the Photoionization Cross Section of an open Shell Atom with Account of Electron Correlations in Two Transitions* Ioffe Institute No. 1079. (Leningrad, 1986) (In Russian)
21. S.K. Semenov, N.A. Cherepkov, *Numerical Calculation of the Dipole Dynamic Polarizability of Atoms* (Ioffe Institute, No. 835, Leningrad, 1983) (in Russian)
22. M. Inokuti, Rev. Mod. Phys., **43**, 296 (1971)
23. L.V. Chernysheva, M.Ya. Amusia, S.I Sheftel, *Numerical Calculation of Atomic Generalized Oscillator Strength with Wave Functions in the Hartree-Fock Approximation and Taking into Account Many-Electron Correlations in One Transition* (Ioffe Institute, No. 493, Leningrad, 1975) (In Russian)
24. M.Ya. Amusia, L.V. Chernysheva, E.Z. Liverts, JETP Lett. **94**(6), 466 (2011)
25. M.Ya. Amusia, L.V. Chernysheva, V.G. Yarzhemsky, *Photon Absorption, Electron Scattering, Vacancy Decay* (Nauka, Saint-Petersburg division, St. Petersburg, 2010) (In Russian)
26. M.Ya. Amusia, L.V. Chernysheva, E.Z. Liverts, Phys. Rev. **A85**, 042722 (2012)
27. M.Ya. Amusia, L.V. Chernysheva, V.G. Yarzhemsky, *Handbook of Theoretical Atomic Physics, Data for Photon Absorption, Electron Scattering, Vacancies Decay* (Springer, Heidelberg, 2012)
28. M.Ya. Amusia, V.K. Ivanov, N.A. Cherepkov, L.V. Chernysheva, *Processes in Multielectron Atoms* (Nauka, St. Petersburg, 2006). ((In Russian))
29. L.V. Chernysheva, M.Ya. Amusia, A.S. Kheifets, *Numerical Calculation of Differential Cross Sections for Excitation and Ionization of An Atom by Electron Impact* (Ioffe Institute, No. 1531, Leningrad, 1990) (In Russian)

30. F.H. Read, Radiat. Res. **64**, 23 (1975)
31. G.F. Drukarev, *Collisions of Electrons with Atoms and Molecules* (Nauka, Moscow, 1978). ((In Russian))
32. J.S. Bell, E.J. Squires, Phys. Rev. Lett. **3**, 96 (1959)
33. G.F. Gribakin, V.K. Ivanov, M.Yu. Kuchiev, L.V. Chernysheva. *Many-Body Theory Calculation of Negatively Charged Ions* (Ioffe Institute, No. 1096, Leningrad, 1987) (In Russian)
34. L.V. Chernysheva, M.Ya. Amusia, S.G. Shapiro, *Numerical Calculation of the Diagonal Matrix Element of the Self-energy Part of the One-Particle Green's Function in a Simplified Random Phase Approximation with Exchange* (Ioffe Institute, No. 504, Leningrad, 1976) (In Russian)
35. L.V. Chernysheva, M.Ya. Amusia, K.L. Tsemekhman, *Numerical Calculation of the Cross Sections of Photoabsorption and Single Photoionization of Intermediate Atomic Shells Taking into Account the Self-energy Part of the Photoelectron* (Ioffe Institute, No. 1492, Leningrad, 1990) (In Russian)
36. L.V. Chernysheva, M.Ya. Amusia, V.K. Ivanov, S.G. Shapiro, *Numerical Calculations of the Width of Atomic Levels Relative to Auger Decay with Allowance for Multielectron Correlations in Two Transitions* (Ioffe Institute, No. 536, Leningrad, 1977) (In Russian)
37. W. Wolfli, Ch. Stoller, G. Bonani, M. Suter, M. Stocli, Phys. Rev. Lett. **35,** 656 (1975)
38. V.V. Afrosimov, Yu.S. Gordeev, A.N. Zinoviev, V.M. Dukelskii, A.P. Shergin, Sov. Phys. JETP Lett. **26**, 273 (1976)
39. V.V. Afrosimov, Yu.S. Gordeev, A.N. Zinoviev, D.K. Rasulov, A.P. Shergin, Sov. J. Tech. Phys. Lett. **21**, 535 (1975)
40. M.Ya. Amusia, V.A. Kilin, I.S. Li, Sov. J. Techn. Phys. **54**, 990 (1984)
41. L.V. Chernysheva, M.Ya. Amusia, V.A. Kilin, I.S. Li, *Numerical Calculation of the Probability of Double Auger Decay* (Ioffe Institute, No. 944, Leningrad, 1985) (In Russian)
42. L.V. Chernysheva, M.Ya. Amusia, V.A. Kilin, A.N. Kolesnikova, I.S. Li, *Numerical Calculation of the Probability of One-Photon Decay of Two-hole states* (Ioffe Institute, No. 1054, Leningrad, 1986) (In Russian)
43. M.Ya. Amusia, V.A. Kilin, I.S. Lee, Sov. J. Tech. Phys. Lett. **10**, 1024 (1984)
44. M.Ya. Amusia, V.A. Kilin, A.N. Kolesnikova, I.S. Lee, Sov. J. Tech. Phys. Lett. **10**, 1029 (1984)
45. M.Ya. Amusia, V.A. Kilin, A.N. Kolesnikova, I.S. Lee, Sov. J. Tech. Phys. Lett. **11**, 343 (1985)
46. H. Bransden, K.H. Winters, J. Phys. B: Atom. Mol. Phys. **8**, 1236 (1975)
47. L.V. Chernysheva, M.Ya. Amusia, S.A. Sheinerman, *Numerical Calculation of the Cross Section of the Triplet Levels Excitation by Electrons in the Second Order of the Distorted Wave Method* (Ioffe Institute, No. 776, Leningrad, 1982) (In Russian)
48. V.B. Berestetskii, E.M. Lifshits, L.P. Pitaevskii, *Quantum Electrodynamics* (Butterworth-Heinemann, Oxford, 1982)
49. M.Ya. Amusia, A.S. Baltenkov, A.A. Paizicv, JETP Letters (USSR Acad. Sci.), **24**, 6, 366–369 (1976)
50. L.V. Chernysheva, N.B. Avdonina, M.Ya. Amusia, M.Yu. Kuchiev, *Numerical Calculation of the Spectrum of "Atomic" Bremsstrahlung of Fast Electrons* (Ioffe Institute, No. 865, Leningrad, 1983) (In Russian)
51. L.V. Chernysheva, N.B. Avdonina, M.Ya. Amusia, A.V. Korol, *Numerical Calculation of the Bremsstrahlung Spectrum of Particles of Intermediate Energies* (Ioffe Institute, No. 1314, Leningrad, 1989) (In Russian)
52. L.V. Chernysheva, N.A. Cherepkov, *Numerical Calculation of the Wave Functions of a Meson in the Field of a "Frozen" Core and the Cross Section for the Capture of a Meson by an Atom* (Ioffe Institute, No. 943, Leningrad, 1985) (In Russian)
53. N.A. Cherepkov, L.V. Chernysheva, *In the Book: Mesons in Matter* (Dubna, Moscow, 1977)
54. T. Jahnke, Th. Weber, A.L. Landers, A. Knapp, S. Schössler, J. Nickles, S. Kammer, O. Jagutzki, L. Schmidt, A. Czasch, T. Osipov, E. Arenholz, A.T. Young, R. Díez Muiño, D. Rolles, F.J. García de Abajo, C.S. Fadley, M.A. Van Hove, S.K. Semenov, N.A. Cherepkov, J. Rösch, M.H. Prior, H. Schmidt-Böcking, C.L. Cocke, R. Dörner, Phys. Rev. Lett. **8**, 073002 (2002)

Chapter 3
Mathematical Description of the Properties of Atomic and Molecular Structure

This chapter outlines a general approach to the method developed in this book for constructing mathematical models for describing physical processes in atoms and simple molecules.

To study a physical process or a physical characteristic of an atom or molecule, first of all, a theoretical model must be developed. A numerical model is built on its basis, with the help of which a computer program is created. The mathematical description of the problem includes both models, and they are essentially inseparable. At present, without numerical calculations, most studies of theoretical models are impracticable. The latter, in its turn, must be solvable and not require too large efforts to carry out calculations [1–6].

The accuracy of theoretical models, as a rule, is unknown and is established by means of estimates obtained as a result of roughly approximate calculations. In contrast, computer calculations can be very accurate. A reasonable choice of accuracy is that it should be slightly higher than the expected error of the actual physical result obtained in the experiment. If the error of the experimental data is approximately equal to 5%, then the computational error should not only be less than that already achieved, but also less than the expected experimental error for the coming years. The error of intermediate results is often more important, since the characteristic obtained in a physical experiment is found not rare as the difference between large numbers. It should also be remembered that in the course of calculations, the error accumulates from one step to the next and therefore it is difficult to estimate it. Therefore, the final result of calculations sometimes does not require such good accuracy as is needed for intermediate data and is implemented at intermediate stages. All this must be taken into account when choosing a numerical model to solve a specific problem.

The study of any physical process discussed in this book includes the following sequence of actions:

- choice of approximation for describing a physical process,
- mathematical description of the process within the selected approximation,

M. Ya. Amusia and L. V. Chernysheva, *Computation of Atomic and Molecular Processes*, Springer Series on Atomic, Optical, and Plasma Physics 117,
https://doi.org/10.1007/978-3-030-85143-9_3

- choice of method and calculation of atomic and molecular wave functions in the ground and excited states,
- calculation with their help of matrix elements of various types, which are determined by multidimensional integrals and can be reduced to numerical calculation of two-dimensional and one-dimensional integrals after separating the angular parts and performing analytical integration over the angular variables (see Sec. 1.3 and Ch. 9),
- construction of the amplitudes of physical processes from matrix elements, which often include summation and integration over intermediate states and can have singular regions due to energy denominators, which often pass through zero,
- determination of physical properties or probabilities of processes in terms of corresponding amplitudes.

When calculating one-electron wave functions for atoms or molecules in the Hartree–Fock approximation, which is chosen as zero approximation, each wave function is represented as a product (for atoms) or as a sum of products (for molecules) of the radial, angular, and spin parts. Radial wave functions are solutions of a Hartree–Fock equation (or system of equations) from which angular variables have been analytically removed (see Sect. 1.3 and Chap. 4). Integration of wave functions over angular variables and summation over spins are performed analytically and are included in the expression for matrix elements. If the expression for the amplitude of the process contains a singularity (singular points), then its neighborhood is isolated, and the integral at the singular point is taken analytically. The resulting contribution is added to the contribution of regular regions, which is found as a result of numerical integration. To perform calculations in the vicinity of the singularity, an infinitesimal imaginary quantity is added to the energy denominator, where the rules of the theory of complex variables are applied. Since an imaginary term appears, the amplitude also has an imaginary part.

In the Hartree–Fock and RPAE approximation, the calculation of the physical characteristics and amplitudes of the process probability is reduced to summation over a large number of intermediate (virtual) states, as well as to integration over the states of the continuous spectrum. The latter in numerical calculations is replaced by the summation of a finite series of terms. An important part of the calculation is choosing the correct upper summation limit. The degree of accuracy with which the integration up to an infinite upper limit can be replaced by the sum of a finite number of terms is determined by the rate of decrease of the integrand. Therefore, in each specific case, the choice of the upper limit and the number of intervals into which the region of integration is divided is made on the basis of preliminary calculations. The choice is considered good if a further increase in the upper limit and the number of intervals gives an error less than some predetermined value. The practice of calculations shows that in the physics of atoms and simple molecules, as a rule, one has to deal with matrix elements, which very quickly decrease as the energy of the states entering into them increases. Therefore, to ensure high accuracy, more integration points must be in the relatively low energy range. On the other hand, it is more convenient to have equidistant integration points in the numerical model. If we use the momentum scale $p = \sqrt{E}$ as the integration variable, then many intervals are in the low energy range, and the upper limit of integration is quite high. This choice of scale has a physical basis, since in atomic units p is the electron momentum (if E is given in Rydberg, and not in atomic units a.u., 1 a.u. = 2 Ry), and the main contribution to the integrals is made by relatively small energies E.

Another complication is associated with the summation over the excited states of an electron in the field of an atom or molecule (i.e., atomic or molecular ion), since the number of discrete excited states and states of the continuous spectrum both are infinite. In numerical calculations, the infinite sum over discrete excited states is replaced by a finite sum including only a few terms. For atoms, this is possible for the following reason. If the wave functions of discrete excited states with quantum numbers $n\ \ell$, σ are normalized in the same way as the functions of the continuous spectrum, then the matrix elements of the discrete spectrum in the limit $n \to \infty$ coincide with those corresponding to the wave functions of the continuous spectrum calculated at $E \to 0$. Therefore, matrix elements do not need to be calculated for $n \to \infty$ and $E = 0$, and thus for large n. For molecules, the situation is more complicated because ℓ is not a good quantum number. Therefore, in the case of molecular calculations, the question of the sufficiency of the excited states set used is solved primarily on the basis of the convergence of the results obtained (see Chaps. 6 and 9).

These considerations form a general approach when using a personal computer to study all the physical processes described in Chaps. 4–22, where the corresponding theoretical and numerical models are presented.

References

1. V.F. Bratsev, *Tables of Atomic Wave Functions* (Nauka, Moscow, 1966). ((In Russian))
2. I.P. Grant, Relativistic calculations of atomic structures. Adv. Phys. **19**, 747 (1970)
3. C. Froese Fischer, *The Hartree-Fock Method for Atoms* (Wiley—Interscience, New York, 1977)
4. C. Jablon, J.-C. Simon. *Applications des Modelles Numeriques en Physique.* (Birkhäuser, Basel, 1978)
5. M.Ya. Amusia, L.V. Chernysheva, *Automated System of Investigating of Atomic Structure* (Nauka, Leningrad Division, 1983). ((In Russian))
6. M.Ya. Amusia, L.V. Chernysheva, *Computation of Atomic Processes* (Institute of Physics Publishing, Bristol and Philadelphia, 1997)

Chapter 4
Wave Functions of the Ground State of Atoms and Simple Molecules in the Hartree–Fock Approximation

This chapter proposes a method for numerical solution of nonrelativistic Hartree–Fock equations for atoms, simple molecules, and endohedrals. The need to solve these equations is determined by the fact that many of the processes under consideration are nonrelativistic, while the solution of relativistic equations for many calculations related to intermediate and outer atomic shells is much more complicated and, in fact, is not necessary, although it gives high accuracy.

A number of programs are known that allow calculating the wave functions of an atom in the Hartree–Fock approximation [1–3]. However, the presented here programs have a number of advantages. One of them is that they are used to obtain atomic wave functions of excited states in discrete and continuous spectra. The specificity of the solution method used in the programs is also presented, namely the change of variables, acceleration of convergence, spin-polarized version of the method, etc.

4.1 Theoretical Model

4.1.1 Spherically Symmetric Potential

As mentioned in Chap. 1, one-electron wave functions $\phi(x)$, $x = (r, \sigma)$ are written as the product of the coordinate and spin parts (see (1.15)):

$$\phi_i(x) = \varphi_i(\mathbf{r})X_\sigma, \tag{4.1}$$

where, in the case of atoms, the coordinate wave functions are in a spherical coordinate system and are of the form:

M. Ya. Amusia and L. V. Chernysheva, *Computation of Atomic and Molecular Processes*, Springer Series on Atomic, Optical, and Plasma Physics 117,
https://doi.org/10.1007/978-3-030-85143-9_4

$$\varphi_i(\mathbf{r}) = \frac{1}{r} P_i(r) Y_{lm}(\vartheta, \varphi),\ i \equiv n_i l_i,\ l \equiv l_i. \tag{4.2}$$

In Formula (4.2), it is taken into account that in the case of spherical symmetry, the energies are degenerate in the quantum number m. This means that at $l_i > 0$, the radial wave functions $P_i(r)$ are the same for all $|m| \leq l_i$.

Substituting (4.1)–(4.2) into (1.8), integrating over angular variables ϑ, φ, summing over the spin variable and over the angular momentum projection m, we obtain the following system of nonlinear integral–differential equations [4, 5] for the radial Hartree–Fock functions $P_i(r)$, $i = 1, \ldots, s$, where s is the number of subshells in an atom:

$$\left(\frac{d^2}{dr^2} + \frac{2}{r}[Z - Y_i(r)] - \frac{l_i(l_i+1)}{r^2} - E_i\right) P_i(r) = X_i(r) + T_i, \tag{4.3}$$

$$Y_i(r) = \sum_{j=1}^{s} \left(N_j - \delta_{ij}\right) Y_{jj}^0(r) - \frac{2}{N_j} \sum_{j=1}^{s} \sum_{k>0} \gamma'_k(ij) Y_{jj}^k(r), \tag{4.4}$$

$$X_i(r) = \frac{2}{rN_i} \sum_{j \neq i}^{s} \sum_{k>0} \gamma_k(ij) Y_{ij}^k(r) P_j(r), \tag{4.5}$$

$$\begin{aligned} Y_{ij}^k(r) &= r \int_0^\infty \frac{r_<^k}{r_>^{k+1}} P_i(r') P_j(r') dr' \\ &= \left\{ \int_0^r \left(\frac{r'}{r}\right)^k + \int_r^\infty \left(\frac{r}{r'}\right)^{k+1} \right\} P_i(r') P_j(r') dr', \end{aligned} \tag{4.6}$$

$$T_i = -\sum_{i \neq j}^{s} \delta_{l_i l_j} \delta_{\sigma_i \sigma_j} E_{ij} P_i(r). \tag{4.7}$$

Here, $(n_i \ell_i)^{N_{ei}}$ denotes, respectively, the principal quantum number, orbital angular momentum, and the number of electrons in the ith subshell; r is the radius of an electron measured from the center of the nucleus; $\delta_{\ell_i \ell_j}$ is the Kronecker symbol, E_i, E_{ij}—diagonal and off-diagonal energy parameters, respectively. In calculations, $r_{\max}$ is usually taken equal to 60–100 atomic units.

Representation (4.3)–(4.7) of the Hartree–Fock equations implies that $E_i > 0$. After calculations, the sign E_i changes to the opposite, since the energy of the occupied states is actually negative. Radial wave functions meet the boundary conditions:

$$P_i(0) = P_i(\infty) = 0 \tag{4.8}$$

and normalization conditions:

$$\langle P_i|P_i\rangle \equiv \int_0^\infty P_i^2(r)dr = 1 \tag{4.9}$$

Representation (4.1) is strictly fulfilled only for atoms with closed and semi-closed shells. In other cases, a different restriction of the Hartree–Fock approximation, called the self-consistent central field approximation, is required.

Let us consider how the coefficients γ in Eqs. (4.4)–(4.5) are determined. In the LS-coupling scheme for atoms with closed subshells or containing a subshell with one electron in addition to the closed ones, the wave function is one Slater determinant. For closed subshells, such a wave function, in its transformation properties with respect to arbitrary rotations of the atomic system in the coordinate and spin spaces correspond to the state with $L = S = 0$.

Let, in addition to the closed subshells, the atom has only one electron with quantum numbers (l, m, σ). Then the total wave function corresponds to the state with $L = l, M_L = m, S = 1/2$ and $M_S = \sigma$.

In this case, the direct coefficients in the Hartree–Fock Eq. (4.3), which describe the interaction of an electron inside the subshell, are

$$\gamma'_k(nl) = N_{nl}(N_{nl} - 1)\frac{2l+1}{2(4l+1)}\begin{pmatrix} l & k & l \\ 0 & 0 & 0 \end{pmatrix}^2 . \tag{4.10}$$

Here N_{nl} is the number of electrons in the nl-subshell. k- an integer number that can take values from 2 to 2ℓ; the term with $k = 0$, i.e., $\gamma_0(nl) = -N_{nl}(N_{nl} - 1)/2$ is taken into account in (4.3); if k odd, $3j$-the symbol vanishes.

The exchange coefficients in (4.4), corresponding to the intershell electron interaction, are determined by the relation:

$$\gamma_k(nl, n'l') = \frac{1}{2}N_{nl}N_{n'l'}\begin{pmatrix} l & k & l' \\ 0 & 0 & 0 \end{pmatrix}^2 . \tag{4.11}$$

This formula is also valid for any term of an atom, provided that at least one of the subshells ($n\ell$ or $n'\ell'$), or both, are closed. For the averaged term (see Sect. 2.4), Formula (4.10) is also true in the case when both subshells are open.

For the state of an atom with an excited electron $l_p m_p \sigma_p$ and a hole $l_q m_q \sigma_q$, i.e., if an electron is absent only in one of the closed subshells, the one-determinant wave function, in its transformation properties with respect to arbitrary rotations of the atomic system, behaves like the following product of spin and angular parts:

$$Y_{l_p m_p}(\vartheta_1, \varphi_1)Y_{l_q - m_q}(\vartheta_2, \varphi_2)X_{\sigma_p}X_{-\sigma_q}.$$

From a set of such functions for different $m_p\, m_q\, \sigma_p\, \sigma_q$ ones, we construct a linear combination corresponding to a system with fixed total angular momentum and spin L и S of both particles and holes. Substituting this combination into the energy functional (1.7) and varying it with respect to ϕ, we obtain, after separating the angular parts, the direct coefficients corresponding to the interaction of open subshells:

$$\gamma'_{k\, k\neq 0}(n_pl_p, n_ql_q, LS) = (-1)^L(2l_p+1)(2l_q+1)$$
$$\begin{pmatrix} l_p & k & l_p \\ 0 & 0 & 0 \end{pmatrix}\begin{pmatrix} l_q & k & l_q \\ 0 & 0 & 0 \end{pmatrix}\begin{Bmatrix} l_p & l_q & L \\ l_q & l_p & k \end{Bmatrix}, \quad (4.12)$$

Here k is an even number that varies within $k = 2, \ldots, min\{2\ell_p, 2\ell_q\}$. Exchange coefficients are given by the following relations:

$$\gamma_k(n_pl_p, n_ql_q, LS) = (2l_q+1)(1-2\delta_{SO}\delta_{Lk})\begin{pmatrix} l_q & k & l_p \\ 0 & 0 & 0 \end{pmatrix}^2. \quad (4.13)$$

For the interaction of closed subshells, as well as subshells containing only one electron, except for closed ones, the coefficients are calculated by Formulas (4.9), (4.10). Solving the self-consistent system of equations for all electrons (4.3)–(4.7), we obtain the radial parts of the wave functions of atomic electrons that accounts for the mutual influence of the closed and open shells on each other.

4.1.2 Axially Symmetric Potential

As already mentioned in Chap. 1 (see (1.15)–(1.17)), the one-electron wave functions are found in the form (4.1), i.e., as products of the coordinate and spin wave functions. If the potential has axial symmetry, the solution of the Hartree–Fock Eq. (1.10) is sought in elongated spheroidal coordinates (ξ, η, φ):

$$\begin{aligned} x &= R(\xi^2-1)^{1/2}(1-\eta^2)^{1/2}\cos\varphi, \\ y &= R(\xi^2-1)^{1/2}(1-\eta^2)^{1/2}\sin\varphi, \\ z &= R\xi\eta, \end{aligned} \quad (4.14)$$

where R is some constant that determines the "elongation" of the coordinates.

It is taken into account, that for the given type of symmetry the orbital angular momentum ℓ is no longer a good quantum number. At the same time, for different projections of the angular moment m, in this case there is no degeneracy in the value of energy. Thus, in this case it is incorrect to use only one spherical harmonic in the representation of the coordinate wave function (as in (4.2) for atoms), and instead we use an expansion in spherical harmonics with a fixed m:

$$\varphi_i(\mathbf{r}) = \sum_{\lambda=|m_i|}^{L} \mathrm{P}_{i\lambda}(\xi) Y_{\lambda m_i}(\eta, \varphi), \tag{4.15}$$

Functions $\mathrm{P}_{i\lambda}$ below, similar to the atomic case, will also be called radial. It should be remembered that, in contrast to the atomic case, here one coordinate function φ_i corresponds to a set of radial functions $\mathrm{P}_{i\lambda}$. Since calculations in elongated spheroidal coordinates, in contrast to the case of spherical coordinates, are not yet so widespread, the description in spheroidal coordinates is presented here in more detail. The Laplace operator in these coordinates is known to be written in the form (see, e.g., [6]):

$$\Delta = \frac{1}{R^2(\xi^2 - \eta^2)} \left\{ \frac{\partial}{\partial \xi}(\xi^2 - 1)\frac{\partial}{\partial \xi} + \frac{\partial}{\partial \eta}(1 - \eta^2)\frac{\partial}{\partial \eta} + \left[\frac{1}{\xi^2 - 1} + \frac{1}{1 - \eta^2} \right] \frac{\partial^2}{\partial \varphi^2} \right\} \tag{4.16}$$

For diatomic molecules, the axis z is chosen so that the nuclei of the molecule lie on this axis at the points

$$z_{\pm} = R_z \equiv R \tag{4.17}$$

where $2R_z = D_z$ is the internuclear distance. In other words, for the case of diatomic molecules, the constant R in Eq. (4.14) is assumed to be equal to half the internuclear distance. In this case, the potential of the nuclei of the molecule with charges Z_1, Z_2 is obviously independent of φ and is written in the form:

$$V(\xi, \eta) = \frac{2(a_s \xi + a_d \eta)}{R^2(\xi^2 - \eta^2)}, \tag{4.18}$$

where

$$a_s = 2R(Z_1 + Z_2),\ a_d = 2R(Z_2 - Z_1). \tag{4.19}$$

The Coulomb integrals from (1.9), after summation over spins, are written in terms of coordinate functions φ in the form:

$$J_{i,j}(\mathbf{r}) = \int \varphi_i^*(\mathbf{r}')|\mathbf{r}' - \mathbf{r}|^{-1} \varphi_j(\mathbf{r}') d\mathbf{r}', \tag{4.20}$$

Taking into account (4.15), these integrals are also found in the form of expansions in spherical harmonics:

$$J_{ij}(\mathbf{r}) = \sum_{\lambda=m_{ek}}^{L} Z_{ij,\lambda}(\xi) Y_{\lambda, m_{ij}}(\eta, \varphi),\ m_{ij} = m_i - m_j. \tag{4.21}$$

Substituting (4.1), (4.15), (4.16), (4.18), and (4.21) into the Hartree–Fock Eq. (1.11), we can reduce the solution of these equations to the solution of a system of equations for functions $P_{i\lambda}(\xi)$. To do this, we must additionally use the following equalities, which correspond to the definition of spherical functions $Y_{\lambda m}(\eta, \varphi)$,

$$\left[\frac{\partial}{\partial\eta}(1-\eta^2)\frac{\partial}{\partial\eta}+\frac{1}{(1-\eta^2)}\frac{\partial^2}{\partial\varphi^2}\right]Y_{\lambda m}(\eta,\varphi)=l(l+1)Y_{\lambda m}(\eta,\varphi),$$
$$\frac{\partial^2}{\partial\varphi^2}Y_{\lambda m}(\eta,\varphi)=-m^2Y_{\lambda m}(\eta,\varphi). \tag{4.22}$$

Further, as in the case of atoms, it is necessary to perform integration over angular variables and summation over spins. As a result, we get [6]:

$$\begin{aligned}&\left[\frac{d}{d\xi}(\xi^2-1)\frac{d}{d\xi}-\frac{m_i^2}{\xi^2-1}-\lambda(\lambda+1)+a_s\xi+\varepsilon_i\frac{R^2}{2}(\xi^2-d^i_{\lambda\lambda})\right]P_{i\lambda}(\xi)\\&-\frac{R^2}{2}\sum_k P_{ik}(\xi)\sum_{j=1}^{N}(a_{ij}-\delta_{ji}b_{ii})\sum_{\lambda'}Z_{jj,\lambda'}(\xi)\left[\xi^2A^{\lambda',k,\lambda}_{0,m_i,m_i}-B^{\lambda',k,\lambda}_{0,m_i,m_i}\right]\\&+\frac{R^2}{2}\sum_{\substack{j=1\\(j\neq i)}}^{N}b_{ij}\sum_{k'}P_{jk'}(\xi)\sum_{\lambda'}Z_{ij,\lambda'}(\xi)\left[\xi^2A^{\lambda',k',\lambda}_{m_{ij},m_j,m_i}-B^{\lambda',k',\lambda}_{m_{ij},m_j,m_i}\right]\\&=\varepsilon_i\frac{R^2}{2}\sum_{k\neq\lambda}d^i_{\lambda k}P_{ik}(\xi)-a_d\sum_k c^i_{\lambda k}P_{ik}(\xi);\quad m_{ij}\equiv m_i-m_j,\ \lambda=|m_i|,\ldots,L.\end{aligned} \tag{4.23}$$

Here

$$c^i_{\lambda\lambda'}=\iint Y_{\lambda m_i}(\eta,\phi)\eta Y^*_{\lambda' m_i}(\eta,\phi)d\eta d\phi,\ d^i_{\lambda\lambda'}=\iint Y_{\lambda m_i}(\eta,\phi)\eta^2Y^*_{\lambda' m_i}(\eta,\phi)d\eta d\phi, \tag{4.24}$$

and the coefficients in front of the direct and exchange terms, a_{ji}, b_{ji} respectively, for doubly occupied spin–orbitals, are known to be equal:

$$a_{ji}=2,\ b_{ji}=1. \tag{4.25}$$

The system of Eq. (4.23) is a linear system of L equations for L unknown functions. The normalization of wave functions per unit corresponds to the condition

$$\int_1^\infty d\xi\int_{-1}^{1}d\eta(\xi^2-\eta^2)\sum_{\lambda=m}^{L}X_{j\lambda}(\xi)Y_{\lambda m}(\eta,\phi)\sum_{\lambda'=m}^{L}X_{j\lambda'}(\xi)Y_{\lambda' m}(\eta,\phi)=1 \tag{4.26}$$

For the ground state, the wave functions of the various shells are automatically orthogonal in accordance with the general properties of the Hartree–Fock equations.

When ionizing inner molecular shells, the rearrangement (or relaxation) of the core should be taken into account (see Chap. 1). In this case, the following coefficients were used in molecular calculations instead of coefficients (4.25):

$$\begin{aligned} a_{i'i} &= 2, \quad b_{i'i} = 1, \\ a_{ii'} &= 2 - z_e, \quad b_{ii'} = 0.5(2 - z_e), \quad i \neq i'; \\ a_{i'i'} &= 2(1 - z_e), \quad b_{i'i'} = 1 - z_e. \end{aligned} \tag{4.27}$$

Here i' is the number of the ionized spin–orbital, i.e., states containing one electron, and z_e $(0 < z_e < 1)$ is a fractional parameter. This parameter is selected from the condition that the obtained results coincide with the experimental data of the position of the shape resonance maximum. It turned out that other numerical results obtained in this way are also in good agreement with experiment.

4.2 Numerical Model

4.2.1 General Consideration

Here we consider the case of spherical symmetry of the potential. The axisymmetric case will be considered separately in Sect. 2.8. We represent the Hartree–Fock Eq. (1.8) in the following form:

$$\frac{d^2 P_i(r)}{dr^2} = [V_i(r) + E_i]P_i(r) + \int\limits_0^\infty U_i(r, r')P_i(r')dr' + T, \tag{4.28}$$

where

$$V_i(r) = \frac{2}{r}[Y_i(r) - Z] + \frac{l_i(l_i + 1)}{r^2} \tag{4.29}$$

is the local part of the potential; $2Y_i(r)/r$ is the direct Hartree–Fock potential, and

$$U_i(r, r') = \frac{2}{N_i} \sum_{j=1, j\neq i}^{s} \sum_k \gamma_k(ij) \frac{r_<^k}{r_>^{k+1}} P_j(r)P_j(r') \tag{4.30}$$

is the core of the nonlocal part of the Hartree–Fock potential; T is a term that includes off-diagonal parameters.

When solving Eqs. (4.3)–(4.7), one should keep in mind the following features.

1. Hartree–Fock equations are nonlinear. The function $Y_i(r)$ in (4.4) depends on the solutions of all equations; they, in turn, are linked to each other through $Y_i(r)$. The exchange term $X_i(r)$ (4.5) also depends on the solutions of all equations.
   The method of successive approximations allows us to solve a linear homogeneous equation at each iteration. The functions $Y_i(r)$ and $X_i(r)$ are calculated with the wave functions obtained at the previous iterative step; i.e., wave functions enter into expressions for $Y_i(r)$ and $X_i(r)$ as free parameters.
   When using the method of successive approximations, one should set the zero approximation, on which the rate of convergence depends. The iterative process starts with calculating $V_i(r)$ and $U_i(r, r')$ with zero approximation for radial wave functions being.

$$P_i(r) = 0, \tag{4.31}$$

   and a zero estimate for the energy eigenvalue

$$E^{(0)}_{i,\text{est}} = \frac{1}{n_i^2}\left(Z - \sum_{i=1, i\neq j} N_j\right)^2. \tag{4.32}$$

   Radial wave functions $P_i(r)$ are ordered in descending order of energy E_i.
2. Due to the presence of exchange interaction with electrons of other subshells, the wave function behaves asymptotically as $\exp(-\alpha r)$, where $\alpha = (I)^{1/2}$ and I is the ionization potential of the atom, and not as $\exp(-\alpha_i r)$, where $\alpha_i = (I_i)^{1/2}$ and I_i is the ionization potential of the shell, $I \leq I_i$.
   For the same reason, the wave function can sometimes have additional zeros, located, however, far from the core, i.e., where the wave function is close to zero [7]. The program does not take them into account, and it is assumed that each wave function $P_i(r)$ has $(n_i - \ell_i - 1)$ zeros. This additional constraint allows us to assign a solution to a specific configuration.
3. The term T_{i} in (4.7), which includes off-diagonal parameters, ensures the orthogonality of wave functions with equal orbital angular momenta. These factors are zero for configurations with closed shells [4] and small for the ground states of almost all atoms [1]. That is why we neglect them in the program.
4. The coefficients $\gamma_k(ij)$ and $\gamma'_k(ij)$ in (4.4) and (4.5) are the coefficients of the Slater integrals G_k and F_k [4]. This program takes into account the coefficients for $k = 0$, $\gamma'_0(ij) = -N_i(N_j - \delta_{ij})$. In addition to them, the ***scfhf*** program calculates (depending on the specification of the attribute *ksu*) the coefficients in the approximation - the *LS*-coupling ($ksu = 1$) or the excited state ($ksu = 2$) of the atom.

In the *LS*-coupling, the ground state of the atom is determined by the direct γ'_1 (4.10) and exchange γ_1 (4.11) coefficients. The coefficients are calculated for an averaged term of an arbitrary configuration or a fixed term (specified L and S in the case of the *LS*-coupling approximation, or *L* in the case of the spin-polarized method) for atoms with closed subshells, for atoms with only one electron above the closed subshells, and also for atoms with all closed subshells and one hole.

The number of coefficients *k* in the Hartree–Fock equation is determined by the following formulas.

For $ksu = 1$ or 2, the number of γ coefficients is

$$K_\gamma = \sum_{i,j,i<j}^{s} In\left(\frac{l_i + l_j - |l_i - l_j|}{2} + 1\right), \tag{4.33}$$

where *In* is the integer part of the number, *i* and *j* are the subshells numbers. Then the number of γ'_1 coefficients is determined as:

$$K_{\gamma_1} = \int_{i=1}^{s} l_i, \quad q_i \neq 1. \tag{4.34}$$

The total number of the coefficient at $ksu = 1$ equals to

$$K_{ksu=1} = K_\gamma + K_{\gamma'}. \tag{4.35}$$

For $ksu = 2$, for $N_i = 1$ or $N_j = 1$ and

$$N_j(N_i) = \begin{cases} 4\ell_j + 1 \text{ for LS-coupling} \\ 2\ell_j \text{ for polarized spin} \end{cases}$$

and an additional number of coefficients γ'_3 is calculated equal to.

$$K_{\gamma'_3} = \min(\ell_i, \ell_j),\ \ell_i = 1, \ldots, s,\ \ell_j = 1, \ldots, s. \tag{4.36}$$

The total number of coefficients for $ksu = 2$ is

$$K_{ksu=2} = K_\gamma + K_{\gamma'_1} + K_{\gamma'_3}. \tag{4.37}$$

It will be shown below that the residual interaction can be partially taken into account by an appropriate choice of coefficients in the Hartree–Fock equation.

5. The required computer time, and sometimes even the very possibility of solving Eq. (4.3), fundamentally depends on the order of enumeration when solving linear integral–differential Eq. (4.28) for various shells. This is especially important for heavy atoms, since the solution procedure for the wave functions of the outer shell is unstable and even a small variation of the wave functions of the inner shell can cause a significant change in the solution of Eq. (4.28) for the

outer shells. Therefore, the iterative procedure is divided into three cycles. In the first cycle, Eqs. (4.28)–(4.29) are solved in several steps. At the first step, equations are chosen for the two innermost subshells. At the next step, two more subshells are added to those for which the wave function has already been obtained. The procedure is repeated until all subshells have been taken into account. For example, the following order applies to an atom with five subshells (= 5): 1.2; 1,2,3,4; 1,2,3,4,5. At the $(m+1)$-th iterative step, the radial functions that determine the direct and exchange Hartree–Fock potential and the eigenvalue of energy, and the one obtained at the previous step (the m-th iteration), are taken as estimates:

$$
\begin{aligned}
P_{i,\mathrm{est}}^{(m)}(r) &= P_i^{(m)}(r),\\
E_{i,\mathrm{est}}^{(m)}(r) &= E_i^{(m)}(r).
\end{aligned}
\tag{4.38}
$$

At the end of the first cycle, the wave function of the inner subshell is found with better accuracy than the wave function of the outer subshell. In other words, the maximum relative discrepancy between the absolute value of the ith wave function of the subshell at m and $(m+1)$ -iterative steps

$$
\eta_i = \max\left(\left|1 - \frac{P_{i,\mathrm{est}}^{(m)}(r)}{P_{i,\mathrm{est}}^{(m+1)}(r)}\right|, \left|1 - \frac{E^{(m)}}{E^{(m+1)}}\right|\right), \tag{4.39}
$$

is less for inner subshells than for outer subshells. In the second cycle, Eq. (4.28) is also solved in several steps, but this time the order is reversed. If the absolute value η_i is less than the specified precision ε, the ith equation is skipped at this step. The procedure is repeated until the absolute values of all η_i are less than ε. Usually $\varepsilon = 10^{-5} \div 10^{-6}$.

As a rule, Eq. (4.28) is solved in this cycle for the least stable outer subshells. To make the iteration converge quickly, the wave functions and energies are calculated using the following expressions:

$$
\begin{aligned}
P_{i,\mathrm{est}}^{(m)}(r) &= (1 - a_1)P_i^{(m)}(r) + a_1 P_i^{(m-1)}(r),\\
E_{i,\mathrm{est}}^{(m)}(r) &= (1 - a_1)E_i^{(m)} + a_1 E_i^{(m-1)}.
\end{aligned}
\tag{4.40}
$$

The parameter a_1 is assumed to be zero if

$$
\mathrm{sign}(E_i^{(m)} - E_i^{(m-1)}) = \mathrm{sign}(E_i^{(m-1)} - E_i^{(m-2)})
$$

In this case, (4.40) turns into (4.38). If at the last three iterative steps $[m-2, m-1, m]$ the corresponding energy values converge to their optimal value, oscillating, i.e.,

$$
\mathrm{sign}(E_i^{(m)} - E_i^{(m-1)}) \neq \mathrm{sign}(E_i^{(m-1)} - E_i^{(m-2)}),
$$

then the parameter a_1 is calculated by Formula [8]

$$a_1 = \frac{E_i^{(m)} - E_i^{(m-1)}}{E_i^{(m)} - 2E_i^{(m-1)} + E_i^{(m-2)}} \tag{4.41}$$

The introduction of the parameter a_1 either accelerates the process of convergence of the procedure (e.g., in the cases of argon or neon atoms), or makes it possible to solve Eq. (4.28) for those atoms, for which the iterative procedure diverges at $a_1 = 0$ (e.g., the ground state of zinc or palladium atoms).

6. Potential $V_i(r)$ (4.29) has a singularity at $r = 0$. In the process of numerically solving the HF equations, r is replaced by a new variable ρ, so that in the vicinity $r = 0$ there is a fairly frequent series of subintervals, and the subintervals by ρ would be the same. In the program for transition to ρ, the following change of variables is introduced [2]:

$$\rho = \alpha r + \beta \ln r \tag{4.42}$$

where α and β are some constants.

At $r \to \infty$, $\rho \sim r$ and both functions $P_i(r)$ and $P_i(\rho)$ decrease exponentially. At $r \to 0$ the function $P_i(r) \sim r^{\ell_i+1}$, and $\rho \sim \ln(r) < 0$. Therefore, the function $P(\rho)$ behaves like $\exp[(\ell_i + 1)\rho]$, i.e., also decreases exponentially. Thus, the step size in the new variable ρ can be taken constant everywhere. The infinite interval of ρ is replaced by a finite one:

$$-\infty < \rho_0 \le \rho \le \rho_N < \infty \tag{4.43}$$

It is convenient to use Newton's method to determine the values r corresponding to the values ρ. In accordance with the variable change (4.42), we solve the equation

$$y(r) = \alpha r + \beta \ln r - \rho = 0. \tag{4.44}$$

The sequence

$$\begin{aligned} r_i &= r_{i-1} - \frac{y(r_{i-1})}{y'(r_{i-1})} = r_{i-1} - \frac{\alpha r_{i-1} + \beta l n r_{i-1} - \rho) r_{i-1}}{\alpha r_{i-1} + \beta} \\ &= r_{i-1}\left(1 - \frac{\alpha r_{i-1} + \beta l n r_{i-1} - \rho}{\alpha r_{i-1} + \beta}\right) = r_{i-1}(1 - \varepsilon) \end{aligned} \tag{4.45}$$

converges to a solution within a given accuracy ε if the initial approximation is in the solution region. The initial value ρ_0 is chosen according to the empirical formula

$$\rho_0 = -\beta(10 + lnZ), \tag{4.46}$$

The final ρ is defined as

$$\rho_N = \rho_0 + Nh. \tag{4.47}$$

The parameter α is calculated by the formula

$$\alpha = (\rho_N - \beta ln r_N)\frac{1}{r_N}. \tag{4.48}$$

Here the parameter β, the integration step h, and the number of points N are the input data of the program.

After performing the change of variables, we look for the wave function in the form

$$f_i(\rho) = \sqrt{\frac{\alpha + \beta}{r}} P_i(r), \quad r = r(\rho). \tag{4.49}$$

From relation (4.28), we obtain, which $f_i(\rho)$ is a solution of the following equation:

$$\frac{d^2 f_i(\rho)}{d\rho^2} = p_i(\rho) f_i(\rho) + q_i(\rho), \tag{4.50}$$

where

$$p_i(\rho) = (\alpha r + \beta)^{-2}\left[r^2(V_i(r) - E_i) + \frac{\beta(\alpha r + \beta/4)}{(\alpha r + \beta)^2}\right], \tag{4.51}$$

$$q_i(p) = \left(\frac{r}{\alpha r + \beta}\right)^{3/2} X_i(\rho), \quad r = r(\rho). \tag{4.52}$$

The procedure we have described quickly converges to a satisfactory solution when applied to the ground states of atoms and positive ions. However, the procedure is different for *negative ions*. The physical reason for the divergence of this procedure is the low binding energy of the negative ion. Because of this, the binding energy and the wave function of the outermost electron itself are strongly affected by the relatively small variations in the wave function of other atomic electrons.

To ensure convergence, an increased value of the atomic charge $Z_i = Z + \Delta Z > Z$ should be chosen as an initial approximation. Then, in the course of iterations, this value is decreased by an amount equal to $\Delta Z/n$, where n is a given number of steps to gradually decrease Z. As a result, the correct value of the atomic charge Z is achieved. Usually, for negative ions, the values $\Delta Z = 0.3$, $n = 1$. For neutral atoms and positive ions, ΔZ is assumed to be zero. Sometimes ΔZ can be chosen based on the requirement that the energy of the outer level coincide with the affinity energy of the negative ion.

4.2.2 Calculation of Y^k and X^k

In the case of spherical symmetry of the potential, the definition of the functions $p_i(\rho)$ and $q_i(\rho)$ is reduced to calculation $Y_{ij}^k(r)$ from (4.6). We denote the first term in (4.6) as

$$Z_{ij}^k(r) = r^{-k} \int_0^r (r')^k P_i(r')P_j(r')dr'. \tag{4.53}$$

The expressions for $Y_{ij}^k(r)$ and $Z_{ij}^k(r)$ use wave functions as free parameters and satisfy the following differential equations [4]:

$$\frac{d}{dr} Z_{ij}^k(r) = -\frac{k}{r} Z_{ij}^k(r) + P_i(r)P_j(r), \tag{4.54}$$

$$\frac{d}{dr} Y_{ij}^k(r) = \frac{k+1}{r} Y_{ij}^{-k}(r) - \frac{2k+1}{r} Z_{ij}^k(r) \tag{4.55}$$

with boundary conditions

$$\begin{aligned} &Z_{ij}^k(r) = 0, \\ &Y_{ij}^k(r) - Z_{ij}^k(r) \to 0, \\ &r \to \infty \end{aligned} \tag{4.56}$$

Therefore, the calculation of $Y_{ij}^k(r)$ is reduced to the successive integration of two first-order differential Eqs. (4.54) and (4.55). The first is integrated from 0 to ∞ with a boundary condition $Z_{ij}^k(0) = 0$, while the second is integrated from ∞ to 0 with a boundary condition $Y_{ij}^k(\infty) = Z_{ij}^k(\infty)$. After replacing the variable r by ρ, we have

$$\frac{d}{d\rho} Z_{ij}^k(\rho) = -\frac{k}{\alpha r + \beta} Z_{ij}^k(\rho) + \left(\frac{r}{\alpha r + \beta}\right)^2 f_i(\rho) f_j(\rho) \tag{4.57}$$

$$\frac{d}{d\rho} Y_{ij}^k(\rho) = \frac{k+1}{\alpha r + \beta} Y_{ij}^k(\rho) - \left(\frac{2k+1}{\alpha r + \beta}\right)^2 Z_{ij}^k(\rho) \tag{4.58}$$

Since at $\rho \to \infty$ the function $Z_{ij}^k(\rho)$ decreases much faster than $f(\rho)$, then it can be set $Z_{ij}^k(\rho_0)$ equal to 0. Then the boundary condition for $Y_{ij}^k(\rho)$ is $Y_{ij}^k(\rho_N) = Z_{ij}^k(\rho_N)$. A numerical solution to a first-order differential equation can be found according to the formulas

$$Z_m = -Z_{m+1} + \frac{h}{12}(5\,Z'_m + 8\,Z'_{m-1} - Z'_{m-2}) \tag{4.59}$$

$$Y_m = -Y_{m+1} - \frac{h}{12}(5\,Y'_m + 8\,Y'_{m-1} - Y'_{m-2}) \tag{4.60}$$

$$\begin{aligned} Y_m &\equiv Y_{ij}^k(\rho_m), \qquad Z_m \equiv Z_{ij}^k(\rho_m), \\ Y'_m &\equiv -\rho Y_{ij}^k|_{\rho=\rho_m},\ Z'_m \equiv -\rho Z_{ij}^k|_{\rho=\rho_m} \end{aligned}$$

with boundary conditions

$$Z_{-2} = Z_{-1} = 0, \quad Y_{N+2} = Y_{N+1} = 0, \quad Y_N = Z_N \tag{4.61}$$

4.2.3 Numerical Procedure for a Differential Equation

To solve linear differential equations in the case of spherical symmetry of the potential, we use the so-called chasing technique. In the case of axial symmetry, to solve systems of linear differential equations, the matrix chasing technique is used, accordingly (see Sect. 2.8).

One of the most effective methods for the numerical solution of a second-order linear differential equation that does not contain the first derivative is Numerov's method based on the expression

$$f_{m+1} - 2f_m + f_{m-1} = \frac{h^2}{12}\left(f''_{m+1} + 10f''_m + f''_{m-1}\right) + R(f), \tag{4.62}$$

where f_m is the value of the function at the point m, f'' is the second derivative of the function. Ignoring the remainder $R(f)$ and substituting the second derivative (4.50) into (4.62), we obtain the following equation:

$$f_i(\rho_{m+1})A_{m+1} - 2f_i(\rho_m)B_m + f_i(\rho_{m-1})A_{m-1} = \omega_m, \tag{4.63}$$

where

$$\begin{aligned} A_m &= 1 - \frac{h^2}{12}p_i(\rho_m), \\ B_m &= 1 + \frac{5h^2}{12}p_i(\rho_m), \\ \omega_m &= \frac{h^2}{12}\left[q_i(\rho_{m+1}) + 10q_1(\rho_m) + q_i(\rho_{m-1})\right]. \end{aligned} \tag{4.64}$$

According to (4.29) and (4.51), at $r \to 0$ the function $p_i(\rho)$ tends to $1/\beta^2[\ell_i(\ell_i + 1) + \beta^2/4]$, since $Y_i(0) = 0$. Similarly, for $r \to \infty$ one has $p_i(\rho) > 0$ for positive energy E_i. A qualitative study of the term $V_i(r)$ in (4.29) shows that it increases smoothly from 0 to its maximum equal to Z. This allows us to estimate the magnitude and sign of the derivative $p_i(r)$ and see that the function passes through zero twice. Let us denote these points as ρ_1 and ρ_2. The range of integration can then be divided into three regions: $[\rho_0, \rho_1]$, $[\rho_1, \rho_2]$, $[\rho_2, \rho_N]$ in which the solution $f(\rho)$ increases, oscillates, and decreases, respectively.

In the intervals $[\rho_0, \rho_1]$ and $[\rho_2, \rho_N]$, where $p_i(\rho) > 0$, to solve Eq. (4.63), the chasing technique is applied. The method is based on the successive elimination of one unknown value f_i from (4.63) through the boundary condition $f(\rho_0) = 0$ or $f(\rho_N) = 0$. Using the induction method, one can show that the values of the functions $f_i(\rho_m)$ and $f_i(\rho_{m+1})$ are connected by the relation:

$$f_i(\rho_m) = U_m + V_m f_i(\rho_{m+1}), \tag{4.65}$$

where

$$U_m = \frac{A_{m-1}U_{m-1} + \omega_m}{2B_m - A_{m-1}V_{m-1}},$$
$$V_m = \frac{A_{m+1}}{2B_m - A_{m-1}V_{m-1}}. \tag{4.66}$$

In a direct run, they are determined starting from the value

$$U_0 = V_0 = 0, \tag{4.67}$$

which corresponds to the boundary conditions $f_i(\rho_0) = 0$. Then, if a value $f_i(\rho_1)$ is available, the values at other points $f_i(\rho_m)$, $\rho_m = \rho_1, \ldots, 0$ are sequentially backtracked using Eq. (4.65). Since the integral–differential equation that is solved at each iteration step is linear and homogeneous, the normalization during the solution can be arbitrary. After the solution has been found, it can be normalized again, in accordance with (4.9). Therefore, an arbitrary value $f_i(\rho_1)$ can be chosen at the normalization point ρ_1. In the program, we assume $f_i(\rho_1) = 1$. In the interval $[\rho_1, \rho_2]$, the function $f_i(\rho)$ can be obtained from (4.63)

$$f_i(\rho_m) = \frac{\omega_{m-1} + 2B_{m-1}f_i(\rho_{m-1}) - A_{m-2}f(\rho_{m-2})}{A_m}, \tag{4.68}$$

and in this area the number of roots of this function is determined.

In the interval $[\rho_2, N]$, the values $f_i(\rho_m)$ are determined again by the chasing technique using a different boundary condition $f_i(\rho_N) = 0$. The corresponding equations in this area are

$$f_i(\rho_m) = \tilde{U}_m + \tilde{V}_m f_i(\rho_{m-1}), \tag{4.69}$$

$$\tilde{U}_m = \frac{A_{m+1}\,\tilde{U}_{m-1} - \omega_m}{2B_m - A_{m+1}\,\tilde{V}_{m+1}},$$
$$\tilde{V}_m = \frac{A_{m-1}}{2B_m - A_{m+1}\,\tilde{V}_{m+1}}. \tag{4.70}$$

with boundary conditions:

$$\tilde{U}_N = \tilde{V}_N = 0 \tag{4.71}$$

The calculation $f_i(\rho)$ is performed by reverse chasing in the direction of increasing ρ, starting from $f_i^{\infty}(\rho_2+h)$, where the solutions are "stitched" together. The last value $f_i^0(\rho_2+h)$ obtained by Numerov's method is used to check if the first derivatives are "stitched" together at the point $\rho_i = \rho_2$. The "jump" of the first derivative at this point can be roughly presented as follows:

$$\frac{d}{d\rho}\left[f_i^{\infty}(\rho_2+h) - f_i^0(\rho_2+h)\right] \approx \frac{f_i^{\infty}(\rho_2+h) - f_i^0(\rho_2+h)}{h} \tag{4.72}$$

This expression is used to correct the energy's eigenvalue.

4.2.4 Energy Eigenvalue Correction

Each Eq. (4.28) is solved for one value, and then the behavior of the solution is studied as a function of the variable E_{nl}. The correct eigenvalue of energy is obtained if the solution has $n_i - l_i - 1$ zeros and if the "jump" of the first derivative at the "stitching" point $\rho_2 + h$ does not exceed a given value ε, which determines the accuracy of the solution. If in course of the solution the number of zeros ζ is not equal to the desired value, the energy is recalculated using the formula:

$$E_{nl}^{(m)} = \left[1 + \frac{0.1}{n}\mathrm{sign}(\zeta - n_i + l_i + 1)\right] \cdot E_{nl}^{(m)}. \tag{4.73}$$

If the solution has the required number of zeros, further improvement E_{nl} is performed depending on the sign of the function $\tau(E_{nl})$ given by:

$$\tau(E_{nl}^{(m)}) = \frac{f_i^{\infty}(\rho_2) - f_i^0(\rho_2)}{h}. \tag{4.74}$$

The value $\tau(E_{nl}) = 0$ corresponds to its eigenvalue E_{nl}. The quantity $\tau(E_{nl})$ shown as a function E_{nl} is presented in Fig. 4.1:

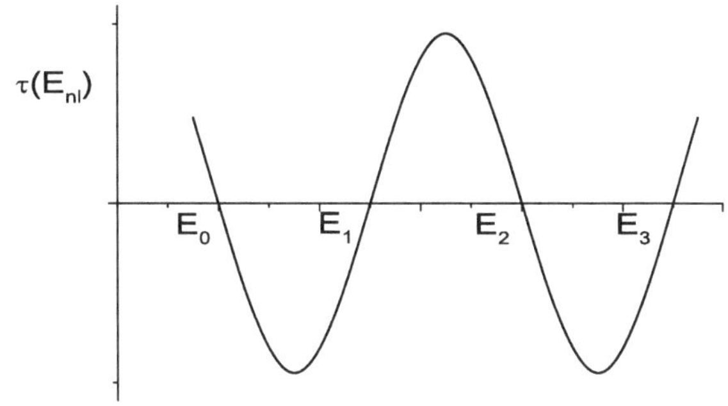

Fig. 4.1 Search for discrete energies in the Hartree-Fock equations

Therefore, for $\tau < 0$ and an even number of zeros, the trial value E_{nl} must be increased, while for $\tau > 0$ and odd number of zeros, it must be decreased, i.e., the energy is recalculated using the expression

$$E_{nl}^{(m+1)} = E_{nl}^{(m)}\left\{1 + (-1)^{n-l}\frac{0.1}{n}\mathrm{sign}\tau(E_{nl}^{(m)})\right\}. \tag{4.75}$$

After the root is isolated, it is corrected using the chord method:

$$E^{(m+1)} = \frac{\tau(E^{(m)})E^{(m-1)} - \tau(E^{(m-1)})E^{(m)}}{\tau(E^{(m)}) - \tau(E^{(m-1)})} \tag{4.76}$$

where $E^{(m-1)}$ and $E^{(m)}$ are two energy estimates, respectively, for the increments $\tau(E^{(m-1)})$ and $\tau(E^{(m)})$, $\tau(E^{(m-1)}) \cdot \tau(E^{(m)}) < 0$. The energy values are corrected repeatedly until the absolute difference between the energy values obtained in two successive iterations becomes smaller than ε. Usually, the accuracy ε is taken equal to 10^{-5}.

After the solutions of equations $f_i(\rho)$ (4.50)–(4.52) with boundary conditions (4.8) are normalized in accordance with (4.9), we divide $f_i(\rho)$ by the value $\left(f_i^2(\rho)d\rho\right)^{1/2}$ and then pass from $f_i(\rho)$ to $P_i(\rho)$ using (4.49).

4.2.5 Total Energy of an Atom in the Hartree–Fock Approximation

Many tasks require knowledge of the total energy of the atom (ion) E_0^{HF}. For example, the ionization potential is equal to the energy difference between an ion with a given vacancy (or vacancies) and an atom. To find E_0^{HF}, use expression [9]:

$$E_0^{HF} = \frac{1}{3}\sum_{i=1}^{s}\left(E_i - Z\left\langle\frac{1}{r}\right\rangle_i\right), \tag{4.77}$$

where $< 1/r >$ is the matrix element of the inverse coordinate:

$$\left\langle \frac{1}{r} \right\rangle_i = \int_0^{\infty} P_i^2(r) \frac{1}{r} dr = \int_{\rho_0}^{\rho} \frac{r}{(\alpha r + \beta)^2} f_i^2(\rho) d\rho, \tag{4.78}$$

Energy is measured in Rydberg (Ry). The total energy of the atom E_0^{HF} is determined in the program after the wave functions $P_i(r)$ and energies E_i are calculated for all atomic subshells.

4.2.6 Spin-Polarized HF Approximation

According to Hund's semiempirical rule, the total spin of the subshell in the ground state reaches the maximum value allowed by the Pauli principle. Therefore, in a semi-closed shell, the spin vectors of all electrons are collinear, and their projections onto an arbitrary fixed direction σ are equal. These electrons interact with other electrons in two different ways, namely with or without exchange, depending on the direction of the spins. Therefore, it is convenient to split each subshell into two levels determined by the spin, describing them with quantum numbers n, l, σ. These spin levels are completely closed with respect to the projections of the orbital angular momentum l.

Since the Coulomb interaction does not affect spin projections, electrons with opposite spin can be considered as particles of two different types, and additional levels, with spins directed "up" and "down," respectively, can be considered completely occupied. We will use the following subscripts $\uparrow$ and $\downarrow$ for the "up" and "down" states of electrons, respectively. HF Eq. (1.8) for a system containing such two types of particles are written in the form

$$\begin{aligned} & -\frac{\Delta}{2}\phi_{j\uparrow(\downarrow)}(x) - \frac{Z}{r}\phi_{j\uparrow(\downarrow)}(x) \\ & + \sum_{k\uparrow(\downarrow)=1}^{N_{e\uparrow(\downarrow)}} \int \phi^*_{k\uparrow(\downarrow)}(x') \frac{dx'}{|\mathbf{r}-\mathbf{r}'|} \left[\phi_{k\uparrow(\downarrow)}(x')\phi_{j\uparrow(\downarrow)}(x) - \phi_{j\uparrow(\downarrow)}(x')\phi_{k\uparrow(\downarrow)}(x)\right] \\ & + \sum_{k\downarrow(\uparrow)=1}^{N_{e\downarrow(\uparrow)}} \int \phi^*_{k\downarrow(\uparrow)}(x') \frac{dx'}{|\mathbf{r}-\mathbf{r}'|} \phi_{k\downarrow(\uparrow)}(x')\phi_{j\uparrow(\downarrow)}(x) = E_{j\uparrow(\downarrow)}\phi_{j\uparrow(\downarrow)}(x), \end{aligned} \tag{4.79}$$

Here $N_{e\uparrow} + N_{e\downarrow} = N_e$ is the total number of electrons in the system; i.e., (4.79) turns into two systems of interconnected equations for $\uparrow$ and $\downarrow$ electrons.

Within the spin-polarized HF approximation, we represent the one-electron wave functions in the following form:

$$\phi_{nlm\uparrow(\downarrow)} = \frac{1}{r} P_{nl\uparrow,\downarrow}(r) Y_{lm}(\vartheta, \phi) \quad (4.80)$$

which differs from (4.2) in the explicit dependence of the radial function on the spin projection.

Thus, since each subshell splits into two spin levels, the number of radial wave functions doubles. Radial Hartree–Fock functions $P_i(r)$ ($i \equiv n_i \ell_i \sigma_i, i = 1, \ldots, s$ where s is the doubled number of subshells in an atom) are solutions of the system of Eqs. (4.3)–(4.7). The difference, however, is that in (4.4) a coefficient $\delta_{\sigma_i \sigma_j}$ appears under the summation sign. The absence of exchange between the electrons of the split subshells and other electrons with the same spin direction with electrons having opposite spin projections is taken into account by choosing the appropriate coefficients.

Let the atomic wave function be one determinant of one-electron spin wave functions and, in addition to closed subshells, there are several more semi-closed ones. Such an electronic system, from the point of view of its transformation properties with respect to rotation in ordinary space, corresponds to the total angular momentum $L = 0$. At the same time, the total spin S of such a system is not generally defined. The total spin is equal to $S = (1/2)\Sigma N_{n\ell\sigma}$ with summation over all spin subshells with the number of electrons $N_{n\ell\sigma}$ in each, only if the spins of all these subshells are turned in the same direction.

If, in addition to the closed spin subshells, there is additionally only one electron in the state $n\ell\sigma$, then the one-determinant wave function describes the system with $L = l\, M_L = m$, and in the general case with an undefined S. For such a wave function, the Hartree–Fock equations have the following coefficients:

$$\gamma'_k(nl\sigma) = N_{nl\sigma}(N_{nl\sigma} - 1)\frac{2l+1}{4l}\begin{pmatrix} l & k & l \\ 0 & 0 & 0 \end{pmatrix}^2 . \quad (4.81)$$

The corresponding term with $k = 0$ has already been taken into account in relations (4.4), (4.5) in an explicit form, since

$$\gamma'_0(n\ell\sigma) = -\frac{N_{n\ell\sigma}(N_{n\ell\sigma} - 1)}{2} \quad (4.82)$$

The coefficients in the exchange term of the Hartree–Fock equations are determined by the relation:

$$\gamma_k(nl\sigma, n'l'\sigma') = \delta_{\sigma\sigma'} N_{nl\sigma} N_{n'l'\sigma'} \begin{pmatrix} l & k & l' \\ 0 & 0 & 0 \end{pmatrix}^2 . \quad (4.83)$$

This formula is also used for an arbitrary atomic term if at least one of the spin subshells ($n\ell\sigma$ or $n'\ell'\sigma'$) or both are closed. In the case of an averaged term, Eq. (4.83) describes the interaction of two open spin subshells.

We construct the wave function for an excited spin-polarized state of an atom of the "particle-hole" type in the same way as in the case of a *LS*-coupling. As one determinant, this function from the point of view of its transformation properties with respect to rotations in space behaves like $Y_{\ell_p m_p} Y_{l_q(-m_q)}$, where $\ell_p m_p$ refer to an excited electron (particle), and $\ell_q(-m_q)$- to a hole. From a set of such wave functions for different m_p, m_q ones, we construct a linear combination corresponding to a system with a fixed total momentum L and a defined spin S.

By varying the total energy functional with such a function, we obtain the Hartree–Fock equation. Direct coefficients corresponding to the interaction of open shells will be

$$\gamma'_k(n_p l_p \sigma_p, n_q l_q \sigma_q, S) = (-1)^L (2l_p+1)(2l_q+1) \begin{pmatrix} l_p & k & l_p \\ 0 & 0 & 0 \end{pmatrix} \begin{pmatrix} l_q & k & l_q \\ 0 & 0 & 0 \end{pmatrix} \begin{Bmatrix} l_p & l_q & L \\ l_q & l_p & k \end{Bmatrix}, \quad (4.84)$$

i.e., γ'_k does not depend on σ_p and σ_q coinciding with (4.12), and the exchange terms are

$$\gamma_k(n_p l_p \sigma_p, n_q l_q \sigma_q, L) = \delta_{\sigma_p \sigma_q} (2l_q+1) \left(1 - \delta_{LK} \cdot \frac{2l_p+1}{2L+1}\right) \begin{pmatrix} l_q & k & l_p \\ 0 & 0 & 0 \end{pmatrix}^2 . \quad (4.85)$$

Solving the self-consistent system of Eqs. (4.2)–(4.7) for all atomic electrons with these coefficients, we obtain electronic wave functions taking into account the mutual influence of open and closed spin subshells.

4.2.7 *Inclusion of the Endohedral's Potential*

In order to take into account the potential of the endohedral, in which an atom or a simple molecule is placed inside a fullerene that is inside an empty shell formed by carbon atoms, Eq. (4.3) is transformed into the form

$$\left(\frac{d^2}{dr^2} + \frac{2}{r}[Z - Y_i(r)] + W_c(r) - \frac{l_i(l_i+1)}{r^2} - E_i\right) P_i(r) = X_i(r) + T, \quad (4.86)$$

Here $W_c(r)$ is the attractive potential of finite width Δ and depth, equal to U_0

$$W_c(r) = U_0 \text{ at } R_0 \le r \le R_0 + \Delta \quad (4.87)$$

The solution of Eq. (4.86) is carried out according to the algorithm described above. In this case, specific calculations were carried out with the following potential $W_c(r)$ parameters chosen to best fit the negative ion of fullerene C_{60}:

$$U_0 = 0.844\,\text{Ry}, R_o = 6.01\,\text{a.u}, \Delta = 1.25\,\text{a.u}$$

4.2.8 Numerical Procedure for HF Equations with Axially Symmetric Potential

For the case of a potential with axial symmetry, the general principle of enumerating subshells when solving the system of Eq. (4.23) is similar to the atomic case—from the inner subshell to the outer one. The numerical solution of the equations is also to a noticeable degree similar to the atomic case, but there are also significant differences associated primarily with the type of symmetry.

Equation (4.23) that needs to be solved include functions $Z_{\lambda,ij}(\xi)$ (projections of the Coulomb integrals onto spherical harmonics), which are similar to the functions Y_{ij}^k from (4.6) for atoms. The functions $Z_{\lambda,ij}(\xi)$ are defined by Eqs. (4.20), (4.21), and these functions must be found before solving the equations themselves. The calculation of these functions, on the one hand, is in many respects similar to the solution of the equations themselves; on the other hand, the solution process is much easier to understand using the example of these functions, since in this case it is not a system of equations that is solved, but individual equations.

From (4.20), we have:

$$\Delta J_{ij}(\mathbf{r}) = -4\pi\varphi_i(\mathbf{r})\varphi_j^*(\mathbf{r}). \tag{4.88}$$

Using (4.15), (4.16), (4.21), and (4.22), we obtain from (4.88):

$$\begin{aligned} &\left[\frac{d}{d\xi}(\xi^2 - 1)\frac{d}{d\xi} - \frac{m^2}{\xi^2 - 1} - \lambda(\lambda + 1)\right] Z_{ij,\lambda}(\xi) Y_{\lambda m}(\eta, \phi) \\ &= -\pi R^2 \left[(\xi^2 - \eta^2) \sum_{\lambda=m}^{L} X_{j\lambda}(\xi) Y_{\lambda m}^*(\eta, \phi) \sum_{\lambda'=m}^{L} X_{i\lambda'}(\xi) Y_{\lambda' m}(\eta, \phi)\right]. \end{aligned} \tag{4.89}$$

By projecting the resulting expression sequentially on harmonics $Y_{\lambda m}(\eta, \phi)$ with different values λ, we obtain a set of independent equations for calculating functions $Z_{ij,\lambda}(\xi)$:

$$\left[\frac{d}{d\xi}(\xi^2-1)\frac{d}{d\xi}-\frac{m^2}{\xi^2-1}-\lambda(\lambda+1)\right]Z_{ij,\lambda}(\xi)$$
$$=-\pi R_z^2(-1)^{m_i}\sum_{k,k'}\mathrm{P}_{i,k}(\xi)\mathrm{P}_{j,k'}(\xi)\left(\xi^2A^{k',k,\lambda}_{m_j,-m_i,m}-B^{k',k,\lambda}_{m_j,-m_i,m}\right),\quad \lambda=1\div L',\tag{4.90}$$

where

$$A^{l,l',l''}_{m,m',m''}=\iint Y_{l,m}Y_{l',m'}Y^*_{l'',m''}d\eta d\varphi,$$
$$B^{l,l',l''}_{m,m',m''}=\iint Y_{l,m}Y_{l',m'}Y^*_{l'',m''}\eta^2 d\eta d\varphi,\tag{4.91}$$

The resulting Eq. (4.90) is solved numerically, sequentially for all λ. The details of the procedure for this numerical solution are described below.

Solutions of Eq. (4.90) are found numerically on a limited interval:

$$1\le\xi\le\xi_{\max},\ \xi_{\max}\approx r_{\max}\cong 60\,a.e.\tag{4.92}$$

Near the singularity point $\xi=1$ (the axis of the molecule), for $1\le\xi\le\xi^*$ ($\xi^*\simeq 1.02$), the solutions of Eq. (4.90) are sought on the basis of the asymptotic expansion of functions $Z_{ij,\lambda}(\xi)$ in the vicinity of this point,

$$Z_{ij,\lambda}(\xi)\simeq(\xi^2-1)^{|m|/2}\sum_{k=0}^{n}c_k(\xi-1)^k,\tag{4.93}$$

which follows directly from (4.90) as a necessary condition for eliminating the singularity at $\xi=1$.

The number of included expansion terms is limited, naturally, to some given $n (n\simeq 5)$ value. In this case, n equidistant integration points are set on the interval $[1,\xi^*]$:

$$\xi_i=1+(i-1)\Delta^*;\ \Delta^*=(\xi^*-1)/(n^*-1),\ i=1\div n^*.\tag{4.94}$$

For the first integration point, to move away from the singularity point, a negligible small shift $\xi_1=1+\delta_1,\ \delta_1<10^{-9}$ is made. The differentiations in (4.90), using the representation (4.93), in this case are performed analytically. As a result, we have:

$$\left[\frac{d}{d\xi}(\xi^2-1)\frac{d}{d\xi}-\frac{m^2}{\xi^2-1}\right]Z_{ij,\lambda}(\xi)=(\xi^2-1)^{m/2}$$
$$\times\left[\sum_{k=0}^{n^*}c_k(\xi-1)^k a_k^m+\sum_{k=0}^{n^*-1}c_{k+1}(\xi-1)^k b_k^m\right],\tag{4.95}$$

where

$$a_k^m = (m+k)(m+k+1);$$
$$b_k^m = 2(k+1)(m+k+1). \tag{4.96}$$

Further, Eq. (4.90) is written in the form:

$$\left[\sum_{k=0}^{n^*} c_k(\xi-1)^k a_k^m + \sum_{k=0}^{n^*-1} c_{k+1}(\xi-1)^k b_k^m \right.$$
$$\left. + p(\xi)\sum_{k=0}^{n} c_k(\xi-1)^k \right] = q(\xi), \tag{4.97}$$

where

$$p(\xi) = -\lambda(\lambda+1),$$
$$q(\xi) = -\pi R_z^2(-1)^{m_i}\sum_{k,k'} \mathrm{P}_{i,k}(\xi)\mathrm{P}_{j,k'}(\xi)\left(\xi^2 A^{k',k,\lambda}_{m_j,-m_i,m} - B^{k',k,\lambda}_{m_j,-m_i,m}\right)/(\xi^2-1)^{m/2}. \tag{4.98}$$

Although p here does not depend on ξ, a general procedure is used for the solution, which is also used below for solving Eq. (4.23), and which assumes the presence of such a dependence. Next, a matrix is formed:

$$\mathbf{w} : w_{ik} = (\xi_i - 1)^k, \tag{4.99}$$

and the inverse matrix $\mathbf{w}^{-1}$ is found. Then the coefficients of the expansion of the functions $p(\xi)$, $q(\xi)$ in power series

$$p(\xi) = \sum_{k=0}^{n^*} p_k(\xi-1)^k, \quad q(\xi) = \sum_{k=0}^{n^*} q_k(\xi-1)^k, \tag{4.100}$$

are defined by the formulas:

$$p_k = \sum_{k'=0}^{n^*} p(\xi_{k'})w^{-1}_{k'k}, \quad q_k = \sum_{k'=0}^{n^*} q(\xi_{k'})w^{-1}_{k'k}. \tag{4.101}$$

Equating in (4.97) the coefficients at the same degrees of $(\xi - 1)$, we obtain the system of equations

$$c_k a_k^m + c_{k+1} b_k^m + \sum_{k'=0}^{n} c_{k'} p_{k-k'} = q_k, \; k = 1 \div n - 1, \tag{4.102}$$

in which the coefficients c_k are unknown, and the number of equations is one less than the number of unknowns. To close the system, it is necessary to consider a solution going from infinity. Before looking for these solutions, to formulate the boundary conditions for large ξ, we note that solutions (4.90) decreasing at infinity have the asymptotic:

$$Z_{ij,\lambda} \sim \xi^{-\lambda-1} \tag{4.103}$$

The calculations use the first three terms of this asymptotic, written in the form:

$$Z_{ij,\lambda} \cong c(\xi^2 - 1)^{-(\lambda+1)/2}[1 + a_1/(\xi^2 - 1) + a_1/(\xi^2 - 1)^2]. \tag{4.104}$$

The coefficients a_1, a_2 are determined recursively and are found by substituting (4.104) in (4.90) and equating the coefficients at the same degrees of ξ:

$$\begin{aligned} a_1 &= -[m^2 + 1 + \lambda(\lambda + 2)]/(4\lambda + 6) \\ a_2 &= -a_1[m^2 + 1 + (\lambda + 4)(\lambda + 2)]/(8\lambda + 20) \end{aligned} \tag{4.105}$$

Further, for the numerical solution on the interval $[\xi^*, \xi_{\max}]$, a change of variables is made, which provides a high density of integration points near the molecular axis (singularity point):

$$\rho = \rho(\xi) = \xi - \frac{c_1}{\xi - c_2}. \tag{4.106}$$

Here c_1, c_2 are some given constants ($c_1 = 10, c_2 = 0.9$). The inverse transform of (4.106) is easy to find:

$$\xi = \xi(\rho) = \frac{(\rho + c_2) + \sqrt{(\rho + c_2)^2 - 4(c_1 + \rho c_2)}}{2}. \tag{4.107}$$

Simultaneously with the change of coordinates, the change of the required functions is done

$$\begin{aligned} F_{ij,\lambda}(\rho) &= Z_{ij,\lambda}(\xi)g(\xi), \\ g(\xi) &= \sqrt{\rho'(\xi)(\xi^2 - 1)} \\ &= \frac{\sqrt{(\xi^2 - 1)[(\xi - c_2)^2 + c_1]}}{(\xi - c_2)}, \end{aligned} \tag{4.108}$$

which ensures the absence of the first derivative of the new functions $dF_{ij,\lambda}/d\rho$ in the transformed Eq. (4.90). To write the transformed equations, we calculate the first three derivatives of the function $\rho(\xi)$ from (4.106):

$$\rho'(\xi) = 1 + \frac{c_1}{(\xi - c_2)^2}, \ \rho''(\xi) = -\frac{2c_1}{(\xi - c_2)^3}, \ \rho'''(\xi) = -\frac{3\rho''(\xi)}{(\xi - c_2)}, \tag{4.109}$$

and define, using them, functions:

$$\begin{aligned} g_1(\xi) &= \frac{1}{(\xi^2 - 1)[\rho'(\xi)]^2}, \\ g_2(\xi) &= -\frac{2\rho'''(\xi)\rho'(\xi) - 3[\rho''(\xi)]^2}{4(\xi^2 - 1)[\rho'(\xi)]^4} \\ &+ \frac{1}{(\xi^2 - 1)^2[\rho'(\xi)]^2}. \end{aligned} \tag{4.110}$$

Then the equations for the functions $F_{ij,\lambda}$ can be written in the form:

$$\frac{d^2F_{ij,\lambda}(\rho)}{d\rho^2} + p'(\xi)F_{ij,\lambda}(\rho) = q'(\xi), \tag{4.111}$$

where the functions $p'(\xi)$, $q'(\xi)$ are defined in terms of the functions $p(\xi)$, $q(\xi)$ from (4.98) as follows:

$$\begin{aligned} p'(\xi) &= p(\xi) - \frac{m^2}{\xi^2 - 1}g_1(\xi) + g_2(\xi), \\ q'(\xi) &= q(\xi)g_1(\xi). \end{aligned} \tag{4.112}$$

The following description of the numerical algorithm for solving Eq. (4.111), as well as radial HF equations, up to notation and small details in the boundary conditions coincides with the description given above for atoms (see Sects. 2.2 and 2.3). The authors consider it necessary to reproduce these differences for those who intend to use the programs presented in this book for some other, more advanced calculations. The primes on the functions from (4.112) are omitted below.

The total number of integration points N is set. The interval of numerical integration with respect to the variable ρ is determined as $[\rho^*, \rho_{\max}]$,

$$\rho^* = \rho(\xi^*), \ \rho_{\max} = \rho(\xi_{\max}), \tag{4.113}$$

the integration step is

$$h = (\rho_{\max} - \rho^*)/(N - n^*), \tag{4.114}$$

and the points of integration are:

$$\rho_i = \rho^* + h \cdot (k - n^* - 1),\ k = n^* + 1, \ldots, N. \tag{4.115}$$

For the subsequent description of the programs, it should be noted here that all equations were initially written in the variable ξ. Therefore, the array of integration points in the variable ξ is decisive for all calculations. In this case, at the point ξ^*, you need to have both of function, namely $Z_{ij,\lambda}(\xi^*)$ for integration over ξ along the region near axis of the molecule, and a transformed function $F_{ij,\lambda}(\rho^*)$ for integration over the variable ρ. The same is for other functions. Therefore, the point ξ^* is written into the array of the variable twice $\xi_n = \xi_{n+1} = \xi^*$, and the point with the number $n + 1$ is the first point for solving equations with the transformed variable ρ and the corresponding substitutions of functions of the form (4.108). Thus, for the required function $Z_{ij,\lambda}(\xi)$ we have:

$$Z_{ij,\lambda}(\xi_n) \equiv Z_{ij,\lambda}(\xi^*) = F_{ij,\lambda}(\rho_{n+1})/g(\xi_{n+1}) \equiv F_{ij,\lambda}(\rho^*)/g(\xi^*) \tag{4.116}$$

According to Numerov's method (see Formulas (4.62)–(4.64)), we write Eq. (4.111) in finite-difference form:

$$\begin{aligned} &F_{ij,\lambda}(\rho_{k+1})[c_h + p(\rho_{k+1})] - 2F_{ij,\lambda}(\rho_k)[c_h + 5p(\rho_k)] \\ &\quad + F_{ij,\lambda}(\rho_{k-1})[c_h + p(\rho_{k-1})] = q(\rho_{k+1}) + 10q(\rho_k) \\ &\quad + q(\rho_{k-1}) \equiv \omega_k, \quad c_h = 12/h^2. \end{aligned} \tag{4.117}$$

Further, we apply the chasing technique. Namely, we write the relationship between the values of the function at two adjacent points in the form

$$F_{ij,\lambda}(\rho_k) = V_k + U_k F_{ij,\lambda}(\rho_{k-1}). \tag{4.118}$$

Substituting this relation into (4.117), we find the recurrence relations for U_k, V_k:

$$\begin{aligned} U_k &= -[c_h + p(\rho_{k-1})]/z, \\ V_k &= \{\omega_k - V_{k+1}[c_h + p(\rho_{k+1})]\}/z, \\ z &= U_{k+1}[c_h + p(\rho_{k+1})] - 2[c_h + 5p(\rho_k)]. \end{aligned} \tag{4.119}$$

The values U_N, V_N are determined from Eqs. (4.104) and (4.108):

$$\begin{aligned} V_N &= 0, \\ U_N &= \frac{F_{ij,\lambda}(\rho_N)}{F_{ij,\lambda}(\rho_{N-1})} \\ &= \frac{g(\xi_N)(\xi_N^2 - 1)^{-(\lambda+1)/2}[1 + a_1/(\xi_N^2 - 1) + a_1/(\xi_N^2 - 1)^2]}{g(\xi_{N-1})(\xi_{N-1}^2 - 1)^{-(\lambda+1)/2}[1 + a_1/(\xi_{N-1}^2 - 1) + a_1/(\xi_{N-1}^2 - 1)^2]}. \end{aligned} \tag{4.120}$$

Sequentially calculating the values U_k, V_k using the recurrence relations, we reach the point ξ^* for which from (4.118) we have:

$$F_{ij,\lambda}(\rho_{n+2}) = V_{n+2} + U_{n+2} F_{ij,\lambda}(\rho_{n+1}). \tag{4.121}$$

Using (4.108), we first replace the functions $F_{ij,\lambda}$ here with functions $Z_{ij,\lambda}$ and then replace $Z_{ij,\lambda}$ with a representation in the form of expansion (4.93), which describes these functions well near the axis of the molecule. As a result, we get:

$$g(\xi_{n+2})(\xi_{n+2}^2 - 1)^{|m|/2} \sum_{k=0}^{n} c_k(\xi_{n+2} - 1)^k = V_{n+2}$$
$$+ U_{n+2} g(\xi_{n+1})(\xi_{n+1}^2 - 1)^{|m|/2} \sum_{k=0}^{n} c_k(\xi_{n+1} - 1)^k. \tag{4.122}$$

This equation, together with the equations from (4.102), form a closed linear system of equations for determining the necessary coefficients.

After these coefficients are found, we calculate the values $Z_{ij,\lambda}(\xi)$ at the points of integration in the interval $[1, \xi^*]$:

$$Z_{ij,\lambda}(\xi_j) = (\xi_j^2 - 1)^{|m|/2} \sum_{k=0}^{n} c_k(\xi_j - 1)^k,\ j = 1 \div n, \tag{4.123}$$

and then

$$F_{ij,\lambda}(\rho_{n+1}) \equiv F_{ij,\lambda}(\rho^*) = Z_{ij,\lambda}(\xi^*) g(\xi^*) \equiv Z_{ij,\lambda}(\xi_n) g(\xi_n). \tag{4.124}$$

Further, from (4.118), we find the rest of the solution.

The solutions of Eq. (4.23) are found numerically similarly to the solution of Eq. (4.90), with the difference that instead of independent equations, it is now necessary to solve their system. Integration points are determined in the same way as when calculating functions $Z_{ij,\lambda}$. Near the point of singularity $\xi = 1$, i.e., on the axis of the molecule, at $1 \le \xi \le \xi^*$, the solution is sought on the basis of the asymptotic expansion of the functions $\mathrm{P}_{i\lambda}$ from (4.23) in the vicinity of this point:

$$\mathrm{P}_{i\lambda}(\xi) \simeq (\xi^2 - 1)^{|m_i|/2} \sum_{k=0}^{n^*} c_{\lambda,k}(\xi - 1)^k, \tag{4.125}$$

which follows directly from (4.23). The number of expansion terms taken into account is also limited, naturally, to some given $n^*(n^* \simeq 5)$. As follows again from (4.23), the coefficients $c_{\lambda,0}$ are not defined. In this case, obviously, one of the coefficients $c_{\lambda,0}$ can be specified arbitrarily, determining the normalization of the sought function. Therefore, in calculations we take

$$c_{\lambda,0} = 1\ (\lambda = m_i). \tag{4.126}$$

The subsequent coefficients $c_{\lambda,i}(i \geq 1)$ must be determined, naturally, from (4.23). For each value λ, a ratio similar to (4.95) is written:

$$\left[\frac{d}{d\xi}(\xi^2 - 1)\frac{d}{d\xi} - \frac{m^2}{\xi^2 - 1}\right]\mathrm{P}_{i,\lambda}(\xi) = (\xi^2 - 1)^{m/2}$$
$$\times\left[\sum_{k=0}^{n^*} c_{\lambda,k}(\xi - 1)^k a_k^m + \sum_{k=0}^{n^*-1} c_{\lambda,k+1}(\xi - 1)^k b_k^m\right], \tag{4.127}$$

where the coefficients a_k^m, b_k^m are defined in (4.96). Similarly to (4.97), the system of Eq. (4.23) is written in the form:

$$\left[\sum_{k=0}^{n^*} c_{\lambda,k}(\xi - 1)^k a_k^m + \sum_{k=0}^{n^*-1} c_{\lambda,k+1}(\xi - 1)^k b_k^m\right.$$
$$\left. + \sum_{\lambda'} p_{\lambda\lambda'}(\xi) \sum_{k=0}^{n} c_{\lambda',k}(\xi - 1)^k\right] = q_\lambda(\xi),\ \lambda = |m_i|, \ldots, L, \tag{4.128}$$

where

$$p_{\lambda\lambda'}(\xi) = \delta_{\lambda,\lambda'}\left[-\lambda(\lambda+1) + a_s\xi + \varepsilon_i R^2(\xi^2 - d_{\lambda\lambda}^i)/2\right]$$
$$- \frac{R^2}{2}\sum_{j=1}^{N}(a_{ij} - \delta_{ij}b_{ii})\sum_{k'} Z_{jj,k'}(\xi)\left[\xi^2 A_{0,m_i,m_i}^{k',\lambda',\lambda} - B_{0,m_i,m_i}^{k',\lambda',\lambda}\right]$$
$$+ \varepsilon_i \frac{R^2}{2} d_{\lambda\lambda'}^i + a_d c_{\lambda\lambda'}^i, \tag{4.129}$$

and

$$q_\lambda(\xi) = \frac{R^2}{2}\sum_{\substack{j=1 \\ (j\neq i)}}^{N} b_{ij}\sum_{k'} \mathrm{P}_{jk'}(\xi)\sum_{\lambda'} Z_{ij,\lambda'}(\xi)\left[\xi^2 A_{m_{ij},m_j,m_i}^{\lambda',k',\lambda} - B_{m_{ij},m_j,m_i}^{\lambda',k',\lambda}\right]$$
$$- \varepsilon_i \frac{R^2}{2}\sum_{k\neq\lambda} d_{\lambda k}^i \mathrm{P}_{ik}(\xi) - a_d \sum_{k} c_{\lambda k}^i \mathrm{P}_{ik}(\xi). \tag{4.130}$$

Then, expansions are performed similar to (4.100), (4.101):

$$p_{\lambda\lambda'}(\xi) = \sum_{k=0}^{n^*} p_{\lambda\lambda',k}(\xi - 1)^k,\ q_\lambda(\xi) = \sum_{k=0}^{n^*} q_{\lambda,k}(\xi - 1)^k, \tag{4.131}$$

where

$$p_{\lambda\lambda',k} = \sum_{k'=0}^{n^*} p_{\lambda\lambda'}(\xi_{k'}) w_{k'k}^{-1}, \quad q_{\lambda,k} = \sum_{k'=0}^{n^*} q_\lambda(\xi_{k'}) w_{k'k}^{-1}. \tag{4.132}$$

Based on (4.128), (4.131), and (4.132), we present the system of equations for determining the coefficients $c_{\lambda,k}$, similarly to (4.102):

$$c_{\lambda,k} a_k^m + c_{\lambda,k+1} b_k^m + \sum_{\lambda'} \sum_{k'=0}^{n} c_{\lambda',k'} p_{\lambda\lambda',k-k'} = q_{\lambda,k},$$
$$k = 1 \div n-1, \ \lambda, \lambda' = |m_i|, ..., L. \tag{4.133}$$

The missing equations in this system must be obtained from the condition of smooth matching at the point ξ^* of solutions of the form (4.125) with solutions coming from the final point of integration $\xi_{\max}$, at which zero boundary conditions are imposed.

For the numerical solution on the interval $[\xi^*, \xi_{\max}]$ change of variables (4.106) is made and all the transformations described in Formulas (4.107)–(4.110) are performed. As a result, for the transformed functions we obtain

$$F_\lambda(\rho) = \mathrm{P}_{i\lambda}(\xi) \frac{\sqrt{(\xi^2-1)[(\xi-c_2)^2+c_1]}}{(\xi-c_2)}, \tag{4.134}$$

Equations (4.23) can be presented in the vector form:

$$\frac{d^2\mathbf{F}(\rho)}{d\rho^2} + \mathbf{F}(\rho)\mathbf{p}(\rho) = \mathbf{q}(\rho). \tag{4.135}$$

Here $\mathbf{F}$ and $\mathbf{q}$ are vectors of dimension L

$$\mathbf{F} = [F_1(\rho_j), ..., F_L(\rho_j)],$$
$$\mathbf{q} = [q_1(\rho_j), ..., q_L(\rho_j)], \tag{4.136}$$

and $\mathbf{p}$ is a square matrix with dimensions $L \times L$:

$$\mathbf{p} = \{p_{\lambda\lambda'}\}. \tag{4.137}$$

Further, to solve the vector Eq. (4.135), the matrix chasing method is used, which is completely analogous to the usual vector chasing described above when solving Eq. (4.111). Namely, Eqs. (4.118) and (4.119) are written in vector–matrix form:

$$\mathbf{F}(\rho_k) = \mathbf{V}_k + \mathbf{U}_k \mathbf{F}(\rho_{k-1}), \tag{4.138}$$

where

$$\begin{aligned}\mathbf{U}_k &= -[c_h + \mathbf{p}(\rho_{k-1})]\mathbf{z}^{-1},\\ \mathbf{V}_k &= \{\boldsymbol{\omega}_k - \mathbf{V}_{k+1}[c_h + \mathbf{p}(\rho_{k+1})]\}\mathbf{z}^{-1},\\ \mathbf{z} &= \mathbf{U}_{k+1}[c_h + \mathbf{p}(\rho_{k+1})] - 2[c_h + 5\mathbf{p}(\rho_k)].\end{aligned} \tag{4.139}$$

Zero boundary conditions at the end of the integration interval correspond to the equalities:

$$\mathbf{U}_N = 0, \mathbf{V}_N = 0. \tag{4.140}$$

Further, the chasing is carried out according to these formulas in the usual way.

Conditions for smooth matching of the required functions at a point ξ^*, similar to (4.122), are written for all $\lambda \neq |m_i|$ in the following form:

$$\begin{aligned}&g(\xi_{n+2})(\xi_{n+2}^2 - 1)^{|m|/2}\sum_{k=0}^{n} c_{\lambda,k}(\xi_{n+2} - 1)^k\\ &= \{\mathbf{V}_{n+2}\}_\lambda + g(\xi_{n+1})(\xi_{n+1}^2 - 1)^{|m|/2}\sum_{\lambda'}\{\mathbf{U}_{n+2}\}_{\lambda,\lambda'}\sum_{k=0}^{n} c_{\lambda',k}(\xi_{n+1} - 1)^k.\end{aligned} \tag{4.141}$$

These relations are necessary additional equations to the system (4.133) for calculating the coefficients $c_{\lambda,k}$ and, thus, for finding the entire solution of Eq. (4.23). In this case, the smoothness of the first harmonic, with $\lambda = m_i$, similar to the atomic case, is achieved by choosing the eigenvalue of the energy.

The wave functions of the ground state of an atom in the HF and SP HF approximations are calculated in the ***scfhf*** program, for an atom in a fullerene shell in the ***scfhfw*** program, and for molecules—in the ***scfhfm*** program.

A more detailed description of some of the computational procedures used can be found in [10–12].

References

1. C. Froese Fischer, *The Hartree-Fock Method for Atoms* (Wiley—Interscience, New York, 1977)
2. V.F. Bratsev, *Tables of Atomic Wave Functions* (Nauka, Moscow, 1966). ((In Russian))
3. I.P. Grant, *Relativistic calculations of atomic structures/* Adv. Phys. **19.** 747 (1970)
4. D. Hartree, *The Calculation of Atomic Structure* (John Wiley & Sons Inc., New York, 1957)
5. V.A. Fock, Z. Phys., **61,** 126; **62**, 795 (1930)
6. E.A. McCullough Jr., J. Chem. Phys. **62**, 3991 (1977)
7. V.A. Dzuba, V.V. Flambaum, O.P. Silvestrov, J. Phys. B: Atom. Mol. Phys. **15**, L585 (1982)
8. J.H. Wegstein, Commun. ACM **1**, 9 (1958)
9. V.L. Yakhontov, Y.M. Kuchiev, J. Phys. B: Atom. Mol. Phys. **16,** L37 (1983)
10. L.V. Chernysheva, N.A. Cherepkov, *Numerical Calculation of the Hartree-Fock Wavefunctions* (Ioffe Institute, No 337, Leningrad, 1971) (in Russian)

11. M.Ya. Amusia, L.V. Chernysheva, *Automated System for Studying the Structure of Atoms* (Nauka, Leningrad Division, 1983). ((In Russian))
12. M.Ya. Amusia, L.V. Chernysheva, *Computation of Atomic Processes* (Institute of physics Publishing, Bristol and Philadelphia, 1997)

Chapter 5
Wave Functions of the Ground State of an Atom in the Hartree–Fock–Dirac Approximation

This chapter discusses a method for solving simplified Hartree–Fock–Dirac (HFD) equations, which are obtained from exact HFD equations by dropping parametrically small terms. We take into account that even in heavy atoms electrons of only the inner and partly intermediate shells are relativistic. For the rest, relativistic corrections are appropriate to take into account as both the corrections in powers of small parameters $\alpha Z_{эфф}$ and α, where α и $Z_{эфф}$ are the fine structure constant and the effective charge of the central Coulomb field acting on the subshell under consideration, respectively. The numerical solution of such a system of equations is simpler, and the results obtained coincide within tenths of a percent with the results obtained using the exact HFD equations [1–6].

We present in this chapter a numerical procedure for solving these equations. Questions of changing variables, accelerating convergence, and correcting the values of the energy levels are discussed.

Taking into account relativistic corrections and solving the corresponding equations is a rather difficult task, which significantly increases the time required for numerical calculations. That is why the program described in this chapter is used only for the inner shells of atoms or for heavy atoms, combining *these results in one calculation with the nonrelativistic functions of* the outer electrons.

M. Ya. Amusia and L. V. Chernysheva, *Computation of Atomic and Molecular Processes*, Springer Series on Atomic, Optical, and Plasma Physics 117,
https://doi.org/10.1007/978-3-030-85143-9_5

5.1 Theoretical Model

We present the one-electron relativistic wave function $\phi_i(\mathbf{r}, \sigma)$($\sigma$—spin variable) of the ith bound state of the atom in the following form[1]

$$\phi_i(\mathbf{r}, \sigma) = \frac{1}{r}\begin{pmatrix} F_{nk}(r)\Omega_{jlj_z}(\mathbf{n}) \\ i\alpha G_{nk}(r)\Omega_{jl j_z}(\mathbf{n}) \end{pmatrix} \tag{5.1}$$

Here, $F_{nk}(r)(G_{nk}(r))$ and $\Omega_{jljz}(n)$ $(\Omega_{jlj_z}(\mathbf{n}) = -(\sigma\mathbf{n})\Omega_{jlj_z}(\mathbf{n}))$ are the radial and angular parts of the large (small) components of the Dirac bi-spinor (5.1), respectively; $\alpha = 1/c \simeq (137.036)^{-1}$ is the fine structure constant; $\mathbf{n}$ is the coordinate; $\mathbf{r}$ is the direction vector; n is the principal quantum number of the electron state, j, l, l_z are its total and orbital angular momenta and the projection of the total momentum onto the axis z. The quantity κ called hereinafter the *effective orbital moment* is related to j and l as follows:

$$k = \begin{cases} -(j+1/2) = -(l+1), \; l = j - 1/2 \\ j + 1/2 = l, \qquad\qquad\quad l = j + 1/2 \end{cases} \tag{5.2}$$

and $\ell' = 2j - \ell$.

Applying the condition of the total energy minimum with respect to the one-electron wave functions (5.1) variation, we obtain from [7] the following basic system of equations for $F_i(r)$ and $G_i(r)$ ($i = nk, i = 1,..., s$, where s is the number of atomic subshells):

$$\frac{d}{dr}F_i(r) + \frac{k_i}{r}F_i(r) - [2 + \alpha^2(\tilde{\varepsilon}_i - U_i(r))]F_i(r)$$
$$= -\frac{\alpha^2}{r}\sum_{j,k(j\neq i)}^{s} D(ijk)\tilde{Y}_{ij}^k(r)G_j(r), \tag{5.3}$$

$$\frac{d}{dr}G_i(r) + \frac{k_i}{r}G_i(r) + [\tilde{\varepsilon}_i - U_i(r)]G_i(r)$$
$$= -\frac{1}{r}\sum_{j,k(j\neq i)}^{s} D(ijk)\tilde{Y}_{ij}^k(r)F_j(r). \tag{5.4}$$

Here, $\tilde{\varepsilon}_i$ is the single-particle energy of an electron in the ith subshell ($\varepsilon_i \equiv \varepsilon_{ii}$ is a diagonal Lagrange multiplier). $D(ijk)$ и $C(ijk)$ are the angular coefficients of the radial Slater integrals $F^k(ij)$ and $G^k(ij)$ [7] in the expression for the total interelectronic Coulomb energy of the atom [8]:

---

[1] This representation of one-particle functions is valid only for atoms with closed or semi-closed subshells. In other cases, an additional assumption is used that lies beyond the Hartree–Fock–Dirac approximation, called the central field approximation.

$$\left\langle \frac{1}{2}\sum_{ij}\frac{1}{|r_i - r_j|}\right\rangle = \frac{1}{2}\sum_{ijk} q_i[C(ijk)F^k(ij) + D(ijk)G^k(ij)] \qquad (5.5)$$

Here q_i is the number of electrons in the subshell with $i = nk$. The function $U_i(r)$ is a self-consistent potential that determines the motion of electrons in the ith shell.

$$U_i(r) = \frac{1}{r}\left[\sum_{jk}^{s} C(ijk)\tilde{Y}_{jj}^k(r) + \sum_{jk}^{s} D(ijk)\tilde{Y}_{jj}^k(r) - Z\right], \qquad (5.6)$$

where Z is the charge of the atomic nucleus, and $\tilde{Y}_{ij}^k(r)$ is determined by the following relation

$$\tilde{Y}_{ij}^k(r) = r\int_0^\infty \frac{r_<^k}{r_>^{k+1}}\left[F_i(r')F_j(r') + \alpha^2 G_i(r')G_j(r')\right]dr' \qquad (5.7)$$

Radial wave functions $F_i(r)$ and $G_i(r)$ satisfy the boundary conditions.

$$\begin{aligned} F_i(0) &= F_i(\infty) = 0, \\ G_i(0) &= G_i(\infty) = 0, \end{aligned} \qquad (5.8)$$

and the normalization condition

$$\int_0^\infty \left(F_i^2(r) + \alpha^2 G_i^2(r)\right)dr = 1. \qquad (5.9)$$

In addition, we assume that each function $F_i(r)$ must have at least $(n_i - l_i - 1)$ zeros at a finite distance [9, 10].

In Eqs. (5.3)–(5.4), we omitted the off-diagonal energy parameters $\varepsilon_{ij}(i \neq j)$ responsible for the orthogonality of the bi-spinor radial components for the shells i and j:

$$\int_0^\infty \left(F_i(r)F_j(r) + \alpha^2 G_i(r)G_i(r)\right)r = \delta_{ij} \qquad (5.10)$$

The basis for the formulation of this relation for $i \neq j$ is the fact that in configurations with closed subshells these parameters are equal to zero [1, 7]. They are very small for the ground states of almost all atoms [10].

In the main system of Eqs. (5.3)–(5.4), the following simplification is possible. The right-hand side of Eq. (5.3) is responsible for the correction to single-particle electron energies and wave functions due to the so-called *exchange of small components*. Simple estimates show that this correction is parametrically small. Indeed, for the innermost shells, the energy shift caused by it is equal to $\Delta\tilde{\varepsilon}_i \sim Z\alpha^2\tilde{\varepsilon}_i$, while a similar effect on the right-hand side of Eq. (5.4) (the so-called *exchange of large components*) leads to an order $(Z\alpha)^2\tilde{\varepsilon}_i$ correction. Therefore, it is quite natural to ignore the right-hand side of Eq. (5.3). Thus, the system of *2 s*-Eqs. (5.3)–(5.4) takes the form:

$$\frac{d}{dr}F_i(r) + \frac{\kappa_i}{r}F_i(r) - \left[2 + \alpha^2(\tilde{\varepsilon}_i - U_i(r))\right]F_i(r) = 0 \tag{5.11}$$

$$\frac{d}{dr}G_i(r) - \frac{\kappa_i}{r}G_i(r) + \left[\tilde{\varepsilon}_i - U_i(r)\right]G_i(r) = \frac{1}{r}\sum_{j,k(j\neq i)} D(ijk)\,\tilde{Y}^k_{ik}(r)F_j(r) \tag{5.12}$$

The single-particle wave functions and electron energies found by solving systems (5.3)–(5.4) and (5.11)–(5.12), respectively, are in good agreement with each other (within 0.1% for the innermost shells and 0.01% for the outer shells). This statement is true for all elements from 2 (He) to 80 (Hg). This numerical result is an additional evidence that the simplification made in passing from (5.3)–(5.4) to (5.11)–(5.12) is quite accurate.

Let us make the following substitution:

$$F_i(r) = \xi_i(r)P_i(r) \tag{5.13}$$

where

$$\xi_i(r) = \left[1 + \frac{\alpha^2}{2}\left(\frac{\varepsilon_i}{2} - U_i(r)\right)\right]^{1/2} \tag{5.14}$$

Here $\varepsilon_i = -2\,\tilde{\varepsilon}_i, \varepsilon_i > 0$ is measured in Rydberg. As a result, the system of *2 s*-equations of the first order is transformed into the system of *s*-equations of the second order

$$\frac{d^2}{dr^2}P_i(r) - \left\{2U_i(r)\left(1 - \frac{1}{2}\alpha_0^2\varepsilon_i\right) - \alpha^2U_i^2(r) + \frac{\kappa_i(\kappa_i+1)}{r^2} + 2\,\tilde{W}_i(r) + \varepsilon_i - \frac{1}{4}\alpha^2\varepsilon_i\right\}$$

$$P_i(r) = \xi_i(r)\,\tilde{X}_i(r), \tag{5.15}$$

where

$$U_i(r) = \frac{1}{r}\Big(\tilde{Y}_i(r) - Z\Big), \tag{5.16}$$

$$\tilde{Y}_i(r) = \sum_j (q_j - \delta_{ij})\,\tilde{Y}_i^0\, j(r) - \frac{2}{N_i}\sum_{j=0}^{s}\sum_{k>0} \gamma'_k(ij)\,\tilde{Y}_{ij}^k(r), \tag{5.17}$$

$$\tilde{W}_i(r) = \frac{\alpha^2}{8\xi_i^2(r)}\left\{U''_i(6) + \frac{3}{4}\cdot\frac{\alpha^2}{\xi_i^2(r)}\left(U'_i(r)\right)^2 - \frac{2\kappa_i}{r}\,U'_i(r)\right\}, \tag{5.18}$$

$$\tilde{X}_i(r) = \frac{2}{N_i}\frac{1}{r}\sum_{j\neq i}^{s}\sum_{k>0}\gamma_k(ij)\tilde{Y}_{ij}^k(r)\xi_j(r)P_j(r), \tag{5.19}$$

$$\tilde{Y}_{ij}^k(r) = \left\{\int_0^r\left(\frac{r'}{r}\right)^k + + \int_r^\infty\left(\frac{r}{r'}\right)^{k+1}\right\}$$
$$\Big(\xi_i(r')\xi_j(r')P_i(r')P_j(r') + \alpha_0^2 G_i(r)G_j(r)\Big)dr', \tag{5.20}$$

$$G_i(r) = \frac{1}{2}\xi_i^{-2}(r)\Big(F'_i(r) + \frac{\kappa_i}{r}F_i(r)\Big). \tag{5.21}$$

Here the function $F_i(r)$ is related to $P_i(r)$ by Eq. (5.13). The coefficients $\gamma'_k(i, j)$ and $\gamma_k(i, j)$ in formulas (5.17) and (5.19) are related to $D(ijk)$ and $C(ijk)$ as follows:

$$\gamma'_k(i, j) = -\frac{1}{2}q_i D(ijk), \quad k > 0,$$
$$\gamma_k(i, j) = -q_i D(ijk), \tag{5.22}$$

and the coefficient $C(ijk) = q_j\delta_{k_0}$(see (5.5)) ($q_i = 2j_i + 1$, j_i is the total angular momentum of the electron ith subshell) enters into definition $\tilde{Y}_i(r)$ (5.17). If the interacting subshells a and b with moments j_a and j_b, respectively, are closed ($q_a = 2j_a+1, q_b = 2j_b+1$, or semi-closed ($q_a = (1/2)(2j_a+1)$), the coefficients $\gamma'_k(a, b)$ and $\gamma_k(a, b)$ are determined by the relations.

$$\gamma'_k(ab) = \begin{cases} \frac{1}{2}q_a q_b\begin{pmatrix} j_a & k & j_b \\ -1/2 & 0 & 1/2\end{pmatrix} & \text{for even } k \\ 0 & \text{for odd k}\end{cases} \quad k > 0, \tag{5.23}$$

$$\gamma_k(ab) = \begin{cases} q_a q_b\begin{pmatrix} j_a & k & j_b \\ -1/2 & 0 & 1/2\end{pmatrix} & \text{for even } k \\ 0 & \text{for odd k}\end{cases}, \tag{5.24}$$

Here $\gamma_k'(ab) = 0$ for $a \neq b$. When one of the subshells, a or b, is closed or semi-closed, the values $\gamma_k'(ab)$ and $\gamma_k(ab)$ depend on how the atomic state is determined. Coefficients $\gamma_0'(ab)$ are included in the definition of $\tilde{Y}_i(r)$.

To calculate the total energy of an atom (ion), a relativistic generalization of the virial theorem, well known in nonrelativistic theory, is used. As shown in [11, 12], the total relativistic energy of an atom (ion), the one-particle wave functions $\phi_i(r, \sigma)$ (5.1) of which were found using the variational principle, is determined by the relation

$$E = -4\sum_{a=1}^{s} q_a \int_0^{\infty} G_a^2(r)dr \tag{5.25}$$

where E is measured in Rydberg.

It follows from this expression that the calculation of the energy requires knowledge of all s-functions $G_\alpha(r)$, each of which is a "small" component of the radial part of the bi-spinor, which describes the states of electrons in the corresponding shells. Therefore, the calculation of E in accordance with (5.25) is performed at the last stage of the calculation, after all the energy levels in the system have been found.

5.2 Numerical Model

5.2.1 General Consideration

The system of equations in the Hartree–Fock–Dirac approximation is solved by the self-consistent field method. This is an iterative method ("large iteration") for solving the corresponding linearized equations obtained by substituting (5.15)–(5.19) into the system of Schrödinger equations with a non-local potential:

$$P''_i(r) = (V_i(r) + \varepsilon_i)P_i(r) + \int_0^{\infty} \Pi(r, r')dr' \tag{5.26}$$

where

$$V_i(r) = 2\left(1 - \frac{1}{2}\alpha^2\varepsilon_i\right)U_i(r) - \alpha^2 U_i^2(r) + \frac{\kappa_i(\kappa_i + 1)}{r^2} + 2\tilde{W}(r) - \frac{1}{4}\alpha^2\varepsilon_i^2 \tag{5.27}$$

is a local, energy-independent part of the potential, and $U_i(r)$, $\tilde{W}_i(r)$ are determined by Eqs. (5.16)–(5.18). Non-local, i.e., the exchange part of the Hartree–Fock–Dirac potential $\Pi(r, r')$ is given by the expression

$$\Pi_i(r,r') = -\frac{2}{q_i}\sum_{j\neq i}\sum_k \gamma_k(i.j)\xi_i(r)\xi_j(r)P_i(r)\frac{r_<^k}{r_>^{k+1}}$$
$$\times\left(\xi_i(r')\xi_j(r')P_i(r')P_j(r') + \alpha^2 G_i(r')G_j(r')\right) \quad (5.28)$$

where the functions $\xi_i(r)$ and $G_i(r)$ are defined by formulas (5.14) and (5.21).

In the program, iteration begins by computing $V_i(r)$ and $\Pi_i(r, r\prime)$. Zero approximation for any $i = 1, 2, 3, \ldots s$ is

$$P_i(r) = G_i(r) = U_i(r) = 0 \quad (5.29)$$

The functions $\xi_i(r)$ included in definition $\tilde{Y}_i(r)$ (5.19) and $\tilde{W}_i(r)$ are assumed to be equal to unity:

$$\xi_i(r) = 1, \quad i = 1, \ldots, s \quad (5.30)$$

and zero estimates of the energy eigenvalues are determined by the equation

$$\varepsilon_{i,\text{est}}^{(0)} = \frac{Z - \sum_{j<i} N_j^2}{n_i^2} \quad (5.31)$$

At this stage, the radial wave functions $P_i(r)$ are constructed in ascending order in energy (and, accordingly, values $G_i(r)$).

Integral–differential Eq. (5.26) are solved using the ***rscfhf*** program (see Sect. 2.2) by an iterative procedure that is divided into three cycles for each subshell. In the first cycle, we start by solving the equations for the two innermost subshells, and at the next step, one or two other subshells are added. Two subshells are added if they constitute a fine structure doublet $j = l \pm 1/2$, otherwise only one.

Such a sequence in consideration is associated with the fact that the closely lying states of the fine structure doublet strongly influence each other. Therefore, it is natural to solve the equations for these states together. The procedure is repeated until all subshells are considered. For example, the following order 1,2; 1,2,3,4; 1,2,3,4,5; 1,2,3,4,5,6,7 (order of subshells starting with the innermost) should be applied to Ar atom with the configuration $1s_{1/2}^2 2s_{1/2}^2 2p_{1/2}^2 2p_{3/2}^4 3s_{1/2}^2 3p_{1/2}^2 3p_{3/2}^4$.

After the wave function $P_i(r)$ and energy for the ith shell are found, the ***rscfhf*** program calculates the radial part of the small component $G_i(r)$ (5.21). At each iteration, the estimates of the radial functions determining the direct and exchange Hartree-Fock-Dirac potential (see $V_i(r)$ (5.27) and $\Pi(r, r\prime)$ (5.28)) as well as the energy eigenvalue are obtained in the previous iteration:

$$P_{i,\text{est}}^{(m)}(r) = P_i^{(m)}(r),$$
$$G_{i,\text{est}}^{(m)}(r) = G_i^{(m)}(r),$$
$$\varepsilon_{i,\text{est}}^{(m)}(r) = \varepsilon_i^{(m)}(r). \quad (5.32)$$

At the end of the first cycle, the wave functions of the inner shell are found with better accuracy than those of the outer subshells; current error for the ith shell

$$\eta_i = max\left\{\left|1 - \frac{P_i^{(m)}(r)}{P_i^{(m+1)}(r)}\right|, \left|1 - \frac{\varepsilon_i^{(m)}}{\varepsilon_i^{(m+1)}}\right|\right\} \tag{5.33}$$

turns out to be less for inner subshells than for outer subshells.

In the second cycle, Eq. (5.24) is solved again, but now the procedure starts with the outermost subshell. Moreover, if the current error for the ith subshell is less than the specified accuracy ε, the ith equation is not solved at this stage.

It should be noted that in this cycle we usually obtain a solution for the parameters of the outer subshells that are less stable with respect to the calculation. To improve convergence, the wave functions and energies are recalculated using the formulas.

$$\begin{aligned} P_{i,\mathrm{est}}^{(m)}(r) &= (1-f)P_i^{(m)}(r) + fP_i^{(m-1)}(r) \\ \varepsilon_{i,\mathrm{est}}^{(m)}(r) &= (1-f)\varepsilon_i^{(m)}(r) + f\varepsilon_i^{(m-1)}(r) \end{aligned} \tag{5.34}$$

The parameter f is nonzero if at the three previous iterations the corresponding energies changed sign, i.e.,

$$\mathrm{sign}\left(\varepsilon_i^{(m)} - \varepsilon_i^{(m-1)}\right) \neq \mathrm{sign}\left(\varepsilon_i^{(m-1)} - \varepsilon_i^{(m-2)}\right). \tag{5.35}$$

The parameter f is calculated on the assumption that the energy converges to its limit value as a geometric progression:

$$c = \frac{\varepsilon_i^{(m)} - \varepsilon_i^{(m-1)}}{\varepsilon_i^{(m)} - 2\varepsilon_i^{(m-1)} + \varepsilon_i^{(m-2)}} \tag{5.36}$$

Otherwise, if

$$\mathrm{sign}\left(\varepsilon_i^{(m)} - \varepsilon_i^{(m-1)}\right) = \mathrm{sign}\left(\varepsilon_i^{(m-1)} - \varepsilon_i^{(m-2)}\right), \tag{5.37}$$

then the parameter f is set equal to zero and (5.34) is identical to (5.32). A nonzero parameter f either leads to a faster convergence of the process or provides a solution in atoms for which the procedure would not converge at all for $f = 0$.

5.2.2 *Change of Variables*

The linear homogeneous integral–differential Eq. (5.26) is solved using an iterative procedure known as *small iteration*, which corresponds to solving the linear inhomogeneous differential equation

$$P''_i(r) = (V_i(r) - \varepsilon_i)P_i(r) + \xi_i(r)\,\tilde{X}_i(r) \tag{5.38}$$

where

$$\xi_i(z)\,\tilde{X}_i(r) \equiv \int_0^{\infty} \Pi_i(r, r')r'. \tag{5.39}$$

When solving Eq. (5.38), it is convenient to use a grid of equally spaced points of the independent variable. Meanwhile, in view of the fact that the potential $V_i(r)$ has a singularity at $r = 0$, the mesh should be relatively dense in the vicinity of this point, while there is no such need at large r. Therefore, it is convenient to introduce the alteration of variables $\rho = \alpha r + \beta \ln r$ (4.42), where α and β are some constants that depend on the charge of the atomic nucleus and the input parameters of the problem.

This replacement is also very convenient for calculating the wave functions of the excited states, since the infinite range of integration over them is replaced by a finite range:

$$-\infty < \rho_0 \leq \rho \leq \rho_N < +\infty, \tag{5.40}$$

where $\rho_0 = -\beta(10+lnz_N)$, which corresponds to $r_{min} \approx 4.5 \cdot 10^{-5}/Z$. Subintervals are assumed to be equal: $\rho_i = \rho_0 + ih$, $i = 0, 1, 2, \ldots, N$. The coefficient $\tilde{\alpha}$ (not to be confused with the fine structure constant, traditionally denoted α!) is defined as $\tilde{\alpha} = (\rho_N - \beta lnz_N)/z_N$. Coefficient β, subinterval h, number of points N, and maximum radius r_N are inputs to the ***rscfhf*** program.

In order that, after the change of variables, Eq. (4.41) does not contain the first derivative, the desired wave function must be transformed as follows:

$$f_i(\rho) = \sqrt{\frac{\tilde{\alpha}r + \beta}{r}}\,P_i(r). \tag{5.41}$$

As a result of this, we obtain from (5.38) the following equation:

$$f''_i(\rho) = \tilde{P}_i(\rho)f_i(\rho) + \tilde{Q}_i(\rho), \tag{5.42}$$

where

$$\begin{aligned}\tilde{P}_i(\rho) =& \frac{\kappa_i(\kappa_i+1)}{(\tilde{\alpha}r+\beta)^2} + \frac{\tilde{\alpha}\beta r + \beta^2/4}{(\tilde{\alpha}r+\beta)^4} + \frac{2r^2}{(\tilde{\alpha}r+\beta)^2}U_i(\rho)\\ &+ \frac{r^2}{(\tilde{\alpha}r+\beta)^2}\left[\varepsilon_i - \alpha^2(\varepsilon_i U_i(\rho)) + \frac{1}{4}\varepsilon_i^2\right] - \alpha^2\frac{r^2}{(\tilde{\alpha}r+\beta)^2}(U_i(\rho))^2\\ &+ \frac{\alpha_0^2}{4\xi_i^2(\rho)}\left[U''_i(\rho) + \alpha_0^2\frac{3}{4\xi_i^2(\rho)}(U_i(\rho))^2 - \frac{2\kappa_i(\alpha r+\beta)+\beta}{(\alpha r+\beta)^2}U_i(\rho)\right],\end{aligned} \tag{5.43}$$

$$\tilde{Q}_i(\rho) = -\frac{2}{q_i}\frac{r}{(\tilde{a}r+\beta)^2}\xi_i(\rho)\sum_{j\neq i}\sum_k \gamma_k(i.j)\,\tilde{Y}^k_{ij}(\rho)\xi_j(\rho)f_j(\rho), \tag{5.44}$$

$$U_i(\rho) = \frac{1}{r}\left(\tilde{Y}_i(\rho) - z\right), \tag{5.45}$$

$$\tilde{Y}_i(\rho) = \sum_j \left(N_j - \delta_{ij}\right)\tilde{Y}^0_{jj}(\rho) - \frac{2}{q_i}\sum_j\sum_{k>0}\gamma'_k(ij)\,\tilde{Y}^k_{jj}(\rho), \tag{5.46}$$

$$\tilde{Y}^k_{jj}(\rho) = \left\{\int_0^r\left(\frac{r'}{r}\right)^k + \int_r^\infty\left(\frac{r}{r'}\right)^{k+1}\right\}\frac{r}{\tilde{a}r+\beta}$$
$$\times\left[\xi_i(\rho')\xi_j(\rho')f_i(\rho')f_j(\rho') + \alpha^2 g_i(\rho')g_j(\rho')\right]dr', \tag{5.47}$$

$$\xi_i(\rho) = \left[1 - \frac{1}{2}\alpha^2\left(\frac{1}{2}\varepsilon_i + U_i(\rho)\right)\right]^{1/2}, \tag{5.48}$$

$$g_i(\rho) = \frac{\tilde{a}r+\beta}{r}\frac{1}{2\xi(\rho)}\left\{f'_i(\rho) + \left[\frac{2\kappa_i(\tilde{a}r+\beta)+\beta}{2(\tilde{a}r+\beta)^2} - \frac{\alpha^2}{4\xi_i^2(\rho)}U_i(\rho)\right]f_i(\rho)\right\} \tag{5.49}$$

For a correct solution of the problem, the boundary conditions for the function $f_i(\rho)$ have to by the following:

$$f_i(\rho_0) = f_i(\rho_N) = 0. \tag{5.50}$$

In this case, the normalization condition is written as:

$$\int_0^{r_N}\frac{r}{\tilde{a}r+\beta}\left(\xi_i^2(\rho)f_i^2(\rho) + \alpha_0^2 g_i^2(\rho)\right)dr = 1. \tag{5.51}$$

5.2.3 Functions $\tilde{P}_i(\rho)$ and $\tilde{Q}_i(\rho)$

The solution of Eq. (5.42) for the ith shell requires the calculation of the functions $\tilde{P}_i(\rho)$ and $\tilde{Q}_i(\rho)$. As follows from Eqs. (5.43)–(5.49), these functions can be found through the sequential calculation of the functions: $g_i(\rho)$, $\tilde{Y}_{ij}(\rho)$, and $U_i(\rho)$ at $i = 1, 2, \ldots, s$.

It should be remembered that to solve Eq. (5.42) for the ith subshell at each *small iteration*, knowledge of the potentials $U_i(\rho)$ and functions $g_i(\rho)$ $(i = 1, 2, \ldots, s)$ obtained at the previous iteration is required. These values, as well as functions $f_i(\rho)$ for all atomic subshells, are stored in the computer memory during the calculation.

5.2.4 Numerical Procedure for Differential Equation

One of the most effective methods for the numerical solution of a second-order linear differential equation is Numerov's method based on the expression

$$f_{m+1} - 2f_m + f_{m-1} = \frac{h^2}{12}\left(f''_{m+1} + 10f''_m + f''_{m-1}\right) + o(h^6) \tag{5.52}$$

Ignoring $0(h^6)$ (5.52) and using (5.42), we get the following expression:

$$f_i(\rho_{m+1})A_{m+1} - f_i(\rho_m)B_m + f_i(\rho_{m-1})A_{m-1} = \omega_m, \tag{5.53}$$

where

$$\begin{aligned} A_m &= 1 - \frac{h^2}{12}\tilde{P}_i(\rho_m), \\ B_m &= 1 + \frac{5h^2}{12}\tilde{P}_i(\rho_m), \\ \omega_m &= \frac{h^2}{12}\cdot\left[Q_i(\rho_{m+1}) - 10\tilde{Q}(\rho_m) + \tilde{Q}(\rho_{m-1}\right]. \end{aligned} \tag{5.54}$$

The procedure for the numerical solution of Eq. (5.53) is unstable under the condition $\tilde{P}_i(\rho) > 0$. Therefore, the region of integration over ρ is divided into several intervals, depending on the sign $\tilde{P}_i(\rho) > 0$.

For each bound electronic state, there are two points where the function $\tilde{P}_i(\rho)$ vanishes, i.e., where $\tilde{P}_i(\rho) = 0$. Therefore, the region of argument $\rho_0 \leq \rho \leq \rho_N$ can be divided into three intervals: two with $\tilde{P}_i(\rho_m) > 0$ $(m = 0, 1, 2, \ldots, j_1 - 1,$ $m = j_2, j_2 + 1, \ldots, N$, $j_1 < j_2)$ and the third with $\tilde{P}_i(\rho_m) < 0$.

In the intervals where $\tilde{P}_i(\rho) > 0$, to solve Eq. (5.53), the chasing technique is used, based on the successive elimination of one unknown value f_i in (5.53) under the boundary conditions $f_i(\rho_0) = 0$ or $f_i(\rho_N) = 0$. In the interval $[0, j_1 - 1]$, the equation has the form

$$f_i(\rho_m) = U_m + V_m f_i(\rho_{m+1}). \tag{5.55}$$

Substituting (5.55) into (5.53), we obtain the following recurrence relations for U_m и V_m

$$U_m = \frac{A_{m-1}U_{m-1} - \omega_m}{2B_m - A_m V_{m-1}},$$
$$V_m = \frac{A_{m+1}}{2B_m - A_{m-1}V_{m-1}}, \tag{5.56}$$

by means of which they are determined (straight chaise), starting with $U_0 = V_0 = 0$, which corresponds to the boundary conditions $f_i(\rho) = 0$. Once $f_i(\rho_{j_1})$ is already known, the values $f_i(\rho_m)$, $m = j_1 - 1, \ldots, 1, 0$ are sequentially obtained (backward chaise) from Eq. (5.55). Since the integral–differential Eq. (5.26), which must be solved at each "large" iteration, is linear and homogeneous, an arbitrary normalization is applied during the solution. After the solution has been found, it can be normalized again through condition (5.51). The program assumes that $f_i(\rho_{j_1}) = 1$ in the interval $[j_1, j_2]$. The function f_i is found using the relation (5.53):

$$f_i(\rho_m) = \frac{\omega_{m-1} + 2B_{m-1}f_i(\rho_{m-1}) - A_{m-2}f_i(\rho_{m-2})}{A_m} \tag{5.57}$$

The number of zeros is also in this region of ρ. The values $f_i(\rho_m)$ in the range $[j_2, N]$ are also determined by the chasing technique. This time, a different boundary condition is used: $f_i(\rho_N) = 0$. The corresponding equations in this region are:

$$f_i(\rho_m) = \tilde{U}_m + \tilde{V}_m f_i(\rho_{m-1}), \tag{5.58}$$

$$\tilde{U}_m = \frac{A_{m-1}\tilde{U}_{m+1} - \omega_m}{2B_m - A_{m+1}\tilde{V}_{m+1}},$$
$$\tilde{V}_m = \frac{A_{m-1}}{2B_m - A_{m+1}\tilde{V}_{m+1}}, \tag{5.59}$$

$$\tilde{U}_N = \tilde{V}_N = 0. \tag{5.60}$$

The values $f_i(\rho)$ are calculated in ascending order in ρ_m (reverse chasing), starting at the value $f_i^{\infty}(\rho_{j_2-1})$ where the functions are "stitched" together. The last value $f_i^0(\rho_{j_2})$ obtained by Numerov's method is used to check if the first derivatives are "stitched" together at a point $\rho = \rho_{j_2-1}$. The "jump" of the first derivative at this point can be roughly estimated in the following way:

$$\frac{d}{d\rho}\left[f_i^{\infty}(\rho_{j_2-1}) - f_i^0(\rho_{j_2-1})\right] \approx \frac{f_i^{\infty}(\rho_{j_2-1}) - f_i^0(\rho_{j_2-1})}{h}. \tag{5.61}$$

This relation is used to correct its own value.

5.2.5 Energy Eigenvalue Correction

The correct eigenvalue of the energy is obtained if the solution $f_i(\rho)$ has $(n_i - l_i - 1)$ roots and the "jump" of the first derivative at the point $\rho = \rho_{j_2-1}$ does not exceed the given computation error ε.

If at some step m of calculating the number of zeros k is not equal to the required number, the energy is corrected using the following formula:

$$\varepsilon_{n\kappa}^{(m+1)} = \left(1 + \frac{\mu}{n}\text{sign}(k - n + l + 1)\right)\varepsilon_{n\kappa}^{(m)}, \tag{5.62}$$

where μ is a constant that determines the amount of change in sample values ε_{nx}.

If the number of zeros is correct, then the increment in the estimate of the eigenvalue of the energy is expressed through the "jump" of the first derivative at the "stitching" point ρ_{j_2-1}, and the energy is recalculated by the formula

$$\varepsilon_{n\kappa}^{(m+1)} = \varepsilon_{n\kappa}^{(m)}\left[1 + (-1)^{n-l}\frac{\mu}{n}\text{sign}\left(\tau(\varepsilon_{n\kappa}^{(m)})\right)\right], \tag{5.63}$$

where

$$\tau(\varepsilon_{n\kappa}^{(m)}) = \frac{1}{h}\left[f_i^{\infty}(\rho_{j_2-1}) - f_i^{0}(\rho_{j_2-1})\right]. \tag{5.64}$$

The equation $\tau(\varepsilon_{n\kappa}) = 0$ corresponds to its eigenvalue $\varepsilon_{n\kappa}$. Equation (5.63) implies that for $\tau > 0$ an odd number of zeros of the function, the trial value $\varepsilon_{n\kappa}$ must be decreased. After the root is located, the energy is corrected using the chord method:

$$\varepsilon^{(m+1)} = \frac{\tau(\varepsilon^{(m)})\varepsilon^{(m+1)} - \tau(\varepsilon^{(m-1)})\varepsilon^{(m)}}{\tau(\varepsilon^{(m)}) - \tau(\varepsilon^{(m-1)})} \tag{5.65}$$

where $\varepsilon^{(m-1)}$ и $\varepsilon^{(m+1)}$ are the energy estimates for the increments $\tau(\varepsilon^{(m)})$ и $\tau(\varepsilon^{(m+1)})$, accordingly, und $\tau(\varepsilon^{(m-1)}) \cdot \tau(\varepsilon^{(m)}) < 0$. The energy is repeatedly corrected until the difference between the values obtained in two successive iterations does not exceed the specified value ε.

The wave functions of the ground state of the atom in the HFD approximation are calculated using the ***rscfhf*** program employing all the procedures described above.

References

1. I.P. Grant, *Relativistic calculations of atomic structures.* Adv. Phys. **19**, 747 (1970)
2. M.A. Goulthard, Proc. Phys. Soc. (London) **9**, 144 (1967)
3. F.C. Smith, W.R. Johnson, Phys. Rev. **160**, 36 (1967)
4. J.P. Desclaux, D.F. Mayers, and O'Brien, J. Phys. B: Atom. Mol. Phys. **4**, 631 (1971)
5. I.M. Band, V.I. Fomichev, *The system of Computer Programs RAINE* (Leningrad Institute for Nuclear Physics, No 498, Leningrad, 1979) (in Russian)
6. V.A. Dsuba, O.P. Sushkov, V.V. Flambaum. *The Computer Package for Calculations of Atomic Wavefunctions and Level Energies* (The Institute for Nuclear Physics of the USSR Academy of Sciences, Siberian division, No 82–89, Novosibirsk, 1982) (in Russian)
7. I. Lindgren, A. Rosen, Case Stud. Atomic Phys. **4**, 10 (1974)
8. D. Hartree, *The Calculation of Atomic Structure* (John Wiley & Sons Inc., New York, 1957)
9. V.A. Dzuba, V.V. Flambaum, O.P. Silvestrov, J. Phys. B: Atom. Mol. Phys. **15**, L585 (1982)
10. C. Froese Fischer, Comp. Phys. Comm. **1**, 151 (1970)
11. Y.K. Kim, Phys. Rev. **154**, 17 (1967)
12. Y.K. Kim, Phys. Rev. **159**, 190 (1967)

Chapter 6
Wave Functions of Excited States of Atoms and Simple Molecules in the Hartree–Fock Approximation

Consideration of any process involving an atom or a simple molecule requires knowledge of the wave functions of the excited states. This chapter proposes the simplest approach to calculating one-particle wave functions in discrete and continuous spectra. In this case, as a rule, the so-called frozen core approximation is used, in which it is assumed that the excitation of one electron does not affect the wave functions of the rest.

Numerical procedures [1–3] are given for finding these wave functions, for accelerating convergence when taking into account the exchange term, for turning on diagonal parameters that ensure the orthogonality of the wave functions of excited electrons to unexcited, core ones, for obtaining phase shifts in elastic and inelastic scattering processes.

The features of the solution of the HF equations for the electron, meson, and positron are described.

6.1 Theoretical Model

6.1.1 General Consideration

Single-electron excited states of an atom and a simple molecule in the Hartree–Fock approximation can be found (see Sect. 1.3) as states of an electron in a fixed field, the so-called field of the "frozen core." The core is understood as a set of wave functions of the initial (ground) state of the system, which are found by solving systems of Eqs. (1.8) and (1.9):

$$\phi_k,\ k = 1 \div N_e. \tag{6.1}$$

M. Ya. Amusia and L. V. Chernysheva, *Computation of Atomic and Molecular Processes*, Springer Series on Atomic, Optical, and Plasma Physics 117,
https://doi.org/10.1007/978-3-030-85143-9_6

The Hartree–Fock Eq. (1.8) for the function of an excited electron ϕ_ε can be represented as:

$$-\frac{\Delta}{2}\phi_\varepsilon(r) + V(r) - a_p^\varepsilon \phi_\varepsilon(r) \sum_{k=1}^{N_e} J_{k,k}(r) + b_p^\varepsilon \sum_{k=1}^{N_e} \phi_k(r) J_{\varepsilon,k}(r) = \varepsilon\phi_\varepsilon(r) + \sum_{k=1}^{N_e} \varepsilon_k^\varepsilon \phi_k(r), \tag{6.2}$$

where $J_{k,k}(\mathbf{r})$ and $J_{\varepsilon,k}(\mathbf{r})$ are direct and exchange Coulomb integrals:

$$\begin{aligned} J_{k,k}(\mathbf{r}) &\equiv \int \phi_k^*(\mathbf{r}')|\mathbf{r}' - \mathbf{r}|^{-1}\phi_k(\mathbf{r}')d\mathbf{r}', \\ J_{\varepsilon,k}(\mathbf{r}) &\equiv \int \phi_k^*(\mathbf{r}')|\mathbf{r}' - \mathbf{r}|^{-1}\phi_\varepsilon(\mathbf{r}')d\mathbf{r}'. \end{aligned} \tag{6.3}$$

The values of the coefficients a_k^ε, b_k^ε depend on the configuration of the excited state, and the potential $V(\mathbf{r})$ for an atom or diatomic molecule can be written explicitly, as Z/r and $(Z_1/|\mathbf{r} - \mathbf{R}_1| + Z_2/|\mathbf{r} - \mathbf{R}_2|)$ in (1.8) and (1.9), respectively. The last term in the summation on the right-hand side of (6.2) ensures the orthogonality of the wave functions of the excited states to the wave functions of the ground state (core) by choosing the numerical values of the so-called off-diagonal energy parameters $\varepsilon_k^\varepsilon$.

Equation (6.2) describes the motion of (N_e+1)-electron in the field of N_e-electrons of an atom or molecule. Thus, it directly allows one to describe the scattering of an electron by these objects. If the excitation of an atom or a molecule is of interest, the summation over N_e in (6.2) is carried out over all electronic states except for the one that has left the excited electron.

6.1.2 The Case of an Atom and an Endohedral

Let us consider the simplest case, when all wave functions ϕ_k in the initial (ground) state of the atom correspond to semi-closed spin orbitals. An excited state arises due to the transition (excitation) of an electron from the shell with a number i', we have the following expressions for the coefficients in (6.2):

$$\begin{aligned} a_k^e &= 1, \\ b_{i'}^e &= -1;\ b_k^e = 1, k \neq i'. \end{aligned} \tag{6.4}$$

For the numerical solution of Eq. (6.2) in the case of an atom, the same transformations are used as in Chap. 4 to find the HF functions of the ground state. For a potential $V(\mathbf{r})$ with central symmetry, the radial wave function of an excited electron $P_{s+1}(r)$ in the Hartree–Fock approximation, similarly to (4.2), is found by solving the equation:

$$\left(\frac{d^2}{dr^2}+\frac{2}{r}[Z-Y(r)]-\frac{l_{s+1}(l_{s+1}+1)}{r^2}-E_{s+1}\right)P_{s+1}(r)=X(r)+T, \quad (6.5)$$

where

$$Y(r)=\sum_{i=1}^{s}N_iY_{ii}^0(r)-\frac{2}{N_{s+1}}\sum_{i=1}^{s}\sum_{k>0}\gamma'_k(s+1,i)Y_{ii}^k(r), \quad (6.6)$$

$$X(r)=\frac{2}{rN_{s+1}}\sum_{i=1}^{s}\sum_{k}\gamma_k(s+1,i)Y_{s+1,i}^k(r)P_i(r), \quad (6.7)$$

$$Y_{ij}^k(r)=r\int_0^{\infty}\frac{r_<^k}{r_>^{k+1}}P_i(r')P_j(r')dr', \quad (6.8)$$

$$T=-\sum_{i=1}^{s}\delta_{l_{s+1},l_i}E_{s+1,i}P_i(r), \quad (6.9)$$

are calculated with fixed radial wave functions $P_i(r)$, $i=1,\ldots,s$. These wave functions are the solution to the Hartree–Fock equations for the ground state of the atom (4.2)–(4.7) and are used as input to this program. The parameters $E_{s+1,i}$ included in (6.9) are off-diagonal parameters that ensure the orthogonality of wave functions with the same orbital angular momentum $l_i=l_{s+1}$.

We neglect the influence of an excited electron on the core, and therefore, Eq. (6.5) is linear. The coefficients γ_k and γ'_k in the case of a fixed term are calculated using Formulas (4.9)–(4.12); i.e., for an excited electron in the field of a frozen core with closed subshells or with one subshell with a hole, except for closed ones, only the interaction of this electron with all subshells is specified.

The solution $P_{s+1}(r)$ of the Hartree–Fock Eq. (6.5) must satisfy the boundary conditions at $r=0$

$$P_{s+1}(0)=0, \quad (6.10)$$

orthogonality condition

$$\int_0^\infty P_{s+1}(r)P_i(r)rdr = 0\, i = 1, \ldots, s \tag{6.11}$$

and a number of others, which are different for the discrete spectrum and states of the continuous spectrum.

For a discrete state $E_{s+1} < 0$, the conditions are as follows:

boundary condition at $r \to \infty$ is

$$P_{s+1}(\infty) = 0, \tag{6.12}$$

and normalization condition is

$$\|P_{s+1}\|^2 = \int_0^\infty P_{s+1}^2(r)dr = 1 \tag{6.13}$$

Here P_{s+1} must also have $n_{s+1} - l_{s+1} - 1$ zeros. For the continuous spectrum state ($E_{s+1} \equiv k^2 > 0$), the normalization is determined by the asymptotic

$$P_{s+1}(r) \approx \frac{1}{\sqrt{\pi p}} sin\left(pr + \frac{1}{p}\left(Z - \sum_{i=1}^{s} N_i\right) ln2pr - \frac{\pi l_{s+1}}{2} + \delta_{l_{s+1}}\right), \tag{6.14}$$

where $\delta_{l_{s+1}}$ is the phase of the wave function of an elastically scattered electron with momentum k. This function is normalized by the following condition:

$$\int_0^\infty P_{p,l_{s+1}}(r)P_{p',l_{s+1}}(r)dr = \delta(p - p') \tag{6.15}$$

We consider atoms with semi-closed shells separately. In Sect. 4.2.6, it was noted that such atoms can be conveniently interpreted as containing two types of electrons, or two types of different particles—"up" and "down" electrons, denoted by the indices $\uparrow$ and $\downarrow$, respectively. There is no exchange between them. The equations for the ground state have the form (4.79). The equation for the incident $(N_e + 1)$-electron, which also belongs to the types either "up" or "down," is similar.

The fact that in the ground state each nl-subshell is characterized by the type of electron, $\uparrow$ or $\downarrow$, significantly modifies the equations for radial functions (6.5) and relations (6.6)–(6.9), and the wave functions themselves acquire indices $\uparrow$ or $\downarrow$, turning into $P_{s+1\uparrow(\downarrow)}(r)$.

Changes in Eq. (6.5) are determined by the fact that for electrons of, say, the "up" type, both the Hartree and Fock terms of the self-consistent field created by atomic electrons of the "up" type are taken into account. The interaction with "down" electrons contains only the Hartree term from these electrons.

The presence of an endohedral is taken into account in the framework of an approximate approach called the pseudo-potential approximation. In it, an atom stuffed into a fullerene is considered to be placed at its center, and the fullerene itself is spherically symmetric. Within the pseudo-potential approximation, the action of fullerene is described by a static potential, similar to that described in Sect. 4.2.7. This potential $W_c(r)$ is simply added to $V(\mathbf{r})$ in (6.2) and (6.5). Of course, it does not have to be rectangular, as described by Formula (4.87).

6.1.3 Diatomic Molecule

As mentioned in Chap. 4, here we are considering diatomic molecules in which in the initial state all the shells are semi-closed; i.e., in each shell, there is both an electron with a spin "up" and an electron with a spin "down." After summation over the spins in (6.2), only the coordinate wave functions remain in the HF equations. As in Chap. 4, in the case of axial symmetry of the potential $V(\mathbf{r})$, the coordinate wave function of the excited (final) state of the electron $\varphi_f(\mathbf{r})$ is, similarly to (4.15), in the form of an expansion in spherical harmonics in elongated spheroidal coordinates:

$$\varphi_f(\mathbf{r}) = \sum_{\lambda=m_f}^{L} \mathrm{P}_{f\lambda}(\xi) Y_{\lambda m_f}(\eta, \varphi). \tag{6.16}$$

Also, the Coulomb integrals (6.3) are written in the form of similar expansions:

$$\begin{aligned} J_{k,k}(\mathbf{r}) &= \sum_{\lambda=0}^{L} Z_{kk,\lambda}(\xi) Y_{\lambda,0}(\eta, \phi), \\ J_{f,k}(\mathbf{r}) &= \sum_{\lambda=|m'|}^{L} Z_{fk,\lambda}(\xi) Y_{\lambda m'}(\eta, \phi), \ m' = m_f - m_k. \end{aligned} \tag{6.17}$$

Then the radial functions $\mathrm{P}_{f\lambda}$ should be found as a solution to the following system of equations:

$$\left[\frac{d}{d\xi}(\xi^2-1)\frac{d}{d\xi}-\frac{m_f^2}{\xi^2-1}-\lambda(\lambda+1)+a_s\xi+\varepsilon_f\frac{R^2}{2}(\xi^2-d^i_{\lambda\lambda})\right]\mathrm{P}_{f\lambda}(\xi)$$
$$-\frac{R^2}{2}\sum_{\lambda'}\mathrm{P}_{f\lambda'}(\xi)\sum_{j=1}^{N}a_j^f\sum_k Z_{jj,k}(\xi)\left[\xi^2A^{k,\lambda',\lambda}_{0,m_i,m_i}-B^{k,\lambda',\lambda}_{0,m_i,m_i}\right]$$
$$+\frac{R^2}{2}\sum_{\substack{j=1\\(j\neq i)}}^{N}b_j^f\sum_{\lambda'}\mathrm{P}_{j\lambda'}(\xi)\sum_{k'}Z_{fj,k'}(\xi)\left[\xi^2A^{k',\lambda',\lambda}_{m',m_j,m_i}-B^{k',\lambda',\lambda}_{m',m_j,m_i}\right] \tag{6.18}$$
$$=\varepsilon_i\frac{R^2}{2}\sum_{k\neq\lambda}d^i_{\lambda k}\mathrm{P}_{fk}(\xi)-a_d\sum_k c^i_{\lambda k}\mathrm{P}_{fk}(\xi)+\sum_{j=1}^{N}\varepsilon_j^f\mathrm{P}_{j\lambda}(\xi);$$
$$m'\equiv m_f-m_j,\ \lambda=\left|m_f\right|,\ldots,L.$$

where

$$a_k^f=2,\ b_k^f=1,\ k\neq i';$$
$$a_{i'}^f=1;\ b_{i'}^f=-1, \tag{6.19}$$

Here i' is the number of the excited (ionized) shell. The derivation of these equations is similar to the derivation of Eqs. (4.23) for the ground state. The last additional term provides the orthogonalization of the considered wave function $\varphi_f(\mathbf{r})$ to the wave functions of the ground state:

$$\int\varphi_f(\mathbf{r})\varphi_i(\mathbf{r})d\mathbf{r}=0. \tag{6.20}$$

For discrete states, we have the boundary conditions:

$$\mathrm{P}_{f\lambda}(\xi)\big|_{\xi\to\infty}=0, \tag{6.21}$$

and the wave function must be normalized to unity:

$$\int\varphi_f(r)\varphi_f(r)dr=\iint d\xi d\eta\sum_{\lambda,\lambda'=m_f}^{L}\mathrm{P}_{f\lambda}(\xi)\mathrm{P}_{f\lambda'}(\xi)Y_{\lambda m_f}(\eta,\varphi)Y_{\lambda' m_f}(\eta,\varphi)(\xi^2-\eta^2)$$
$$=1. \tag{6.22}$$

In this case, for the discrete spectrum, successively (with decreasing energy modulus) are found all the solutions that are possible under the condition $\left|\varepsilon_f\right|<|\varepsilon_N|$.

A more essential difference from the case of central symmetry is the problem of calculating the set of functions of the continuous spectrum [4]. This difference is due to the fact that in the case of axial symmetry, the angular momentum is not a good quantum number. The normalization conditions are not problematic. They are based on the well-known asymptotic behavior of continuous spectrum functions:

$$\varphi_{\varepsilon lm}(\mathbf{r})|_{r\to\infty} \propto \frac{1}{r}\left(\frac{2}{\pi p}\right)^{1/2} \sum_{l'} c_{ll'}^{\varepsilon m} Y_{l'm}(\Omega_{\mathbf{r}}) \sin\left[\vartheta(r) + \delta_{ll'}^{\varepsilon m}\right], \tag{6.23}$$

where

$$\vartheta(r) = pr + p^{-1} \ln 2kr \tag{6.24}$$

and the coefficients $c_{ll'}^{\varepsilon m}$ for the correct normalization of the functions $\varphi_{\varepsilon lm}$ must satisfy the relation

$$\sum_{l'} (c_{ll'}^{\varepsilon m})^2 = 1. \tag{6.25}$$

However, due to the degeneracy in the quantum number l, for each energy value there is a countable set of functions with asymptotic of the form (6.23). Various methods can be used to construct orthogonal sets (6.23). Historically, the original is, apparently, the phase method [5], in which solutions with different phases are found. This method goes back to solving a one-electron problem in which all variables are separated. Since, in the general case, such solutions are not easy to find due to poor convergence, then, as a rule, other approaches are used, for example, the K-matrix method [6].

This book uses what appears to be the most general method. Namely, first we find a set of linearly independent solutions of the form (6.23). Linear independence is achieved by the fact that when solving the system (6.18), linearly independent boundary conditions are sequentially set. Further, their mutual orthogonalization is performed in the usual way. As shown in [4], the orthogonality conditions for functions (6.23) have the form:

$$\sum_{l} c_{ll'}^{\varepsilon m} c_{ll''}^{\varepsilon m} \exp[i(\delta_{ll'}^{\varepsilon m} - \delta_{ll''}^{\varepsilon m})] = \delta_{l'l''} \tag{6.26}$$

A numerical procedure for determining the orthogonal sets of continuous spectrum functions will be described in 6.2.7.

6.2 Numerical Model

6.2.1 General Consideration

For the case of both spherical and axial symmetry of the potential, it is common that the angular variables are eliminated analytically from the original three-dimensional HF Eqs. (6.2). Thus, the original HF equations can be reduced to the so-called radial equations in one variable. Although in the case of axial symmetry a variable ξ is used instead of a radial variable r, here the equations obtained for this case are also, by analogy, called radial. These equations, for the case of diatomic molecules, are presented in (6.18).

For atoms, the radial Hartree–Fock Eq. (6.5) can be expressed in the form of the inhomogeneous Schrödinger equation with a nonlocal potential:

$$\frac{d^2 P_{s+1}(r)}{dr^2} = \left[V(r) + E_{s+1}\right] P_{s+1}(r) + \int\limits_0^\infty K(r, r') P_{s+1}(r') dr' + T, \qquad (6.27)$$

where

$$V(r) = \frac{2}{r}[Y(r) - Z] + \frac{l_{s+1}(l_{s+1}+1)}{r^2} \qquad (6.28)$$

is a local part of the potential, while

$$K(r, r') = \frac{2}{r N_{s+1}} \sum_{i=1}^{s} \sum_{k} \gamma_k(s+1, i) \frac{r_<^k}{r_>^{k+1}} P_i(r) P_i(r') \qquad (6.29)$$

is the kernel of the nonlocal potential, and

$$T = -\sum_{i=1}^{s} \delta_{l_{s+1} l_i} \delta_{\sigma_{s+1} \sigma_i} E_{s+1,i} P_i(r) \qquad (6.30)$$

is a term with off-diagonal parameters $E_{s+1,i}$ required to satisfy the orthogonality condition (6.11). The ***fchf*** program uses two independent procedures to solve Eq. (6.27) for discrete states and for continuous spectrum states. In any procedure, the same substitution (4.27) is used, and the same calculation technique for $Y_{ij}^k(r)$ is applied as described in Sect. 4.2.2.

In the case of a fixed term for an excited electron in the field of a frozen core with closed subshells or with one subshell with a hole, except for closed ones, only the interaction of this electron with all subshells is specified; that is, the coefficients (4.10)–(4.13) are calculated.

The numerical procedure for the discrete spectrum is similar to that described in Sect. 4.2.3.

6.2.2 *Numerical Procedure for Continuous Spectrum States*

Although the procedure for solving Eq. (6.27) for continuous spectrum states is based on the same methods as for discrete states, it still has several specific features. Namely,

1. In this task, you do not need to find the eigenvalue, since $E_{s+1} = \kappa^2 > 0$ is specified,
2. The solution is found in accordance with Formulas (4.38)–(4.40). It is easy to see that in this case the function $p(\rho)$ from the differential equation

$$f''_{s+1}(\rho) + p(\rho)f_{s+1}(\rho) + q(\rho) = 0 \tag{6.31}$$

has one root inside the region of integration, so that the latter is split into two intervals $[\rho_0, \rho_1]$ and $[\rho_1, \rho_N]$. For the same reason, as in the case of discrete states, the chasing method is used in the interval $[\rho_0, \rho_1]$, and Numerov's method is used in the interval $[\rho_1, \rho_N]$. The estimates of successive approximations accuracy for the wave function are compared in amplitude and if

$$max\left|\frac{f_{s+1}^{(m+1)}(\rho_j) - f_{s+1}^{m}(\rho_j)}{f_{s+1}^{(m+1)}(\rho_j)}\right|, \qquad j = 1, \ldots, N \tag{6.32}$$

does not exceed the specified accuracy ε, the iterative process ends.

3. The resulting solution is normalized to the δ-function of energy in accordance with the asymptotic (6.14).

6.2.3 *Convergence Procedure for Calculating the Exchange Term for Atoms*

As our calculations show, when the exchange term in expression (6.27) is taken into account, the convergence of the iterative process for some electronic states is too slow. Therefore, convergence acceleration is introduced in the program.

At each step (e.g., the m-th step), after solving the corresponding differential equation, the current mismatch of the iterative process is calculated, which is equal to the maximum absolute value of the relative mismatch calculated at the m-th and

(m-1)-th step of the iteration:

$$\eta^{(m)} = max\left|\frac{1 - f_{s+1}^{(m-1)}(\rho_j)}{f_{s+1}^{m}(\rho_j)}\right|, \qquad j = 1, \ldots, N \tag{6.33}$$

The convergence acceleration procedure is applied only when the value η oscillates on the last 3 iterations:

$$\text{sign}(\eta^{(m)} - \eta^{(m-1)}) \neq \text{sign}(\eta^{(m-1)} - \eta^{(m-2)}) \tag{6.34}$$

In this case, if η still exceeds the given error, i.e., $|\eta^{(m)} - \eta^{(m-1)}| > \varepsilon$, the new solution is calculated by the formula:

$$f_{s+1}^{(m)}(\rho) = (1 - a_1) f_{s+1}^{(m)}(\rho) + a_1 f_{s+1}^{(m-1)}(\rho) \tag{6.35}$$

and for the discrete state a new energy eigenvalue is introduced:

$$E_{s+1}^{(m)} = (1 - a_1) E_{s+1}^{(m)} + a_1 E_{s+1}^{(m-1)}, \tag{6.36}$$

where a_1 is the convergence acceleration parameter

$$a_1 = \frac{\eta^{(m)} - \eta^{(m-1)}}{\eta^{(m)} - 2\eta^{(m-1)} + \eta^{(m-2)}}. \tag{6.37}$$

If

$$\text{sign}\left(\eta^{(m)} - \eta^{(m-1)}\right) = \text{sign}\left(\eta^{(m-1)} - \eta^{(m-2)}\right), \tag{6.38}$$

or

$$|\eta^{(m)} - \eta^{(m-1)}| < \varepsilon, \tag{6.39}$$

then the parameter a_1 is assumed to be zero.

6.2.4 Off-Diagonal Parameters

As it was said, the off-diagonal parameters here ensure orthogonality of the wave functions of the excited states to the wave functions of the ground state (core). Since the wave function of any excited state is found as a result of some iterative process, then the off-diagonal parameters, naturally, are also determined iteratively, in the course of this process.

In the case of atoms, the off-diagonal parameters $E_{s+1,i}$ ensure the orthogonality of the radial wave functions with equal ℓ. As the first option, these parameters are determined by the following method of successive approximations. First, Eq. (6.27) is solved with zero off-diagonal parameters $E^{(0)}_{s+1,i} = 0$. After each iterative step, the overlap integral is calculated using the formula:

$$I_i^{(m)} = \int P^{(m)}_{s+1} P_i(r) r, \qquad i = 1, \ldots, s \tag{6.40}$$

If the sum of the absolute values of the overlap integrals $I_1, \ldots, I_s$ does not exceed the specified accuracy ε, the procedure for determining the off-diagonal parameters ends. Otherwise, the new values of the off-diagonal parameters are calculated in two ways, depending on the values of the overlap integral in two subsequent iterations. If they have the same sign $\mathrm{sign}(I_i^{(m)}) = \mathrm{sign}(I_i^{(m-1)})$, the off-diagonal parameters are adjusted by an amount proportional to the overlap integral:

$$E^{(m+1)}_{s+1,i} = E^{(m)}_{s+1,i} - a_i^{(m)} I_i(m), \tag{6.41}$$

where parameters

$$a_i^{(m)} = \left| \frac{I_i^{(m)}}{I_i^{(m)} - I_i^{(m-1)}} \right| \tag{6.42}$$

are introduced to speed up the convergence of the iterative process. Estimation for them is

$$a_i^{(0)} = 0. \tag{6.43}$$

If $\mathrm{sign}(I_i^{(m)}) \neq \mathrm{sign}(I_i^{(m+1)})$, then the new value of the off-diagonal parameter is determined by the chord method:

$$E^{(m+1)}_{s+1,i} = \frac{I_i^{(m)} E^{(m-1)}_{s+1,i} - I_i^{(m-1)} E^{(m)}_{s+1,i}}{I_i^{(m)} - I_i^{(m-1)}} \tag{6.44}$$

Convergence is achieved if $E_{s+1,i}, i = 1, \ldots, s$ are small. In most cases, this condition is met.

Consider the following option for accelerating convergence. It is used when the excited state wave function is calculated with a large negative exchange coefficient and the iteration does not converge. Although the wave function is fixed at $\rho = \rho_1$, its value to the right of this point increases indefinitely at each iteration step. The reason for this behavior is that the exchange potential $X_{s+1}(r)$ is, in this case, positive and abnormally large. This enhances the mutual influence of the wave function and exchange potential on each other, leading to a further increase in both. Obviously, if the calculated wave function were orthogonal to the wave functions of the core, the exchange integral

$$Y^{(0)}_{El,n,} = r \int_0^\infty \frac{1}{r_>} P_{El}(r') P_{nl}(r') dr', \tag{6.45}$$

which comes in $X_{s+1}(r)$ with a large negative coefficient $\gamma_0(El, nl^{-1}, s+1)$ would be much less. However, the iteration diverges at $E_{s+1,i} = 0$ when the desired function $P_{El}(r)$ is not orthogonal to the wave functions of the core.

To avoid this difficulty, a second algorithm for solving Eq. (6.5) with conditions (6.11) was developed. This algorithm is also a method of successive approximations and can be called "*at first orthogonalization, then exchange,*" in contrast to the first, "*at first exchange, then orthogonalization*" described above. In the new algorithm, the initial estimate for the wave function in Eq. (6.5) is the zero wave function, and therefore, the exchange potential is also zero. At the first iteration step, the Hartree wave function is calculated. With a proper choice of off-diagonal parameters, the function meets condition (6.11). The orthogonalized Hartree function is then used to compute the exchange potential. Equation (6.5) is solved again with this potential, and new orthogonalization parameters $E_{s+1,i}$ are chosen. The exchange potential is determined according to the new orthogonalized wave function, and the calculation is repeated again. It is considered that Eq. (6.5) is solved when the difference between two sequentially calculated orthogonalized wave functions does not exceed ε. Since the wave function is orthogonalized during the iterative procedure, the exchange potential does not reach too high values and the iterative convergence is fast. This usually takes for $\varepsilon = 10^{-5}$ about 7–10 steps.

Since the orthogonalization procedure takes a relatively long time, and the off-diagonal parameters for ^1S configurations are large, the described algorithm is slightly changed for small values of the overlap integrals ($|I_i| < 10^{-2}$). This allows the chord method to be used despite the sign of I_i. The convergence rate, therefore, turns out to be significantly increased, since the dependence $E_{s+1,i}$ on the overlap integrals in this range is almost linear.

The third option for accelerating the convergence for atoms is that the energy parameters $E_{s+1,i}$ are determined as follows:

$$E_{s+1,i} = \sum_k \Big\{ a^k_{s+1,j} \int Y^k_{jj}(r') \frac{2}{r'} P_{s+1}(r') P_i(r') dr' \\ - b^k_{s+1,j} \int Y^k_{s+1,j}(r') \frac{2}{r'} P_i(r') P_j(r') dr' \Big\}, \tag{6.46}$$

where the index j corresponds to a subshell $(n_j l_j)^N$ with N-electrons and an excited electron in the state $(n_{s+1} l_{s+1})$. The coefficients $a^k_{s+1,j}$ and $b^k_{s+1,j}$ in (6.46) are usually derived from the condition of orthogonality of the wave function of the excited state and the wave functions of the occupied states:

$$\int P_s(r)P_i(r)dr = 0, \qquad i = 1, \ldots, s \tag{6.47}$$

$$\begin{aligned}
a^k_{s+1,j} &= -N(2l_j+1)(2l_{s+1}+1)\begin{pmatrix} l_j & k & l_j \\ 0 & 0 & 0 \end{pmatrix}\begin{pmatrix} l_{s+1} & k & l_{s+1} \\ 0 & 0 & 0 \end{pmatrix} \\
&\times \sum_{\bar{L}\bar{S}} (2\bar{L}+1)(2\bar{S}+1)(2S'+1)^{-1}(-1)^{L_1+\bar{L}} \\
&(l_j^{N-1}L'S', l_j|\}l_j^N \bar{L}\bar{S})^2 \begin{Bmatrix} k & L' & L' \\ \bar{L} & l_j & l_j \end{Bmatrix}\begin{Bmatrix} k & L' & L' \\ L_1 & l_{s+1} & l_{s+1} \end{Bmatrix}, \\
b^k_{sj} &= -N(2l_j+1)(2l_s+1)\sum_{\bar{L}\bar{S}} \delta_{\bar{S}S_1}(2\bar{L}+1) \\
&(l_j^{N-1}L'S', l_j|\}l_j^N \bar{L}\bar{S})^2 \begin{pmatrix} l_j & l_k & l_{s+1} \\ 0 & 0 & 0 \end{pmatrix}\begin{Bmatrix} L_1 & k & \bar{L} \\ l_j & L' & l_{s+1} \end{Bmatrix},
\end{aligned} \tag{6.48}$$

Here quantum numbers $L'S'$ and L_1S_1 describe, respectively, the state of the ion and the whole system (ion and photoelectron) in the final state; $(l_j^{n-1}L'S', |\}l_j^N \overline{LS})$ are the genealogical coefficients [IS]. In the case of excitation of atoms with closed shells, Eq. (6.48) is simplified:

$$\begin{aligned}
a^k_{s+1,j} &= \delta_{k0}(4l_j + 2 - q_j)' \\
b^k_{s+1,j} &= (2l_j+1)\begin{pmatrix} l_j & k & l_{s+1} \\ 0 & 0 & 0 \end{pmatrix} - \gamma_k(s+1, j).
\end{aligned} \tag{6.49}$$

With the version of the RPAE for atoms with an open shell [7], conditions (6.47) are not satisfied, and the coefficients $a^k_{s+1,j}$ and $b^k_{s+1,j}$ are derived from the condition that the photoionization cross section in the form of length and velocity are equal. In the case of ionization of an atom with an unfilled shell, these coefficients are calculated as

$$\begin{aligned}
a^k_{s+1,j} &= (2l_j+1)(2l_{s+1}+1)\begin{pmatrix} l_j & k & l_j \\ 0 & 0 & 0 \end{pmatrix}\begin{pmatrix} l_{s+1} & k & l_{s+1} \\ 0 & 0 & 0 \end{pmatrix}\begin{Bmatrix} k & l_j & l_j \\ 1 & l_{s+1} & l_{s+1} \end{Bmatrix} \\
&- N(2l_j+1)(2l_{s+1}+1)\begin{pmatrix} l_j & k & l_j \\ 0 & 0 & 0 \end{pmatrix}\begin{pmatrix} l_{s+1} & k & l_{s+1} \\ 0 & 0 & 0 \end{pmatrix} \\
&\sum_{\bar{L}\bar{S}} (2\bar{L}+1)(2\bar{S}+1) \times \Big[(2S'+1)^{-1}(-1)^{L_1+\bar{L}}
\end{aligned}$$

$$\times\,(l_j^{N-1}L'S', l_j|\}l_j^N\bar{L}\bar{S})^2\begin{Bmatrix} k & L' & L' \\ \bar{L} & l_j & l_j \end{Bmatrix}\begin{Bmatrix} k & L' & L' \\ L_1 & l_{s+1} & l_{s+1} \end{Bmatrix}$$

$$+\sum_{l'S'}\sum_{l''S''}\delta_{s'+1,s''+1}(2l_j+1)(2S'+1)^{-1}\,(-1)^{L_1+L'}$$

$$(l_j^{N-1}L'S', l_j|\}l_j^N LS)(l_j^{N-1}L'S', l_j|\}l_j^N\bar{L}\bar{S})$$

$$\times\left(l_j^{N-1}L'S', l_j\|\}l_j^N LS\right)\left(l_j^{N-1}L'S', l_j\|\}l_j^N\bar{L}\bar{S}\right)\Big]$$

$$\begin{Bmatrix} k & L' & L'' \\ \bar{L} & l_j & l_j \end{Bmatrix}\begin{Bmatrix} k & L' & L'' \\ L & l_j & l_j \end{Bmatrix}\begin{Bmatrix} k & l_j & l_j \\ 1 & l_{s+1} & l_{s+1} \end{Bmatrix}\Bigg], \tag{6.50}$$

$$b_{s+1j}^k = -N(2l_j+1)(2l_{s+1}+1)$$

$$\times\sum_{\bar{L}\bar{S}}\delta_{\bar{S}S_1}(2\bar{L}+1)(l_j^{N-1}L'S', l_j|\}l_j^N\bar{L}\bar{S})^2$$

$$\begin{Bmatrix} L_1 & k & \bar{L} \\ l_j & L' & l_{s+1} \end{Bmatrix}\begin{pmatrix} l_j & l_k & l_{s+1} \\ 0 & 0 & 0 \end{pmatrix} \tag{6.51}$$

with quantum numbers $L'S'$ and L_1S_1 defined as in (6.48).

6.2.5 *Normalization and Phase Shift of the Radial Wave Function in Continuous Spectrum*

The radial wave function of the continuous spectrum is normalized to the δ-function in energy. This results in appearance of a factor $(\pi p)^{-1/2}$ in the asymptotic expression (6.14). The normalization constant for the solution of Eq. (6.27) is derived from the semiclassical approximation for the radial wave function

$$P_{s+1}(r) \approx F(r)|^{-1/4} sin\int_0^r \sqrt{F(r')}dr',$$

$$F(r) = p^2 - V(r), \tag{6.52}$$

which is true for sufficiently large values of the radius r, with the asymptotic behavior described by (6.14). The function defined in this way oscillates as sin pr.

To obtain the normalization factor and the phase shift, Eq. (6.27) is solved by Numerov's method at equidistant points outside the specified range.

The size of the step is chosen to be equal to the difference between two successive values r, $\Delta r = r_N - r_{N-1}$, provided that there are at least five nodes within one half-wave, that is $\Delta r < 1/5p$. Otherwise, the difference is halved until this condition is met. Integration starts from the penultimate value r_{N-1}.

If the step is defined as $(r_N - r_{N-1})/2$, the next value of the function required for integration is found by quadratic interpolation. Only the last three values of the function needed for interpolation are stored in memory, when solving the equation in this area. The total number of zeros n is determined in course of the calculation. If in the procedure for solving (6.27) in the outer region the difference between two successive maxima of the expression $\sqrt[4]{F(r)}P(r)$ [4] is less than ε, the normalization parameter A for the solution, in accordance with (6.14), is determined by the expression

$$A = \frac{1}{\sqrt{\pi} max \left| P_{s+1}(r)\,(F(r))^{1/4} \right|}. \tag{6.53}$$

The phase shift is determined starting from the normalization point ρ_1 (see Sect. 6.2.2) at the first and every sixth root as follows:

$$\delta = n\pi - pr_n - \frac{1}{\kappa}(Z - \sum_{j=1}^{s} N_j) \ln 2pr_n$$
$$+ \frac{\pi l_{s+1}}{2} - \frac{1}{2p}\left(\left[\left(Z - \sum_{j=1}^{s} N_j\right)/p\right]^2 + l_{s+1}(l_{s+1}+1)\right),$$
$$r = r(\rho), \qquad p = \sqrt{E}, \tag{6.54}$$

where r_n is the n-th zero of the solution. Each calculated phase value is compared with the value obtained in the previous step. If the difference between them is less than ε, the last value is taken as the phase shift, and the integration ends.

The disadvantage of this method is that the error accumulates in the course of calculations at small values κ_{s+1}, $\kappa_{s+1}^2 = E_{s+1}$, when the interval at the maximum $\sqrt[4]{F(r)}P_{s+1}(r)$ [4] is quite large. In this case, there is a more convenient expression for the phase difference [8]

$$\delta_l = arcsin\left(-\frac{\sqrt{\pi p}}{p} \int_0^\infty j_l(pr) V_i(r) P_{s+1}(r) dr\right), \tag{6.55}$$

where $j_e(pr)$ is the spherical Bessel function, $V_i(r) = 2/r[Z - Y_i(r)]$. However, for large values of the argument p, when the functions $j_e(pr)$ and $P_{s+1}(r)$ oscillate with high frequency, the number of roots within the half-wave for these functions in (6.55) is too small for this expression to be still valid.

6.2.6 Radial Hartree Equation for a Meson and a Positron in the Field of a Frozen Core of an Atom and an Endohedral

The radial wave function in the Hartree–Fock approximation for a meson in the field of a frozen atomic core is a solution of the Hartree equation:

$$\left(\frac{d^2}{dr^2}+\frac{2m_1}{r}[Z-Y(r)]-\frac{l_{s+1}(l_{s+1}+1)}{r^2}-E_{s+1}m_1\right)P^M_{s+1}(r)=0,$$

$$Y(r)=\sum_{i=1}^{s}N_iY_ii^0(r)-\frac{2}{N_{s+1}}\sum_{i=1}^{s}\sum_{k>0}\gamma'_k(s+1,i)Y^k_{ii}(r),$$

$$Y^k_{ii}(r)=r\int_0^\infty\frac{r^k_<}{r^{k+1}_>}P_i(r')P_i(r')dr', \qquad (6.56)$$

where m_1 is the product of the ratio of the particle mass to the electron mass multiplied by the particle charge, taken with a minus sign, for a positron $m_1=-1$, $m_1=\mu$ - for a negatively charged meson $m_1=\mu$, where μ is the meson mass in units of electron masses; for a positively charged meson $m_1=-\mu$.

In contrast to the HF equations for an excited electron (6.5)–(6.9), in this equation the exchange term $X(r)$ and the term T, including the off-diagonal parameters, are absent, since positron and meson are not identical to an electron. In addition, the meson mass is taken into account, i.e., $E_{s+1}=p^2/|m|_1>0$. The coefficients γ_k in this case are calculated by Formula (4.11).

For a meson in a discrete spectrum, the eigenvalue problem (6.56) must be solved with the same boundary conditions (6.10), (6.12) and the normalization condition $P^M_{s+1}(r)$ (6.13). The meson mass m_1 should be taken into account in (4.73):

$$E_{s+1}=E_{s+1}\left[1+\frac{0.01m_1}{n_{s+1}}\text{sign}(\zeta-n_{s+1}+l_{s+1}+1)\right]. \qquad (6.57)$$

For a meson in the continuous spectrum ($E_{s+1}=p^2/|m_1|>0$), in Eq. (6.56), the energy E_{s+1} is known. The radial wave function $P^M_{s+1}(r)$ is the solution of the Hartree equation for the meson. The function $P^M_{s+1}(r)$ is normalized to the δ- function in energy, and its asymptotic behavior is as follows:

$$P^M_{s+1}(r)\approx\sqrt{\frac{m_1}{\pi p}}sin\left(pr+\frac{m_1Z_0}{p}ln2pr-\frac{\pi l}{2}+\delta_l\right), \qquad (6.58)$$

where Z_0 is the electric charge of the core ($Z_0=0$ for a neutral atom). When calculating the phase shift, the normalization constant A (6.53) is multiplied by $\sqrt{m_1}$.

Equation (6.56) is solved similarly to (6.5). The zero estimate of the energy eigenvalue for a meson in the bound s-state is determined accounting for the mass m_1:

$$E^{(0)}_{s+1,\text{est}} = m_1 \left[\left(Z - \sum_{j=1}^{s} N_j \right) \Big/ n_{s+1} \right]^2 . \tag{6.59}$$

Hartree equation for the positron, where the electric charge of the positron is taken into account, is:

$$\begin{aligned}
&\left(\frac{d^2}{dr^2} + \frac{2}{r}[Z - Y(r)] - \frac{l_{s+1}(l_{s+1}+1)}{r^2} - E_{s+1} \right) P^P_{s+1}(r) = 0, \\
&Y(r) = \sum_{i=1}^{s} N_i Y_i i^0(r) - 2 \sum_{i=1}^{s} \sum_{k>0} \gamma'_k(s+1, i) Y^k_{ii}(r), \\
&Y^k_{ii}(r) = r \int_0^\infty \frac{r^k_<}{r^{k+1}_>} P_i(r') P_i(r') dr'
\end{aligned} \tag{6.60}$$

In the case of a positron, the exchange term $X_i(r)$ and the term T (which includes the off-diagonal parameters) do not exist, as well as for the meson. The wave functions of both the meson and the positron are calculated similarly to excited states of electrons (see Sect. 2.3), using the same program that takes into account the specific features in Eqs. (6.5)–(6.9), (6.56), and (6.60). The exchange term for the positron and meson $X(r)$, and, if necessary, for electrons removed from the outer shell is omitted; i.e., exchange coefficients $\gamma_k(s+1, i)$ are set equal to zero. For the meson and positron in the field of the frozen core, only the coefficients (4.11) are calculated. The condition of orthogonality of the wave function of the excited state to all other functions (6.11) is introduced by a parameter ort_i, $i = 1, \ldots, s$ that is equal to 1 if the condition is satisfied, otherwise this parameter is equal to 0.

In order to account for the potential of the endohedral, i.e., for an atom or simple molecule placed inside a fullerene, Eq. (6.60) is transformed to the form

$$\left(\frac{d^2}{dr^2} + \frac{2}{r}[Z - Y_i(r)] + W_c(r) - \frac{l_i(l_i+1)}{r^2} - E_i \right) P_i(r) = X_i(r) + T, \tag{6.61}$$

Here $W_c(r)$ is the attractive potential of finite width Δ and depth U_0, equal to

$$W_c(r) = U_0,\ R_0 \le r \le R_0 + \Delta \tag{6.62}$$

The solution to Eq. (6.61) is carried out according to the algorithm described above.

6.2.7 Numerical Procedure for HF Equations with Axially Symmetric Potential

The numerical procedure to determine the wave functions of excited states of the discrete spectrum is practically the same as the procedure for calculating the wave functions of the ground state. For excited states, the solution of the HF equations is even somewhat simpler, since the procedure of self-consistency of various solutions is not required. Naturally, in this case, in the algorithm described in Chap. 4, the value of the angular momentum projection m_i must be replaced everywhere by the value m_f. In particular, Eqs. (4.133) and (4.141) for determining the coefficients of asymptotic expansions $c_{\lambda,k}$ are then written in the form:

$$c_{\lambda,k}a_k^m + c_{\lambda,k+1}b_k^m + \sum_{\lambda'}\sum_{k'=0}^{n} c_{\lambda',k'}p_{\lambda\lambda',k-k'} = q_{\lambda,k},$$
$$c_{\lambda,0} = 1\ (\lambda = m_f);\ k = 1 \div n-1,\ \lambda, \lambda' = \left|m_f\right|, \ldots, L. \tag{6.63}$$

and

$$g(\xi_{n+2})(\xi_{n+2}^2 - 1)^{|m|/2}\sum_{k=0}^{n} c_{\lambda,k}(\xi_{n+2} - 1)^k$$
$$= \{\mathbf{V}_{n+2}\}_\lambda + g(\xi_{n+1})(\xi_{n+1}^2 - 1)^{|m|/2}\sum_{\lambda'}\{\mathbf{U}_{n+2}\}_{\lambda,\lambda'}\sum_{k=0}^{n} c_{\lambda',k}(\xi_{n+1} - 1)^k;\ \lambda \neq |m_f|. \tag{6.64}$$

Solutions in the discrete spectrum are found sequentially in descending order of energy modulus; i.e., for a given value of the projection of the moment m_f, a smooth solution of Eq. (6.18) is found with the maximum possible magnitude of the energy value $\varepsilon_f^{(1)}$, which, naturally, will always be less in modulus than the value of the energy for the outermost shell in the ground state. Then, the next solution is found, again with the maximum energy value $\varepsilon_f^{(2)}$, possibly large in absolute value, different from $\varepsilon_f^{(1)}$, etc. Naturally, it is always correct that $\left|\varepsilon_f^{(n+1)}\right| < \left|\varepsilon_f^{(n)}\right|$.

The entire basic numerical algorithm for solving Eq. (6.18) for the continuous spectrum remains the same as for the discrete spectrum (or, the same as for the ground state in Chap. 4), and differs in only a few details. Namely, when solving Eq. (6.18) for the case of a continuous spectrum, in contrast to the case of a discrete spectrum, a smooth solution of Eq. (6.18) exists for any positive energy value $\varepsilon_f (\varepsilon_f > 0)$.

To take this into account in the solution algorithm, we consider the closed system of Eqs. (6.63) and (6.64) to determine the coefficients of the asymptotic expansions $c_{\lambda,k}$. The condition $\lambda \neq |m_f|$ in (6.64), as noted in Chap. 4, means that we are looking for a solution with smooth harmonics, except for the harmonic number $\lambda = |m_f|$. By removing this condition, we thereby make the solution completely smooth. However, in this case, an additional equation appears in system (6.64). For the new system of Eqs. (6.63), (6.64) to have a solution, we additionally remove the equation $c_{\lambda,0} = 1$ $(\lambda = m_f)$ in (6.63). As a result, instead of (6.63) and (6.64) we get:

$$\begin{gathered} c_{\lambda,k}a_k^m + c_{\lambda,k+1}b_k^m + \sum_{\lambda'}\sum_{k'=0}^{n} c_{\lambda',k'}p_{\lambda\lambda',k-k'} = q_{\lambda,k}, \\ k = 1 \div n-1,\ \lambda,\lambda' = \left|m_f\right|, \ldots, L. \end{gathered} \tag{6.65}$$

and

$$\begin{gathered} g(\xi_{n+2})(\xi_{n+2}^2 - 1)^{|m|/2}\sum_{k=0}^{n} c_{\lambda,k}(\xi_{n+2} - 1)^k \\ = \{\mathbf{V}_{n+2}\}_\lambda + g(\xi_{n+1})(\xi_{n+1}^2 - 1)^{|m|/2}\sum_{\lambda'}\{\mathbf{U}_{n+2}\}_{\lambda,\lambda'}\sum_{k=0}^{n} c_{\lambda',k}(\xi_{n+1} - 1)^k. \end{gathered} \tag{6.66}$$

In the new system of linear Eqs. (6.65), (6.66), the number of equations again coincides with the number of unknowns, and Eq. (6.66) ensures the smoothness of the wave function of the continuous spectrum.

Another significant difference from the case of a discrete spectrum is that the boundary conditions at infinity are nonzero. On the contrary, as our calculation experience shows, it is only by the boundary conditions at infinity that it is possible in the particular solution of Eqs. (6.18) to provide in the asymptotic (6.23) the predominant contribution of one harmonic with $l' = l$. For this, when constructing a set of linearly independent functions $\varphi_{f,l}(l = 1, 2, \ldots)$, when calculating each function $\varphi_{f,l}$, the value of its harmonic with a number ℓ on the upper boundary of the integration interval is fixed, i.e., at $\xi = \xi_{\max}$ $(\rho = \rho_{\max})$:

$$\mathrm{P}_{f,l}(\rho_{\max}) = 1. \tag{6.67}$$

In this case, for all other harmonics it is assumed:

$$\mathrm{P}_{f,\lambda}(\rho_{\max}) = 0,\ \lambda \neq l. \tag{6.68}$$

Naturally, linearly independent boundary conditions provide a linearly independent set of functions.

It is known that the wave functions of the continuous spectrum for a fixed energy form a linear space; i.e., any linear combination of them is also a solution to the Hartree–Fock equation. Therefore, from an initially non-orthogonal set of wave functions of the continuous spectrum, it is easy to construct an orthogonal set satisfying conditions (6.26). For example, if two non-orthogonal functions $\varphi_{f,l'}$ and $\varphi_{f,l''}$ $(l' \neq l'')$ are found numerically and for which, contrary to (6.26),

$$\sum_l c_{ll'}^{\varepsilon m} c_{ll''}^{\varepsilon m} \exp[i(\delta_{ll'}^{\varepsilon m} - \delta_{ll''}^{\varepsilon m})] = a_{l'l''} \neq 0, \tag{6.69}$$

then the function

$$\varphi'_{f,l''} = \varphi_{f,l''} - a_{l'l''}\varphi_{f,l'} \tag{6.70}$$

will, on the one hand, also be a solution to the Hartree–Fock equation. On the other hand, it will be orthogonal to the function $\varphi_{\varepsilon l'm}$. It remains only to properly normalize it in accordance with condition (6.25). This procedure is repeated for all functions from the set, resulting in a set of orthonormal functions of the continuous spectrum.

For diatomic molecules, off-diagonal energy parameters are always expressed in terms of the Coulomb matrix elements by the formula

$$\begin{aligned} \varepsilon_j^f = {} & \sum_{k=1}^N a_{jk}\langle \varphi_f, \varphi_k|V|\varphi_j, \varphi_k\rangle - \sum_{k=1}^N b_{jk}\langle \varphi_f, \varphi_k|V|\varphi_k, \varphi_j\rangle \\ & - \sum_{k=1}^N a_k^f\langle \varphi_j, \varphi_k|V|\varphi_f, \varphi_k\rangle + \sum_{k=1}^N b_k^f\langle \varphi_j, \varphi_k|V|\varphi_k, \varphi_f\rangle, \end{aligned} \tag{6.71}$$

where a_{jk}, b_{jk} are the coefficients in the HF equations of the ground state, and a_k^f, b_k^f are the coefficients in the HF equations for excited states. This formula is similar to Formula (6.46) for atoms. It ensures the orthogonality of the wave functions of the excited states to the wave functions of the core by virtue of the HF equations themselves. Calculation of the Coulomb matrix elements is described in Chap. 9 (see (9.45)–(9.52)).

Using all the procedures described above, the ***fchf*** program [1–3] allows finding the wave functions of an electron, meson, and positron in discrete and continuous spectra, and the ***fchfw*** program gives wave functions for an atom placed in a fullerene shell. The same for molecules is achieved using the programs ***fchfmd*** (for discrete spectrum) and **fchfmc** for continuous spectrum.

The scattering of a positron and a meson on an endohedral is easy to describe in the pseudo-potential approximation, if, as described in Sect. 4.2.7, one simply adds the potential $W_c(r)$ to Eqs. (6.56) or (6.60).

References

1. L.V. Chernysheva, N.A. Cherepkov, and V. Radojevič, *Program for calculating the Hartree-Fock wave functions of discrete and continuous spectra in the field of a "frozen" core* (Physical-Technical Institute, No. 487, Leningrad, 1975). (In Russian).
2. M.Ya. Amusia, L.V. Chernysheva, *Automated System for Studying the Structure of Atoms* (Nauka, Leningrad Division, 1983). ((In Russian))
3. M.Ya. Amusia, L.V. Chernysheva, *Computation of Atomic Processes* (Institute of physics Publishing, Bristol and Philadelphia, 1997)
4. S.K. Semenov, N.A. Cherepkov, G.H. Fecher, and G. Schönhense, Phys. Rev. A, **61**, 032704 (2000).
5. D.I. Abramov, I.V. Komarov, Teor. Mat. Fiz. **22**, 253 (1975). ((In Russian))
6. H. Park, R.N. Zare, J. Chem. Phys. **104**, 4554 (1996)
7. V.F. Orlov, N.A. Cherepkov, L.V. Chernysheva, Opt. Spectrosc. **64**, 409 (1988). ((In Russian))
8. F. Calogero, *Variable approach to potential scattering* (Academic Press, New York – London, 1967).

Chapter 7
Wave Functions of an Excited State of an Atom in the Hartree–Fock–Dirac Approximation

Consideration in the Hartree–Fock–Dirac approximation of any process in which atom participates require knowledge of the wave functions of the excited states. This chapter proposes the simplest approach to calculating one-particle wave functions in discrete and continuous spectra. In this case, as a rule, the frozen core approximation is used, in which it is assumed that the excitation of one electron does not affect the others.

Numerical procedures are presented that permit: to calculate the wave functions of an electron, to accelerate the convergence when taking into account the exchange term, to include off-diagonal parameters that ensure the orthogonality of the wave functions of excited states to the wave functions of the core, to obtain phase shifts of elastic and inelastic scattering.

7.1 Theoretical Model

The radial parts of the relativistic wave functions in the "frozen" core approximation in an excited state (discrete and continuous spectrum) are found by a direct generalization of the corresponding non-relativistic approaches [1–5]. The wave function of an arbitrarily charged particle is still determined by Eq. (5.1). In the same way as in Chap. 5, it can be shown that the radial part $P_s(r)$,

$$F_s(r) = \xi_s(r) P_s(r) \tag{7.1}$$

$$\xi_s(r) = \left[1 - \frac{\alpha^2}{2}\left(\frac{\varepsilon_s}{2} - \frac{|m_s|}{m_s} U_s(r)\right)\right]^{1/2}, \tag{7.2}$$

satisfies the linear equation without self-consistency:

M. Ya. Amusia and L. V. Chernysheva, *Computation of Atomic and Molecular Processes*, Springer Series on Atomic, Optical, and Plasma Physics 117,
https://doi.org/10.1007/978-3-030-85143-9_7

$$\frac{d^2}{dr^2}P_s(r) + \left\{ |m_s|\varepsilon_s + \frac{1}{4}\alpha^2\varepsilon_s^2 - 2U_s(r)\left(m_s + \frac{1}{2}\alpha^2\varepsilon_s\right) + \alpha^2 U_s^2(r) - \frac{\kappa_s(\kappa_s+1)}{r^2} - \alpha^2 \frac{1}{4\xi_s^2(r)} \frac{d^2}{dr^2}U_s(r) + \alpha^2 \frac{k_s}{2\xi_s^2(r)} \frac{1}{r}\frac{d}{dr}U_s(r) - \alpha^4 \frac{3}{16\xi_s^4(r)}\left(\frac{d}{dr}U_s(r)\right)^2 \right\} P_s(r) = \xi_s(r)\left[\tilde{X}_s(r) + T_s(r)\right]. \quad (7.3)$$

Here the following relations are used

$$U_s(r) = \frac{1}{r}[Y_s(r) - Z], \quad (7.4)$$

$$Y_s(r) = \sum_{j \neq s}^{s-1} q_j\, Y_{jj}^0(r) - \frac{2}{q_s}\sum_{j \neq s}^{s-1}\sum_{k>0} \gamma'_k(sj)\, Y_{jj}^k(r), \quad (7.5)$$

$$X_s(r) = -\frac{2}{q_s}\frac{1}{r}\sum_{j \neq s}^{s-1}\sum_{k>0} \gamma_k(sj)\, Y_{sj}^k(r)\xi_j(r)P_j(r), \quad (7.6)$$

$$Y_{sj}^k(r) = r\int_0^\infty \frac{r_<^k}{r_>^{k+1}}\left(\xi_s(r')\xi_j(r')P_s(r')P_j(r') + \alpha^2 G_s(r')G_j(r')\right)dr\prime, \quad (7.7)$$

$$G_s(r) = \frac{1}{2}\xi_s^{-2}(r)\left(F'_s(r) + \frac{\kappa_s}{r}F_s(r)\right), \quad (7.8)$$

$$T_s(r) = -\sum_{j \neq s} \delta_{k_j k_s}\, \varepsilon_{sj}\xi_j(r)P_j(r). \quad (7.9)$$

Here $m_s = -QM$, where Q and M are the charge and mass of the particle ($m_s = 1$ for an electron), $s \equiv (n_s k_s)$ denotes the exited subshell for which electron wave function is calculated, while $j \equiv (n_j k_j)$ corresponds to all sublevels of the core. The corresponding radial parts $F_j(r), G_j(r), j = 1 \ldots s-1$, as well as are the self-consistent potentials $U_i(r)$, $i = 1 \ldots s-1$, are the same and already defined in advance. Quantity ε_s, the energy eigenvalue, is the diagonal Lagrange multiplier, $\varepsilon_s < 0$—for discrete, and $\varepsilon_s \equiv p^2 > 0$—for continuous spectrum.

Off-diagonal Lagrange multipliers ε_{sj}, $s \neq j$ are introduced in (7.3) to ensure the orthogonality of the calculated wave functions to the core ones for $k_j = k_s$. Angular coefficients are determined by the formulas

$$\gamma'_k(aa) = \begin{cases} \frac{1}{2}\begin{pmatrix} j_a & k & j_a \\ -1/2 & 0 & 1/2 \end{pmatrix} \text{for even } k \\ 0 \text{ for odd k} \end{cases} \quad k > 0 \quad (7.10)$$

$$\gamma'_k(aa) = \begin{cases} \begin{pmatrix} j_a & k & j_b \\ -1/2 & 0 & 1/2 \end{pmatrix} \text{ for even } k \\ 0 \text{ for odd k} \end{cases} \tag{7.11}$$

Solution (7.3) corresponding to the discrete spectrum must satisfy the same boundary condition for $r = 0$ and $r = \infty$

$$\begin{aligned} F_{n_i k_i}(0) &= F_{n_i k_i}(\infty) = 0, \\ G_{n_i k_i}(0) &= G_{n_i k_i}(\infty) = 0. \end{aligned} \tag{7.12}$$

The same are the boundary conditions for $r = 0$.

On the other hand, one should distinguish discrete and continuous spectra by introduction proper boundary condition at $r = \infty$, and the normalization condition. In the case of discrete excitation $\varepsilon_s < 0$, the boundary conditions are still given by Eq. (7.12) and the normalization conditions are

$$\int_0^\infty \left(F^2_{n_i k_i}(r) + \alpha_0^2 G^2_{n_i k_i}(r)\right) dr = 1 \tag{7.13}$$

These relations unambiguously determine the desired solution, provided that they meet the requirement that the large radial component $F_{n_i k_i}(r)$ must have at least $(n_i - l_i - 1)$ nodes at finite distances [2, 4]. In the case of a continuous spectrum, the asymptotic behavior of the wave function at $r \to \infty$ simultaneously determines its normalization. Here the asymptotic has the form:

$$P_s(r) \sim \sqrt{\frac{|m_s|}{\pi \tilde{p}}} sin\left(pr + \frac{m_s Z_{eff}}{\tilde{p}}\left(1 + \frac{\alpha^2 p^2}{2}\right) ln 2\tilde{p}r - \frac{\pi l_s}{2} + \delta_k(p)\right), \tag{7.14}$$

where $\tilde{p} \equiv p\sqrt{1 + \frac{\alpha^2 p^2}{4}}$ is the relativistic momentum of an electron, $Z_{eff} = Z - \sum_{j=1}^{s-1} q_j = Z - N - 1$ is the effective charge acting upon an atomic electron at large distances from the core, ℓ_s is the orbital moment of the excited state, $\delta_k(p)$ is the phase of elastic scattering. The asymptotic behavior of the small radial component $G_s(r)$ can be easily obtained from Eqs. (7.9), (7.14):

$$G_s(r) \sim \sqrt{\frac{|m_s|}{\pi p}} \frac{p}{2} \cos\left(pr + \frac{m_s Z_{eff}}{p}\left(1 + \frac{\alpha^2 p^2}{2}\right) ln 2pr - \frac{\pi l_s}{2} + \delta_k(p)\right), \tag{7.15}$$

Equations (7.14) and (7.15) determine the normalization of the function $\varphi_s(\mathbf{r}) \equiv \varphi_{k\kappa m}(\mathbf{r}),\ \varepsilon_s \equiv p^2 > 0$ on the energy scale (in Ry):

$$\int_0^\infty \left(\xi_k(r)\xi_{k'}(r)P_k(r)P_{k'}(r) + \alpha^2 G_p(r)G_{p'}(r)\right)dr = \delta(p^2 - p'^2)\delta_{\kappa\kappa'}\delta_{mm'} \quad (7.16)$$

7.2 Numerical Model

The radial Eq. (7.3) for determining one wave function $P_s(r)$ can be written in a form similar to Eq. (5.26), except for the additional term $T(r)$ defined by Formula (7.9). The latter includes off-diagonal Lagrange multipliers, which ensure the orthogonality of the wave function $P_s(r)$, which we define, to the wave functions of the ground state $P_i(r),\ i = 1 \ldots s-1$. Because of these similarities, the structure of the program ***rfchf*** is close in general to the ***rscfhf*** program (Chap. 5) and includes two separate procedures. The first is intended to obtain solutions describing discrete states, and the second—for states of the continuous spectrum. Both procedures are based on the corresponding nonrelativistic analogs (see [3]), have common parts, and use the same radial grids (see Formulas (5.34) and (5.35)). In addition, in both cases, the same auxiliary transformations of radial functions (5.41) are used. The procedure in discrete spectrum is almost identical to the ***rscfhf*** program (Chap. 5), with the exception of a few expansion opportunities presented in it:

(a) additional acceleration of convergence in the exchange term,
(b) determination of off-diagonal parameters,
(c) the inclusion of the dependence on the parameter m_s, which allows us to consider particles other than electrons.

As in the nonrelativistic case, the procedures for the continuous spectrum differ from the discrete in the following way:

(a) they include phase shift $\delta_\kappa(p)$, which is calculated using the asymptotic relations (7.14) and (7.15). They determine at the same time the normalization of the wave functions of the continuous spectrum to the energy δ-function given by Formula (7.16),
(b) the one-particle energy $\varepsilon_s = p^2 > 0$ is determined in advance, since it is not an eigenvalue.

As with the ground state, this program starts by setting the potentials that is achieved by putting $P_s^{(0)}(r) = G_s^{(0)}(r) = U_s^{(0)}(r) = 0,\ \xi_s^{(0)}(r) = 1$. The initial approximation for the eigenvalue in the discrete spectrum is given by the simplest hydrogen-like formula:

$$\varepsilon_s^{(0)} = -\left[\left(Z - \sum_{j=1}^{s-1} q_j \Big/ n_s\right)\right]^2. \tag{7.17}$$

For continuous spectrum, this is a given value $\varepsilon_s^{(0)} = \varepsilon_s \equiv p_s^2$.

By definition, the off-diagonal parameters $\varepsilon_{sj},\ j = 1...s-1$ values that enter Eqs. (7.8) must be such that the overlap integrals

$$I_j \equiv \int_0^\infty \left(\xi_s(r)\xi_j(r)P_s(r)P_j(r) + \alpha^2 G_s(r)G_j(r)\right)dr' \tag{7.18}$$

be close to zero with a given accuracy for all pairs of subshells $(n_s\kappa_s, n_j\kappa_j)$, where $\kappa_s = \kappa_j,\ j = 1...s-1$. Note that, $\varepsilon_{sj}, \kappa_s \neq \kappa_j,\ j = 1...s-1$ are set equal to zero, since one-particle wave functions that differ in κ_j are mutually orthogonal (see (7.16)). At the beginning of the programs, $\varepsilon_j \equiv \varepsilon_{sj},\ j = 1...s-1$ are set as

$$\varepsilon_{j,1}^{(0)} = 0,\ \varepsilon_{j,\nu}^{(0)} = 0.9\varepsilon_{j,\nu-1}^{(0)},\ \nu = 2, 3, 4,...K_1 + K_2 \tag{7.19}$$

Here ν means $K1$ or $K2$, for wave functions in discrete and continuous spectrum, respectively, according to their energies, going from lowest to highest. A successful iterative value for $\varepsilon_{j,\nu}^{(1)}$ can be chosen as

$$\varepsilon_{j,\nu}^{(1)} = \varepsilon_{j,\nu}^{(0)} + I_j^{(1)} \tag{7.20}$$

Starting from the second iteration $\varepsilon_{j,\nu}^{(m)}$ $(m \geq 2)$

$$\varepsilon_{j,\nu}^{(m)} = \frac{I_j^{(m)}\varepsilon_{j,\nu}^{(m-1)} - I_j^{(m-1)}\varepsilon_{j,\nu}^{(m)}}{I_j^{(m)}\varepsilon_{j,\nu}^{(m-1)}},\ m = 2, 3, 4... \tag{7.21}$$

Equation (7.3) is solved using previously found values of the off-diagonal parameters, while the overlap integrals are calculated at the end of each iteration. This process continues until all the overlap integrals become less in absolute value than the specified accuracy $|I_j|/\|\varphi_s\| < \varepsilon, \quad j = 1, 2, \ldots, s-1$.

Calculations show that the iterative procedure for determining the off-diagonal parameters does not always converge. In particular, this can happen when these parameters are of the same order of magnitude (or greater) than the corresponding diagonal Lagrange multiplier, i.e., eigenvalues of energy. As noted above, there are two options for resolving this difficulty: One should either automatically control the calculation of off-diagonal parameters or introduce an acceleration of convergence.

According to the first variant, the calculation is performed as described above, i.e., using numerical iterative elaboration of ε_j. If convergence cannot be achieved using this method in a given number of iterations (usually ≤ 6), then the second option is automatically included. In the latter case, the off-diagonal parameters are expressed in accordance with [6] through the matrix elements of the Coulomb interaction. At the mth iteration, these matrix elements are expressed via the function $P_s^{(m-1)}(r)$ found at the previous iteration. As a result, we determine the wave functions of excited states (discrete and continuous spectra) for those configurations for which such a calculation using the first option did not give a result.

Apart from the features highlighted above, this ***rfchf*** program is similar to the ***fchf*** program [3]. Common characteristics of these programs include the same in:

(a) calculating of the exchange terms,
(b) determination of the normalization term and phase shift for the wave functions of the continuous spectrum,
(c) numerical integration outside the radial grid.

A detailed account of this can also be found in Chap. 6, as well as in Refs. [3, 6–9].

References

1. D. Hartree, *The Calculation of Atomic Structure* (Wiley, New York, 1957)
2. L.V. Chernysheva, N.A. Cherepkov, V. Radojevič, Self-consistent field Hartree-Fock program for atoms. Comp. Phys. Comm. **11**, 57 (1976)
3. L.V. Chernysheva, N.A. Cherepkov, V. Radojevič, Frozen-core Hartree-Fock program for atomic discrete and continuum states. Comp. Phys. Comm **18**, 87 (1979)
4. C. Froese Fischer, Comp. Phys. Comm. **1**, 151 (1970)
5. V.A. Dsuba, O.P. Sushkov, V.V. Flambaum, *The Computer Package for Calculations of Atomic Wavefunctions and Level Energies* (The Institute for Nuclear Physics of the USSR Academy of Sciences, Siberian division, No 82–89, Novosibirsk, 1982) (in Russian)
6. L.V. Chernysheva, M.Yu. Kuchiev, V.L. Yakhontov, *Numerical Calculations of Ground-state Atomic Wavefunctions in the Hartree-Fock-Dirac Approximation* (Ioffe Institute, No 1015, Leningrad, 1986) (in Russian)
7. L.V. Chernysheva, M.Ya. Amusia, V.F. Orlov, S.K. Semenov, N.A. Cherepkov, *Numerical Calculations of Electron Wavefunctions in the Frozen-core Field with Various Input of the Off-diagonal Energy Parameters* (Ioffe Institute, No 1319, Leningrad, 1989) (in Russian)
8. M.Ya. Amusia, L.V. Chernysheva, *Automated System for Studying the Structure of Atoms* (Nauka, Leningrad Division, 1983). ((In Russian))
9. L.V. Chernysheva, V.L. Yakhontov, Two-program package to calculate the ground and excited state wave functions in the Hartree-Fock-Dirac approximation. Comput. Phys. Commun. **119**, 2–3, 232 (1999); Preprint: Germany, Heidelberg, HD-PY 96/06.

Chapter 8
Choice of Wave Functions

Each specific physical problem requires the choice of the initial basis of wave functions for its consideration. Obtaining the characteristics of a physical process requires knowledge of the wave functions in the ground and excited states. The field in which the excited particle is located can be described in various ways. In order to take into account the residual interaction V_{res} in the framework of perturbation theory or the random phase approximation with exchange (RPAE), the one-electron wave functions of occupied and empty states must represent a complete set of mutually orthogonal normalized functions. This chapter discusses some possible options for choosing the wave functions of excited states of atoms and simple molecules for different ways of accounting for the electron–electron interaction [1–3].

8.1 Theoretical Model

8.1.1 Spherical Symmetry (Atoms)

The wave functions of the ground state of an atom or ion are calculated in the Hartree–Fock approximation by formulas (4.2)–(4.7) or in the Hartree–Fock–Dirac approximation by formulas (5.1)–(5.6). The wave functions of excited (unbound) states can be found in the Hartree–Fock approximation, or in a self-consistent field by formulas (4.2)–(4.7), as well as in the field of the "frozen core" in the Hartree–Fock approximation for an electron (6.1)–(6.9), meson (6.60), and positron (6.60) or in the Hartree–Fock–Dirac approximation, solving Eqs. (7.3–7.9).

Two approaches are used to select the wave functions of the excited state of an atom. In the "frozen core" approximation, it is assumed that during the excitation process, only the state of the electron directly involved in the process changes, while the states of the rest of the electrons remain unchanged, i.e., "frozen." The second

M. Ya. Amusia and L. V. Chernysheva, *Computation of Atomic and Molecular Processes*,
Springer Series on Atomic, Optical, and Plasma Physics 117,
https://doi.org/10.1007/978-3-030-85143-9_8

approach permits to include rearrangement effects, i.e., all changes in the state of atomic electrons (or electrons of an ion) that occur due to the excitation or ionization of an atom.

The ground state in the frozen core approximation is obtained from Eqs. (4.2)–(4.7). If i in (4.3) corresponds to the occupied states ($i \le F$), then δ_{ij} can be equal to 1, and the second term on the left in (4.3) actually describes the field of $N_e - 1$ atomic electrons, i.e., all but the considered i-electron. The reason for this is that the electron's action on itself (i.e., self-action) must be excluded. Therefore, at large distances from the atom, the self-consistent field depends on r as $(Z - N_e + 1)/r$.

The wave functions of the excited state of an atom in a frozen core are solution of one Eq. (6.2), in which it is taken into account that the excited electron moves in the field of N_e-electrons. Therefore, its wave function is denoted as $P_i^{N+1}(r)$ (or $\phi_i^{n+1}(x)$), if we substitute functions $P_i^{N+1}(r)$ in relation (4.2)). In this case, the field at large distances from the atom behaves like $-(Z - N_e)/r$. So, for a neutral atom, the $(N_e + 1)$th electron turns out to be in a short-range field, of the order of the atom size radius, and, as a rule, does not have discrete excited states. The wave functions $P_i^{N+1}(r)$, at $i > F$, and $P_i(r)$, at $i \le F$, are solution of the same system of equations. Therefore, they are mutually orthogonal and form a complete set in the sense that

$$\sum_{i \le F} P_i(r)P_i(r') + \sum_{i>F} P_i^{N+1}(r)P_i^{N+1}(r') = \delta(r - r'). \tag{8.1}$$

Accounting for residual interactions is based on these functions. However, it is not convenient to use ϕ^{N+1} in numerical calculations, since the matrix elements of the Coulomb interaction $\langle \nu_1 \nu_4 | V | \nu_3 \nu_2 \rangle$, diagonal in hole states, diverge at $\nu_4 = \nu_2 \le F$ and $\nu_3 = \nu_1 > F$. Indeed, in a matrix element of the form $\langle \nu_1 \nu_4 | V | \nu_3 \nu_2 \rangle$ (for small in comparison with the ionization potential and positive E_1, E_3), we have

$$\langle \nu_1 \nu_4 | V | \nu_3 \nu_2 \rangle \equiv \int \phi_{\nu_1}^{*N+1}(x) \phi_{\nu_2}^{*}(x') \frac{1}{|r - r'|} \phi_{\nu_3}^{N+1}(x) \phi_{\nu_2}(x') dx dx' \tag{8.2}$$

The main contribution comes from the region of large values $r (r \gg r')$, so that

$$\langle \nu_1 \nu_4 | V | \nu_3 \nu_2 \rangle \approx \int \phi_{\nu_1}^{*N+1}(x) \frac{1}{r} \phi_{\nu_3}^{N+1}(x) dx = \int_0^\infty P_{\nu_1}^{*N+1}(r) P_{\nu_3}^{N+1}(r) \frac{dr}{r} \tag{8.3}$$

This integral diverges at $E_1 = E_3 > 0$, since at large r the function $P_{\nu_1}^{N+1}(r)$ behaves like (6.14), i.e., $|P_{\nu_1}^{N+1}(r)|^2$ does not decrease with growth r. The indicated divergence can be avoided by summing up all the "forward in time" diagrams diagonal in hole states. This is achieved by choosing those one-electron wave functions of the excited state that are solutions of Eqs. (6.2)–(6.5), where the first sum (6.2) does not contain the term corresponding to the state of the hole, so that the field acting on the excited electron is created by the $(N_e - 1)$ electron of the residual ion.

Defined in such a way, the wave functions of the excited state are not orthogonal to the wave functions of the ground state. We denote these functions as $P_i^N(r)$, or $\phi_i^N(r)$, if substituted $P_i^N(r)$ in (4.1), where i is a level from which the electron is removed. At large distances from the atom, the field acting on an electron with a wave function P_i^N behaves like $-(Z - N_e + 1)/r$, i.e., the same as for the wave function of the ground state.

By choosing the proper one-electron wave functions, one can take into account a much broader class of diagrams, namely all those in which the hole interacts with the electron, and its state does not change its energy, but only the projections of the spin σ and angular momentum m can change. This is achieved by appropriately specifying the coefficients in (6.2). We denote such radial functions as $P^{N(LS)}$ or $\phi^{N(LS)}$, if substituting $P^{N(LS)}$ in (4.1). The index $N(LS)$ emphasizes that the total angular momentum and spin of the electron–hole pair L and S are fixed, while the projections of angular momentum and spin for the hole and the electron are not specified. The functions $\phi^{N(LS)}$ differ from ϕ^N, since in the latter the state of the hole, given by all four quantum numbers, is fixed. At large distances from the atom, the field acting on the electron, described by the wave functions P^N or $P^{N(L,S)}$, decreases as $-(Z - N_e + 1)/r$. For a neutral atom ($N_e = Z$) at large distances, this field is purely Coulomb $(-1/r)$. In such a field, in contrast to the case with P^{N+1}, there are infinite number of discrete levels "condensing" to energy $E = 0$ at $n \to \infty$.

In general, $P^{N(L,S)}$ are not orthogonal to the wave functions of the ground state, but they can be orthogonalized in the same way as P^N using the off-diagonal parameters in (6.2).

Within RPAE frame, the use of orthogonalized functions is equivalent to the summation of all "time-forward" diagrams diagonal along the hole, for example (1.50a, c), at $\nu_6 = \nu_4 = \nu_2$.

The choice of the orthogonal functions $\phi^{N(LS)}$ corresponds to the summation of a wider class of "time-forward" diagrams, namely simplified RPAE (SRPAE) diagrams, in which the hole states differ in angular momentum m and spin projections σ and have the same energy. As, for example, in (1.50a, c) with $E_4 = E_6 = E_2$. We emphasize once again that the functions ϕ^N and $\phi^{N(LS)}$ account for all the divergent matrix elements of the Coulomb interaction of an electron and a hole. However, in the calculations with them, a certain caution is required in order to avoid double counting of the same correlation corrections.

All the above-mentioned one-electron wave functions of excited states are calculated in a fixed field of all other atomic electrons. In reality, the removal of an electron or the excitation of an atom leads to alteration in the states of all electrons, and not just the one that is directly removed. This can be to large extent included in the framework of the Hartree–Fock approximation with an "unfrozen" core. In this case, the problems of ionization and excitation are solved in different ways.

In the case of ionization, the wave functions of the residual ion in a self-consistent field are calculated by solving the system of Eqs. (4.2)–(4.7), where the actual ionization is taken into account by specifying the configuration. The obtained one-particle functions of the electrons of the residual ion are used in Eqs. (6.2)–(6.9) to specify the field, in which the removed electron moves. The wave function of the removed electron corresponds to a certain total angular momentum and spin of an electron–hole pair, if the projections of the moment and spin of this electron are not fixed. This function is analogous to $\phi^{N(LS)}$ in the case of a "frozen" core, and the field in which the electron is located at large distances from the atom behaves like $-(Z-N+1)/r$.

In the case of excitation of an atom, the rearrangement within the framework of the generalized RPAE (GRPAE) ("unfreezing" of the core) is taken into account by solving the system of Eqs. (4.3) for an atom in an excited state, where the actual excitation is given by the configuration of the atom.

In both cases considered, the one-electron wave functions, after accounting for the rearrangement turn out to be not orthogonal to the wave functions of the initial state. The orthogonality condition can be essential in some problems and is ensured by inclusion of the off-diagonal parameters in (6.2).

Considering the process that changes the charge of the atomic nucleus, i.e., β-decay, the corresponding atomic wave functions should be found by solving Eq. (4.3) with a new charge.

8.1.2 Axial Symmetry (Diatomic Molecules)

The wave functions of the ground state of diatomic molecules, as well as for atoms, are obtained in the Hartree–Fock approximation (see Chap. 4). As already indicated in Sect. 4.1.2, the wave functions of the ground state obtained by solving Eqs. (4.23 and 4.24) with coefficients (4.25) are automatically orthogonal for molecules with closed shells.

To determine the wave functions of excited states in the Hartree–Fock approximation, two approaches are used, similar to the case of atoms. However, in both approaches, these functions are found from Eqs. (6.18) with coefficients (6.19). The difference lies in the possibility of choosing the functions of the core.

If in Eq. (6.18) the wave functions of the ground state are used as the core functions, then this corresponds to the approximation of the "frozen" core. Such excited states wave functions are orthogonalized to the functions of the ground state using off-diagonal parameters. Thus, the wave functions of the ground and excited states form a complete orthogonal set required to obtain results in the RPAE. However, as calculations showed, the "frozen core" approximation does not give good agreement with experiment for photoionization of the inner shells of a molecule ($1s$ shells of atoms in a molecule) even with the subsequent use of RPAE.

To account for the rearrangement of the core during ionization of the inner shells, we used the following approach (see [4]), which generalizes the core rearrangement method, or the Rearrenged Core Hartree–Fock approximation (RCHF). In addition to the wave functions of the ground state, the wave functions of the rearranged core are found as solutions of Eqs. (4.23 and 4.24), where the following values are used instead of the values of the coefficients (4.25):

$$a_{i',i'} = 2(1 - z_e), \quad b_{i',i'} = 2(1 - z_e);$$
$$a_{i,i'} = 2 - z_e, \quad b_{i,i'} = 0.5(2 - z_e), \quad i \neq i'; \tag{8.4}$$
$$a_{i',j} = 2, \quad b_{i',j} = 1, \quad j \neq i'.$$

Here, $z_e(0 < z_e < 1)$ is an adjustable parameter that can be determined based on ensuring the best agreement with experiment. In our calculations, it was usually determined from the condition that the position of the shape resonance peak (σ^* shape resonance). coincided with the experiment. Note that $z_e = 0$ corresponds to the "frozen core" approximation, while $z_e = 1$ corresponds to the RCHF approximation. The core wave functions with coefficients (8.4) are orthogonalized using off-diagonal energy parameters. These wave functions are further used as core functions in relations (6.23) when calculating the wave functions of excited states. In this case, with the help of off-diagonal energy parameters, the functions of excited states are orthogonalized to the functions of the ground state of the molecule. Thus, the wave functions of the ground and excited states again form a complete orthonormal set, which is further used, if necessary, for the RPAE.

All Hartree–Fock coefficients are input data for programs ***scfhfm***, ***fchfm***. The calculation variant is specified by one integer parameter (see Chap. 22).

8.2 Numerical Model

When carrying out calculations, the choice of certain wave functions of an electron (positron, meson, proton) is determined by specifying the coefficients in the Hartree–Fock equations through a number of parameters. Let us consider the possible options.

1. The ground state of the atom, programs ***scfhf***, ***scfhfw*** (when specifying data, we put $ksu = 1$). Calculation formulas in *LS*-coupling for direct coefficients γ' (indicated in programs F_1) are determined by formula (4.11), for exchange coefficients γ (indicated in programs G_1)—(4.12). Within the framework of the spin-polarized method, we have, respectively, $F_4 = F_1$, $G_4 = 2G_1$. In this case, calculations make it possible to find the coefficients of the mean term of an arbitrary configuration or a fixed term (L and S are given in the *LS* coupling method or L in the spin-polarized method) for the occupied shells. The approach is similar for closed subshells plus one electron or for closed subshells plus one subshell with a hole.
2. Excited state is calculated by the ***scfcf***, ***scfhfw*** programs (sign $ksu = 2$). Possible cases: closed subshells; closed subshells plus one electron; closed subshells plus a single hole subshell; closed subshells plus two open ones: one with a hole, the other with one electron. In all cases, the term is fixed. In this case, the coefficients F_1 and F_4 describe the interaction inside the closed subshell, and G_1 and G_4 describe the interaction of any subshell with the closed one. In the *LS*-coupling scheme, the interaction of a particle with a hole is characterized by

Table 8.1 Coefficients and formulas used when calculating wave functions

Sign *ksu*	*LS*-coupling $S_1 = 2S + 1, L_1 = L$	Spin-polarized method ($S_1 = 111$)
1	F_1—(4.10), G_1—(4.11)	$F_4 = F_1, G_4 = 2G_1$
2	F_1—(4.10), G_1—(4.11) F_3—(4.12), G_3—(4.13)	$F_4 = F_1, G_4 = G_1$ $F_5 = F_3,$ G_5—(4.83)
3	F_3—(4.12), G_1—(4.11), G_3—4.13)	$F_5 = F_3, G_4 = 2G_1$ G_5—4.83)
3 (meson, positron)	F_3—(4.12)	

the coefficients F_3 and G_3, and in the spin-polarized method, by the coefficients $F_5 = F_3$ and G_5.

3. Excited state in the field of the frozen core, programs *fchf, fchfw* ($ksu = 3$). Possible cases: closed subshells plus one electron; closed subshells plus one with a hole and one electron in the other. In both cases, the term is fixed. Only the interaction of one excited electron with all subshells is considered; therefore, only the coefficients G_1, F_3, G_3 are calculated, and for the polarized spin G_4, $F_5 = F_3$, G_5. For mesons and positrons, only the coefficients F_3 are calculated. For the convenience of using the formulas, everything stated is summarized in the table. The relationship with the parameters introduced in Chap. 4 is indicated by the formulas in Table 8.1.

In the HF approximation for the averaged term (***scfhfaver*** program), for any distribution of electrons over subshells, the coefficients are specified in the ***coefav*** procedure, with the parameter $S_1 = 1$. For the averaged term, the HF wave functions of the excited state are found using the ***fchfaver*** program. In this case, the coefficients are determined by the formulas given in the table, but with an additional factor proportional to the number of electrons of the subshell, and also the parameter $S_1 = 1$.

When calculating the wave functions of particles (meson, positron, proton), everything described above remains valid. Additionally, the parameter *mu* is set, taking into account the mass and charge of the particle: $mu = 1$ for an electron, $mu = -1$ for a positron, $mu \simeq 201.1$ for a negative meson, $mu \simeq -201.1$ for a positive meson, and $mu \simeq -1836$ for a proton. This parameter in the program provides the solution of the Hartree equation for the meson, positron, and proton, i.e., in equations where there are no exchange coefficients.

In the GRPAE, parameters L_1 and S_1 acquire the same values as described above. Taking into account the rearrangement in the ground state, an electron is removed from the selected shell. The wave function of an electron in an excited state in a fixed field is calculated in the field of a core with a new configuration using the ***fchf*** program.

In the Hartree–Fock–Dirac approximation for an atom with closed shells, L_1 is the total orbital angular momentum and S_1 is the total spin in the initial data to the ***rscfhf*** program, in which the coefficients are calculated by formulas (5.20)–(5.22), and for excited states by ***rfchf*** program according to the same formulas.

References

1. L.V. Chernysheva, N.A. Cherepkov, S.I. Sheftel, *Program for calculating the coefficients in the Hartree-Fock equations* (Ioffe Institute, No. 760, Leningrad, 1982). (In Russian)
2. M.Ya. Amusia, L.V. Chernysheva, *Automated System for Studying the Structure of Atoms* (Nauka, Leningrad Division, 1983). ((In Russian))
3. M.Ya. Amusia, L.V. Chernysheva, *Computation of Atomic Processes* (Institute of physics Publishing, Bristol and Philadelphia, 1997)
4. S.K. Semenov, N.A. Cherepkov, T. Jahnke, R. Dorner, J. Phys. B. **37**, 1331 (2004)

Chapter 9
Photoionization Cross Section, Oscillator Strengths of Discrete Transitions, Polarizability

This chapter contains the basic formulas that are used in calculating the characteristics of the process of atoms and simple molecules interaction with electromagnetic radiation. The necessary details of the calculation technique are given for objects with closed and open subshells and levels, as well as with semi-closed subshells. Formulas are given for the numerical integration of matrix elements, including expressions containing first-order poles in the integrand. The latter is especially important when considering discrete excitations.

9.1 Theoretical Model

In accordance with the general approach presented in Chap. 1, the initial approximation for calculating the photoionization cross sections and oscillator strengths is the Hartree–Fock method. A subsequent and very significant refinement of the results is achieved by using the random phase approximation with exchange (RPAE).

There is not only a terminological difference, but also a difference in essence between the random phase approximation (RPA), widely used in the theory of metals or condensed matter, and RPAE as applied to atoms and molecules. The difference is that within the framework of the RPAE, the exchange of atomic electrons is properly accounted for, and RPAE is a consistent generalization of the HF approximation to the case of weak external fields.[1]

In the Hartree–Fock method [1, 2], the wave functions of the initial and final states are represented in the simplest case as a single Slater determinant, and in the general case as a combination of several such determinants [2]. The total photoionization cross section is determined by the expression:

[1] Sometimes, unfortunately, the abbreviation RPA is used along with RPAE, even when it means RPAE.

M. Ya. Amusia and L. V. Chernysheva, *Computation of Atomic and Molecular Processes*,
Springer Series on Atomic, Optical, and Plasma Physics 117,
https://doi.org/10.1007/978-3-030-85143-9_9

$$\sigma(\omega) = \sum_{\nu_1 \nu_2} \sigma_{\nu_1 \nu_2}(\omega), \tag{9.1}$$

where $\sigma_{\nu_1 \nu_2}(\omega)$ is the *partial* cross section for the transition of an electron from a discrete state ν_1 to a continuous spectrum state ν_2. In the general case, the partial cross section for photoionization is:

$$\sigma_{\nu_1 \nu_2}(\omega) = 4\pi\alpha a_0^2 \omega \sum_{\mu=-1,0,1} \left| e_\mu \langle \Phi_f | d_\mu | \Phi_i \rangle \right|^2, \tag{9.2}$$

where e_μ is the spherical projections of the photon polarization vector. The corresponding projections of the dipole operator in the length form are defined as

$$d_\mu^L = \begin{cases} \frac{1}{\sqrt{2}}(x \pm iy), \ \mu = \pm 1, \\ z = r\cos\theta, \ \mu = 0, \end{cases} \tag{9.3}$$

or, in another notation,

$$d_\mu^L = r\sqrt{\frac{4\pi}{3}} Y_{1\mu}(\theta, \varphi). \tag{9.4}$$

The projections of the dipole operator in the velocity form are determined by the relations:

$$d_\mu^V = \begin{cases} \frac{1}{\omega\sqrt{2}}\left(\mp\frac{\partial}{\partial x} - i\frac{\partial}{\partial y}\right), \ \mu = \pm 1 \\ \frac{1}{\omega}\frac{\partial}{\partial z}, \ \mu = 0 \end{cases} \tag{9.5}$$

9.1.1 Spherical Symmetry (Atoms)

In the case of spherical symmetric potential, it is sufficient to use only one projection of the dipole operator on the z-axis, then multiplying the result by 3. Additionally, summation is performed over the projections of the orbital angular momentum. In this case, formula (9.2) is simplified and written in the form:

$$\sigma_{\nu_1 \nu_2}(\omega) \simeq 2.689 \cdot \frac{\omega N_1}{2l+1} |d_{\nu_1 \nu_2}|^2, \quad \text{in Mb}, \tag{9.6}$$

where ω is the photon energy, $E_2 = \omega + E_1$. Note that ν corresponds to only two quantum numbers $\nu \equiv n(E), l$.

Excitation of an electron by a photon to a discrete state is characterized by the oscillator strength

$$f_{\nu_1\nu_2} = \frac{\omega_{\nu_1\nu_2} N_1}{3(2l+1)} |d_{\nu_1\nu_2}|^2, \tag{9.7}$$

where $\omega_{\nu_1\nu_2} = E_2 - E_1$, N_1 is the number of electrons in the ν_1-subshell, $d_{\nu_1\nu_2}$ is the reduced dipole matrix element in the Hartree–Fock or RPAE approximation. The constant 2.689 in (9.6) is the numerical value of the quantity $(4\pi^2\alpha a_0^2/3) \times 10^{18}$ in CGSE units, where $\alpha = e^2/\hbar c \approx 1/137$ is the fine structure constant, and $a_0 = \hbar^2/me^2$ is the Bohr radius ($\alpha = 1/c$ and $a_0 = 1$ in atomic units accepted in this book).

A significant difference arises when the photoionization cross section for atoms with an open shell $(n_1 l_1)^N$ is considered, where N is the number of electrons in this shell, $(N < 4l_1 + 2)$. In the *LS*-coupling approximation, the ground state of an atom can be described by the total angular momentum L and total spin S, while the state of the ion and the entire system "ion + photoelectron" in the final state can be described by quantum numbers $L'S'$ and $L_1 S_1$, respectively. Then the photoionization cross section for the process

$$(n_1 l_1)^N (LS) + h\nu \rightarrow (n_1 l_1)^{N-1} (L'S') E_2 l_2 (L_1 S_1)$$

is defined as follows:

$$\sigma(\omega) = \frac{8\pi^2 \alpha a_0^2 \omega}{3[L]} \left| \langle L'S', \nu_2 L_1 S_1 \| d \| \nu_1 LS \rangle \right|^2 \left(10^{-18}\ \text{cm}^2\right), \tag{9.8}$$

where $[L] \equiv 2L + 1$.

The photoionization cross section, oscillator strengths, and polarizability are calculated in the single-configuration Hartree–Fock approximation and in the RPAE in the length and velocity forms. Since these two forms within the RPAE should coincide, when the influence of all atomic subshells is accounted for, their difference can be used to assess both the total numerical error of the results obtained and the influence of transitions from other subshells, except for the ionizable one.

In the framework of the RPAE, the photoionization cross section is calculated including the intra- and intershell correlations. The transition is characterized by the specification of a vacancy $n_1 l_1$ and an excited state $n_2(E_2)l_2$, where n_2—denotes a discrete level and E_2—a state in the continuous spectrum.

The dipole dynamic polarizability is defined as

$$\alpha(\omega) \approx \sum_{\nu_1\nu_2} \frac{4f_{\nu_1\nu_2}}{\omega^2_{\nu_1\nu_2} - \omega^2} + \frac{1}{\pi^2 \alpha a_0^2} \int\limits_I^\infty \frac{\sigma(\omega')\omega'}{\omega'^2 - \omega^2}. \tag{9.9}$$

where $I = |E_0|$ is the ionization threshold.

9.1.2 Axial Symmetry (Diatomic Molecules)

In the case of axial symmetry, formulas (9.1) and (9.2) correspond to two quantum numbers—the projection of angular momentum m and energy E. For continuous spectrum, a number λ is added to this, which enumerates the functions of the continuous spectrum at a given energy E. The number λ is analogous to orbital angular momentum ℓ, which is not a good quantum number for the case of axial symmetry. In what follows, we will assume that $\nu \equiv (m, E, \lambda)$, having in mind that λ is not used for the discrete spectrum.

For the case of axial symmetry, it is necessary to use all three components of the dipole operator. In this case, formulas (9.4) are written in elongated spheroidal coordinates in the form:

$$\begin{aligned} d^L_{\pm 1} &= \frac{R_z}{2}\sqrt{\frac{4\pi}{3}}(\xi^2 - 1)^{1/2} Y_{1,\pm 1}(\eta, \varphi), \\ d^L_0 &= \frac{R_z}{2}\xi\eta. \end{aligned} \tag{9.10}$$

Instead of (9.5) in spheroidal coordinates, we have:

$$\begin{aligned} d^V_{\pm 1} &= \frac{2}{R_z(\xi^2 - \eta^2)} Y_{1,\pm 1}\left[\xi(\xi^2 - 1)^{1/2}\frac{\partial}{\partial \xi} - \eta(\xi^2 - 1)^{1/2}\frac{\partial}{\partial \eta}\right] \\ &\quad + \frac{1}{(1-\eta^2)(\xi^2 - 1)^{1/2}}(\xi^2 - \eta^2) Y_{1,\pm 1}\frac{\partial}{\partial \varphi}, \\ d^V_0 &= \frac{2}{R_z(\xi^2 - \eta^2)}\left[\eta(\xi^2 - 1)\frac{\partial}{\partial \xi} + \xi(1 - \eta^2)\frac{\partial}{\partial \eta}\right]. \end{aligned} \tag{9.11}$$

Accordingly, the dipole matrix element has an additional index $\mu = -1, 0, 1$ characterizing the projection of the dipole operator. In this case, the formula for the partial ionization cross sections is written as:

$$\sigma_{\nu_1\nu_2}(\omega) = 2.689 \cdot \omega \sum_\mu |d_{\mu,\nu_1\nu_2}|^2, \quad \text{(M6)}, \tag{9.12}$$

where ω is the photon energy. For the oscillator strengths, respectively, we have:

$$f_{\nu_1\nu_2} = \frac{\omega}{3}\sum_{\mu} |d_{\mu,\nu_1\nu_2}|^2. \tag{9.13}$$

Formulas (9.12), (9.13) differ from the corresponding formulas for atoms (9.6), (9.7) by the absence of the term $N_1/(2l+1)$ and by the presence of additional summation over μ.

9.2 Numerical Model

9.2.1 *Dipole Matrix Elements in the Hartree–Fock Approximation*

9.2.1.1 Spherical Symmetry (Atoms)

The given matrix elements included in (9.6) and (9.7) are related to the photoionization amplitude by the following relation

$$d_{\nu_1\nu_2} \equiv \langle \nu_1 | d_{lm} | \nu_2 \rangle = (-1)^{l_1-m_1} \begin{pmatrix} l_1 & l & l_2 \\ -m_1 & m & m_2 \end{pmatrix} \langle \nu_1 \| d_l \| \nu_2 \rangle, \tag{9.14}$$

where

$$\begin{pmatrix} l_1 & l & l_2 \\ -m_1 & m & m_2 \end{pmatrix}$$

is the $3j$-symbol, $k_1, k_2 \equiv (nlm\sigma)$, $\nu_1\nu_2 \equiv [n(E)l]$. For partially closed shells, the reduced matrix element is determined by the relation

$$\left\langle L'S', \nu_2 L_1 S_1 \| d \| \nu_1 LS \right\rangle = \delta_{SS_1} C^{l_1 L_1 S_1}_{L'S', l_2 L_1 S_1} \langle \nu_2 \| d \| \nu_1 \rangle, \tag{9.15}$$

where

$$C^{l_1 LS\kappa}_{L'S', l_2 L_1 S_1} = (-1)^{l_1+l_2+L_1+L'+\kappa} \sqrt{\frac{N}{2}[L, L_1]}(l_1^{N-1} L'S', l_1 |\} \, l_1^N LS) \begin{Bmatrix} L & \kappa & L_1 \\ l_2 & L' & l_1 \end{Bmatrix} \tag{9.16}$$

and $(l_1^{N-1} L'S', l_1 |\} l_1^N LS)$ is the so-called genealogical coefficient.

The reduced dipole matrix element in the one-configuration Hartree–Fock approximation is defined in the length form as

$$d^{(r)}_{\nu_1\nu_2} \equiv \left\langle \nu_1 \left\| d^{(r)} \right\| \nu_2 \right\rangle = \sqrt{(2l_1+1)(2l_2+1)} \begin{pmatrix} l_1 & 1 & l_2 \\ 0 & 0 & 0 \end{pmatrix} \int\limits_0^\infty P_{\nu_1}(r) P_{\nu_2}(r) r dr, \quad (9.17)$$

and in the velocity form (∇-form) as

$$d^{(\nabla)}_{\nu_1\nu_2} \equiv \frac{2}{\omega} \left\langle \nu_1 \left\| d^{(\nabla)} \right\| \nu_2 \right\rangle = \frac{2}{\omega} \sqrt{(2l+1)(2l_2+1)} \begin{pmatrix} l_1 & 1 & l_2 \\ 0 & 0 & 0 \end{pmatrix}$$
$$\int\limits_0^\infty P_{\nu_1}(r) \left(\frac{d}{dr} \pm \frac{l_1+l_2+1}{2r} \right) P_{\nu_2}(r) \mathrm{d}r \quad (9.18)$$

Here, the (+) sign refers to $l_2 = l_1 + 1$, and the (−) sign refers to $l_2 = l_1 - 1$.

It is convenient to define the matrix in the electron–hole transition so that a certain state of the electron–hole pair $\alpha \equiv (\nu_1\nu_2)$, $\nu_1 \le F(\nu_2 > F)$ or $\nu_2 > F(\nu_1 \le F)$ would correspond to each row and (or) column of this matrix. Thus, the single-particle matrix elements of the dipole type $d_{\nu_1\nu_2}$ in the Hartree–Fock approximation or $D_{\nu_1\nu_2}(\omega)$ in the RPAE form a row (or column) in the matrix

$$d_\alpha \equiv d_{\nu_1\nu_2} = \langle \nu_1 \| d \| \nu_2 \rangle \quad (9.19)$$

The dipole matrix elements in formulas (9.15) and (9.16) are calculated with one-particle Hartree–Fock wave functions obtained by the ***scfhf*** program for the ground state of an atom ϕ^N, ϕ^{N+1}, $\phi^{N(LS)}$ and the ***fchf*** program for excited atomic states (see Chap. 6). It is possible to use the wave functions of excited states obtained by the rearrangement method. In the latter case, the ionization potential of the ith electron (i.e., the energy of removing an electron from the ith subshell) is not equal to its Hartree–Fock energy $|E_i|$, but it is taken either its experimental value or the Hartree–Fock energy, adjusted by accounting for the rearrangement of the atom due to its ionization (see Chap. 15).

In the ***scfhf*** and ***fchf*** programs, the functions are calculated at equidistant points using a new variable ρ that replaces r. Therefore, expressions (9.17) and (9.18) for the dipole matrix elements, taking into account the change of variables (4.42), have the form:

$$\left\langle \nu_1 \left\| d^{(r)} \right\| \nu_2 \right\rangle = (-1)^{l_>} \sqrt{l_>} \int\limits_{-\infty}^{\infty} \frac{r^3}{(\alpha r+\beta)^2} f_2(\rho) f_1(\rho) d\rho, \quad (9.20)$$

$$\left\langle \nu_1 \left\| d^{(\nabla)} \right\| \nu_2 \right\rangle = (-1)^{l_>} \sqrt{l_>} \int\limits_0^\infty \mathrm{d}\rho \frac{r^2 f_2(\rho)}{\alpha r+\beta} \left(\frac{d}{d\rho} \pm \frac{l_>}{\alpha r+\beta} - \frac{\beta/2}{(\alpha r+\beta)^2} \right) f_1(\rho), \quad (9.21)$$

where $l_> = max(l_1, l_2)$. Note that the coefficient in front of the radial integral in expressions (9.20) and (9.21) is

$$(-1)^{l_>}\sqrt{l_>} = \sqrt{(2l_1+1)(2l_2+1)}\begin{pmatrix} l_2 & 1 & l_1 \\ 0 & 0 & 0 \end{pmatrix}. \tag{9.22}$$

The values of the integrals in Eqs. (9.20) and (9.21) are calculated using the generalized Simpson formula:

$$\int_c^d y(x)x = \frac{d-c}{6m}\left[y_0 + 4y_1 + 2y_2 + 4y_3 + 2y_4 + \cdots + 4y_{2m-1} + y_{2m}\right] + R(y), \tag{9.23}$$

where $R(y)$ is the remainder that we neglect. The interval $[c, d]$ includes $2\ m$ equal line segments.

The derivative $f_j' \equiv df/d\rho|_{\rho=\rho_j}$ in formula (9.21) is expressed in terms of the function $f(\rho)$ values at five consecutive points:

$$\begin{aligned}
f_0' &= \frac{1}{12h}(-25f_0 + 48f_1 - 36f_2 + 16f_3 - 3f_4) + R(f), \\
f_1' &= \frac{1}{12h}(-3f_0 - 10f_1 + 18f_2 - 6f_3 - 3f_4) + R(f), \\
f_j' &= \frac{1}{12h}\left(f_{j-2} - 8f_{j-1} + 8f_{j+1} - f_{j+2}\right) + R(f),\ j = 2, 3, \ldots, N-2 \qquad (9.24) \\
f_{N-1}' &= \frac{1}{12h}(-f_{N-4} + 6f_{N-3} - 18f_{N-2} + 10f_{N-1} + 3f_N) + R(f), \\
f_{N-2}' &= \frac{1}{12h}(3f_{N-4} - 16f_{N-3} + 36f_{N-2} - 48f_{N-1} + 25f_N) + R(f),
\end{aligned}$$

where h is the constant step of integration in the variable ρ.

For a subshell containing only one electron ($N_1 = 1$), the calculation ends with the determination of the HF dipole matrix elements, through which the cross section and the oscillator strength are expressed.

9.2.1.2 Axial Symmetry (Diatomic Molecules)

In the case of axial symmetry, the variant of photoionization of diatomic molecules with closed shells in the ground state was considered, i.e., the case of spin–orbital singlets. In this case, the dipole matrix element in the length or velocity form is written in spheroidal coordinates as:

$$
\begin{aligned}
d_{l,\mu}^{L,V} &\equiv \langle f|d_{\mu}^{L,V}|i\rangle \\
&= \int \sum_{\lambda=m_i}^{L} X_{i,\lambda}(\xi)Y_{\lambda m_i}(\eta,\varphi)d_{\mu}^{L,V} \sum_{\lambda'=m_f}^{L} X_{f,l,\lambda'}(\xi)Y_{\lambda' m_f}(\eta,\varphi)(\xi^2-\eta^2)d\mathbf{r} \\
&= \sum_{\lambda=m_i}^{L}\sum_{\lambda'=m_f}^{L} \int \big[\langle l,\lambda|d_{\mu}^{L,V}|\lambda' m_f\rangle \xi^2 X_{i,\lambda}(\xi)X_{f,l,\lambda}(\xi) \\
&\quad - \langle l,\lambda|d_{\mu}^{L,V}\eta^2|\lambda' m_f\rangle X_{i,\lambda}(\xi)X_{f,l,\lambda}(\xi)\big]d\xi.
\end{aligned} \tag{9.25}
$$

Here operators $d_{\mu}^{L,V}$ are determined by formulas (9.10) and (9.11), and

$$
|i\rangle \equiv |\nu_1\rangle,\ |f\rangle \equiv |\nu_2\rangle \tag{9.26}
$$

corresponds to initial and final state of a molecule, respectively.

In the programs for calculating the wave functions of the ground and excited states, ***scfhfm*** and ***fchfm***, the values of the functions are found at equidistant points in the variable ρ (see (4.106)). After changing the variable, as well as performing other standard transformations, instead of formula (9.25), we get the following final expressions for the dipole matrix elements. For the length form we have:

$$
\begin{aligned}
d_{l,\mu}^{L} &= \frac{2}{\sqrt{3\pi}} \sum_{\lambda,\lambda'} \big[\langle \lambda' m_f|Y_{1,\mu}|\lambda m_i\rangle \xi^2 - \langle km|Y_{1,\mu}\eta^2|\lambda m_i\rangle\big] \\
&\quad \times \int \frac{X'_{i,\lambda}(\rho)X'_{f,l,\lambda'}(\rho)(\xi-c_2)^2}{(\xi^2-1)\big[(\xi-c_2)^2-c_1\big]} F_{\mu}(\rho)\mathrm{d}\rho,
\end{aligned} \tag{9.27}
$$

where

$$
F_0(\rho) = \xi = \frac{(\rho+c_2)+\sqrt{(\rho+c_2)^2-4(c_1+\rho c_2)}}{2}, \tag{9.28}
$$

$$
F_{\pm 1}(\rho) = \sqrt{\xi^2-1}. \tag{9.29}
$$

For the velocity form, we find:

$$
\begin{aligned}
d_{l,0}^{V} &= \frac{2}{\sqrt{3\pi}} \sum_{\lambda,\lambda'} (\lambda m_i|Y_{1,0}|\lambda' m_f) \int \frac{X'_{f,l,\lambda'}(\rho)(\xi-c_2)^4}{(\xi^2-1)\big[(\xi-c_2)^2-c_1\big]^2} \\
&\quad \times \big\{X'_{i,\lambda}(\rho)\big[2\xi + (\lambda(\lambda+1)-\lambda'(\lambda'+1))\xi\big] \\
&\quad + 2(\xi^2-1)\frac{(\xi-c_2)^2-c_1}{(\xi-c_2)^2}\frac{d}{d\rho}X'_{i,\lambda}(\rho)\big\}\mathrm{d}\rho,
\end{aligned} \tag{9.30}
$$

$$
\begin{aligned}
d_{l,\pm 1}^{V} = \sum_{\lambda,\lambda'} \langle \lambda m_i | Y_{1,\mp 1} | \lambda' m_f \rangle \int \frac{X'_{f,l,\lambda'}(\rho)(\xi - c_2)^4}{(\xi^2 - 1)\left[(\xi - c_2)^2 - c_1\right]^2} \\
\Big\{ X'_{i,\lambda}(\rho) \Big[2\sqrt{\xi^2 - 1} + 1/\sqrt{\xi^2 - 1} \\
+ \big(\lambda(\lambda + 1) - \lambda'(\lambda' + 1) + (m_i^2 - m_f^2)/(\xi^2 - 1) \big) \sqrt{\xi^2 - 1} \\
+ 2\xi \sqrt{\xi^2 - 1} \frac{\left[(\xi - c_2)^2 - c_1\right]}{(\xi - c_2)^2} \frac{d}{d\rho} X'_{i,\lambda}(\rho) \Big\} d\rho.
\end{aligned} \tag{9.31}
$$

The variable ξ in formulas (9.27), (9.29)–(9.31) is replaced by variable ρ in accordance with (9.28). Integration in the above formulas with respect to ρ in programs is carried out numerically using the Simpson formula.

9.2.2 Coulomb Matrix Elements

9.2.2.1 Spherical Symmetry (Atoms)

The reduced Coulomb matrix element of the electron–electron interaction is obtained from the Coulomb matrix element after separating the factor that depends on the orbital quantum numbers:

$$
\langle \alpha | V | \beta \rangle = \sum_{\lambda=0}^{\infty} \sum_{m=-\lambda}^{\lambda} (-1)^{m_1 + m_3 + m} \begin{pmatrix} l_1 & \lambda & l_2 \\ -m_1 & m & m_2 \end{pmatrix} \begin{pmatrix} l_3 & \lambda & l_4 \\ -m_3 & m & m_4 \end{pmatrix} V_{\alpha\beta}^{\lambda}. \tag{9.32}
$$

It is determined by the relation

$$
\begin{aligned}
V_{\alpha\beta}^{\lambda} = \langle \nu_1 \nu_4 \| V_\lambda \| \nu_2 \nu_3 \rangle - \sqrt{(2l_1 + 1)(2l_2 + 1)(2l_3 + 1)(2l_4 + 1)} \times \\
\begin{pmatrix} l_1 & \lambda & l_2 \\ 0 & 0 & 0 \end{pmatrix} \begin{pmatrix} l_3 & \lambda & l_4 \\ 0 & 0 & 0 \end{pmatrix} R_\lambda(\nu_1 \nu_2 \nu_3 \nu_4),
\end{aligned} \tag{9.33}
$$

where $\nu \equiv nl (\nu \leq F)$, or $\nu \equiv [n(E)l]\ (\nu > F)$.

The total reduced Coulomb matrix element is equal to the difference between the direct and exchange ones:

$$
\begin{aligned}
U_{\nu_1 \nu_2 \nu_3 \nu_4}^{\lambda} = \langle \nu_1 \nu_4 \| U_\lambda \| \nu_2 \nu_3 \rangle = \frac{N_1}{(2\lambda + 1)} V_{\nu_1 \nu_2 \nu_3 \nu_4}^{\lambda} \\
- (2\lambda + 1) \sum_{\lambda'} (-1)^{\lambda + \lambda'} \begin{Bmatrix} l_1 & \lambda & l_2 \\ l_4 & \lambda' & l_3 \end{Bmatrix} V_{\nu_1 \nu_3 \nu_2 \nu_4}^{\lambda'},
\end{aligned} \tag{9.34}
$$

where

$$\begin{Bmatrix} l_1 & \lambda & l_2 \\ l_4 & \lambda' & l_3 \end{Bmatrix}$$

is the $6j$-symbol (see, for example, [1]) and

$$R_\lambda(\nu_1\nu_2\nu_3\nu_4) = \int_0^\infty P_{\nu_1}(r)P_{\nu_2}(r)dr \int_0^r \frac{r_<^\lambda}{r_>^{\lambda+1}} P_{\nu_3}(r')P_{\nu_4}(r')dr'. \tag{9.35}$$

While calculating the Coulomb matrix element, it is convenient to pass from the two-dimensional to the one-dimensional integral

$$R_\lambda(\nu_1\nu_2\nu_3\nu_4) = \int_0^\infty P_{\nu_1}(r)P_{\nu_2}(r)Y^\lambda_{\nu_3\nu_4}(r)\frac{dr}{r} = \int_0^\infty P_{\nu_3}(r)P_{\nu_4}(r)Y^\lambda_{\nu_1\nu_2}(r)\frac{dr}{r}, \tag{9.36}$$

in which the function $Y^\lambda(r)$ defined by (4.5), (4.6) is found numerically by solving two first-order differential equations. The method of solution is described in Sect. 4.2.2. The one-dimensional radial integral (9.36) is calculated using the Simpson formula.

If one transition is taken into account, then of the two-particle matrix elements of the Coulomb interaction

$$V_{\alpha\beta} = \langle \nu_1\nu_3 \| V \| \nu_2\nu_4 \rangle, \; \alpha \equiv (\nu_1\nu_2), \; \beta \equiv (\nu_3\nu_4), \tag{9.37}$$

a matrix of size $2m \times 2m$ is formed, where m is the number of electron—hole pairs. Recall that a discrete row of points is taken in the continuous spectrum, as described in Sect. 4.2.3. The structure of the matrix of Coulomb matrix elements can be represented as

$$U = \begin{pmatrix} U_{\alpha\alpha} & U_{\alpha\overline{\alpha}} \\ U_{\overline{\alpha}\alpha} & U_{\overline{\alpha}\overline{\alpha}} \end{pmatrix} \tag{9.38}$$

where $U_{\alpha\alpha}$ is the submatrix $m \times m$. Indexes are defined as follows: $\alpha \equiv (\nu_1\nu_2)$, $\overline{\alpha} \equiv (\nu_2\nu_1)$, $\nu_2 > F$, $\nu_1 \leq F$. This matrix can be represented graphically in the following way:

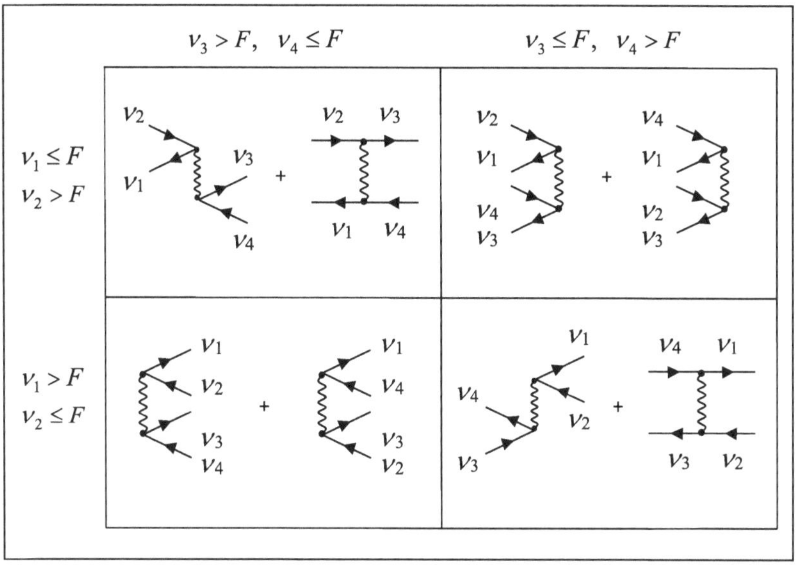

(9.39)

One can see that in each row the initial state $\nu_1\nu_2$ is fixed, while the final state $\nu_3\nu_4$ changes, whereas in the column it is vice versa; for a single transition $\nu_1 = \nu_4\nu_2 = \nu_3$.

For two channels, the structure of the matrix of Coulomb matrix elements can be represented as

$$U = \begin{pmatrix} U_{\alpha\alpha} & U_{\alpha\beta} & U_{\alpha\overline{\alpha}} & U_{\overline{\alpha}\overline{\beta}} \\ U_{\beta\alpha} & U_{\beta\beta} & U_{\beta\overline{\alpha}} & U_{\beta\overline{\beta}} \\ U_{\overline{\alpha}\alpha} & U_{\overline{\alpha}\beta} & U_{\overline{\alpha}\overline{\alpha}} & U_{\overline{\alpha}\overline{\beta}} \\ U_{\overline{\beta}\alpha} & U_{\overline{\beta}\beta} & U_{\overline{\beta}\overline{\alpha}} & U_{\overline{\beta}\overline{\beta}} \end{pmatrix} \tag{9.40}$$

where $U_{\alpha\alpha}$ is a submatrix $m \times m$, $U_{\beta\beta}$ is a submatrix $k \times k$, and indices are defined as $\alpha \equiv (\nu_1\nu_2)$, $\beta \equiv (\nu_3\nu_4)$, $\overline{\alpha} \equiv (\nu_2\nu_1)$, $\overline{\beta} \equiv (\nu_4\nu_3)$, and $\nu_2(\nu_3) > F,\ \nu_1(\nu_4) \le F$.

Usually, the wave functions of excited single-particle states ν_2 are calculated in the field of the remaining $N - 1$ electrons (self-consistent potential V^{N-1} created by $(N - 1)$-electron) with a vacancy ν_1.

Let the excited atomic state correspond to a certain *LS*-term. Then the part of the multielectron correlations of the RPAE described by the "forward in time" Coulomb matrix elements diagonal in the hole state (i.e., matrix elements $U_{\nu_1\nu_2\nu_3\nu_4}$ with $\nu_2 = \nu_4 \le F$ or $\nu_1 = \nu_3 \le F$) is taken into account by the very choice of the wave functions of the excited state (6.2). Therefore, when multielectron correlations are taken into account in the RPAE, in order to avoid double counting, an additional projection factor $(1 - n_{\nu_2}\delta_{\nu_2\nu_4})(1 - n_{\nu_1}\delta_{\nu_1\nu_3})$ should be included in expression (9.34) for the Coulomb matrix element, where n_ν is the Fermi step (1.53). This implies that the diagonal submatrices in (9.38) are equal to zero $U_{\alpha\alpha} = U_{\alpha\alpha} = 0$.

However, if the wave function of the final state ϕ^N is presented in the form of one Slater determinant and does not correspond to a certain *LS*-term, expression (9.34) for the Coulomb matrix element should be replaced by another one (let $\nu_2 \leq F$):

$$
\begin{aligned}
U^{\lambda}_{\nu_1\nu_2\nu_3\nu_4} &= \frac{N_1}{2\lambda+1}\sum_{\lambda'} V^{\lambda'}_{\nu_1\nu_2\nu_3\nu_4} \sum_{\text{all } m}\left(\delta_{\lambda\lambda'} - \frac{1}{2}\delta_{\nu_1\nu_3}\delta_{m_1m_3}\delta_{l_2l_4}\right) \\
&\quad\times \begin{pmatrix} l_1 & \lambda & l_2 \\ -m_1 & -m & m_2 \end{pmatrix}\begin{pmatrix} l_1 & \lambda' & l_2 \\ -m_1 & -m' & m_2 \end{pmatrix} \\
&\quad \begin{pmatrix} l_3 & \lambda & l_4 \\ -m_4 & -m & m_4 \end{pmatrix}\begin{pmatrix} l_3 & \lambda' & l_4 \\ -m_4 & -m' & m_4 \end{pmatrix} \\
&\quad - (2\lambda+1)\sum_{\lambda'}(-1)^{\lambda+\lambda'}\begin{Bmatrix} l_1 & \lambda & l_2 \\ l_3 & \lambda' & l_4 \end{Bmatrix} V^{\lambda'}_{\nu_1\nu_3\nu_2\nu_4} \\
&\quad \sum_{\text{all } m}\left(1 - \delta_{\nu_1\nu_3}\delta_{m_1m_3}\delta_{l_2l_4}\right)\begin{pmatrix} l_1 & \lambda' & l_3 \\ -m_1 & m & m_3 \end{pmatrix}^2 .
\end{aligned} \tag{9.41}
$$

For the off-diagonal submatrix in (9.38) (i.e., $\nu_2(\nu_3) > F$, $\nu_4(\nu_1) \leq F$ or $\nu_2(\nu_3) \leq F$, $\nu_1(\nu_4) > F$), Eq. (9.41) turns into (9.34). Matrix elements $d_{\nu_1\nu_2}$ (9.17), (9.18) and $U_{\nu_1\nu_2\nu_3\nu_4}$ (9.34) are also calculated with functions obtained in the field of the rearranged ion. Substituting them into the RPAE equations gives the amplitude of photoionization or photoexcitation in the so-called *generalized* RPAE (GRPAE). From the symmetry properties of the Coulomb matrix elements, relations follow, which, in fact, reflect the fact that electrons are identical particles:

$$
V_{\nu_1\nu_2\nu_3\nu_4} = V_{\nu_3\nu_4\nu_1\nu_2} \tag{9.42}
$$

and the fact that the matrix elements are real:

$$
V_{\nu_1\nu_2\nu_3\nu_4} = V_{\nu_2\nu_1\nu_4\nu_3} \tag{9.43}
$$

Similar relations are valid for matrix elements (9.34) and (9.41):

$$
\begin{aligned}
U_{\alpha\alpha} &= U_{\overline{\alpha}\overline{\alpha}}; \\
U_{\alpha\overline{\alpha}} &= U_{\overline{\alpha}\alpha}.
\end{aligned} \tag{9.44}
$$

Therefore, it is no need to calculate the complete matrix $2m \times 2m$ (9.38), but it is enough to find its part of size $m \times 2m$ containing only the submatrices $U_{\alpha\alpha}$ and $U_{\overline{\alpha}\alpha}$.. When calculating the elements of these submatrices, the symmetry properties (9.42), (9.43) are taken into account, which saves computer time. It follows directly from these properties that for a fixed hole the submatrices are symmetric, therefore the matrix elements are calculated only on one side of the diagonal, including the diagonal elements. The elements of each submatrix are calculated row by row, starting at the diagonal element. The function $Y_{\nu_1\nu_2}(r)$ under the integral in (9.36) is the same for the entire row. Therefore, it is calculated for a given row once.

9.2.2.2 Axial Symmetry (Diatomic Molecules)

In contrast to the case of central symmetry, with axial symmetry there is no degeneracy in the quantum number m—the projection of the orbital angular momentum. Therefore, the Coulomb matrix elements are calculated in their original notation.

The elements of the Coulomb matrix are written here in the form:

$$\begin{aligned} \langle \nu', \nu | U | i', i \rangle &= \langle \nu', \nu | V | i', i \rangle - \langle \nu', \nu | V | i, i' \rangle, \\ \langle i', \nu | U | \nu', i \rangle &= \langle i', \nu | V | \nu', i \rangle - \langle i', \nu | V | i, \nu' \rangle, \end{aligned} \tag{9.45}$$

where

$$\langle i, j | V | m, n \rangle = \iint \varphi_i^*(\mathbf{r}) \varphi_j^*(\mathbf{r}') \left| \mathbf{r} - \mathbf{r}' \right|^{-1} \varphi_m(\mathbf{r}) \varphi_n(\mathbf{r}') d\mathbf{r} d\mathbf{r}'. \tag{9.46}$$

The indices i, i' in (9.45) denote the sets of quantum numbers needed to identify the wave functions of the shells in the ground state, i.e., possible initial states of the electron,

$$|i\rangle \equiv |m_i, \varepsilon_i\rangle, \ \left|i'\right\rangle \equiv \left|m_i', \varepsilon_i'\right\rangle. \tag{9.47}$$

The indices ν, ν' denote the sets of quantum numbers required to identify the wave functions in excited states or continuous spectrum:

$$|\nu\rangle \equiv \left|\lambda_f, m_f, \varepsilon_f\right\rangle, \ \left|\nu'\right\rangle \equiv \left|\lambda_f', m_f', \varepsilon_f'\right\rangle. \tag{9.48}$$

Naturally, there is no index λ_f for discrete excited states.

The calculation of the integrals (Coulomb matrix elements) in formula (9.46) is carried out as follows. Formula (9.46) is written as:

$$\langle i, j | V | k, n \rangle = \int \varphi_i^*(\mathbf{r}) \varphi_k(\mathbf{r}) J_{nj}(\mathbf{r}) d\mathbf{r}. \tag{9.49}$$

Integral

$$J_{nj}(\mathbf{r}) = \int \varphi_j^*(\mathbf{r}') \left| \mathbf{r} - \mathbf{r}' \right|^{-1} \varphi_n(\mathbf{r}') d\mathbf{r}' \tag{9.50}$$

calculation method by expanding it in spherical harmonics in spheroidal coordinates has already been considered in Chaps. 4 and 6. As a result we have, similarly to

(6.17):

$$J_{n,j}(\mathbf{r}) = \sum_{\lambda=m_{nj}}^{L} Z_{nj,\lambda}(\xi) Y_{\lambda m_{nj}}(\eta, \phi), \; m_{nj} = m_n - m_j. \tag{9.51}$$

Since all wave functions are also calculated as expansions in spherical harmonics, formula (9.49) can be rewritten as

$$\begin{aligned}\langle i, j|V|k, n\rangle &= \int \sum_{\lambda=|m_i|}^{L} X_{i,\lambda}(\xi) Y^{*}_{\lambda m_i}(\eta, \phi) \sum_{\lambda'=|m_k|}^{L} X_{k,\lambda'}(\xi) Y_{\lambda' m_k}(\eta, \phi) \\ &\times \sum_{\lambda''=|m_{nj}|}^{L} Z_{nj,\lambda''}(\xi) Y_{\lambda'' m_{nj}}(\eta, \phi)(\xi^2 - \eta^2) d\mathbf{r} \\ &= \int \sum_{\lambda=|m_i|}^{L} X_{i,\lambda}(\xi) \sum_{\lambda'=|m_k|}^{L} X_{k,\lambda'}(\xi) \\ &\sum_{\lambda''=|m_{nj}|}^{L} Z_{nj,\lambda''}(\xi) \left(\xi^2 A^{l_{ij}, l_j, \lambda}_{m_{ij}, m_j, m_i} - B^{l_{ij}, l_j, \lambda}_{m_{ij}, m_j, m_i} \right) d\xi\end{aligned} \tag{9.52}$$

where the coefficients *A, B* are defined in (4.91) and are calculated via *3j*-symbols. The one-dimensional integral in (9.52) is found numerically after replacing ξ by ρ according to (9.28).

The formation of the Coulomb matrix in the axial-symmetric case is carried out in the same way as in the case of atoms, with insignificant differences only. Let for a given initial state i_1 and a given Δm—the difference between the values of the angular momentum projection in the final and initial states, the sets of wave functions of the continuous spectrum with a given step in momentum $k_j = k_0 + \Delta k \cdot (j-1), \; j = 1 \div N$ are calculated.

The calculations were carried out for a number of values of $\lambda (\lambda = 1 \div L)$. Let also several functions of discrete excited states be calculated (see Chap. 6). It is in this order that the Coulomb matrix is formed. At first, the Coulomb matrix elements are calculated for the continuous spectrum, for $\lambda = 1, j = 1, 2, ..., N$,, etc. up to $\lambda = L, j = 1, 2, ..., N$. After that, the Coulomb matrix elements for the discrete spectrum are added. The rest of the formation of the Coulomb matrix coincides with the atomic case, i.e., formulas (9.38), (9.40), (9.44) are used.

9.2.3 RPAE Equation for Dipole Matrix Elements

9.2.3.1 Closed Shells (Atom)

Dipole matrix elements $D_\alpha(\omega)$ in RPAE satisfy Eq. (1.52) and take into account diagrams (1.51):

$$D_\alpha(\omega) = d_\alpha + \int\limits_\beta U_{\alpha\beta}\,\Pi^0_\beta(\omega) D_\beta(\omega), \tag{9.53}$$

where

$$\Pi^0_\beta(\omega) = \frac{1}{3}\frac{\zeta_\beta}{\omega - \omega_\beta + i\zeta_\beta\eta} \tag{9.54}$$

is the so-called polarization operator in the Hartree–Fock approximation. The factor 3 in the denominator (9.54) appears from the summation over the angular variables; $\omega_\beta = E_3 - E_4$ is the energy of the electron–hole pair $\beta \equiv (\nu_3\nu_4)$, $\nu_3 > F$, $\nu_4 \le F$, or $\nu_3 \le F$, $\nu_4 > F$ and $\zeta_\beta = \mathrm{sign}(\omega_\beta)$, $\eta \to +0$.

The integral in formula (9.53) includes summation over discrete excited states and integration over states of the continuous spectrum of an electron above the ionization threshold. Integral Eq. (9.53) is solved for electron–hole excitations $\alpha \equiv (\nu_1\nu_2)$, $\beta \equiv (\nu_3\nu_4)$ belonging to one (or more) electron–hole transitions.

Since the polarization operator $\Pi^0_\beta(\omega)$ has a pole in the continuous spectrum, then (9.53) is a complex equation, and the direction of traversing the pole is determined by the sign ζ_β, $\zeta_\beta = \pm 1$, depending on ω_β. It is considered that $\eta \to +0$. Integral Eq. (9.53) can be rewritten as follows:

$$P\int\limits_\beta \left[\delta_{\alpha\beta} - \frac{2}{3}U_{\alpha\beta}\frac{\zeta_\beta}{\omega - \omega_\beta}\right] D_\beta(\omega) + i\frac{2\pi}{3}U_{\alpha\lambda}D_\lambda(\omega) = d_\alpha,$$

$$\omega = \omega_\lambda \tag{9.55}$$

where $\delta_{\alpha\beta}$ is the Kronecker symbol or the delta function, depending on whether ω_β is the energy of a discrete excitation or a continuous spectrum, λ is one of the electron–hole excitations β, and P means that the principal value of the integral in (9.55) is taken. If λ belongs to the discrete spectrum, the last term on the left-hand side of Eq. (9.55) $i(2\pi/3)U_{\alpha\lambda}D_\lambda(\omega)$ is absent, since in this case (9.55) is a real equation.

Equation (9.55) can be written for the reduced matrix elements:

$$\langle\nu_1\|D(\omega)\|\nu_2\rangle = \langle\nu_1\|d\|\nu_2\rangle$$

$$+\left(\sum_{\nu_3\le F,\nu_4>F}-\sum_{\nu_3>F,\nu_4\le F}\right)\frac{\langle\nu_4\|D(\omega)\|\nu_3\rangle\langle\nu_3\nu_1\|U_1\|\nu_4\nu_2\rangle}{3\left[\omega-E_{\nu_4}+E_{\nu_3}+i\delta(1-2n_{\nu_4})\right]} \quad (9.56)$$

9.2.3.2 Open Shells (Atom)

The RPAE equation for the photoionization amplitude for open shell atoms is much more complicated. This equation also includes summation over the total moment L'' and spin S'' of the intermediate state of the ion:

$$\begin{aligned}\left\langle L'S',\nu_2L_1S_1\|D(\omega)\|\nu_1LS\right\rangle &= \left\langle L'S',\nu_2L_1S_1\|d\|\nu_1LS\right\rangle\\ &+\sum_{\nu_3>F}\sum_{L''S''}\left[\left\langle L''S'',\nu_3L_1S_1\|D(\omega)\|\nu_1\right\rangle\right.\\ &\left\langle\nu_1;L'S',\nu_2L_1S_1\|U\|L''S'',\nu_3L_1S_1;\nu_1\right\rangle\\ &\times\left([L_1](\omega-\varepsilon_{\nu_3}+\varepsilon_{\nu_1}+i\delta)\right)^{-1}\\ &-\sum_{L_2}\left\langle\nu_1LS\|D(\omega)\|L''S'',\nu_3L_2S_1\right\rangle\\ &\times\left\langle L''S'',\nu_3L_2S_1;L'S'\nu_2L_1S_1\|U\|\nu_1;\nu_1\right\rangle\\ &\left.\left([L_1](\omega+\varepsilon_{\nu_3}-\varepsilon_{\nu_1})\right)^{-1}\right],\end{aligned} \quad (9.57)$$

Here, summation over $\nu_3 > F$ implies both summation over excited discrete states and integration over states of the continuous spectrum. The Coulomb matrix elements included in this equation are defined as

$$\begin{aligned}\left\langle\nu_1;L'S',\nu_2L_1S_1\|U\|L''S'',\nu_3L_1S_1;\nu_1\right\rangle &= 2\sum_l\left[a_l\langle\nu_1\nu_2\|V_l\|\nu_3\nu_1\rangle\right.\\ &-[l]b_l\sum_{l'}(-1)^{l+l'}\begin{Bmatrix}l_1 & l & l_3\\ l_2 & l' & l_1\end{Bmatrix}\\ &\left.\langle\nu_1\nu_2\|V_{l'}\|\nu_1\nu_3\rangle\right],\end{aligned} \quad (9.58)$$

$$a_l=\sum_{\overline{LS}}C^{l_1\overline{LS}l}_{L'S',l_2L_1S_1}C^{l_1\overline{LS}l}_{L''S'',l_3L_1S_1}\delta_{\overline{S}S_1}, \quad (9.59)$$

$$b_l=\sum_{\overline{LS}}C^{l_1\overline{LS}l}_{L'S',l_2L_1S_1}C^{l_1\overline{LS}l}_{L''S'',l_3L_1S_1}\frac{[\overline{S}]}{[S']}\delta_{S'S''}, \quad (9.60)$$

$$\langle L''S'', \nu_3 L_2 S_1; L'S', \nu_2 L_1 S_1 \| U \| \nu_1; \nu_1 \rangle = 2 \sum_l \Big[a_l \langle \nu_3 \nu_2 \| V_l \| \nu_1 \nu_1 \rangle \delta_{L_1 L_2} - \beta [L_1] \delta_{l1} \sum_{l'} (-1)^{l+l'} \begin{Bmatrix} l_1 & l & l_2 \\ l_1 & l' & l_3 \end{Bmatrix} \langle \nu_3 \nu_2 \| V_{l'} \| \nu_1 \nu_1 \rangle \Big], \quad (9.61)$$

$$\beta = C^{l_1 LS1}_{L''S'', l_3 L_2 S_1} C^{l_1 LS1}_{L'S', l_2 L_1 S_1} \frac{[l_1]}{N[L]}, \quad (9.62)$$

9.2.3.3 Semi-closed Shells (Atom)

One can think of an atom with semi-closed shells in the ground state as having two sets of electrons with different spin directions $\downarrow\uparrow$ (see Sect. 4.2.6). An electron–hole pair can be characterized by one index $\uparrow$ or $\downarrow$, since the excitation of a photon does not change the direction of the spin.

The combinations of electron–hole pairs $\uparrow$ and $\downarrow$, taken into account in RPAE, can change in the initial, intermediate, and final states. As a result, different equations can be obtained. If we ignore spin-dependent interactions (as we do all the time in this book) and keep in mind that the contribution of RPAE exchange diagrams associated with states with different spin projections must be eliminated, then we get a system of four equations for effective interaction $\Gamma(\omega)$. This system is represented in matrix form as

$$\begin{pmatrix} \Gamma^{\uparrow\uparrow}_{\alpha\beta} & \Gamma^{\uparrow\downarrow}_{\alpha\beta} \\ \Gamma^{\downarrow\uparrow}_{\alpha\beta} & \Gamma^{\downarrow\downarrow}_{\alpha\beta} \end{pmatrix} = \begin{pmatrix} U^{\uparrow\uparrow}_{\alpha\beta} & V^{\uparrow\downarrow}_{\alpha\beta} \\ V^{\downarrow\uparrow}_{\alpha\beta} & U^{\downarrow\downarrow}_{\alpha\beta} \end{pmatrix} + \begin{pmatrix} U^{\uparrow\uparrow}_{\alpha\alpha'} & V^{\uparrow\downarrow}_{\alpha\alpha'} \\ V^{\downarrow\uparrow}_{\alpha\alpha'} & U^{\downarrow\downarrow}_{\alpha\alpha'} \end{pmatrix} \begin{pmatrix} \hat{\chi}^{\uparrow}_{\alpha'} & 0 \\ 0 & \hat{\chi}^{\downarrow}_{\alpha'} \end{pmatrix} \begin{pmatrix} \Gamma^{\uparrow\uparrow}_{\alpha'\beta} & \Gamma^{\uparrow\downarrow}_{\alpha'\beta} \\ \Gamma^{\downarrow\uparrow}_{\alpha'\beta} & \Gamma^{\downarrow\downarrow}_{\alpha'\beta} \end{pmatrix} \quad (9.63)$$

The arrows show the type of particle–hole pairs, between the states of which the matrix elements are calculated; $\Gamma_{\alpha\beta} = \langle \nu_2 \nu_3 | \Gamma(\omega) | \nu_1 \nu_4 \rangle$ (see Eq. (1.56)), where $\alpha \equiv (\nu_1 \nu_2)$, $\beta \equiv (\nu_3 \nu_4)$,$\nu_1 \leq F \nu_2 > F$, the Coulomb matrix $U_{\alpha\beta}$ includes direct and exchange interactions:

$$U_{\alpha\beta} = \langle \nu_1 \nu_4 | V | \nu_2 \nu_3 \rangle - \langle \nu_1 \nu_4 | V | \nu_3 \nu_2 \rangle, \quad (9.64)$$

$\hat{\chi}_{\alpha'}$ is an operator defined in the following way:

$$\hat{U}_{\alpha\alpha'} \hat{\chi}_{\alpha'} \hat{\Gamma}_{\alpha'\beta} \equiv \sum_{\alpha'} \left(\frac{\hat{U}_{\alpha\alpha'} \hat{\Gamma}_{\alpha'\beta}}{E_2 - E_1 - E_6 + E_5 + i\eta} - \frac{\hat{U}_{\alpha\alpha'} \hat{\Gamma}_{\alpha'\beta}}{E_2 - E_1 - E_6 + E_5 - i\eta} \right). \quad (9.65)$$

The last expression is a matrix in indices $\uparrow$ and $\downarrow$. Summation in (9.65) is performed over all intermediate states α' of electron–hole pairs ($\alpha' \equiv (\nu_5\nu_6)\nu_5 > F$, $\nu_6 \leq F$).

The equation for $D(\omega)$ in RPAE can be expressed via $\Gamma(\omega)$ (see (1.57)). Within this approach, the dipole matrix elements $D(\omega)$ depend on what kind of electrons are removed from the atom. Similar to $\Gamma(\omega)$, the following equation is used for $D(\omega)$:

$$(D_\alpha^\uparrow, D_\alpha^\downarrow) = (d_\alpha^\uparrow, d_\alpha^\downarrow) + (d_{\alpha'}^\uparrow \hat{\chi}_{\alpha'}^\uparrow, d_{\alpha'}^\downarrow \hat{\chi}_{\alpha'}^\downarrow) \begin{pmatrix} \Gamma_{\alpha'\alpha}^{\uparrow\uparrow} & \Gamma_{\alpha'\alpha}^{\uparrow\downarrow} \\ \Gamma_{\alpha'\alpha}^{\downarrow\uparrow} & \Gamma_{\alpha'\alpha}^{\downarrow\downarrow} \end{pmatrix}, \tag{9.66}$$

where d_α and D_α are the dipole matrix elements in the Hartree–Fock and RPAE approximations, respectively. The operator $\chi_{\alpha'}$ is determined from a condition similar to (9.65), which here, in accordance with the energy conservation law, is transformed into an equation $E_2 - E_1 = E_6 - E_5 = \omega$.

9.2.3.4 Axially Symmetric Potential (Diatomic Molecule)

In contrast to the case of central symmetry, the RPAE equations for axial symmetry can be written for each of the three projections of the dipole operator. Using the notation introduced at the end of the previous subsection, we write down the RPAE equations for the projection μ ($\mu = -1, 0, +1$) of the dipole operator:

$$\begin{aligned} \langle \nu | D_\mu | i \rangle &= \langle \nu | d_\mu | i \rangle \\ &+ \sum_{i'<F} \sum_{\nu'>F} \left[\frac{\langle \nu' | D_\mu | i' \rangle \langle i', \nu | U | \nu', i \rangle}{\omega - \varepsilon'_f + \varepsilon'_i + i\delta} - \frac{\langle i' | D_\mu | \nu' \rangle \langle \nu', \nu | U | i', i \rangle}{\omega + \varepsilon'_f - \varepsilon'_i} \right], \end{aligned} \tag{9.67}$$

where $\langle \nu | D_\mu | i \rangle$ is the dipole matrix element in RPAE. Summation over $\nu' > F$, as usual, includes also integration over the continuous spectrum. For the conjugate matrix element in RPAE, we additionally have the equations:

$$\begin{aligned} \langle i | D_\mu | \nu \rangle &= \langle i | d_\mu | \nu \rangle \\ &+ \sum_{i'<F} \sum_{\nu'>F} \left[\frac{\langle \nu' | D_\mu | i' \rangle \langle i', i | U | \nu', \nu \rangle}{\omega - \varepsilon'_f + \varepsilon'_i + i\delta} - \frac{\langle i' | D_\mu | \nu' \rangle \langle \nu', i | U | i', \nu \rangle}{\omega + \varepsilon'_f - \varepsilon'_i} \right] \end{aligned} \tag{9.68}$$

According to the selection rules, it is obvious that if on the left side of Eqs. (9.67) and (9.68) we have $\Delta m = 0$ for the difference in the projections of the moments of the initial and final states, then on the right side only the dipole matrix elements with $\Delta m = 0$ should be taken into account. The rest of the terms will simply be preceded by zero Coulomb matrix elements. At the same time, it is obvious that the dipole matrix elements $\Delta m = 1$ and $\Delta m = -1$ enter the equations simultaneously. But in this case, the zero Coulomb matrix elements are in front of the matrix element with $\Delta m = 0$. Thus, the RPAE equations should be solved separately for the case $\Delta m = 0$ and separately for the case $\Delta m = \pm 1$ [3].

9.2.4 Reduction of RPAE Equation to a System of Algebraic Equations

To solve the integral Eqs. (9.53), (9.56) or (9.67), (9.68), the integration interval should be chosen equal to at least several ionization potentials of the subshell under consideration. Then, in accordance with the approximate formula, the final integral is replaced by the sum. However, the usual formulas must be modified, since the integration interval contains a pole (see, for example, (1.56)).

The procedure for finding the principal value of the integral consists in applying the usual method of integration to the difference

$$F(x) = f(x) - \frac{r}{x - x_\lambda}, \tag{9.69}$$

which does not contain a pole at $x = x_\lambda$, if r is the residue of the function $f(x)$ at the point $x = x_\lambda$. As a result, the approximate expression for the integral $F(x)$ is:

$$\int_a^b F(x)x \approx \sum_{j=0}^{n} C_j F(x_j) \tag{9.70}$$

where $x_j (a \le x_j \le b)$ are the points at which the values $F(x_j)$ are set; C_j—coefficients of numerical integration. The function $f(x)$ is evaluated only in a sequence of discrete points $x = x_j$. Its values at other points are determined by applying the Lagrange interpolation formula to a regular function $(x - x_\lambda) f(x)$, after which the residue is easily found. From Eqs. (9.69) and (9.70), it follows that the principal value of the integral of a function $f(x)$ with a pole at $x = x_\lambda (a < x_\lambda < b)$ is

$$\int_a^b f(x)x \approx \sum_{j=0}^{n} (C_j - Q_j) f(x_j). \tag{9.71}$$

Corrections Q_j to the numerical integration coefficients are given by the formula:

$$Q_j = \frac{\varsigma(x_\lambda)}{\varsigma'(x_j)}\left[ln\left|\frac{b - x_\lambda}{x_\lambda - a}\right| - \sum_{j=1}^{n}\frac{C'_j}{x'_j - x_\lambda}\right] \tag{9.72}$$

Here $\varsigma(x) = (x - x_1)(x - x_2)\ldots(x - x_n)$. Relations (9.71) and (9.72) can be applied to the entire integration interval with a sufficiently large number of points n or to a number of subintervals with fewer points. In the latter case, these formulas apply only to one subinterval containing a pole or to two subintervals if the pole is on their border. The chosen option has, as a rule, three or five points, and the generalized Simpson formula is used for numerical integration. To obtain a more accurate value of the integral for a relatively small number of points, the linear momentum of a particle in the continuous spectrum $p = \sqrt{E}$ is taken as the integration variable. After discretization of the interval of integration, (9.55) turns into the following equation:

$$\sum_{\beta}\left[\delta_{\alpha\beta} - U_{\alpha\beta}\chi_\beta(\omega)\right]D_\beta(\omega) + i\frac{2\pi}{3}U_{\alpha\lambda}D_\lambda(\omega) = d_\alpha,$$

$$\omega = \omega_\lambda \tag{9.73}$$

where

$$\chi_\beta(\omega) = \alpha_\beta \Re\Pi^0_\beta(\omega) = \frac{2\alpha_\beta\zeta_\beta}{3(\omega - \omega_\beta)} = \chi^{\mp}_\beta(\omega) \tag{9.74}$$

The coefficients α_β are defined as $\alpha_\beta = 1$ if β corresponds to a discrete spectrum, or as

$$\alpha_\beta = 2\kappa_\beta(C_\beta - Q_\beta) \tag{9.75}$$

if β corresponds to a continuous spectrum of states.

Thus, the system (9.73) is a set of complex linear algebraic equations. In general, they can be written as

$$(A + iB)(x + iy) = u + iv, \tag{9.76}$$

where A and B are real square matrices; x, y, u, v are real vector matrices:

$$A = \begin{pmatrix} \delta_{\alpha\beta} - U_{\alpha\beta}\chi^-_\beta(\omega) & -U_{\alpha\bar{\beta}}\chi^+_\beta(\omega) \\ -U_{\bar{\alpha}\beta}\chi^-_\beta(\omega) & \delta_{\alpha\beta} - U_{\bar{\alpha}\bar{\beta}}\chi^+_\beta(\omega) \end{pmatrix}, \tag{9.77}$$

$$B = \frac{2\pi}{3}U_{\alpha\beta}\delta_{\beta\lambda}, \tag{9.78}$$

if λ belongs to the continuous spectrum and

$$B = 0, \tag{9.79}$$

if λ belongs to the discrete spectrum;

$$u_\alpha = d_\alpha, \tag{9.80}$$

$$v_\alpha = 0. \tag{9.81}$$

9.2.5 Solution of RPAE Equation

After real and imaginary parts of Eq. (9.76) are separated, we obtain a system of algebraic equations, which is conveniently represented in matrix form:

$$\begin{pmatrix} A & -B \\ B & A \end{pmatrix}\begin{pmatrix} x \\ y \end{pmatrix} = \begin{pmatrix} u \\ v \end{pmatrix}. \tag{9.82}$$

Here the upper line corresponds to the real part of (9.76), and the lower line—to its imaginary part. Solution (9.82) can be symbolically represented as

$$\begin{pmatrix} x \\ y \end{pmatrix} = \begin{pmatrix} A & -B \\ B & A \end{pmatrix}^{-1}\begin{pmatrix} u \\ v \end{pmatrix} = \begin{pmatrix} R & S \\ -S & R \end{pmatrix}\begin{pmatrix} u \\ v \end{pmatrix}, \tag{9.83}$$

where

$$R = (A + BA^{-1}B)^{-1}, \tag{9.84}$$

$$S = RBA^{-1} = (A + BA^{-1}B)^{-1}BA^{-1}, \tag{9.85}$$

$$x = \mathrm{Re}D(\omega),\ y = \mathrm{Im}D(\omega). \tag{9.86}$$

Thus, the solution of Eq. (9.76) or (9.82) is reduced to the inversion of two matrices A and $(A + BA^{-1}B)$, half the dimension than the original matrix in (9.82). This significantly reduces machine time and required memory for large matrix sizes. For a fixed state of a hole ν_1, the matrices to be inverted have the size $2(m_1 + m_2) \times 2(m_1 + m_2)$, where m_1 и m_2—are the numbers of wave functions of excited states of the first and second channels, respectively, the interaction of which is taken into account in the framework of the RPAE.

To calculate the cross sections or oscillator strengths, in accordance with (9.2) or (9.7) only the matrix element on the energy shell $D_{\nu_1\nu_2}(\omega)$ is taken from the column matrix $D(\omega)$, $\omega = E_2 - E_1$, where E_1 and E_2 are energies of the ground and excited states of the atom, respectively. Therefore, the RPAE equation must be solved for such a photon energy ω for which the following condition is satisfied:

$$\omega = \omega_\alpha = E_2 - E_1,\ \nu_2 > F, \qquad \nu_1 \le F \tag{9.87}$$

In those cases, when matrix elements outside the energy shell are needed, they are calculated for all photon energies (***pabsout*** program).

9.2.6 *Solution of RPAE Equation for Discrete States*

To describe the excitations of discrete states, Eqs. (9.53), (9.56), or (9.67) cannot be solved directly, since at the photon energy $\omega = \omega_\alpha$ the polarization operator $\Pi^0_\alpha(\omega)$ (9.54) has a pole at this point. Under the influence of the residual interaction, the energy of the discrete excited state changes and is now determined by the poles not of $\Pi^0_\alpha(\omega)$, but $\Gamma(\omega)$. To find this energy, the function χ must be divided into two terms: $\chi_1 = -1/(\omega - \omega_\alpha)$ and the term χ_2, which includes the summation over all states except α. Using (1.60) and (1.61), we obtain

$$\Gamma_{\alpha\lambda}(\omega) = U_{\alpha\lambda} + S_\beta U_{\alpha\beta}\chi_2(\omega)\Gamma_{\beta\lambda}(\omega), \tag{9.88}$$

$$\Gamma(\omega) = \Gamma_{\alpha\alpha}(\omega) + \Gamma_{\alpha\alpha}(\omega)\frac{1}{\omega - \omega_\alpha}\Gamma(\omega) \tag{9.89}$$

For the case of one transition (9.89) is an algebraic equation that can be easily solved:

$$\Gamma(\omega) = \frac{\Gamma_{\alpha\alpha}(\omega - \omega_\alpha)}{\omega - \omega_\alpha - \Gamma_{\alpha\alpha}}, \tag{9.90}$$

so $\Gamma(\omega)$ has a pole at a new energy value

$$\tilde{\omega}_\alpha = \omega_\alpha + \Gamma_{\alpha\alpha}(\tilde{\omega}_\alpha) \tag{9.91}$$

Assuming that $\Gamma_{\alpha\alpha}(\omega)$ weakly depends on ω, this equation can be solved by expanding $\Gamma_{\alpha\alpha}(\omega)$ in the vicinity of ω_α and keeping only the first two terms of the expansion:

$$\Gamma_{\alpha\alpha}(\omega) = \Gamma_{\alpha\alpha}(\omega_\alpha) + (\omega - \omega_\alpha)\frac{\partial\Gamma_{\alpha\alpha}(\omega)}{\partial\omega}|_{\omega=\omega_\alpha}, \tag{9.92}$$

so that

$$\tilde{\omega}_\alpha \approx \omega_\alpha + \Gamma_{\alpha\alpha}(\omega_\alpha)\left(1 - \left.\frac{\partial \Gamma_{\alpha\alpha}(\omega)}{\partial \omega}\right|_{\omega=\omega_\alpha}\right)^{-1} \tag{9.93}$$

From the physical point of view, $\Gamma_{\alpha\alpha}(\omega)$ describes the interaction of an electron and a hole not only directly, as a Coulomb potential, but also through other electron–hole real and virtual excited states. Note also that $\Gamma_{\alpha\alpha}(\omega_\alpha)$ does not diverge, since it does not contain a denominator $\omega - \omega_\alpha$ that vanishes at $\omega = \omega_\alpha$.

Similarly, the amplitude of photoexcitation is also modified, so that in expression (9.7) for the oscillator strength, the square of the dipole matrix element should be replaced by the following expression

$$\left|D'_\alpha(\omega)\right|^2 \left(1 - \frac{\partial \Gamma_{\alpha\alpha}(\omega)}{\partial \omega}|_{\omega=\omega_\alpha}\right)^{-1} \tag{9.94}$$

where $D'_\alpha(\omega)$ is the solution to Eqs. (9.73) and (9.74), in which the term $\chi(\omega)$ diverging at $\omega = \omega_\alpha$ is set equal to zero. For the highest discrete excitations, the energy shift $\tilde{\omega}_\alpha - \omega$ disappears, so that the ionization thresholds in the Hartree–Fock and RPAE approximations coincide.

The RPAE equation for discrete excited states is solved in the following sequence:

(a) the matrices R, S (9.84), (9.85) for the Hartree–Fock value of the photon energy $\omega = \omega_\alpha$ are calculated, with the term $\chi^-(\omega)$ in (9.73) set to zero;
(b) $\Gamma_{\alpha\alpha}(\omega)$ is calculated through the inverse matrix of Eq. (9.82), and, since the derivative $\partial\Gamma_{\alpha\alpha}/\partial\omega$ is usually much less than one, it is calculated in the lowest order in U:

$$\frac{\partial \Gamma_{\alpha\alpha}}{\partial \omega} \approx \int_{\beta\neq\alpha} |U_{\alpha\beta}|^2 \frac{\partial \Pi^0_\beta}{\partial \omega} \approx -\sum_{\beta\neq\alpha} |U_{\alpha\beta}|^2 \frac{\chi_\beta(\omega)}{\omega - \omega_\beta} \tag{9.95}$$

(c) the shifted (new) value of the photon energy $\tilde{\omega}$ is calculated using Eq. (9.93);
(d) matrices R, S are calculated again, but now for the shifted photon energy $\tilde{\omega}$;
(e) after the matrix inversion, $D_\alpha(\tilde{\omega})$ is calculated and with its help the oscillator strength is found according to (9.94) and (9.7).

Note that the matrix $B(\omega)$ in (9.79) is equal to zero for transitions to discrete states, and therefore $\Im D(\omega) = 0$.

9.2.7 *Solution of RPAE Equation for a Continuous Spectrum*

For transitions to the states in continuous spectrum, the pole of the polarization operator $\Pi^0_\alpha(\omega)$ at $\omega = \omega_\alpha$, where α corresponds to the continuous spectrum, is under the integral, which is understood in the sense of the principal value.

According to Eq. (9.79), only one column in the matrix $B(\omega)$ can differ from zero in the case of an excited state in the continuous spectrum. For this column $\beta \equiv (\nu_3\nu_4)$, $\omega = E_3 - E_4$, $\nu_3 > F$, $\nu_4 \leq F$ and ν_3, and belongs to the continuous spectrum. Therefore, the photon energy must be chosen from the condition $\omega = p_j^2 - E_4$, where $p_j = p_0 + j\Delta p$, $j = 0, 1, \ldots, m$ are equidistant points in momentum $p = E_3^{1/2}$.

9.2.8 Photoionization Cross Section

The photoionization cross section and the amplitude D_α are related, as usual, by Eq. (9.2). The partial photoionization cross section in the Hartree–Fock approximation is obtained from formula (9.2) with one-electron wave functions or obtained by the polarized spin method (see Sect. 4.2.6). The total cross section for the transition in the Hartree–Fock approximation is equal to the sum of the partial cross sections of ionization, and in the SP RPAE, the sum of the partial cross sections of ionization for spin subshells with ↓ and ↑ spins. If the states of the electrons ↓ and ↑ are equivalent, then (9.63) and (9.66) lead to the general equations of the RPAE (1.56) and (1.57). In this respect, the equations obtained by the RPAE method apply to atoms with semi-closed subshells. Wave functions of excited states in discrete and continuous spectra are calculated for spin subshells in the ***fchf*** or ***fchfw*** programs.

One-electron functions, being orthogonalized, are input data. The wave functions of excited atomic electrons with the same ℓ but different σ are always orthogonal due to their spin components. The total transition matrix element and the photoionization cross section are obtained by summing the contributions from interactions with other subshells and transitions (see Sect. 2.5). The inputs to this program include information about the transitions in which the exchange should be accounted for. In some transitions, the interacting electrons are described by the same n, l, but different projections σ. Naturally, there is no exchange in such transitions.

9.2.9 Photoionization Quasi-Cross Sections

Quasi-cross section is the oscillator strength of a discrete level *f*, "smeared" over the width of the resolution Δ of discrete levels. It has the dimension of the cross section and is calculated by the ***level*** program that use the formula

$$\sigma(\omega) = \frac{8\pi f \Delta}{c(\omega - \omega_0)^2 + (\Delta/2)^2} \tag{9.96}$$

Here ω_0 is the photon energy of the discrete transition, ω, Δ—are given in Ry. In our specific calculations, we usually assume $\omega_0 - 3\Delta < \omega < \omega_0 + 3\Delta$, $c = 137.02$, $\Delta = 0.008$ [4].

9.2.9.1 Dipole Angular Anisotropy Parameters

The angular distribution of photoelectrons emitted off atoms and non-oriented molecules by non-polarized light is described by the equation

$$\frac{d\sigma}{d\Omega} = \frac{\sigma}{4\pi}\left[1 - \frac{\beta(\omega)}{2} \cdot P_2(cos\vartheta)\right], \tag{9.97}$$

where P_2 is the Legendre polynomial, ϑ is the angle between the direction of electron emission and the direction of ionizing radiation, and β is the dipole parameter of angular anisotropy.

In the case of atoms, the dipole angular anisotropy parameter for the nl-subshell is calculated using the following formula:

$$\begin{aligned}\beta_{nl} = &\frac{1}{(2l+1)\left[|D_{l-1}|^2 + |D_{l+1}|^2\right]}\Big\{(l-1)|D_{l-1}|^2 + (l+2)|D_{l+1}|^2 + 6\sqrt{l(l+1)}\\ &\times \left[(\mathrm{Re}D_{l-1}\mathrm{Re}D_{l+1} + \mathrm{Im}D_{l-1}\mathrm{Im}D_{l+1})cos(\delta_{l+1} - \delta_{l-1})\right.\\ &\left.-(\mathrm{Re}D_{l-1}\mathrm{Im}D_{l+1} - \mathrm{Im}D_{l-1}\mathrm{Re}D_{l+1})sin(\delta_{l+1} - \delta_{l-1})\right]\Big\},\end{aligned} \tag{9.98}$$

where $D_{l\pm1}$ is the dipole matrix elements of the operator in the length or velocity form, taking into account many-electron correlations, and $\delta_{l\pm1}$ is the scattering phase shift of the photoelectron wave with angular momentum $l \pm 1$ in the ion field in the one-particle approximation, where $\Im D_{l\pm1} = 0$.

According to (9.98), the angular distribution of photoelectrons is more sensitive to the transition amplitude $nl \to E, l-1$, than the cross Sect. (9.2). In addition, the angular distribution depends on the phase difference $(\delta_{l+1} - \delta_{l-1})$. Note that the transition $nl \to E, l-1$ is called weak, since its contribution to the cross section is usually much smaller than that of the strong transition $nl \to E, l+1$. The dipole parameters of the angular anisotropy are calculated using the ***beta*** procedure.

In the case of diatomic molecules, the angular anisotropy parameter for the ionizable i-shell and the electron emission energy ε is determined by the formula

$$\begin{aligned}\beta = &\frac{\sqrt{30}}{B}\sum_{l,l'}\sum_{m,m',M}(\mathrm{i})^{l'-l}\exp[\mathrm{i}(\delta_{lm} - \delta_{l'm'})]\sqrt{[l,l']}\times\\ &\begin{pmatrix} l & l' & 2 \\ m & -m' & -M \end{pmatrix}\begin{pmatrix} l & l' & 2 \\ 0 & 0 & 0 \end{pmatrix}\begin{pmatrix} 1 & 1 & 2 \\ m & -m' & -M \end{pmatrix}B,\end{aligned} \tag{9.99}$$

where

$$B = \sum_{\mu,l,m}\left|\langle\varepsilon lm|d_\mu|i\rangle\right|^2. \tag{9.100}$$

In formulas (9.99), (9.100), dipole matrix elements can be used both in the HF approximation and in the RPAE approximation.

9.2.9.2 Dipole Dynamic Polarizability of Atoms

Let us assume that f_n and E_n, are known for the first m_1 discrete excited states, and the photoionization cross section is calculated at an odd number of $m - m_1$ points equidistant in photoelectron's linear momentum. Then from (9.9) the approximate equation follows:

$$\alpha(\omega) \approx \sum_{n=u}^{u+m_1-1} \frac{4f_n}{\omega_n^2 - \omega^2} + 0.49584 \cdot 2h \sum_{j=1}^{m-m_1} \frac{c_j \sigma(\omega_j) p_j}{\omega_j{}^2 - \omega^2}, \tag{9.101}$$

where u is the principal quantum number of the first excited state, h is the step size in momentum $p, \omega_j = p_j^2$. C_j,.—coefficients of numerical integration by the Simpson method. A factor of 0.49584 is a value of $(\pi^2 \alpha a_0^2)^{-1}$, if σ is measured in megabars (Mb).

In our programs, the oscillator strengths are usually calculated only for the first two or three excited states. However, it is known that discrete excited states make a significant contribution to polarizability. To provide a more accurate calculation of the polarizability $\alpha(\omega)$, in our program f_n and E_n for higher discrete excited states are approximated by the following well-known expressions:

$$E_n \approx \frac{a_E}{n^{*2}}, \; n^* = n + \Delta n, \tag{9.102}$$

$$f_n \approx \frac{a_F}{n^3}, \qquad n \to \infty, \tag{9.103}$$

$$\frac{f_n}{E_{n+1} - E_n} \to \frac{\sigma(I)}{4\pi^2 \alpha a_0^2}, \quad n \to \infty, \tag{9.104}$$

where Δn is the so-called *quantum defect*, and a_E, a_{F_1} are constants. If we denote by n_1 the principal quantum number of the highest discrete excited state, the energy E_{n_1} and f_{n_1} of which is known, then using (9.102) the following relations can be obtained, which include a_E and Δn:

$$\begin{aligned} E_{n_1-1} &= \frac{a_E}{(n_1 - 1 + \Delta n)^2} \\ E_{n_1} &= \frac{a_E}{(n_1 + \Delta n)^2} \end{aligned} \tag{9.105}$$

From (9.105) we have

$$\Delta n = \frac{E_{n_1-1} - \sqrt{E_{n_1} E_{n_1-1}}}{E_{n_1-1} - E_{n_1}} - n_1, \tag{9.106}$$

$$a_E = E_{n_1}(n_1 + \Delta n)^2 \tag{9.107}$$

After substituting (9.102) and (9.103) into (9.104), for $n \to \infty$, we obtain

$$\frac{a_F}{a_E} = \frac{\sigma(I)}{2\pi^2 \alpha a_0^2} = \frac{\sigma(I)}{4.0335} \tag{9.108}$$

Hence, using (9.107), we find a_F. In addition, the cross section $\sigma(I)$, which is usually not known, is determined by linear interpolation between the two points closest to the threshold in the continuous spectrum where the photoionization cross section is available. To approximate the oscillator strength at $n > n_1$, the program uses the formula

$$f_n = \frac{a_F}{n^3 + b_F n^2}, \tag{9.109}$$

which for large n corresponds to (9.103). The coefficient b_F is determined from the relation (9.109) at $n = n_1$,

$$f_{n_1} = \frac{a_F}{n_1^3 + b_F n_1^2}, \tag{9.110}$$

that gives

$$b_F = \frac{a_F}{f_{n_1} n_1^2} - n_1 \tag{9.111}$$

The additional term in the denominator (9.109) ensures that the precalculated discrete values f_n, $(n \le n_1)$ at $n > n_1$ smoothly pass to the known approximation of the oscillator strengths (9.103).

To find the dipole polarizability (***polar*** program) for ω above the ionization threshold, the integral in (9.9) should be calculated in accordance with (9.69)–(9.71).

In the discrete spectrum, the ***polar*** program takes into account the contribution of excited states, the oscillator strengths of which are calculated using (9.7) and (9.94), and adds the contribution of a large number (usually 50) of higher excited states calculated by formulas (9.101)–(9.104).

References

1. Messiah, *Quantum Mechanics* (Wiley, North Holland, 1966)
2. R. McWeeney, B. Sutcliffe, *Quantum Mechanics of Molecules* (Academic Press, London, 1969)
3. S.K. Semenov, N.A. Cherepkov, G.H. Fecher, G. Schönhense, Phys. Rev. A**61**, 032704 (2000)
4. M.Ya. Amusia, L.V. Chernysheva, V.G. Yarzhemsky, *Handbook of Theoretical Atomic Physics, Data for Photon Absorption, Electron Scattering, Vacancies Decay* (Springer, Heidelberg, New York, Dordrecht, London, 2012)

Chapter 10
Scattering Cross Section of Fast Electrons, Generalized Oscillator Strengths, and the Compton Effect

Generalized oscillator strengths (GOS) are characteristics that describe the process of interaction of a fast, charged particle with an atom or molecule. This chapter describes how to calculate GOS for electrons in the one-particle Hartree–Fock and RPAE approximations that account for the interaction of various transitions. RPAE equations are presented for the matrix elements of GOS, which are solved by methods similar to those described in Chap. 9. Atoms with closed and semi-closed subshells are considered, as well as atoms with an open shell with electron (or hole) in it.

10.1 Theoretical Model

The differential in angle cross section for inelastic scattering of fast electrons or other charged particles by atoms and molecules in the first Born approximation is determined by the expressions [LL]

$$\frac{\partial \sigma_{\nu_1 \nu_2}}{\partial \Omega} = \frac{4 p_1}{q^2 p_2 \omega} f_{\nu_1 \nu_2}(\mathbf{q}, \omega) \tag{10.1}$$

in the case of a transition of one of the target electrons to a discrete excited state, or

$$\frac{\partial^2 \sigma_{\nu_1 \nu_2}}{\partial \Omega \partial \omega} = \frac{4 p_1}{q^2 p_2 \omega} f_{\nu_1 \nu_2}(\mathbf{q}, \omega) \tag{10.2}$$

in case of its ionization. Here p_1 and p_2 are the impulses of the incident and scattered electrons, respectively, Ω is the emission angle of the scattered electron. ν_2 and ν_1—quantum numbers of the initial and final atomic or molecular states, q—momentum transferred to the target atom $\mathbf{q} = \mathbf{p}_2 - \mathbf{p}_1$, while $\omega = E_{\nu_2} - E_{\nu_1}$ is the transferred

M. Ya. Amusia and L. V. Chernysheva, *Computation of Atomic and Molecular Processes*, Springer Series on Atomic, Optical, and Plasma Physics 117,
https://doi.org/10.1007/978-3-030-85143-9_10

energy; $f_{\nu_1\nu_2}(q,\omega)$ is the generalized oscillator strengths in (10.1) or the generalized oscillator strengths density in (10.2).

Note that the generalized oscillator strengths are dimensionless quantities, while the generalized oscillator strength densities are of inverse energy Ry^{-1} dimension. The reason for the difference is in different normalization of excited states in the discrete (6.13) and continuous (6.14) spectra. In HF, for an atom or a molecule $f_{\nu_1\nu_2}(q,\omega)$ can be written in two different forms [1]: the length form $f^{(r)}$ or velocity $f^{(\nabla)}$ form:

$$f^{(r)}_{\nu_1\nu_2}(\mathbf{q},\omega) = \frac{2\omega}{q^2}\left|\langle\nu_1|g^{(r)}(\mathbf{q})|\nu_2\rangle\right|^2, \tag{10.3}$$

$$f^{(\nabla)}_{\nu_1\nu_2}(\mathbf{q},\omega) = \frac{1}{2q^2\omega}\left|\langle\nu_1|g^{(\nabla)}(\mathbf{q})|\nu_2\rangle\right|^2, \tag{10.4}$$

where

$$\begin{aligned} g^{(r)}(\mathbf{q}) &= exp(i\mathbf{qr}), \\ g^{(\nabla)}(\mathbf{q}) &= exp(i\mathbf{qr})(\mathbf{q}\nabla) - (\mathbf{q}\overleftarrow{\nabla})exp(i\mathbf{qr}), \end{aligned} \tag{10.5}$$

$|\nu_2\rangle$ and $\langle\nu_1|$ are the wave functions of the initial and final states of an electron in the Hartree–Fock approximation for an atom or molecule. The arrow above the operator ∇ indicates that it acts to the left.

The process corresponding to Eqs. (10.3) or (10.4) in the RPAE is depicted by diagrams (1.50) and (1.51):

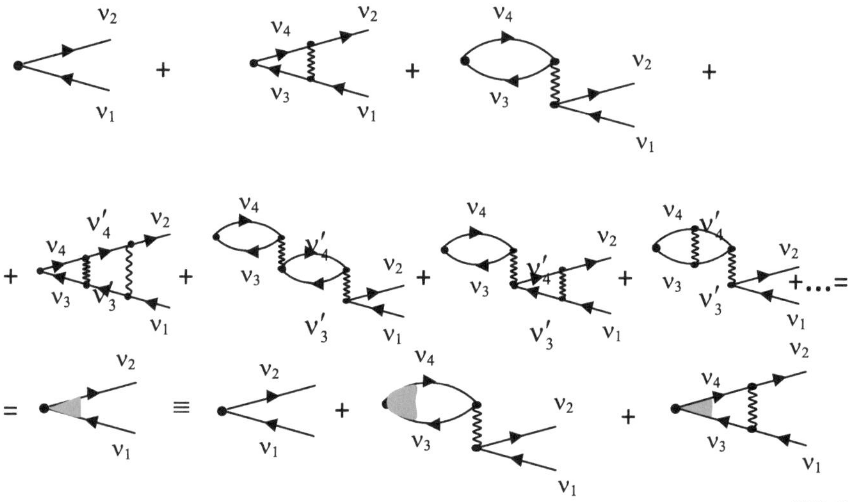

(10.6)

+ the same diagrams with "time reversal" terms (see Chap. 1).

In RPAE, the matrix element of the generalized oscillator strength $\langle \nu_1|G(\mathbf{q},\omega)|\nu_2\rangle$, depicted by the shaded triangle in (10.6), is determined from (1.57), where $D(\omega)$ is replaced by $G(\mathbf{q},\omega)$, and $d(\omega)$ by $g(\mathbf{q},\omega)$. The generalized oscillator strengths in RPAE are obtained by substituting $G(\mathbf{q},\omega)$ in the length or velocity forms in (10.4) and (10.5).

When considering atoms, a significant simplification of the equations is possible if we write the HF wave function in the form (4.2). We use the well-known expansion

$$\exp(i\mathbf{qr}) = \sum (2L+1) j_L(qr) P_L(\cos\varsigma) \tag{10.7}$$

Here $j_L(qr)$ is the spherical Bessel function, $P_L(\cos\varsigma)$ is the Legendre polynomial, and ς is the angle between the vectors $\boldsymbol{q}$ and $\boldsymbol{r}$. Separating the angular and spin parts of the one-electron HF wave function, and using expansion (10.7), we can analytically perform integration over the angles in the matrix elements of Eq. (10.3) or Eq. (10.4). As a result, along with $g_{fi}(\omega,q)$, it is possible to introduce partial one-electron GOS $g^L_{nl,\varepsilon l'}(\mathbf{q},\omega)$, with l', the values of which lie within the following limits $|L-l| \le l' \le L+l$.

The expression for the GOS of the transition *fi* with multipolarity L in the HF approximation $g^L_{fi}(\mathbf{q},\omega^{HF}_{fi})$ follows from Formula (10.3)

$$\begin{aligned} f^L_{fi}(q,\omega^{HF}_{fi}) &= \frac{2\omega^{HF}_{fi}}{q^2}\left|\int \phi^*_f(\mathbf{r}) j_L(qr) P_L(cos\vartheta)\phi_i(\mathbf{r})d\mathbf{r}\right|^2 \\ &= \frac{2\omega^{HF}_{fi}}{q^2}|\langle f|j_L(qr)|i\rangle|^2. \end{aligned} \tag{10.8}$$

To find the GOS with inclusion of the electron correlations within RPAE, the matrix elements of the operator $g^{(r,\nabla)}(\mathbf{q})$ in (10.3) should be replaced by solutions of the equation

$$\begin{aligned} \left\langle f|\hat{G}(\mathbf{q},\omega)|i\right\rangle &= \langle f|g^{(r,\nabla)}(\mathbf{q})|i\rangle \\ &+ \left(\sum_{n'\le F,k'>F} - \sum_{n'>F,k'\le F}\right)\frac{\left\langle k'\left|\hat{G}(\mathbf{q},\omega)\right|n'\right\rangle\langle n'f|U|k'i\rangle}{\omega-\varepsilon_{k'}+\varepsilon_{n'}+i\eta(1-2n_{k'})}. \end{aligned} \tag{10.9}$$

This equation is an obvious generalization of (9.56) for the photoionization amplitude to the case of a GOS. For atoms with spherical symmetry of the field acting on the electron, (10.9) can be reduced to a system of uncoupled equations for the L-components of the GOS

$$\left\langle f|\hat{G}_L(\omega,\mathbf{q})|i\right\rangle = \langle f|g^{L(r,\nabla)}(\mathbf{q})|i\rangle$$

$$+\left(\sum_{n'\le F,k'>F}-\sum_{n'>F,k'\le F}\right)\frac{\left\langle k'\middle|\hat{G}_L(\omega,\mathbf{q})\middle|n'\right\rangle\langle n'f|U_L|k'i\rangle}{\omega-\varepsilon_{k'}+\varepsilon_{n'}+i\eta(1-2n_{k'})}. \tag{10.10}$$

where U_L is the L component of the combination of the direct and exchange Coulomb interelectron interaction.

The Compton scattering cross section under some conditions can be expressed in terms of the GOS. Indeed, the operator describing the interaction of photons with N atomic electrons in the nonrelativistic approximation looks like [BLP]:

$$\hat{\mathrm{K}}=\sum_{i=1}^{N}\left(-\frac{1}{c}\hat{\mathbf{p}}_i\vec{A}(\mathbf{r}_i)+\frac{1}{2c^2}\vec{A}^2(\mathbf{r}_i)\right), \tag{10.11}$$

where $\boldsymbol{A}(\boldsymbol{r}_i)$ is the vector potential of the electromagnetic field, $\boldsymbol{r}_i$ and $\hat{\mathbf{p}}_i$ are the coordinate and momentum of the ith electron, c is the speed of light. The cross section for inelastic (or Compton) scattering of photons is expressed in terms of the second power of the operator $\hat{\mathbf{p}}_i\vec{A}(\mathbf{r}_i)/c$ and the first power $\vec{A}^2(\mathbf{r}_i)/2c^2$. But the contribution of the term $\hat{\mathbf{p}}_i\vec{A}(\mathbf{r}_i)/c$ to Compton scattering by nonrelativistic electrons is small, since it contains a factor $(v/c)^2 << 1$, so in fact the operator $\vec{A}^2(\mathbf{r}_i)/2c^2$ makes the main contribution to the Compton scattering cross section.

For an external electromagnetic field, we have the expression $\vec{A}(\mathbf{r}_i)\sim\mathbf{e}\exp(i\mathbf{kr}_i)$, where $\mathbf{e}$ is the radiation polarization vector [BLP]. Therefore, the cross section of Compton scattering $d\sigma_{if}^{C}(\omega)/d\Omega$, which is differential in the photon scattering angle $d\Omega$, accompanied by an atomic transition from the initial $\langle i|$ to the final $|f\rangle$ state, can be expressed in terms of the matrix elements of the operator $\exp(i\vec{k}\vec{r}_i)$:

$$\frac{\mathrm{d}\sigma_{if}^{C}(\omega)}{\mathrm{d}\Omega}=\left(\frac{\mathrm{d}\sigma}{\mathrm{d}\Omega}\right)_0\frac{E-\omega}{E}\sum_f\left|\left\langle i\middle|\sum_{j=1}^{N}\mathrm{e}^{i\mathbf{qr}_j}\middle|f\right\rangle\right|^2. \tag{10.12}$$

Here E is the incident photon energy and ω is the energy transferred to the atom in the scattering process, $\left(d\sigma/d\Omega\right)_0$ is the classical differential Thomson cross section for light scattering by an electron (see [BLP, 2]). The angle θ determines the direction of emission of the photon. The summation over the final states f is performed having in mind the energy conservation law $\omega=E_f-E_i$, where E_i and E_f are the energies of the atom in the initial state and the system "ionized electron + residual ion" in the final state, respectively.

The contribution of the last term on the right-hand side of (10.11) can be expressed in terms of the GOS, which are given by Formula (10.3) and determine the inelastic scattering cross section of a charged fast particle by an atom. Then (10.12) can be written in the following way:

$$\frac{d\sigma_{if}^{C}(\omega)}{d\Omega} = \left(\frac{d\sigma}{d\Omega}\right)_0 \frac{E-\omega}{E}\frac{q^2}{2\omega}\sum_f f_{fi}(\omega,\mathbf{q}). \tag{10.13}$$

Compton scattering is of interest at high energies E, for medium and heavy atoms at approximately dozens and even hundreds of keV, where it begins to dominate in the total cross section of the "photon–atom" interaction over photoabsorption [2]. Bearing in mind that from the point of view of understanding the structure of atoms, not too large ω, $\omega \leq 100Ry$ are of interest, we get a limitation $\omega/E << 1$. Neglecting corrections of the order of $\sim \omega/E << 1$, we obtain the following relation.

$$q = \frac{2E}{c}\left(1-\frac{\omega}{2E}\right)^{1/2}\sin\vartheta, \tag{10.14}$$

where $\vartheta = \theta/2$ and θ- is the scattering angle of the outgoing photon.

In the HF approximation, the Compton scattering cross section is given by (10.13) with the GOS determined by the relation (10.3):

$$\frac{d\sigma_{HF}^{C}(\omega)}{d\Omega} = \left(\frac{d\sigma}{d\Omega}\right)_0 \frac{E-\omega}{E}\frac{q^2}{2\omega}\sum_{s>F,t\leq F} f_{st}(\omega,q) \tag{10.15}$$

Here, summation is performed over all one-electron states $t \leq F$ occupied in the initial state of the i atom and excited states $s > F$, which satisfy the energy conservation law $\varepsilon_s = \omega + \varepsilon_t$.

The cross section for elastic classical Thomson scattering is given by the following expression [BLP]:

$$\left(\frac{d\sigma}{d\Omega}\right)_0 = \frac{1}{2c^4}(1+\cos^2\theta) \approx \frac{1}{c^4}\left[1-\frac{1}{2}\left(\frac{qc}{E}\right)^2\right], \tag{10.16}$$

where θ is the scattering angle of the photon. The second step of simplification in Formula (10.16) is valid for small scattering angles.

The next step in our consideration of Compton scattering should account for the multielectron correlations within RPAE. The corresponding expression is obtained by replacing f_{st} in Formula (10.15) by the oscillator strengths f_{st}^{RPAE} in RPAE, determined by (10.3) or (10.4), where $g^{(r,\nabla)}(\mathbf{q})$ is replaced by solutions of Eq. (10.9)—matrix elements of the operator $\hat{G}(\omega,\mathbf{q})$.

Equation (10.10) for $\hat{G}(\omega,\mathbf{q})$, based on (10.7), transforms into a system $G^L(\omega,q)$ of independent equations for partial contributions with a given angular momentum L, which must be solved numerically, as described in [AC] and in Chap. 9, resulting in matrix elements $\left\langle nl\middle|\hat{G}^L(\omega,\mathbf{q})\middle|\varepsilon l'\right\rangle$. With the help of these matrix elements, it is possible to obtain the partial in L GOS in the RPAE:

$$f_{nl,\varepsilon l'}^{R,L}(\omega, q) = \frac{2\omega}{q^2}\left|\left\langle nl\left|\hat{G}^L(\omega, \mathbf{q})\right|\varepsilon l'\right\rangle\right|^2. \tag{10.17}$$

Then the values $f_{st}(\omega, \mathbf{q})$ should be replaced in (10.15) by $f_{nl,\varepsilon l'}^{R,L}(\omega, q)$, which leads to the relation

$$\frac{d\sigma_R^C(\omega, q)}{d\Omega} = \left(\frac{d\sigma}{d\Omega}\right)_0 \frac{E-\omega}{E}\frac{q^2}{2\omega}\sum_{s>F,t\leq F.L>0} f_{nl,\varepsilon l'}^{LR}(\omega, q), \tag{10.18}$$

which is the Compton scattering cross sections in RPAE (see [3, 4] and references therein).

10.2 Numerical Model

10.2.1 Matrix Elements of Bessel Functions in the Hartree–Fock Approximation

In the Hartree–Fock approximation, due to non-locality of the exchange potential, (10.3) and (10.4) lead to different results [5]. Representing $exp(i\mathbf{qr})$ from expression (10.7) in the form

$$exp(i\mathbf{qr}) = 4\pi\sum_{L,m} i^L j_L(qr)Y_{Lm}^*(\theta_\mathbf{q}, \varphi_\mathbf{q})Y_{Lm}(\theta_\mathbf{r}, \varphi_\mathbf{r}) \tag{10.19}$$

where $j_l(qr)$ is the Bessel function, and after integrating over the angles, summing over the projections of the spin and orbital angular momentum in the final state, and averaging in the initial state, we obtain that the GOS $f(q, \omega)$ can be expanded into the series in the total angular momentum of the electron–hole pair L in the form:

$$f_{\nu_1\nu_2}(q, \omega) = \sum_L f_{\nu_1\nu_2}^L(q, \omega), \tag{10.20}$$

where

$$f_{\nu_1\nu_2}^{(r,L)}(q, \omega) = (2L+1)\frac{N_1\omega q^{2(L-1)}}{(2l_1+1)}\left|\left\langle \nu_1\left\|g_L^{(r)}(q)\right\|\nu_2\right\rangle\right|^2, \tag{10.21}$$

$$f_{\nu_1\nu_2}^{(\nabla,L)}(q, \omega) = (2L+1)\frac{N_1 q^{2(L-1)}}{\omega(2l_1+1)}\left|\left\langle \nu_1\left\|g_L^{(\nabla)}(q)\right\|\nu_2\right\rangle\right|^2. \tag{10.22}$$

Here N_1 is the number of electrons in the subshell under consideration, $\omega = E_{\nu_2} - E_{\nu_1}$. The GOS matrix elements in the length r and velocity ∇ forms are calculated using the following formulas:

$$\left\langle \nu_1 \left\| g_L^{(r)}(q) \right\| \nu_2 \right\rangle = \sqrt{(2l_1+1)(2l_2+1)} \times \begin{pmatrix} l_1 & L & l_2 \\ 0 & 0 & 0 \end{pmatrix} \int_0^\infty P_{\nu_1}(r) P_{\nu_2}(r) \tilde{j}_L(qr) dr, \tag{10.23}$$

$$\begin{aligned} \left\langle \nu_1 \left\| g_L^{(\nabla)}(q) \right\| \nu_2 \right\rangle &= \frac{q\sqrt{(2l_1+1)(2l_2+1)}}{2(2L+1)} \begin{pmatrix} l_1 & L & l_2 \\ 0 & 0 & 0 \end{pmatrix} \\ &\times \left\{ [l_1(l_1+1) - l_2(l_2+1)] \int_0^\infty P_{\nu_1}(r) P_{\nu_2}(r) \frac{q^{-1}\tilde{j}_{L-1}(qr) + q\tilde{j}_{L+1}(qr)}{r} \mathrm{d}r \right. \\ &+ \int_0^\infty \left(P_{\nu_1}(r) \frac{\mathrm{d}P_{\nu_2}(r)}{\mathrm{d}r} - P_{\nu_2}(r) \frac{\mathrm{d}P_{\nu_1}(r)}{\mathrm{d}r} \right) \\ &\left. \left[(L+1)q\tilde{j}_{L+1}(qr) - Lq^{-1}\tilde{j}_{L-1}(qr) \right] \mathrm{d}r \right\}. \end{aligned} \tag{10.24}$$

Here P_{ν_1} and P_{ν_2} are the radial wave functions of the initial and final states of the electron in the HF approximation. The angular momentum L satisfies the triangle rule $l_1 + l_2 > L > |l_i - l_2|$.

There are no restrictions on the choice of wave functions in the program. They can be ϕ^N, ϕ^{N+1}, $\phi^{N(LS)}$ and functions that include rearrangement effect (see Chap. 8). $\tilde{j}_L(qr)$ is the spherical Bessel function [IS] divided by q^L. Note that the generalized oscillator strengths are dimensionless quantities, and the densities of the generalized oscillator strengths, determined by (10.8) and using (10.9), have the dimension Ry^{-1}.

The value $\tilde{j}_L(qr)$ for $qr \le 1$ is calculated by Formulas [6, 7]:

$$\tilde{j}_L(qr) = r^L \sum_{k=0}^{\infty} \frac{(-1)^k (qr)^{2k} 2^L (L+k)!}{k![2(L+k)+1]!}, \tag{10.25}$$

and for $qr > 1$—by the formula.

$$\begin{aligned} \tilde{j}_L(qr) = \frac{1}{q^L(qr)} &\left[\sin\left(qr - \frac{\pi L}{2}\right) \sum_{k=0}^{E(L/2)} \frac{(-1)^k (L+2k)!}{(2k)!(L-2k)!(2qr)^{2k}} \right. \\ &\left. + \cos\left(qr - \frac{\pi L}{2}\right) \sum_{k=0}^{E\left(\frac{L-1}{2}\right)} \frac{(-1)^k (L+2k+1)!}{(2k+1)!(L-2k-1)!(2qr)^{2k+1}} \right], \end{aligned} \tag{10.26}$$

where $E(x)$ is the integer part of x. Expressions (10.25) and (10.26) are transformed into the following form:

$$qr \le 1,\ \tilde{j}_L(qr) = r^L \sum_{k=0}^{\infty} a_k (qr)^{2k}, \tag{10.27}$$

$$a_o = \frac{2^L L!}{(2L+1)!},\ a_1(qr) = a_0 \frac{(-1)(qr)^2}{2[2L+3]}$$

$$a_k(qr) = a_{k-1}(qr)\frac{(-1)(qr)^2}{2k[2(L+k)+1]}, \tag{10.28}$$

The calculations continue until $|a_i|$ becomes less than a certain small, predetermined number.

For $qr > 1$, the following formula is used

$$\tilde{j}_L(qr) = \frac{1}{q^L(qr)}\left[\sin\left(qr - \frac{\pi L}{2}\right)\sum_{k=0}^{E(L/2)} b_k + \frac{1}{2qr}\cos\left(qr - \frac{\pi L}{2}\right)\sum_{k=0}^{E\left(\frac{L-1}{2}\right)} c_k\right] \tag{10.29}$$

$$b_o = 1,\ b_1(qr) = \frac{(-1)(L-1)L(L+2)(L+1)}{2(2qr)^2},$$

$$b_k = -b_{k-1}\frac{(L+2k)(L+2k-1)}{2k(L-2k)(L-2k-1)(2qr)^2}, \tag{10.30}$$

$$c_o = L(L+1),\ c_1 = c_0 \frac{(-1)(L-2)(L-1)(L+2)(L+3)}{2k(2k+1)(2qr)^2},$$

$$c_k = c_{k-1}\frac{(-1)(L-2k)(L-2k+1)(L+2k)(L+2k+1)}{2k(2k+1)(2qr)^2}. \tag{10.31}$$

When calculating the matrix elements, we pass to the variable ρ through which the wave functions (4.50) are expressed. After changing the variables, we come to the following expressions:

$$\left\langle \nu_1 \left\| g_L^{(r)}(q) \right\| \| \nu_2 \right\rangle = \sqrt{(2l_1+1)(2l_2+1)}\begin{pmatrix} l_1 & L & l_2 \\ 0 & 0 & 0 \end{pmatrix}\int_{-\infty}^{\infty} f_1(\rho) f_2(\rho) \tilde{j}_l(qr)\frac{r^2 d\rho}{(\alpha r + \beta)^2}, \tag{10.32}$$

$$\left\langle \nu_1 \left\| g_L^{(\nabla)} \right\| \nu_2 \right\rangle = \frac{q\sqrt{(2l_1+1)(2l_2+1)}}{2L+1}\begin{pmatrix} l_1 & L & l_2 \\ 0 & 0 & 0 \end{pmatrix}[l_1(l_1+1) - l_2(l_2+1)]$$

$$
\times \left\{ \int_{-\infty}^{\infty} f_1(\rho) f_2(\rho) \left[q^{-1} \tilde{j}_{L-1}(qr) + q \tilde{j}_{L+1}(qr) \right] \frac{r^2 \mathrm{d}\rho}{(\alpha r + \beta)^2} \right.
$$

$$
+ \int_{-\infty}^{\infty} \left(f_2(\rho) \frac{\mathrm{d} f_1(\rho)}{\mathrm{d}r} - f_1(\rho) \frac{\mathrm{d} f_2(\rho)}{\mathrm{d}\rho} \right)
$$

$$
\left. \times \left[q(L+1) \tilde{j}_{L+1}(qr) - L q^{-1} \tilde{j}_{L-1}(qr) \right] \frac{r \mathrm{d}\rho}{(\alpha r + \beta)} \right\}
$$

$$
r = r(\rho). \tag{10.33}
$$

The integrals included in (10.32) and (10.33) are calculated using the Simpson formula. The derivative $df/d\rho$ is found from the formulas of numerical differentiation and is expressed via the values of the function at five successive points [see (9.24)].

10.2.2 Coulomb Multipole Matrix Elements

The complete reduced Coulomb multipole matrix elements are obtained by integrating over the angles and separating the factors containing the projections of the angular momenta [3] and are determined by an expression that includes the difference between the direct and exchange terms:

$$
U^{\lambda}_{\nu_1 \nu_2 \nu_3 \nu_4} = \langle \nu_1 \nu_4 \| U_\lambda \| \nu_2 \nu_3 \rangle = \frac{N_1}{(2\lambda + 1)} V^{\lambda}_{\nu_1 \nu_2 \nu_3 \nu_4}
$$

$$
- (2\lambda + 1) \sum_{\lambda'} (-1)^{\lambda + \lambda'} \begin{Bmatrix} l_1 & \lambda & l_2 \\ l_4 & \lambda' & l_3 \end{Bmatrix} V^{\lambda'}_{\nu_1 \nu_3 \nu_2 \nu_4}, \tag{10.34}
$$

where
$\begin{Bmatrix} l_1 & \lambda & l_2 \\ l_4 & \lambda' & l_3 \end{Bmatrix}$ is the $6j$- symbol [LL, 7].

Four matrices are formed from these matrix elements. Each of them has the structure described above in Sect. 9.2.2. The complete Coulomb matrix is similar to the matrix (9.35) and (9.36). The Coulomb matrix is calculated using the ***mael2*** procedure described in Sect. 9.2.2.

10.2.3 RPAE Equation for Multipole Matrix Elements

The generalized oscillator strengths, that account for the correlations in RPAE, are found by Formulas (10.8) and (10.9), in which the square of the matrix element

(10.12) or (10.13) from the operator (10.19) is substituted by the square modulus of the corresponding matrix element in RPAE. In analytic form, the equation for the matrix element in RPAE (10.6) can be represented in a somewhat more symmetric form than (10.9):

$$\left\langle \nu_1 | G_L(q,\omega) | \nu_2 \right\rangle = \left\langle \nu_1 | g_L^{(r,\nabla)}(q) | \nu_2 \right\rangle + \left(\sum_{\nu_3 \le F, \nu_4 > F} - \sum_{\nu_3 > F, \nu_4 \le F} \right) \frac{\left\langle \nu_4 | G_L(q,\omega) | \nu_3 \right\rangle \langle \nu_3 \nu_1 | U_L | \nu_4 \nu_2 \rangle}{\omega - E_4 + E_3 + i\eta(1 - 2n_{\nu_4})}. \tag{10.35}$$

The numerical solution of this equation for each q is similar to how it is done for the photoionization amplitude (9.53). Instead of an integral equation, we solve a system of linear algebraic equations, which we schematically write as follows:

$$G_i(q,\omega) = g_i^{(r,\nabla)}(q) + \sum_j G_j(q,\omega)\chi_j(\omega)U_{ji} \tag{10.36}$$

Here $g_i^{(r,\nabla)}(q)$ and $G_i(q,\omega)$ are the rows of dipole matrix elements, U_{ji} is the Coulomb matrix, and χ_j is the energy denominator.

The formal solution to this equation is written as:

$$G_i(q,\omega) = \sum_j g_i^{(r,\nabla)}(q)[\delta_{ij} - \chi_j(\omega)U_{ji}]^{-1} \tag{10.37}$$

To find the solution of Eq. (10.37), it is necessary to construct the matrix $[\delta_{ij} - \chi_j(\omega)U_{ji}]$, invert it, and then multiply the row of matrix elements by the column of the inverse matrix corresponding to the given energy ω.

The Coulomb matrix element U in (10.34), as in (9.34), includes direct and exchange interaction terms:

$$\langle \nu_3 \nu_1 | U | \nu_4 \nu_2 \rangle = \langle \nu_3 \nu_1 | V | \nu_4 \nu_2 \rangle - \langle \nu_3 \nu_1 | U | \nu_2 \nu_4 \rangle \tag{10.38}$$

In the Hartree–Fock and RPAE approximation, the generalized oscillator strengths are calculated by the ***gos*** program at each energy value ω for the set q obtained as follows: The initial impulse values q_0, the increment step Δq_0, and the step change factor τ are set. Then

$$q_1 = q_0 + \tau\,\Delta q_0, \ldots\ q_i = q_{i-1} + \tau\,\Delta q_{i-1}, \ldots. \tag{10.39}$$

For a subshell with one electron, calculations are performed only in the Hartree–Fock approximation.

In RPAE, the correlations are taken into account in one $\nu_1\nu_2$ (10.6), two $\nu_1\nu_2$, $\nu_3\nu_4$ or more transitions. In this case, hole states ν_1, ν_3 can coincide, but the linear moments of the particles must be different.

This ***gos*** program is also used when calculating the generalized oscillator strengths of atoms with semi-closed shells. Calculations are performed according to the algorithms described in Sect. 9.2.3.3.

10.2.4 Scattering Cross Section of Fast Electrons in the Logarithmic Approximation

The total cross section for inelastic scattering of fast electrons with ionization of nl—atomic subshell is expressed through the photoionization cross section $\sigma_{nl}(\omega)$ using the following expression [LL]:

$$\sigma^B_{nl,In}(E) = \frac{c}{2\pi E}\int_{I_{nl}}^{E} \frac{\sigma_{nl}(\omega)}{\omega}\ln\left(\varsigma_n\frac{4E}{\omega}\right)\mathrm{d}\omega, \tag{10.40}$$

where the parameter $\ln\varsigma_n \sim 1$ cannot be determined without detailed calculations. Expression (10.40) is called the Bethe cross section of inelastic scattering of an electron on nl-subshell. It is valid with the so-called logarithmic precision, i.e., under the assumption that $\ln(E/\omega) >> 1$.

The total cross section for inelastic scattering of fast electrons with ionization of any atomic electron in the same logarithmic approximation is given by the following expression

$$\sigma^B_{In}(E) = \frac{c}{2\pi E}\sum_{nl}\int_{I_{nl}}^{E} \frac{\sigma_{nl}(\omega)}{\omega}\ln\left(\varsigma_n\frac{4E}{\omega}\right)\mathrm{d}\omega \approx \frac{c}{2\pi E}\int_{I}^{E} \frac{\sigma(\omega)}{\omega}\ln\left(\varsigma\frac{4E}{\omega}\right)\mathrm{d}\omega$$
$$\equiv \frac{c}{2\pi E}\int_{I}^{E} \frac{\sigma(\omega)}{\omega}\ln\left(\frac{4E}{\omega}\right)\mathrm{d}\omega + A\frac{c}{2\pi E}\int_{I}^{E} \frac{\sigma(\omega)}{\omega}\mathrm{d}\omega \equiv \sigma_{In}(E) + A\sigma^{(1)}_{In}(E) \tag{10.41}$$

where $\sigma_{nl}(\omega)$ and $\sigma(\omega)$ are the partial cross section of the nl-subshell and the total cross section of photoionization, respectively; $A \equiv \ln\varsigma > 0$. Here the cross section $\sigma^{(1)}_{In}(E)$ is determined according to (10.41). The introduction $\sigma^{(1)}_{In}(E)$ allows to go beyond the logarithmic approximation by choosing the parameter A. When performing calculations using (10.40) and (10.41), it is best to set $\varsigma_n = \varsigma = 1$. Thus, we get $\sigma_{In}(E)$ and separately $\sigma^{(1)}_{In}(E)$. This allows to obtain the cross section with the so-called logarithmic accuracy $\sigma_{In}(E)$ and the first non-vanishing non-logarithmic correction, which is determined by $\sigma_{In}^{(1)}(E)$. Parameter A is chosen to improve overall agreement with experimental data.

The scattering cross section for fast electrons in the logarithmic approximation is determined using the ***scatlog*** program.

10.2.5 Compton Effect

10.2.5.1 Compton Ionization of Atoms and Ions

In Compton scattering, both the excitation of discrete levels of the target atom and its ionization can occur. The corresponding differential cross section is given by an expression similar to (10.18) [8] for a discrete transition:

$$\frac{\mathrm{d}\sigma^{CR}_{i\to f,L_{if}}(\omega_{if},q)}{\mathrm{d}\Omega} = \left(\frac{\mathrm{d}\sigma}{\mathrm{d}\Omega}\right)_0 \frac{E-\omega_{if}}{E}\frac{q^2}{2\omega_{if}} f^R_{i,f,L_{if}}(\omega_{if},q) \tag{10.42}$$

Here $i \to f$ denotes a discrete transition from state i to f, with excitation energy ω_{if} and angular momentum L_{if}.

For Compton ionization, instead of (10.42), we have a similar equation:

$$\frac{d^2\sigma^{CR}_{i\to\varepsilon_f,L_{if}}(\omega,q)}{\mathrm{d}\Omega\mathrm{d}\omega} = \left(\frac{\mathrm{d}\sigma}{\mathrm{d}\Omega}\right)_0 \frac{E-\omega}{E}\frac{q^2}{2\omega} f^R_{i,\varepsilon_f,L_{if}}(\omega,q), \tag{10.43}$$

where $\omega = \varepsilon_f + I_i$.

In specific calculations of the Compton scattering cross section, we restrict ourselves to monopole, dipole, quadrupole, and octupole transitions ($L = 0, 1, 2, 3$) and calculate $f^L_{nl,\varepsilon l'}(\omega,q)$ and $f^{RL}_{nl,\varepsilon l'}(\omega,q)$ in a rather wide range of q, $0 \le q \le 8$ a.u. and ω up to 100–120 Ry. For reasonable values of E, $E > 1000$ Ry with an accuracy of 1–2%, Expressions (10.42) and (10.43) can be simplified by reducing it to $q = (E/c)\theta$. Then from (10.42) and (10.43), it is obtained:

$$\left(\frac{\mathrm{d}\sigma^{CR}_{nl,\varepsilon_f}(\omega,q)}{\mathrm{d}\Omega}\right)\bigg/\left(\frac{\mathrm{d}\sigma}{\mathrm{d}\Omega}\right)_0 = \frac{q^2}{2\omega} f_{nl,\varepsilon_f}(\omega,q) \equiv \varsigma^R_{nl,\varepsilon_f}(\omega,q) = \sum_{L=0}^{3} \varsigma^R_{nl,\varepsilon_f,L}(\omega,q), \tag{10.44}$$

where

$$f_{nl,\varepsilon_f}(\omega,q) = \sum_{L=0}^{3}\sum_{l'} f^L_{nl,\varepsilon_f l'}(\omega,q)$$

$$\varepsilon = \omega - I_{nl}, q = E\theta/c. \quad (10.45)$$

We will call this quantity $\varsigma^{L}_{nl,\varepsilon}(\omega, q)$ the Compton scattering relative cross section. For a discrete transition, it is convenient to introduce a similar quantity

$$\varsigma^{RL}_{nl,n'l'}(\omega_{nl,n'l'}, q) \equiv q^2 f^{RL}_{nl,n'l'}(\omega_{nl,n'l'}, q)/2\omega_{nl,n'l'}. \quad (10.46)$$

The relative role of RPAE correlations in the ionization of an atom in Compton scattering is determined, in accordance with (10.42) and (10.43), by the following ratio

$$\left(\frac{d\sigma^{C,R}_{nl,\varepsilon_f}(\omega)}{d\Omega}\right) \Big/ \left(\frac{d\sigma^{C,HF}_{nl,\varepsilon_f}}{d\Omega}\right) = f^{RPAE}_{nl,\varepsilon_f}(\omega, q)/f_{nl,\varepsilon_f}(\omega, q). \quad (10.47)$$

We can also calculate the total Compton scattering cross sections. For a given transferred angular momentum L, we have the following formulas in the HF and RPAE (denoted by the superscript R).

$$\frac{d^2\sigma^{CHF,L}(\omega, q)}{d\omega dq} = \frac{q^2}{2\omega c^4}\left[1 - \frac{1}{2}\left(\frac{qc}{E}\right)^2\right]\sum_{nl,l'} f^{L}_{nl,\varepsilon_f l'}(\omega, q), \quad (10.48)$$

$$\frac{d^2\sigma^{CR,L}(\omega, q)}{d\omega dq} = \frac{q^2}{2\omega c^4}\left[1 - \frac{1}{2}\left(\frac{qc}{E}\right)^2\right]\sum_{nl,l'} f^{RL}_{nl,\varepsilon_f l'}(\omega, q). \quad (10.49)$$

The cross section for Compton scattering, which is differential in ω and q, must be summed over all angular momenta L that is included in the calculations, which leads to the relation

$$\frac{d^2\sigma^{CR,HF}(\omega, q)}{d\omega dq} = \sum_{L=0}^{3} \frac{d^2\sigma^{CR,HF,L}(\omega, q)}{d\omega dq} \quad (10.50)$$

The cross section that is differential only in ω is obtained from Formulas (10.49) and (10.50) by integrating them over q:

$$\frac{d\sigma^{CHF}(\omega)}{d\omega} = \frac{1}{2\omega c^2 E^2}\int_0^{q_{\max}} q^3\left[1 - \frac{1}{2}\left(\frac{qc}{E}\right)^2\right]\sum_{nl,l'L} f^{L}_{nl,\varepsilon_f l'}(\omega, q)dq, \quad (10.51)$$

$$\frac{d\sigma^{CR}(\omega)}{d\omega} = \frac{1}{2\omega c^2 E^2}\int_0^{q_{\max}} q^3\left[1 - \frac{1}{2}\left(\frac{qc}{E}\right)^2\right]\sum_{nl,l'L} f^{RL}_{nl,\varepsilon_f l'}(\omega, q)dq. \quad (10.52)$$

Here $q_{\max}$ is the upper limit in q that in our specific calculations was taken equal to 8 a.u., which is large enough. Notice, that $q_{\max}c/E << 1$.

Numerical calculations of the Compton ionization cross section are complicated by the fact that with increase of ω above the ionization threshold, the number of essentially contributing angular momenta of the knocked out electron l' grows very rapidly, which should be taken into account in the corresponding sums in Formulas (10.51) and (10.52). The situation with growth of ω becomes even more complicated, if one wishes to calculate the total cross sections of Compton ionization, which are obtained from (10.51) and (10.52) by integration over ω,:

$$\sigma^{CHF} = \frac{1}{2\omega c^2 E^2} \int_0^{q_{\max}} q^3 \left[1 - \frac{1}{2}\left(\frac{qc}{E}\right)^2\right] \sum_{nl,l'L} \int_{I_{nl}}^{E} f^L_{nl,\varepsilon_f l'}(\omega, q) \mathrm{d}q \mathrm{d}\omega, \quad (10.53)$$

$$\sigma^{CR} = \frac{1}{2\omega c^2 E^2} \int_0^{q_{\max}} q^3 \left[1 - \frac{1}{2}\left(\frac{qc}{E}\right)^2\right] \sum_{nl,l'L} \int_{I_{nl}}^{E} f^{RL}_{nl,\varepsilon_f l'}(\omega, q) \mathrm{d}q \mathrm{d}\omega. \quad (10.54)$$

Note that the momentum $q_{\max}$ also grows with increasing ω, which further complicates the calculation of the total Compton ionization cross sections.

10.2.5.2 The Cross Section of the Compton Excitation of Discrete Levels

This cross section is determined by expressions similar to (10.53) and (10.54), where the energy of the excited atomic electron is determined by the combination of the principal quantum number and angular momentum $n'l'$:

$$\sigma^{CHFL}_{nl,n'l'}(\omega_{nl,n'l'}) = \frac{1}{2\omega_{nl,n'l'} c^2 E^2} \int_0^{q_{\max}} q^3 \left[1 - \frac{1}{2}\left(\frac{qc}{E}\right)^2\right] f^L_{nl,n'l'}(\omega_{nl,n'l'}, q) \mathrm{d}q, \quad (10.55)$$

$$\sigma^{CRL}_{nl,n'l'}(\omega_{nl,n'l'}) = \frac{1}{2\omega_{nl,n'l'} c^2 E^2} \int_0^{q_{\max}} q^3 \left[1 - \frac{1}{2}\left(\frac{qc}{E}\right)^2\right] f^{LR}_{nl,\varepsilon l'}(\omega_{nl,n'l'}, q) \mathrm{d}q, \quad (10.56)$$

Note that the calculation of (10.55) and (10.56) is much simpler than the Compton ionization cross section.

After summing (10.55) and (10.56) over all $n'l'$ и L, we obtain the total contribution to the Compton scattering cross section $\sigma^{CHF,R}_{nl}(\omega)$ from the excitation of all discrete levels of nl-electrons. It is appropriate to compare this cross section with the total classical Thompson cross section for light scattering by an electron σ_0

$$\sigma_0 = \frac{8\pi}{3c^4} \approx 2.3755 \times 10^{-8} \text{at.un.} \quad (10.57)$$

Note that the contribution of the group of discrete levels decreases with increasing E as $1/E^2$, while the total Compton cross section is almost independent of E. This means that with increasing E, the main contribution to the Compton cross section comes not from excitations, but from the ionization of atomic electrons.

We represent the cross sections $\sigma_{nl}^{R}(\omega)$, according to (10.55) and (10.56), in the form [8]:

$$\sigma_{nl}^{CHF,R}(\omega) = \frac{c^2}{E^2}\sigma_0 A_{nl}^{HF,R}(\omega) \tag{10.58}$$

As an example, the factors $A_{nl}^{HF,R}(\varpi)$ were calculated in [8] and turned out to be equal to $A_{2p}^{R} \approx 0.2210$ for Ne, $A_{3p}^{R} \approx 0.1454$ for Ar, $A_{4p}^{R} \approx 0.2084$ for Kr, and $A_{5p}^{R} \approx 0.2413$ for Xe. For energies that are of interest in Compton scattering, namely for E greater than tens of keV, we have $cq_{\max}/E \ll 1$. Recall that we are using the atomic system of units. As can be seen that $A_{nl}(\omega) << Z$, and therefore, the cross section for excitation of discrete levels is much smaller than the total atomic classical Thompson cross section given in (10.57).

To obtain all the characteristics described above, the ***gos*** program [9] is used.

References

1. D.R. Bates, A. Fundaminsky and H.S.W. Massey, Trans. Royal Soc. **A243** (1950)
2. R.H. Pratt, Radiat. Phys. Chem. **74**, 6 (2008)
3. M.Ya. Amusia, L.V. Chernysheva, Z. Felfli, A.Z. Msezane, Surf. Rev. Lett. **9**, 2 (2002)
4. M.Ya. Amusia, L.V. Chernysheva, Z. Felfli, A.Z. Msezane, Phys. Rev. A **65**, 62705 (2002)
5. M.Ya. Amusia, N.A. Cherepkov, S.I. Sheftel, Zh. Zhivanovich, R. Yanev, J. Phys. B: At. Mol. Phys. **6** (1973)
6. I.S. Gradshteyn, I.M. Ryzhik, *Table of Integrals, Series and Products* (Academic Press, 1965)
7. D.A. Varshalovich, A.N. Moskalev, V.K. Khersonskii, *Quantum Theory of Angular Momentum* (World Scientist, Singapore, 1988).
8. M.Ya. Amusia, L.V. Chernysheva, Z. Felfli and A.Z. Msezane, Phys. Rev. A**75**, 062703 (2007)
9. L.V. Chernysheva, M.Ya. Amusia, S.I. Sheftel, *Numerical Calculation of Atomic Generalized Oscillator Strength Allowing for Many-Electron Correlations within Two Transitions* (The Ioffe Institute, No. 495, Leningrad, 1975) (In Russian)

Chapter 11
Angular Distribution of Photoelectrons and Secondary Electrons Outside the Dipole Approximation

In this chapter, we discuss the angular distributions of photoelectrons and secondary electrons knocked out from the target in the process of inelastic scattering of fast electrons. Their general and specific features are investigated, demonstrating a significant difference in the corresponding characteristics of the two processes under consideration, namely photoionization and fast electron scattering.

Specifically, formulas are given and programs are developed for calculating the nondipole parameters of *s*-, *p*-, and *d*-subshells of atoms with closed subshells. It is noted how the anisotropy parameters can be calculated for semi-closed shells.

11.1 Theoretical Model

11.1.1 Angular Distribution of Photoelectrons

Let us start by considering the angular distribution of photoelectrons. As the photon energy ω increases, the dipole approximation, within which the process of photoionization of an atom is usually considered, begins to break down, and non-dipole corrections start to play an increasingly important role. At not too high energies ω, the main non-dipole corrections are those due to the presence of the photon linear momentum. Taking it into account, the operator of the interaction of an electron with an electromagnetic field is determined in the "velocity" form not by an expression $\hat{d}^{\nabla} = i\mathbf{e}\vec{\nabla}$ where e is a photon polarization vector, but by a more general expression [AM], [AICC]

$$\hat{d}^{\nabla} = \mathrm{e}^{i\kappa\mathbf{r}}\left(\mathbf{e}\,\vec{\nabla}\right), \tag{11.1}$$

M. Ya. Amusia and L. V. Chernysheva, *Computation of Atomic and Molecular Processes*, Springer Series on Atomic, Optical, and Plasma Physics 117,
https://doi.org/10.1007/978-3-030-85143-9_11

Likewise, the operator in the "length" form instead of expression $\hat{d}^r = \omega \mathbf{er}$ becomes:

$$\hat{d}^r = \omega e^{i\kappa \mathbf{r}}(\mathbf{er}), \tag{11.2}$$

where $\kappa = \omega/c$ is the photon momentum. The order of magnitude of the non-dipole corrections is characterized by a dimensionless factor κa_i, where a_i is the radius of the ith subshell being ionized. If κa_i is noticeably less than 1, then instead of expressions (11.1) and (11.2), the following expressions for the operators d_κ^∇ and d_κ^r, taking into account the first-order corrections in powers κa_i, become sufficiently accurate:

$$\begin{aligned} \hat{d}_\kappa^\nabla &= \left(\mathbf{e}\vec{\nabla}\right) + i(\kappa \mathbf{r})\left(\mathbf{e}\vec{\nabla}\right) \\ \hat{d}_\kappa^\kappa &= \omega[(\mathbf{er}) + i(\kappa \mathbf{r})(\mathbf{er})], \end{aligned} \tag{11.3}$$

Note that expressions for the operator $\hat{d}$ are given in (11.1) and (11.2) without some numerical coefficients. Exact expressions can be obtained from formulas found in manuals on quantum electrodynamics, for example, [BLP] and [IS]. Additional (in comparison with $\hat{d}^\nabla = i\mathbf{e}\vec{\nabla}$ and $\hat{d}^r = \omega \mathbf{er}$) terms in operators (11.1), (11.2) lead to the fact that not only the amplitudes of dipole transitions with a change in the angular momentum of the electron by unity $l_i \to l_i \pm 1$ are nonzero, but also the quadrupole matrix elements. For them, the angular momentum changes by two or does not change at all, i.e., we have matrix elements with $l_i \to l_i \pm 2, 0$.

In the total photoionization cross sections of the i-subshell, the contributions of dipole and quadrupole terms in (11.3) do not interfere, so that the second terms in (11.3) lead to corrections of the order of $(\kappa a_i)^2$, and at $(\kappa a_i) \ll 1$, its contribution to the cross section is much smaller than in the angular distribution. In the angular distribution of photoelectrons $d\sigma_i/d\Omega$, corrections associated with the presence of terms $(\kappa \mathbf{r})$ in the operators $\hat{d}^\nabla$ and $\hat{d}^r$, are of the order κa_i, i.e., much more than in cross section σ_i. As a result, the study of $d\sigma_i/d\Omega$ makes it possible to investigate the role of electronic correlations not only in dipole, but also in quadrupole transitions.

The expression for the differential cross section with allowance for the non-dipole corrections in (11.2) for unpolarized light can be represented as [1, 2]:

$$\begin{aligned} &\frac{d\sigma_{nl}(\omega)}{d\Omega} \\ &= \frac{\sigma_{nl}(\omega)}{4\pi}\left\{1 - \frac{\beta_{nl}(\omega)}{2}P_2(\cos\theta) + \frac{\omega}{c}\left[\gamma_{nl}(\omega)P_1(\cos\theta) + \eta_{nl}(\omega)P_3(\cos\theta)\right]\right\} \end{aligned} \tag{11.4}$$

Here $\beta_{nl}(\omega)$ is the dipole angular anisotropy parameter described in Sect. 9.2.10, $P_i(\cos\theta)$ with $i = 1, 2, 3$ are the Legendre polynomials, $\gamma_{nl}(\omega)$ and $\eta_{nl}(\omega)$ are the non-dipole angular anisotropy parameters. In the one-electron approximation, $\gamma_{nl}(\omega)$

and $\eta_{nl}(\omega)$ are expressed in terms of the dipole $l_i \to l' = l_i \pm 1$ and quadrupole $l_i \to l' = l_i \pm 2, 0$ matrix elements $\left\langle \varepsilon, l' \middle| \hat{d}_k^{\nabla, r} \middle| n l_i \right\rangle$.

Sources of linearly polarized radiation are usually used in experiments. In this case, instead of (11.4), a slightly different expression is used [3, 4]:

$$\frac{d\sigma_{nl}(\omega)}{d\Omega} = \frac{\sigma_{nl}(\omega)}{4\pi}\left\{1 + \beta_{nl}(\omega)P_2(\cos\vartheta) + \left[\delta_{nl}^C(\omega) + \gamma_{nl}^C(\omega)\cos^2\vartheta\right]\sin\vartheta\cos\Phi\right\} \quad (11.5)$$

Here ϑ is the polar angle between the vectors of the photoelectron velocity $\vec{v}$ and the polarization of the photon **e**, Φ is the azimuthal angle determined by the projection of the velocity $\vec{v}$ in the plane perpendicular to **e** and containing the photon propagation vector, $\delta_{nl}^C(\omega)$ and $\gamma_{nl}^C(\omega)$ are the non-dipole Cooper parameters connected to the parameters $\gamma_{nl}(\omega)$ and $\eta_{nl}(\omega)$ introduced in (11.4) by simple relations

$$\gamma_{nl}^C/5 + \delta_{nl}^C = \kappa\gamma_{nl}, \quad \gamma_{nl}^C/5 = -\kappa\eta_{nl} \quad (11.6)$$

The corresponding expressions for $\gamma_{nl}(\omega)$ and $\eta_{nl}(\omega)$ are in general rather cumbersome:

$$\begin{aligned}\gamma_{nl}(\omega) = &\frac{6}{5\left[l d_{l-1}^2 + (l+1)d_{l+1}^2\right]} \\ &\times\left\{\frac{l+1}{2l+3}\left[3(l+2)q_{l+2}d_{l+1}\cos(\delta_{l+2}-\delta_{l+1}) - l q_l d_{l+1}\cos(\delta_{l+2}-\delta_{l+1})\right]\right. \\ &\left. - \frac{l}{2l-1}\left[3(l-1)q_{l-2}d_{l-1}\cos(\delta_{l-2}-\delta_{l-1}) - (l+1)q_l d_{l-1}\cos(\delta_l-\delta_{l-1})\right]\right\}\end{aligned} \quad (11.7)$$

$$\begin{aligned}\eta_{nl}(\omega) = &\frac{6}{5\left[l d_{l-1}^2 + (l+1)d_{l+1}^2\right]}\left\{\frac{(l+1)(l+2)}{(2l+1)(2l+3)}q_{l+2}\left[5l d_{l-1}\cos(\delta_{l+2}-\delta_{l-1})\right.\right. \\ &\left. - (l+3)d_{l+1}\cos(\delta_{l+2}-\delta_{l-1})\right] - \frac{(l-1)l}{(2l+1)(2l-1)}q_{l-2} \\ &\times\left[5(l+1)d_{l+1}\cos(\delta_{l-2}-\delta_{l+1}) - (l-2)d_{l-1}\cos(\delta_{l-2}-\delta_{l-1})\right] \\ &\left. + 2\frac{l(l+1)}{(2l-1)(2l+3)}q_l\left[(l+2)d_{l+1}\cos(\delta_l-\delta_{l+1}) - (l-1)d_{l-1}\cos(\delta_l-\delta_{l-1})\right]\right\}.\end{aligned} \quad (11.8)$$

Here, in the r-form, the dipole and quadrupole matrix elements $d_{l\pm1}$ are given by the equations

$$d_{l\pm1} = \int_0^\infty P_{nl}(r) r P_{\varepsilon l\pm1}(r) \mathrm{d}r, \; q_{l\pm2,0} = \frac{1}{2}\int_0^\infty P_{nl}(r) r^2 P_{\varepsilon l\pm2,0}(r)\mathrm{d}r. \tag{11.9}$$

To account for multielectron correlations when calculating the non-dipole parameters, one has to replace the HF matrix elements with matrix elements in the RPAE. Quadrupole matrix elements $Q_{l\pm2,0}$ are obtained from the solution of the radial part of the quadrupole RPAE of an equation similar to Eq. (9.56):

$$\langle\nu_2|Q(\omega)|\nu_1\rangle = \langle\nu_2|\hat{q}|\nu_1\rangle + \sum_{\nu_3,\nu_4} \frac{\langle\nu_3|Q(\omega)|\nu_4\rangle\left(n_{\nu_4} - n_{\nu_3}\right)\langle\nu_4\nu_2|U|\nu_3\nu_1\rangle}{\varepsilon_{\nu_4} - \varepsilon_{\nu_3} + \omega + i\eta\left(1 - 2n_{\nu_3}\right)}. \tag{11.10}$$

Here in r-form one has $\hat{q} = r^2 P_2(\cos\theta)$. The numerical procedure for solving Eq. (11.10) is similar to the procedure for solving the equation for the dipole component described in Sect. 9.2.3. This equation can be generalized to allow for rearrangement (also called relaxation) (see Sect. 8.1) in the same way as the equations for the dipole component. This is achieved by an appropriate choice of one-electron wave functions of excited states in the Hartree–Fock approximation.

Expressions for non-dipole parameters were first obtained in [1, 2]. For s-shells, expressions (11.7–11.8) are quite simple, taking the form:

$$\gamma_{ns}(\omega) = -\eta_{ns}(\omega) = \frac{6q_2}{5d_1}\cos\left(\delta_d - \delta_p\right), \tag{11.11}$$

where $d_1(q_2)$ is the dipole (quadrupole) matrix element $\left\langle\varepsilon p, d\middle|\hat{d}(\hat{q})\middle|ns\right\rangle$, and $\delta_{p(d)}$ is the scattering phase of the *p(d)*-photoelectron. In order to account for multielectron correlations, it is necessary to solve Eq. (11.10) separately for dipole and for quadrupole matrix elements.

Let us obtain the value $\gamma_{nl}(\omega)$ determined in (11.11) in the simplest one-particle approximation—the hydrogen-like one. Near the ionization threshold, the ratio q_2/d_1 can be estimated as $q_2/d_1 \sim a_{ns}$, where a_{ns} is the radius of the ionizable subshell, so that at the threshold the contribution of the non-dipole parameter, according to (11.11), is determined by the relation

$$\frac{\omega}{c}\gamma_{ns} \sim \frac{\omega}{c}a_{ns} = \kappa a_{ns}. \tag{11.12}$$

For the outer shells $\omega \approx 1, a_{ns} \approx 1$ and $\gamma_{nl}(\omega) \approx 1/137$. In the simplest hydrogen approximation, i.e., for a purely Coulomb field of a nucleus with a charge Z, we have a much smaller value near the photoionization threshold. Indeed, in this case, we find

$$\frac{q_2^C}{d_1^C} = \frac{2\sqrt{Z^2 + 4v^2}}{Z^2 + v^2}, \tag{11.13}$$

$$\cos(\delta_d^C - \delta_p^C) = \frac{2v}{\sqrt{Z^2 + 4v^2}}, \tag{11.14}$$

where v is the speed of the photoelectron. Hence, we get the expression

$$\kappa\gamma_{1s} = \frac{12}{5}\frac{v}{c}, \tag{11.15}$$

which vanishes at the photoionization threshold, in accordance with [5]. This behavior is a consequence of the existence of a delicate balance between phases and matrix elements q and d in a purely Coulomb field, namely the validity of the relation

$$\delta_d^K - \delta_p^K = \frac{\pi}{2} - \arcsin\frac{2v}{\sqrt{Z^2 + 4v^2}}. \tag{11.16}$$

The presence of a short-range potential along with the Coulomb potential leads to violation of relation (11.15) and returns us to estimate (11.12). However, as evidenced by the results of calculations, the contribution of non-dipole corrections can be much larger and even exceed unity. Qualitatively, the reason for this lies either in the particular smallness of the dipole or in the large value of the quadrupole matrix elements.

From the point of view of elucidating the role of quadrupole multielectron correlations, outer and intermediate shells and low-energy photoelectrons are of particular interest. The expression for the parameters $\gamma_{nl}(\omega)$ and $\eta_{nl}(\omega)$ with allowance for multielectron correlations differs little from (11.7) and (11.8). For example, for s-electrons instead of (11.11) we get:

$$\gamma_{n0}(\omega) - -\eta_{n0}(\omega) = \frac{6\tilde{Q}_2(\omega)}{5\tilde{D}_1(\omega)}\cos(\overline{\delta}_d - \overline{\delta}_p), \tag{11.17}$$

Here $\tilde{D}_1(\omega)$ and $\tilde{Q}_2(\omega)$ are the moduli of the dipole and quadrupole matrix elements in the RPAE—solutions of Eq. (11.10) in the dipole and quadrupole channels, respectively, and the phases are given by the relation $\overline{\delta}_{p(d)} = \delta_{p(d)} + \xi_{p(d)}$. The phase correction $\xi_{p(d)}$ is determined by the presence of the imaginary part in the dipole (quadrupole) matrix element of the RPAE

$$\begin{aligned} D_1(\omega) &= |D_1(\omega)|\exp(i\xi_p) \equiv \tilde{D}_1(\omega)\exp(i\xi_p), \\ Q_2(\omega) &= |Q_2(\omega)|\exp(i\xi_d) \equiv \tilde{Q}_2(\omega)\exp(i\xi_d) \end{aligned} \tag{11.18}$$

As follows from formula (11.17), the parameter $\gamma_{n0}(\omega)$ significantly increases near the zeros of the matrix element $D_1(\omega)$ and the maxima in the amplitude $Q_2(\omega)$.

At zeros $D_1(\omega)$, the photoabsorption cross section itself is small, which for the *ns*-subshell is determined by the expression $\sigma_{ns}(\omega) \sim |D_1(\omega)|^2$. This makes it difficult to observe the effect of increasing the parameter $\gamma_{ns}(\omega)$.

The situation is different with the maximum in the quadrupole amplitude $Q_2(\omega)$. The amplitude $Q_2(\omega)$ can have a significant maximum near the discrete quadrupole transition from the inner shell, i.e., near the autoionization resonance [6, 7]. Significant amplifications $\gamma_{ns}(\omega)$ also arise in the vicinity of other than those considered in these works, quadrupole resonances, for example, shape resonances [6].

In order to simplify the detection of non-dipole corrections, the cross section (11.5) is measured at the so-called magic angle, equal to $\vartheta_m \cong 57.3°$, for which $P_2(\cos\vartheta_m) = 0$, so that there is no contribution from the term with β_{nl}. At an angle ϑ_m, the terms $\delta_{nl}(\omega)$ and $\gamma_{nl}(\omega)$ enter (11.5) in the following "magic" combination

$$\lambda_{nl}^C(\omega) = \gamma_{nl}^C(\omega) + 3\delta_{nl}^C(\omega) \tag{11.19}$$

In other words, at the "magic angle" $d\sigma_{nl}/d\Omega$ is determined by only one parameter.

11.1.2 Angular Distribution of Electrons Knocked Out in the Process of Scattering

Let us consider the angular distribution of electrons knocked out of an atom or ion by a charged fast particle. In this case, the interaction of the "projectile" and the "target" is sufficient to take into account in the first Born approximation. The corresponding cross section, according to (10.1), is expressed in terms of the generalized oscillator strength determined with the help of (10.3) and depending on the momentum $\mathbf{q}$ and energy ω transferred to the target. In this subsection, we consider the angular distribution relative to the vector $\mathbf{q}$ of electrons knocked out of an atom in the course of inelastic scattering.

Let us compare the result obtained for such an angular distribution in the limit of small q with the known formula for the differential cross section for the photoionization of an atom by unpolarized light. As can be seen from formulas (10.8), the generalized oscillator strengths are determined by the matrix element of the Bessel function $j_L(qr)$. At small enough q, one can use the lower-order terms of at least four first spherical Bessel functions:

$$j_0(qr) \cong 1 - \frac{(qr)^2}{6};\; j_1(qr) \cong \frac{qr}{3}\left(1 - \frac{(qr)^2}{10}\right);$$
$$j_2(qr) \cong \frac{(qr)^2}{15}\left(1 - \frac{(qr)^2}{14}\right);\, j_3(qr) \cong \frac{(qr)^3}{105}. \tag{11.20}$$

The matrix elements of the generalized oscillator strengths of nl-subshells of the target atom or ion have the form (10.9) and the corresponding short notation, which implies that it is considered the ionization of nl-subshells:

$$G_{l'L} \equiv G^{L}_{nl,\varepsilon l'}(\omega, q) = \left\langle nl \middle| \hat{G}^{L}(\omega, q) \middle| \varepsilon l' \right\rangle \tag{11.21}$$

Keeping terms of the order of q^4 or less, we have, using the expression from [8] for the angular distribution of s-electrons, the following formula:

$$\begin{aligned}
&\frac{dF_{n0}(q,\omega)}{d\Omega}\\
&= \frac{F_{ns}(q,\omega)}{4\pi}\left\{1 + 2P_2(\cos\theta) + \frac{2}{\tilde{G}_{11}}\left[\tilde{G}_{00}\cos(\bar{\delta}_0 - \bar{\delta}_1) + 2\tilde{G}_{22}\cos(\bar{\delta}_1 - \bar{\delta}_2)\right]\right.\\
&\quad \left.\times\, P_1(\cos\theta) + \frac{6\tilde{G}_{22}}{\tilde{G}_{11}}\cos(\bar{\delta}_1 - \bar{\delta}_2)P_3(\cos\theta)\right\}\\
&\equiv \frac{F_{ns}(q,\omega)}{4\pi}\left\{1 + 2P_2(\cos\theta) + q\left[\gamma^{(in)}_{ns}(q,\omega)P_1(\cos\theta) + \eta^{(in)}_{ns}(q,\omega)P_3(\cos\theta)\right]\right\}.
\end{aligned} \tag{11.22}$$

Here G_{00}, G_{11}, G_{22} are the matrix elements of the generalized oscillator strengths for transitions $s - s(L = 0)$, $s - p(L = 1)$ and $s - d(L = 2)$, correspondingly, related, according to (10.8) and (10.9), to the matrix elements of $j_L(qr)$. As follows from (11.20), the term with the lowest degree of q is $\tilde{G}_{11} \sim q << 1$. Corrections to $\tilde{G}_{11}$ are proportional to q^3. As for $\tilde{G}_{00}$ and $\tilde{G}_{22}$, they are proportional to q^2 with corrections of the order of q^4.

It is necessary to compare relation (11.22) with a similar one for the photoionization of ns-subshell:

$$\begin{aligned}
\frac{d\sigma_{ns}(\omega)}{d\Omega} &= \frac{\sigma_{ns}(\omega)}{4\pi}\left\{1 - P_2(\cos\theta) + \kappa\frac{6\tilde{Q}_2}{5\tilde{D}_1}\cos(\bar{\delta}_1 - \bar{\delta}_2)[P_1(\cos\theta) - P_3(\cos\theta)]\right\}\\
&\equiv \frac{\sigma_{ns}(\omega)}{4\pi}\left\{1 - P_2(\cos\theta) + \kappa\left[\gamma_{ns}(\omega)P_1(\cos\theta) + \eta_{ns}(\omega)P_3(\cos\theta)\right]\right\},
\end{aligned} \tag{11.23}$$

where, $\gamma_{n0}(\omega) = -\eta_{n0}(\omega) = \frac{6\tilde{Q}_2}{5\tilde{D}_1}\cos(\bar{\delta}_1 - \bar{\delta}_2)$.

Expressions (11.22) and (11.23) differ in the sign and magnitude of the dipole parameter and in slightly different expressions for the non-dipole terms. This difference exists and is significant even in the so-called optical limit $q \to 0$. According to (11.7) and (11.20), in the limit $q \to 0$, there are simple relations between the dipole $\tilde{D}_1$ and quadrupole $\tilde{Q}_2$ matrix elements and functions $\tilde{G}_{11}$, $\tilde{G}_{22}$: $\tilde{G}_{11} = q\tilde{D}_1/3$ и $\tilde{G}_{22} = 2q^2\tilde{Q}_2/15$. With the help of relations $\tilde{G}_{00} = -q^2\tilde{Q}_2/3 = -(5/2)\tilde{G}_{22}$, formula (11.22) is transformed into the following expression:

$$\frac{dF_{ns}(q,\omega)}{d\Omega} = \frac{F_{ns}(q,\omega)}{4\pi}\{1 + 2P_2(\cos\theta)$$
$$+ q\left[\frac{2\tilde{Q}_2}{\tilde{D}_1}\left(\frac{4}{5}\cos(\bar{\delta}_1 - \bar{\delta}_2) - \cos(\bar{\delta}_0 - \bar{\delta}_1)\right)P_1(\cos\theta) + 2\gamma_{ns}(\omega)P_3(\cos\theta)\right]\Bigg\}. \quad (11.24)$$

The difference with (11.23) is obvious, since the angular distribution cannot be expressed in terms of a single non-dipole parameter $\gamma_{ns}(\omega)$. This is due to the appearance of a new phase difference $\bar{\delta}_0 - \bar{\delta}_1$. As a result, for very small q, the following relations are valid:

$$\gamma_{ns}^{(in)}(\omega) = \frac{2\tilde{Q}_2}{\tilde{D}_1}\left[\frac{4}{5}\cos(\bar{\delta}_1 - \bar{\delta}_2) - \cos(\bar{\delta}_0 - \bar{\delta}_1)\right],$$
$$\eta_{ns}^{(in)}(\omega) = 2\gamma_{ns}(\omega) = \frac{12}{5}\frac{\tilde{Q}_2}{\tilde{D}_1}\cos(\bar{\delta}_1 - \bar{\delta}_2). \quad (11.25)$$

We see that the study of inelastic scattering even in the limit $q \to 0$ makes it possible to obtain an additional characteristic of the ionization process, namely the phase of the s-wave.

Note that formula (11.24) is valid not only in the limit $q \to 0$, but also at nonzero values of the transferred momentum. To obtain the corresponding value for $q \neq 0$, it is necessary to calculate all the anisotropy and phase parameters included in (11.22) as functions of q.

For $l > 0$, even at very small values of q, the relationship between the non-dipole parameters in photoionization and inelastic scattering of fast electrons is rather complicated.

The similarity of the general structure and the significant difference between (11.23) and (11.24) is obvious. In fact, the contribution of the non-dipole parameters to inelastic scattering can be much larger than to photoionization, since the condition $\omega/c = \kappa << q < R^{-1}$ is easy to achieve. Note that even if the terms with q are neglected, formulas (11.23) and (11.24) differ: in photoionization, the angular distribution is proportional to $\sin^2\theta$ [see (11.23)], while in inelastic scattering it is proportional $\cos^2\theta$ [see (11.24)].

Let us compare the angular distributions of the electrons of the subshells with $l \neq 0$ knocked out from the target atom in inelastic scattering and photoionization. Consider, whether the difference already exists in the limit $q \to 0$, as is the case for s-subshells. In the limit $q \to 0$, according to [8], the following expressions are valid.

For $l = 1$

$$\beta_{np}^{(in)}(q = 0, \omega) = -\frac{4}{\tilde{D}_0^2 + 2\tilde{D}_2^2}[\tilde{D}_2^2 - 2\tilde{D}_0\tilde{D}_2\cos(\bar{\delta}_0 - \bar{\delta}_2)] \quad (11.26)$$

$$\gamma_{np}^{(in)}(q = 0, \omega) = \frac{18}{25[\tilde{D}_0^2 + 2\tilde{D}_2^2]}$$

$$\times\left\{5\tilde{D}_0\tilde{Q}_1\cos(\overline{\delta}_1-\overline{\delta}_0)+2\tilde{D}_2[2\tilde{Q}_3\cos(\overline{\delta}_3-\overline{\delta}_2)-3\tilde{Q}_1\cos(\overline{\delta}_1-\overline{\delta}_2)\right\},\tag{11.27}$$

$$\eta_{np}^{(in)}(q=0,\omega)=\frac{12}{25[\tilde{D}_0^2+2\tilde{D}_2^2]}$$
$$\times\left\{5\tilde{D}_0\tilde{Q}_3\cos(\overline{\delta}_3-\overline{\delta}_0)+2\tilde{D}_2\left[3\tilde{Q}_1\cos(\overline{\delta}_1-\overline{\delta}_2)-2\tilde{Q}_3\cos(\overline{\delta}_3-\overline{\delta}_2)\right]\right\}\tag{11.28}$$

For $l=2$ we have

$$\beta_{nd}^{(in)}(q=0,\omega)=-\frac{4}{5[2\tilde{D}_1^2+3\tilde{D}_3^2]}[\tilde{D}_1^2+6\tilde{D}_3^2-18\tilde{D}_1\tilde{D}_3\cos(\overline{\delta}_1-\overline{\delta}_3)],\tag{11.29}$$

$$\gamma_{nd}^{(in)}(q=0,\omega)=\frac{2}{35[2\tilde{D}_1^2+3\tilde{D}_3^2]}\left\{14\tilde{D}_1\left[7\tilde{Q}_2\cos(\overline{\delta}_1-\overline{\delta}_2)-2\tilde{Q}_0\cos(\overline{\delta}_0-\overline{\delta}_1)\right]\right.$$
$$\left.+\,9\tilde{D}_3\left[8\tilde{Q}_4\cos(\overline{\delta}_4-\overline{\delta}_3)-13\tilde{Q}_2\cos(\overline{\delta}_2-\overline{\delta}_3)\right]\right\},\tag{11.30}$$

$$\eta_{nd}^{(in)}(q=0,\omega)=\frac{12}{35[2\tilde{D}_1^2+3\tilde{D}_3^2]}\left\{2\tilde{D}_1\left[\tilde{Q}_2\cos(\overline{\delta}_2-\overline{\delta}_1)-6\tilde{Q}_4\cos(\overline{\delta}_4-\overline{\delta}_1)\right]\right.$$
$$\left.+\,\tilde{D}_3\left[7\tilde{Q}_0\cos(\overline{\delta}_0-\overline{\delta}_3)-8\tilde{Q}_2\cos(\overline{\delta}_2-\overline{\delta}_3)-6\tilde{Q}_4\cos(\overline{\delta}_4-\overline{\delta}_3)\right]\right\}.\tag{11.31}$$

In order to obtain expressions (11.26)–(11.31), the following relations based on (11.21) were used:

$$\tilde{G}_{l'1}\equiv\frac{q}{3}\tilde{D}_{l'}(l'=l\pm1);\ \tilde{G}_{l'0}\equiv-\frac{q^2}{3}\tilde{Q}_{l'}(l'=l);\ \tilde{G}_{l'2}\equiv\frac{2q^2}{15}\tilde{Q}_{l'}(l'=l,l\pm2)\tag{11.32}$$

To compare the parameters of angular anisotropy in photoionization and scattering of fast electrons, note that the following relations are valid in the HF approximation

$$\tilde{D}_{l'}^r\Rightarrow\tilde{d}_{l'}^r=\int\limits_0^\infty P_{nl}(r)rP_{kl'}(r)\mathrm{d}r;\ \mathrm{Q}_{l'}^r\Rightarrow q_{l'}^r=\frac{1}{2}\int\limits_0^\infty P_{nl}(r)r^2P_{kl'}(r)\mathrm{d}r\tag{11.33}$$

Here, $P_{nl(kl')}(r)=rR_{nl(kl')}(r)$, $R_{nl(kl')}(r)$ is the radial parts of the electron wave function in the initial and final states, respectively, in the HF approximation.

The angular distribution of photoelectrons for any l with the inclusion of corrections of only the lowest order in the photon momentum $\kappa=\omega/c$ is given by the following expression:

$$\frac{d\sigma_{nl}(\omega)}{d\Omega} = \frac{\sigma_{nl}(\omega)}{4\pi}\left\{1 - \frac{\beta_{nl}(\omega)}{2}P_2(\cos\theta) + \kappa\gamma_{nl}(\omega)P_1(\cos\theta) + \kappa\eta_{nl}(\omega)P_3(\cos\theta)\right\}. \quad (11.34)$$

For $l = 1$ the dipole angular anisotropy parameters are determined by the following expression [AM, 6, 7]

$$\beta_{np}(\omega) = \frac{2}{\tilde{D}_0^2 + 2\tilde{D}_2^2}[\tilde{D}_2^2 - 2\tilde{D}_0\tilde{D}_2\cos(\bar{\delta}_0 - \bar{\delta}_2)], \quad (11.35)$$

As can be seen from (11.35), for $l = 1$ the relation $\beta_{np}^{(in)}(q = 0, \omega) = -2\beta_{np}(\omega)$ is the same as for s-subshells.

The following expressions define the non-dipole angular anisotropy parameters [7] for $l = 1$:

$$\gamma_{np}(\omega) = \frac{6}{25[\tilde{D}_0^2 + 2\tilde{D}_2^2]}\left\{5\tilde{D}_0\tilde{Q}_1\cos(\bar{\delta}_1 - \bar{\delta}_0) + \tilde{D}_2\left[9\tilde{Q}_3\cos(\bar{\delta}_3 - \bar{\delta}_2) - \tilde{Q}_1\cos(\bar{\delta}_1 - \bar{\delta}_2)\right]\right\}, \quad (11.36)$$

$$\eta_{np}(\omega) = \frac{6}{25[\tilde{D}_0^2 + 2\tilde{D}_2^2]}\left\{5\tilde{D}_0\tilde{Q}_3\cos(\bar{\delta}_3 - \bar{\delta}_0) + 2\tilde{D}_2\left[3\tilde{Q}_1\cos(\bar{\delta}_1 - \bar{\delta}_2) - 2\tilde{Q}_3\cos(\bar{\delta}_3 - \bar{\delta}_2)\right]\right\}. \quad (11.37)$$

The dipole angular anisotropy parameter for $l = 2$ is determined by the following expression [AM]:

$$\beta_{nd}(\omega) = \frac{2}{5[2\tilde{D}_1^2 + 3\tilde{D}_3^2]}[\tilde{D}_1^2 + 6\tilde{D}_3^2 - 18\tilde{D}_1\tilde{D}_3\cos(\bar{\delta}_1 - \bar{\delta}_3)]. \quad (11.38)$$

Note that, as follows from (11.29), relations similar to those for $l = 0; 1$, $\beta_{nd}^{(in)}(q = 0, \omega) = -2\beta_{nd}(\omega)$, are valid for $l = 2$. It is quite possible that such a relation is also correct for any l.

The following expressions determine the non-dipole angular anisotropy parameters [AM] for $l = 2$

$$\gamma_{nd}(\omega) = \frac{6}{35[2\tilde{D}_1^2 + 3\tilde{D}_3^2]}\left\{7\tilde{D}_1[\tilde{Q}_2\cos(\bar{\delta}_2 - \bar{\delta}_1) - \tilde{Q}_0\cos(\bar{\delta}_0 - \bar{\delta}_1)] + 3\tilde{D}_3[6\tilde{Q}_4\cos(\bar{\delta}_4 - \bar{\delta}_3) - \tilde{Q}_2\cos(\bar{\delta}_2 - \bar{\delta}_2)]\right\}, \quad (11.39)$$

$$\eta_{nd}(\omega) = \frac{6}{35[2\tilde{D}_1^2 + 3\tilde{D}_3^2]}\Big\{2\tilde{D}_1\Big[6\tilde{Q}_4\cos(\overline{\delta}_4 - \overline{\delta}_1) - \tilde{Q}_2\cos(\overline{\delta}_2 - \overline{\delta}_1)\Big]$$
$$- \tilde{D}_3\Big[8\tilde{Q}_2\cos(\overline{\delta}_2 - \overline{\delta}_3) - 6\tilde{Q}_4\cos(\overline{\delta}_4 - \overline{\delta}_3) - 7\tilde{Q}_0\cos(\overline{\delta}_0 - \overline{\delta}_3)\Big]\Big\}. \quad (11.40)$$

It is seen that expressions (11.39) and (11.40) differ significantly from the corresponding non-dipole parameters for inelastic scattering given in (11.27)–(11.31). However, in contrast to the dipole parameters, there are no simple relations independent of the transferred energy ω that relate the corresponding non-dipole parameters for photoionization and inelastic scattering of fast electrons.

Note that the limit $q = 0$ at $\omega \neq 0$ cannot be reached experimentally, since no energy can be transferred from the incident electron to the target without linear momentum transfer. However, with an increase in the speed of the incident particle, an ever-smaller momentum q is sufficient to transfer the given energy ω.

Despite the apparently deep similarity between photoionization and scattering of fast electrons, it appears that there is also a large difference. Indeed, the angular distributions in photoionization and scattering of fast electrons are different even in the limit $q \to 0$. This can be explained by the difference between the transverse (in photoionization) and longitudinal (in the scattering of fast electrons) photons, which ionize the target atom. Analytically, this is reflected in the difference between the operators responsible for ionization by photons and fast electrons, which take into account only the lowest photon momentum in q non-dipole corrections. For photoionization in the length form it is $(\mathbf{er}) + i(\kappa\mathbf{r})(\mathbf{er})$, where $\mathbf{e}$ is the polarization vector of the photon, which is orthogonal to the direction of propagation of the light. As for the scattering of fast electrons, the corresponding operator is $(\mathbf{qr}) + i(\mathbf{qr})(\mathbf{qr})$, which contains only one angle between $\mathbf{q}$ and $\mathbf{r}$, in contrast to the case of photoionization with its two angles—between $\mathbf{r}$, $\mathbf{e}$, and $\boldsymbol{\kappa}$.

Because of this difference, in photoionization, the force that acts on the outgoing electron is orthogonal to the direction of the photon momentum $\boldsymbol{\kappa}/\kappa$ and thus the photon beam. Therefore, photoelectron emission is minimal along $\boldsymbol{\kappa}/\kappa$, while in inelastic electron scattering, the force and maximum field leading to knocking out the electron are directed along q.

So, the reason for the difference in angular distributions in photoabsorption and inelastic scattering of fast electrons is clear. In photoabsorption, an atomic electron is "knocked out" of the atom by the electric field of a photon, which is perpendicular to the direction of the light beam. In inelastic scattering, an electron is knocked out of an atom in the direction of the momentum $\mathbf{q}$, so that the preferred emission of electrons takes place in the direction of the $\mathbf{q}$-vector, with a maximum at $\theta = 0$.

This explains the difference in non-dipole terms. Note that the term with $\cos(\overline{\delta}_0 - \overline{\delta}_1)$ in relation (11.24) arises due to the monopole transition, and therefore is absent in the angular distribution of photoabsorption (11.23). This confirms that the angular distribution of the densities of the generalized oscillator strengths is richer in information on the structure of atoms than the same characteristic of photoionization.

11.2 Numerical Model

11.2.1 Non-dipole Angular Anisotropy Parameters for s-subshells in HF

As mentioned above, the non-dipole angular anisotropy parameters are expressed via the corresponding matrix elements, for example (11.9) or (11.21). In HF approximation for generalized oscillator strengths, one can use dipole matrix elements at small q and quadrupole matrix elements in the r-form, which are calculated in the ***gos*** program by the formula:

$$\langle n'l'|g_L^r(q)|nl\rangle = c_{Lll'}\int P_{nl}\tilde{j}_L(qr)P_{n'l'}dr \tag{11.41}$$

Here $\tilde{j}_L(qr) \equiv j_L(qr)/q^L$ and

$$c_{Lll'} = \sqrt{(2l+1)(2l'+1)}\begin{pmatrix} l\,L\,l' \\ 0\,0\,0 \end{pmatrix}. \tag{11.42}$$

Dipole matrix elements in HF for photoionization can also be determined in the ***phoabs*** program by the formula

$$\langle n'l'|d_1^r|nl\rangle = c_{1ll'}\int_0^\infty P_{n'l'}r\,P_{nl}dr \tag{11.43}$$

We consider mainly dipole and quadrupole excitations. For s-subshell, this corresponds to the following transitions: (a) dipole—$ns \to kp$, i.e., $L = 1, l = 0, l' = 1$. For it, the coefficient c_{101} is 1, (b) quadrupole—$ns \to kd$, i.e., $L = 2, l = 0, l' = 2$. For this transition, we have the coefficient $c_{202} = 1$.

As a result, instead of (11.41) for the dipole matrix elements, we have

$$\langle n'1|g_1^r(q)|n0\rangle = -\int P_{n'1}\tilde{j}_1(qr)P_{n0}dr \tag{11.44}$$

and for quadrupole

$$\langle n'2|g_2^r(q)|nl\rangle = \frac{1}{2}\int P_{n2}\tilde{j}_2(qr)P_{n0}dr \tag{11.45}$$

Now let us look at the Bessel functions. For small values of the momentum q, they have the following form:

$$\tilde{j}_1(qr) = \frac{1}{3}r, \tag{11.46}$$

$$\tilde{j}_2(qr) = \frac{1}{15}r^2. \tag{11.47}$$

Substituting them into integrals (11.43) and (11.44), we obtain the required connection between the matrix elements

$$\langle nl|d_1^r(q)|n'l'\rangle\big|_{q\to 0} = -\frac{1}{3}\int P_{nl}rP_{n'l'}\mathrm{d}r \equiv -\frac{1}{3}d_1^r, \tag{11.48}$$

$$\langle nl|g_2^r(q)|n'l'\rangle\big|_{q\to 0} = \frac{2}{15}\frac{1}{2}\int P_{nl}r^2P_{n'l'}\mathrm{d}r \equiv \frac{2}{15}q_1^{\prime r} \tag{11.49}$$

We now turn to calculating the non-dipole parameter for s-electrons. In the HF approximation, and ω in Ry, it is equal to

$$\gamma = -7.5\alpha\omega\frac{q_2^r}{d_1^r}\cos(\delta_2 - \delta_1) \tag{11.50}$$

This formula is used in the ***anglaniz0*** program.

11.2.2 Non-dipole Angular Anisotropy Parameters for s-subshells in the RPAE

Within the framework of the RPAE with ω in Ry, formula (11.50) should be rewritten as follows:

$$\gamma_{n0} = -7.5\alpha\omega\frac{1}{2|D_1|^2}\left[Q_2D_1^*e^{i(\delta_2-\delta_1)} + Q_2^*D_1e^{-i(\delta_2-\delta_1)}\right], \tag{11.51}$$

where $\alpha \simeq 1/137$ is the fine structure constant.

After that, we make changes in (11.51) in accordance with the following relations

$$|D_1|^2 \to [\mathrm{Re}D_1]^2 + [\mathrm{Im}D_1]^2, \tag{11.52}$$

$$\begin{aligned}\left[Q_2D_1^*\mathrm{e}^{i(\delta_2-\delta_1)} + Q_2^*D_1\mathrm{e}^{-i(\delta_2-\delta_1)}\right] &= [\mathrm{Re}Q_2\mathrm{Re}D_1 + \mathrm{Im}Q_2\mathrm{Im}D_1]\cos(\delta_2-\delta_1)\\ &\quad - [\mathrm{Im}Q_2\mathrm{Re}D_1 - \mathrm{Re}Q_2\mathrm{Im}D_1]\sin(\delta_2-\delta_1)].\end{aligned} \tag{11.53}$$

As a result, from (11.51) we obtain:

$$\gamma = -7.5\frac{\omega}{137}[(\mathrm{Re}Q_2\mathrm{Re}D_1 + \mathrm{Im}Q_2\mathrm{Im}D_1)\cos(\delta_2 - \delta_1) \\ - (\mathrm{Im}Q_2\mathrm{Re}D_1 - \mathrm{Re}Q_2\mathrm{Im}D_1)\sin(\delta_2 - \delta_1)]/[(\mathrm{Re}D_1)^2 + (\mathrm{Im}D_1)^2]. \quad (11.54)$$

Matrix elements are preliminarily calculated in RPAE, and phase shifts are taken from the results of calculations of wave functions in HF approximation.

The non-dipole angular anisotropy parameters in RPAE can also be expressed in terms of the modules of the matrix elements, namely

$$D_{l\pm1}(\omega) \equiv \tilde{D}_{l\pm1}(\omega)\exp[i\xi_{l\pm1}(\varepsilon)], \quad (11.55)$$

where $\tilde{D}_{l\pm1}(\omega)$ and $\xi_{l\pm1}$ are the absolute values (moduli) of the amplitudes of the corresponding transitions and their phases for angular momenta $l \pm 1$, and

$$Q_{l\pm2,0}(\omega) \equiv \tilde{Q}_{l\pm2,0}(\omega)\exp[i\xi_{l\pm2,0}(\varepsilon)], \quad (11.56)$$

where $\tilde{Q}_{l\pm2,0}(\omega)$ and $\xi_{l\pm2,0}(\varepsilon)$ are the absolute values (moduli) of the amplitudes of the corresponding transitions and their phases for angular momenta $l \pm 2, 0$.

The so-called reduced matrix elements are calculated by the formulas

$$\overline{\tilde{Q}}_{l'} = 2\sqrt{(2l+1)(2l'+1)}\begin{pmatrix} l\,2\,l' \\ 0\,0\,0 \end{pmatrix}\tilde{Q}_{l'} \quad (11.57)$$

$$\overline{\tilde{D}}_{l'} = \sqrt{(2l+1)(2l'+1)}\begin{pmatrix} l\,1\,l' \\ 0\,0\,0 \end{pmatrix}\tilde{D}_{l'} \quad (11.58)$$

Based on the general expression of the Clebsch–Gordan coefficients [IS]

$$\begin{pmatrix} l\,l_1\,l' \\ 000 \end{pmatrix} \\ = (-1)^{\frac{l+l_1+l'}{2}}\sqrt{\frac{(l+l_1-l')!(l+l'-l_1)!(l'+l_1-l)!}{(l+l_1+l'+1)!}}\frac{g!}{(g-l)!(g-l')!(g-l_1)!}, \\ 2g = l + l' + l_1;\ l_1 = 1;\ 2. \quad (11.59)$$

we obtain the formula for the non-dipole parameter γ_{nl} in the RPAE:

$$\gamma_{nl} = \frac{3}{10(\overline{\tilde{D}}_{l+1}^2 + \overline{\tilde{D}}_{l-1}^2)}\Big[A_1\overline{\tilde{Q}}_{l+2}\overline{\tilde{D}}_{l+1}\cos(\bar{\delta}_{l+2} - \bar{\delta}_{l+1}) + A_2\overline{\tilde{Q}}_l\overline{\tilde{D}}_{l+1}\cos(\bar{\delta}_l - \bar{\delta}_{l+1}) \\ + A_3\overline{\tilde{Q}}_{l-2}\overline{\tilde{D}}_{l-1}\cos(\bar{\delta}_{l-2} - \bar{\delta}_{l-1}) + A_4\overline{\tilde{Q}}_l\overline{\tilde{D}}_{l-1}\cos(\bar{\delta}_l - \bar{\delta}_{l-1})\Big], \quad (11.60)$$

where $\overline{\delta}_{l'} = \delta_{l'} + \xi_{l'}$ and

$$A_1 = -\sqrt{\frac{6(l+2)}{(2l+3)}},\ A_2 = -\sqrt{\frac{l(2l-1)}{(2l+1)(2l+3)}},$$
$$A_3 = -\sqrt{\frac{6(l-1)}{(2l-1)}},\ A_4 = -\sqrt{\frac{(l+1)(2l+3)}{(2l-1)(2l+1)}}. \tag{11.61}$$

From this we obtain for the s-subshell

$$\gamma_{n0} = \frac{6\bar{\tilde{Q}}_2}{5\bar{\tilde{D}}_1}\cos(\bar{\delta}_2 - \bar{\delta}_1). \tag{11.62}$$

We will do similar transformations for the second parameter

$$\begin{aligned}\eta_{nl} = &\frac{3}{10(\bar{\tilde{D}}_{l+1}^2 + \bar{\tilde{D}}_{l-1}^2)}\Big[B_1\bar{\tilde{Q}}_{l+2}\bar{\tilde{D}}_{l-1}\cos(\bar{\delta}_{l+2} - \bar{\delta}_{l-1})\\
&+ B_2\bar{G}_{l+2}\bar{\tilde{D}}_{l+1}\cos(\bar{\delta}_{l+2} - \bar{\delta}_{l+1}) + B_3\bar{\tilde{Q}}_{l-2}\bar{\tilde{D}}_{l+1}\cos(\bar{\delta}_{l-2} - \bar{\delta}_{l+1})\\
&+ B_4\bar{\tilde{Q}}_{l-2}\bar{\tilde{D}}_{l-1}\cos(\bar{\delta}_{l-2} - \bar{\delta}_{l-1}) + B_5\bar{\tilde{Q}}_l\bar{\tilde{D}}_{l+1}\cos(\bar{\delta}_l - \bar{\delta}_{l+1})\\
&+ B_6\bar{\tilde{Q}}_l\bar{\tilde{D}}_{l-1}\cos(\bar{\delta}_l - \bar{\delta}_{l-1})\Big],\end{aligned} \tag{11.63}$$

where

$$B_1 = \frac{5}{(2l+1)}\sqrt{\frac{2l(l+1)(l+2)}{3(2l+3)}},\ B_2 = \frac{(l+3)}{(2l+1)}\sqrt{\frac{2(l+2)}{3(2l+3)}},$$
$$B_3 = \frac{5}{2l+1}\sqrt{\frac{2l(l\quad 1)(l+1)}{3(2l-1)}},\ B_4 = \frac{l-2}{(2l+1)}\sqrt{\frac{2(l-1)}{3(2l-1)}},$$
$$B_5 = 2(l+2)\sqrt{\frac{l}{(2l-1)(2l+1)(2l+3)}},$$
$$B_6 = 2(l-1)\sqrt{\frac{(l+1)}{(2l-1)(2l+1)(2l+3)}}. \tag{11.64}$$

For the s-subshell, relation (11.63) takes the form

$$\eta_{n0} = -\frac{6\bar{\tilde{Q}}_2}{5\bar{\tilde{D}}_1}\cos(\bar{\delta}_2 - \bar{\delta}_1) = -\gamma_{n0}. \tag{11.65}$$

The described approach is also implemented in the ***anglaniz0*** program and serves as a criterion for the correct assignment of the initial data.

In the same program, using formulas (11.30) and (11.31) the non-dipole angular anisotropy parameters for secondary electrons $\gamma_{ns}^{(in)}(\omega)$, $\eta_{ns}^{(in)}(\omega)$ are calculated for the transferred momentum $q = 0$.

Note that the formulas (11.60) and (11.63) given in this subsection can also be used in approximations other than HF and RPAE. It is only necessary to calculate in a certain approximation the dipole and quadrupole matrix elements and the corresponding phases.

11.2.3 Non-dipole Angular Anisotropy Parameters for p-subshells

For the non-dipole parameters of *p*-subshells, the matrix elements are calculated using the same formulas as for *s*-subshells, but in this case, naturally, other transitions are considered. For all these transitions, we determine the coefficient in front of the integral sign in (11.43), (11.66) and the expressions for the matrix elements.

For a dipole transition $np \to ks$, i.e., $L = 1, l = 1, l' = 0$, we have $c_{101} = -1$, and the corresponding matrix element is found by the formula

$$d_0 \equiv \frac{3}{c_{110}} D_{L=1,l'=0} \tag{11.66}$$

For the transition $np \to kd$, i.e., $L = 1, l = 1, l' = 2$, the coefficient $c_{112} = \sqrt{2}$, and the corresponding matrix element is determined by the formula

$$d_2 \equiv \frac{3}{c_{112}} D_{L=1,l'=2} \tag{11.67}$$

For the quadrupole transition $np \to kp$, i.e., $L = 2, l = 1, l' = 1$, the coefficient in front of the integral sign is $c_{211} = \sqrt{9} \cdot (0.365148) = 1.09544$, so the matrix element is

$$q_1 = \frac{15}{2c_1} Q_{L=2,l'=1}. \tag{11.68}$$

For the quadrupole transition $np \to kf$, i.e., $L = 2, l = 1, l' = 3$, we have $c_{213} \cong -\sqrt{21} \cdot 0.29277 \cong -1.34164$, and the matrix element is

$$q_3 = \frac{15}{2c_{213}} Q_{L=2,l'=3}. \tag{11.69}$$

We now turn to the non-dipole parameters, which are determined by formula (11.5) and are expressed in terms of the dipole and quadrupole matrix elements in the following way

$$\gamma_p^C = -\frac{3\kappa}{5[d_0^2 + 2d_2^2]}\{5d_0q_3\cos(\delta_3 - \delta_0) - 2d_2 \\ [2q_3\cos(\delta_3 - \delta_2) + 3q_1\cos(\delta_1 - \delta_2)]\}, \tag{11.70}$$

$$\delta_p^C = \frac{3\kappa}{5[d_0^2 + 2d_2^2]}[d_0q_1\cos(\delta_1 - \delta_0) + d_0q_3\cos(\delta_3 - \delta_0) \\ + q_1d_2\cos(\delta_1 - \delta_2) + q_3d_2\cos(\delta_3 - \delta_2)]. \tag{11.71}$$

where, $\kappa = \alpha\omega(\mathrm{Ry})/2$.

The combination of these coefficients $(\gamma^C + 3\delta^C)$ (11.19) is called magic, since it determines $d\sigma_{nl}(\omega, \vartheta)/d\Omega$ at the magic angle of electron emission at which $P_2(\cos\vartheta) = 0$, i.e., the contribution of the dipole corrections to the angular distribution vanishes. In each approximation (HF or RPAE), matrix elements calculated in the corresponding approximation are used, as well as phase shifts—δ_l in HF and $\bar{\delta}_l$ in RPAE.

In RPAE, the non-dipole angular anisotropy parameters are expressed as follows:

$$\gamma_{n1} = \frac{3}{5\left[(l+1)\tilde{\bar{D}}_{l+1}^2 + l\tilde{\bar{D}}_{l-1}^2\right]}\left\{\frac{(l+1)}{2l+3}\tilde{\bar{D}}_{l+1}\left[3(l+2)\tilde{\bar{Q}}_{l+2}\cos(\bar{\delta}_{l+2} - \bar{\delta}_{l+1})\right.\right. \\ \left. - l\tilde{\bar{Q}}_l\cos(\bar{\delta}_l - \bar{\delta}_{l+1})\right] + \frac{l}{(2l-1)}\tilde{\bar{D}}_{l-1}\left[(l+1)\tilde{\bar{Q}}_l\cos(\bar{\delta}_l - \bar{\delta}_{l-1})\right. \\ \left.\left. - 3(l-1)\tilde{\bar{Q}}_{l-2}\cos(\bar{\delta}_{l-2} - \bar{\delta}_{l-1})\right]\right\} \\ = \frac{3}{5\left[2\tilde{\bar{D}}_{l+1}^2 + \tilde{\bar{D}}_{l-1}^2\right]}\left\{\frac{2}{5}\tilde{\bar{D}}_{l+1}\left[9\tilde{\bar{Q}}_{l+2}\cos(\bar{\delta}_{l+2} - \bar{\delta}_{l+1}) - \tilde{\bar{Q}}_l\cos(\bar{\delta}_l - \bar{\delta}_{l+1})\right]\right. \\ \left. + \frac{2}{(2l-1)}\tilde{\bar{D}}_{l-1}\tilde{\bar{Q}}_l\cos(\bar{\delta}_l - \bar{\delta}_{l-1})\right\} \\ \Rightarrow \frac{6}{25\left[2\tilde{\bar{D}}_2^2 + \tilde{\bar{D}}_0^2\right]}\left\{\tilde{\bar{D}}_2\left[9\tilde{\bar{Q}}_3\cos(\bar{\delta}_3 - \bar{\delta}_2) - \tilde{\bar{Q}}_1\cos(\bar{\delta}_1 - \bar{\delta}_2)\right]\right. \\ \left. + 5\tilde{\bar{D}}_0\tilde{\bar{Q}}_1\cos(\bar{\delta}_1 - \bar{\delta}_0)\right\}, \tag{11.72}$$

$$\eta_{n1} = \frac{3}{5\left[(l+1)\tilde{\bar{D}}_{l+1}^2 + l\tilde{\bar{D}}_{l-1}^2\right]}\left\{\frac{(l+1)(l+2)}{(2l+1)(2l+3)}\tilde{\bar{Q}}_{l+2}\left[5l\tilde{\bar{D}}_{l-1}\cos(\bar{\delta}_{l+2} - \bar{\delta}_{l-1})\right.\right. \\ \left. - (l+3)\tilde{\bar{D}}_{l+1}\cos(\bar{\delta}_{l+2} - \bar{\delta}_{l+1})\right]$$

$$
\begin{aligned}
&- \frac{l(l-1)}{(2l+1)(2l-1)} \bar{\tilde{Q}}_{l-2} \Big[5(l+1) \bar{\tilde{D}}_{l+1} \cos(\bar{\delta}_{l-2} - \bar{\delta}_{l+1}) \\
&- (l-2) \bar{\tilde{D}}_{l-1} \cos(\bar{\delta}_{l-2} - \bar{\delta}_{l-1}) \Big] \\
&+ \frac{2l(l+1)}{(2l-1)(2l+3)} \bar{\tilde{Q}}_{l} \Big[(l+2) \bar{\tilde{D}}_{l+1} \cos(\bar{\delta}_{l} - \bar{\delta}_{l+1}) \\
&- (l-1) \bar{\tilde{D}}_{l-1} \cos(\bar{\delta}_{l} - \bar{\delta}_{l-1}) \Big] \Big\} \\
&\Rightarrow \frac{6}{25 \Big[2\bar{\tilde{D}}_2^2 + \bar{\tilde{D}}_0^2 \Big]} \Big\{ \bar{\tilde{Q}}_3 \Big[5\bar{\tilde{D}}_0 \cos(\bar{\delta}_3 - \bar{\delta}_0) - 4\bar{\tilde{D}}_2 \cos(\bar{\delta}_3 - \bar{\delta}_2) \Big] \\
&+ 6\bar{\tilde{Q}}_1 \bar{\tilde{D}}_2 \cos(\bar{\delta}_1 - \bar{\delta}_2) \Big\}. \qquad (11.73)
\end{aligned}
$$

Anglaniz1 program implements these formulas.

For fast electrons colliding with an atom, at transferred linear momentum $q = 0$, the dipole parameter $\beta_{np}^{(in)}(\omega)$ is calculated by the formula (11.26), while the non-dipole angular anisotropy parameters $\gamma_{np}^{(in)}(\omega)$, $\eta_{np}^{(in)}(\omega)$ are determined by the formulas (11.27)–(11.28).

11.2.4 Non-dipole Angular Anisotropy Parameters for d*-subshells*

When calculating the non-dipole parameters for d-subshells, the matrix elements are determined in essentially the same way as for s- and p-subshells, but, of course, other transitions are considered. For all these transitions, we determine the coefficient in front of the integral sign in (11.43), (11.74), and the required matrix elements.

For the dipole transition $nd \to kp$, i.e., $L = 1, l = 2, l' = 1$,, we have $c_{121} = \sqrt{2}$, and the matrix element is equal to

$$d_1 = \frac{3}{c_{121}} D_{L=2,l'=1}. \qquad (11.74)$$

For the dipole transition $nd \to kf$, i.e., $L = 1, l = 2, l' = 3$, we obtain $c_{123} = -\sqrt{3}$, and the matrix element is

$$d_3 = \frac{3}{c_{123}} G_{L=2,l'=3}. \qquad (11.75)$$

For the quadrupole transition $nd \to ks$, i.e., $L = 2, l = 2, l' = 0$, the coefficient in front of the integral sign is $c_{220} = 2/15$, and the matrix element is

$$q_0 = \frac{15}{2c_{220}} G_{L=2,l'=0}. \qquad (11.76)$$

For the quadrupole transition $nd \to kd$, i.e., $L = 2$, $l = 2$, $l' = 2$, we have $c_{222} = -4/3\sqrt{70}$, and the matrix element is

$$q_2 = \frac{15}{2c_{222}} Q_{L=2,l'=2}. \tag{11.77}$$

For the quadrupole transition $nd \to kg$, i.e., $L = 2$, $l = 2$, $l' = 4$, we obtain $c_{222} = -4/5\sqrt{14}$ and the matrix element is

$$q_4 = \frac{15}{2c_{224}} Q_{L=2,l'=4}. \tag{11.78}$$

Non-dipole parameters, which are determined by the formula (11.6), are expressed through the dipole and quadrupole matrix elements in the following way:

$$\begin{aligned}\gamma_{n2}^{C} = & \frac{30\kappa}{35(2\bar{\tilde{D}}_1^2 + 3\bar{\tilde{D}}_3^2)}\Big\{2\bar{\tilde{D}}_1\Big[\bar{\tilde{Q}}_2\cos(\bar{\delta}_2 - \bar{\delta}_1) - 6\bar{\tilde{Q}}_4\cos(\bar{\delta}_4 - \bar{\delta}_1)\Big] \\ & + \bar{\tilde{D}}_3\Big[7\bar{\tilde{Q}}_0\cos(\bar{\delta}_0 - \bar{\delta}_3) - 8\bar{\tilde{Q}}_2\cos(\bar{\delta}_2 - \bar{\delta}_3) + 6\bar{\tilde{Q}}_4\cos(\bar{\delta}_4 - \bar{\delta}_3)\Big]\Big\},\end{aligned} \tag{11.79}$$

$$\begin{aligned}\eta_{n2}^{C} = & \frac{6\kappa}{35\Big[3\bar{\tilde{D}}_3^2 + 2\bar{\tilde{D}}_1^2\Big]} \\ & \times \Big\{\bar{\tilde{D}}_3\Big[12\bar{\tilde{Q}}_4\cos(\bar{\delta}_4 - \bar{\delta}_3) + 5\bar{\tilde{Q}}_2\cos(\delta_2 - \delta_3) \\ & + 6\bar{\tilde{Q}}_4\cos(\bar{\delta}_4 - \bar{\delta}_1) - 7\bar{\tilde{Q}}_0\cos(\bar{\delta}_0 - \bar{\delta}_3)\Big] \\ & + \bar{\tilde{D}}_1\Big[5\bar{\tilde{Q}}_2\cos(\bar{\delta}_2 - \bar{\delta}_1) - 7\bar{\tilde{Q}}_0\cos(\bar{\delta}_0 - \bar{\delta}_1)\Big]\Big\},\end{aligned} \tag{11.80}$$

where, ω is in Ry and therefore

$$\kappa = \alpha\omega/2 \tag{11.81}$$

Now we turn to the non-dipole parameters, which are calculated by formulas (11.7) and (11.8). In the RPAE approximation, they have the following form:

$$\begin{aligned}\gamma_{n2} = & \frac{3}{5\Big[(l+1)\bar{\tilde{D}}_{l+1}^2 + l\bar{\tilde{D}}_{l-1}^2\Big]}\Big\{\frac{(l+1)}{2l+3}\bar{\tilde{D}}_{l+1}\Big[3(l+2)\bar{\tilde{Q}}_{l+2}\cos(\bar{\delta}_{l+2} - \bar{\delta}_{l+1}) \\ & -l\bar{\tilde{Q}}_l\cos(\bar{\delta}_l - \bar{\delta}_{l+1})\Big] + \frac{l}{(2l-1)}\bar{\tilde{D}}_{l-1}\Big[(l+1)\bar{\tilde{Q}}_l\cos(\bar{\delta}_l - \bar{\delta}_{l-1}) \\ & -3(l-1)\bar{\tilde{Q}}_{l-2}\cos(\bar{\delta}_{l-2} - \bar{\delta}_{l-1})\Big]\Big\}\end{aligned}$$

$$\Rightarrow \frac{6}{35\left[3\bar{\tilde{D}}_3^2 + 2\bar{\tilde{D}}_1^2\right]}\left\{3\bar{\tilde{D}}_3\left[6\bar{\tilde{Q}}_4\cos(\bar{\delta}_4 - \bar{\delta}_3) - \bar{\tilde{Q}}_2\cos(\bar{\delta}_2 - \bar{\delta}_3)\right]\right.$$
$$\left. +7\bar{\tilde{D}}_1\left[\bar{\tilde{Q}}_2\cos(\bar{\delta}_2 - \bar{\delta}_1) - \bar{\tilde{Q}}_0\cos(\bar{\delta}_0 - \bar{\delta}_1)\right]\right\}, \tag{11.82}$$

$$\eta_{n2} = \frac{6}{35\left[3\overline{\tilde{D}}_2^{\,2} + 2\overline{\tilde{D}}_1^{\,2}\right]}\left\{2\overline{\tilde{D}}_1\left[6\overline{\tilde{Q}}_4\cos(\overline{\delta}_4 - \overline{\delta}_1) - \overline{\tilde{Q}}_2\cos(\overline{\delta}_2 - \overline{\delta}_1)\right]\right.$$
$$\left. + \overline{\tilde{D}}_3\left[8\overline{\tilde{Q}}_2\cos(\overline{\delta}_2 - \overline{\delta}_3) - 6\overline{\tilde{Q}}_4\cos(\overline{\delta}_4 - \overline{\delta}_3) - 7\overline{\tilde{Q}}_0\cos(\overline{\delta}_0 - \overline{\delta}_3)\right]\right\}. \tag{11.83}$$

These formulas are materialized in the ***anglaniz2*** program. It also calculates the dipole parameter for secondary electrons $\beta_{nd}^{(in)}(\omega)$ at the transferred moment $q = 0$ by formula (11.28) and angular anisotropy non-dipole parameters $\gamma_{nd}^{(in)}(\omega)$, $\eta_{nd}^{(in)}(\omega)$ by formula (11.29)–(11.30).

11.2.5 Non-dipole Angular Anisotropy Parameters for Semi-Closed Shell Atoms

The non-dipole parameters of the angular anisotropy of photoelectrons with nl "up" ("down") levels in the SP RPAE are obtained by generalizing Eqs. (11.7) and (11.8), similar to that described in Sect. 9.2.3.3 and in previous chapters:

$$\gamma_{nl\uparrow(\downarrow)}(\omega) = \frac{6}{5\left[(l+1)\tilde{D}_{l+1\uparrow(\downarrow)}^2 + l\tilde{D}_{l-1\uparrow(\downarrow)}^2\right]}$$
$$\times\left\{\frac{(l+1)}{2l+3}\tilde{D}_{l+1\uparrow(\downarrow)}\left[3(l+2)\tilde{Q}_{l+2\uparrow(\downarrow)}\cos\left(\tilde{\delta}_{l+2\uparrow(\downarrow)} - \tilde{\delta}_{l+1\uparrow(\downarrow)}\right)\right.\right.$$
$$\left. - l\tilde{Q}_{l\uparrow(\downarrow)}\cos\left(\tilde{\delta}_{l\uparrow(\downarrow)} - \tilde{\delta}_{l+1\uparrow(\downarrow)}\right)\right]$$
$$- \frac{l}{2l-1}\tilde{D}_{l-1\uparrow(\downarrow)}\left[3(l-1)\tilde{Q}_{l-2\uparrow(\downarrow)}\cos\left(\tilde{\delta}_{l-2\uparrow(\downarrow)} - \tilde{\delta}_{l-1\uparrow(\downarrow)}\right)\right.$$
$$\left.\left. - (l+1)\tilde{Q}_{l\uparrow(\downarrow)}\cos\left(\tilde{\delta}_{l\uparrow(\downarrow)} - \tilde{\delta}_{l-1\uparrow(\downarrow)}\right)\right]\right\}, \tag{11.84}$$

$$\eta_{nl\uparrow(\downarrow)}(\omega) = \frac{6}{5\left[(l+1)\tilde{D}_{l+1\uparrow(\downarrow)}^2 + l\tilde{D}_{l-1\uparrow(\downarrow)}^2\right]} \times \left\{\frac{(l+1)(l+2)}{(2l+1)(2l+3)}\tilde{Q}_{l+2\uparrow(\downarrow)}\right.$$
$$\times\left[5l\tilde{D}_{l-1\uparrow(\downarrow)}\cos\left(\tilde{\delta}_{l+2\uparrow(\downarrow)} - \tilde{\delta}_{l-1\uparrow(\downarrow)}\right) - (l+3)\tilde{D}_{l+1\uparrow(\downarrow)}\cos\left(\tilde{\delta}_{l+2\uparrow(\downarrow)} - \tilde{\delta}_{l+1\uparrow(\downarrow)}\right)\right]$$
$$- \frac{(l-1)l}{(2l+1)(2l-1)}\tilde{Q}_{l-2\uparrow(\downarrow)} \times \left[5(l+1)\tilde{D}_{l+1\uparrow(\downarrow)}\cos\left(\tilde{\delta}_{l-2\uparrow(\downarrow)} - \tilde{\delta}_{l+1\uparrow(\downarrow)}\right)\right.$$

$$
\begin{aligned}
&- (l-2)\tilde{D}_{l-1\uparrow(\downarrow)}\cos\Big(\tilde{\delta}_{l-2\uparrow(\downarrow)} - \tilde{\delta}_{l-1\uparrow(\downarrow)}\Big)\Big] + 2\frac{l(l+1)\tilde{Q}_{l\uparrow(\downarrow)}}{(2l-1)(2l+3)} \\
&\times \Big[(l+2)\tilde{D}_{l+1\uparrow(\downarrow)}\cos\Big(\tilde{\delta}_{l\uparrow(\downarrow)} - \tilde{\delta}_{l+1\uparrow(\downarrow)}\Big) \\
&- (l-1)\,D_{l-1\uparrow(\downarrow)}\cos\Big(\tilde{\delta}_{l\uparrow(\downarrow)} - \tilde{\delta}_{l-1\uparrow(\downarrow)}\Big)\Big]\Big\}. \qquad (11.85)
\end{aligned}
$$

where

$$
Q_{l\pm 2,0\uparrow(\downarrow)}(\omega) \equiv \tilde{Q}_{l\pm 2\uparrow(\downarrow)}(\omega)\exp[i\xi_{l\pm 1,0\uparrow(\downarrow)}(\varepsilon)]; \qquad (11.86)
$$

and $\tilde{\delta}_{l_j\uparrow(\downarrow)} = \delta_{l_j\uparrow(\downarrow)} + \xi_{l_j\uparrow(\downarrow)}$.

It is interesting to note that the energy distance between the "up" and "down" levels with the same quantum numbers n and l is much larger than the spin–orbit splitting in all but the heaviest atoms. This is understandable, since the exchange interaction is usually greater than the relativistic spin–orbit interaction.

In contrast to atoms with closed subshells, the angular anisotropy parameters for semi-closed subshells are calculated using the presented formulas with the wave functions calculated in the SP RPAE approximation.

The non-dipole angular anisotropy parameters for photoelectrons in the SP RPAE are calculated using the same programs ***anglaniz0***, ***anglaniz1***, and ***anglaniz2*** with the wave functions previously calculated in the SP HF approximation.

References

1. M.Ya. Amusia, A.S. Baltenkov, A.A. Grinberg, S.G. Shapiro, JETP **68** (1975)
2. M.Ya. Amusia, P.U. Arifov, A.S. Baltenkov, A.A. Grinberg, S.G. Shapiro, Phys. Lett. **A47**, 66 (1974)
3. J.W. Cooper, Phys. Rev. A**42**, 6942 (1990); Phys. Rev. A**45**, 3362 (1992); Phys. Rev. A**47**, 1841 (1993)
4. A. Bechler, R.H. Pratt, Phys. Rev. A. **42**, 6400 (1990)
5. A. Sommerfeld, *Wave Mechanics* (Mathuen, London, 1930)
6. M.Ya. Amusia, V.K. Dolmatov, V.K. Ivanov, JETP Lett. **6**, 23 (1980)
7. M.Ya. Amusia, A.S. Baltenkov, Z. Felfli and A.Z. Msezane, Phys. Rev. A**63**, 052506 (2001)
8. M.Ya. Amusia, L.V. Chernysheva, E.Z. Liverts. Phys. Rev. A**85**, 4, 042722 (2012)

Chapter 12
Spin Polarization Parameters of Photoelectrons

This chapter describes the procedures for calculating the spin polarization parameters in the dipole approximation and with account of the lowest non-dipole corrections. The results of these calculations are presented in [AICC].

12.1 Theoretical Model

12.1.1 Spin Polarization Parameters in the Dipole Approximation

Consideration of the spin polarization of photoelectrons in photoionization processes is of great theoretical and practical interest. In the general case, the angular distribution of photoelectrons with a certain spin orientation knocked out from unpolarized atoms, as well as non-oriented molecules, by circularly polarized radiation is given by the following expression [1–3]:

$$I_j^{\pm}(\mathbf{k},\mathbf{s}) = \frac{\sigma(\omega)}{8\pi}\left\{1 - \frac{\beta}{2}P_2(\mathbf{k}\cdot\mathbf{k}_\gamma) + A^j(\mathbf{s}\cdot\mathbf{k}_\gamma)\right.$$
$$\left. -\alpha^j\left[\frac{3}{2}(\mathbf{k}\cdot\mathbf{k}_\gamma)(\mathbf{k}\cdot\mathbf{s}) - \frac{1}{2}(\mathbf{s}\cdot\mathbf{k}_\gamma)\right] - 2\xi^j\left(\mathbf{s}\cdot\left[\mathbf{k}\times\mathbf{k}_\gamma\right]\right)(\mathbf{k}\cdot\mathbf{k}_\gamma)\right\} \quad (12.1)$$

Here, $\mathbf{k}$, $\mathbf{s}$, $\mathbf{k}_\gamma$ are the unit vectors in the directions of the photoelectron momentum, photoelectron spin, and photon momentum, respectively, j is the total angular and spin moment of the electron in the initial state, P_2 is the Legendre polynomial. As follows from the above expression, the angular distribution of photoelectrons is determined by five parameters, namely the photoionization cross section $\sigma(\omega)$, the

M. Ya. Amusia and L. V. Chernysheva, *Computation of Atomic and Molecular Processes*, Springer Series on Atomic, Optical, and Plasma Physics 117,
https://doi.org/10.1007/978-3-030-85143-9_12

angular anisotropy parameter $\beta(\omega)$, and three spin parameters $A^j(\omega)$, $\alpha^j(\omega)$, and $\xi^j(\omega)$. In the case of photoionization of a shell with an angular momentum $l \geq 1$, the spin parameters in the HF approximation are given by the following relations:

$$A^j(\omega) = \frac{(-1)^{j-l-1/2}}{(d_{l+1}^2 + d_{l-1}^2)(2j+1)}\left(ld_{l+1}^2 - (l+1)d_{l-1}^2\right),$$

$$\alpha^j(\omega)$$

$$= \frac{2(-1)^{j-l-1/2}l(l+1)}{(d_{l+1}^2 + d_{l-1}^2)(2j+1)(2l+1)}\left[l(l+2)d_{l+1}^2 - (l^2-1)d_{l-1}^2 - 3\sqrt{l(l+1)}d_{l+1}d_{l-1}\cos(\delta_{l+1} - \delta_{l-1})\right],$$

$$\xi^j(\omega) = \frac{3(-1)^{j-l-1/2}\sqrt{l(l+1)}}{(d_{l+1}^2 + d_{l-1}^2)(2j+1)}d_{l+1}d_{l-1}\sin(\delta_{l+1} - \delta_{l-1}). \tag{12.2}$$

The values of the parameters $A^j(\omega)$, $\alpha^j(\omega)$, and $\xi^j(\omega)$ in the RPAE can be obtained from formula (12.2) using the following substitutions:

$$d_{l\pm1}^2 \to |D_{l\pm1}|^2,$$
$$d_{l+1}d_{l-1}\cos(\delta_{l+1} - \delta_{l-1}) \to \mathrm{Re}\left[D_{l+1}D_{l-1}^* e^{i(\delta_{l+1}-\delta_{l-1})}\right],$$
$$d_{l+1}d_{l-1}\sin(\delta_{l+1} - \delta_{l-1}) \to \mathrm{Im}\left[D_{l+1}D_{l-1}^* e^{i(\delta_{l+1}-\delta_{l-1})}\right]. \tag{12.3}$$

It should be noted that upon absorption of linearly polarized and unpolarized radiation, the terms with parameters $A^j(\omega)$ and $\alpha^j(\omega)$ in (12.1) disappear, as shown in [2]. It is clear that in these cases there is a significant reduction in the information that can be obtained by comparing theory with experiment. Therefore, the authors do not consider it in any way useful to investigate here additionally these cases of radiation polarization.

12.1.2 Spin Polarization Parameters with Allowance for Quadrupole Terms

The angular distribution of photoelectrons with a certain spin orientation with allowance for non-dipole corrections is described in [4, 5]. Non-dipole corrections can make a significant contribution not only to the dipole angular distribution (see Chap. 11), but also to the spin polarization of photoelectrons. When non-dipole corrections are accounted for in the lowest approximation in photon momentum, the angular distribution of photoelectrons with a certain spin orientation becomes much more complex than that given by formula (12.1·):

$$I_j^{\pm}(\mathbf{k}, \mathbf{s}) = \frac{\sigma(\omega)}{8\pi} \times \left\{1 - \frac{\beta}{2}P_2(\mathbf{k}\cdot\mathbf{k}_\gamma) \pm A^j(\mathbf{s}\cdot\mathbf{k}_\gamma) \mp \alpha^j\left[\frac{3}{2}(\mathbf{k}\cdot\mathbf{k}_\gamma)(\mathbf{k}\cdot\mathbf{s}) - \frac{1}{2}(\mathbf{s}\cdot\mathbf{k}_\gamma)\right]\right.$$
$$- 2\xi^j(\mathbf{s}\cdot[\mathbf{k}\times\mathbf{k}_\gamma])(\mathbf{k}\cdot\mathbf{k}_\gamma) - 3\sqrt{6/5}\,(\mathbf{k}\cdot\mathbf{k}_\gamma)\mathrm{Re}b_{110}^j - 2\sqrt{21/5}\,P_3(\mathbf{k}\cdot\mathbf{k}_\gamma)\mathrm{Re}b_{330}^j$$

$$+\frac{9}{\sqrt{5}}(\mathbf{s}\cdot[\mathbf{k}\times\mathbf{k}_\gamma])\mathrm{Im}b^j_{111}-3[3(\mathbf{k}\cdot\mathbf{k}_\gamma)(\mathbf{s}\cdot\mathbf{k}_\gamma)-(\mathbf{k}\cdot\mathbf{s})]\mathrm{Re}b^j_{211}$$
$$\left.-3\sqrt{\frac{3}{2}}\left[2(\mathbf{k}\cdot\mathbf{k}_\gamma)(\mathbf{s}\cdot\mathbf{k}_\gamma)-(\mathbf{k}\cdot\mathbf{s})\left[5(\mathbf{k}\cdot\mathbf{k}_\gamma)^2-1\right]\right]\mathrm{Re}b^j_{231}\right\} \tag{12.4}$$

where the parameters b are determined by the formula [AICC]

$$b^j_{ILJ}(\Lambda,\Lambda')=\frac{\sqrt{6}}{N}\sum_{l_1l_2}\hat{l}\sqrt{\hat{l}_1\hat{l}_2\hat{L}}\,(i)^{l_2-l_1}(-1)^{l-j-\frac{1}{2}+l_2-J+\Lambda+\Lambda'}\exp[i(\delta_{l_1}-\delta_{l_2})]$$
$$\times\begin{pmatrix} l_1 & l_2 & L\\ 0 & 0 & 0\end{pmatrix}\begin{Bmatrix} 1/2 & 1/2 & x\\ l & l & j\end{Bmatrix}\begin{Bmatrix} I & L & J\\ \Lambda' & l_2 & l\\ \Lambda & l_1 & l\end{Bmatrix}\langle\varepsilon l_1\|H_\Lambda\|nlj\rangle\langle nlj\|H^*_{\Lambda'}\|\varepsilon l_2\rangle \tag{12.5}$$

Here, $\hat{l}=2l+1$, and Λ, Λ' can take the values 1 or 2. In this case,

$$H_1\equiv\hat{d},$$
$$H_2\equiv\hat{q}, \tag{12.6}$$

H_1. corresponds to the electric dipole operator and H_2—to the electric quadrupole operator. Also, in formula (12.5), the notations are used

$$\hat{l}\equiv 2l+1,\ N=\sum_{l_1}|\langle\varepsilon l_1\|H_1\|nlj\rangle|^2\equiv\sum_{l_1}\left|\left\langle\varepsilon l_1\left\|\hat{d}\right\|nlj\right\rangle\right|^2 \tag{12.7}$$

and the following definitions for the reduced dipole and quadrupole matrix elements are employed

$$d_{l'}=\langle\varepsilon l'\|H_1\|nlj\rangle=(-1)^{l'}\frac{\omega}{c}\begin{pmatrix} l' & 1 & l\\ 0 & 0 & 0\end{pmatrix}\sqrt{\hat{l}\hat{l'}}\int_0^\infty r^3R_{\varepsilon l'}(r)R^j_{nl}(r)dr$$
$$q_{l''}=\langle\varepsilon l''\|H_2\|nlj\rangle=i\frac{\omega}{c}(-1)^{l''}\begin{pmatrix} l'' & 2 & l\\ 0 & 0 & 0\end{pmatrix}\frac{\sqrt{\hat{l}\hat{l''}}}{2\sqrt{3}}\int_0^\infty r^4R_{\varepsilon l''}(r)R^j_{nl}(r)dr \tag{12.8}$$

Substituting in (12.5)–(12.8) the values of 3j-, 6j-, and 9j-symbols, for the case of photoionization of $d_{3/2}$-shell, we obtain the following formulas for the coefficients b (similar formulas for the case of photoionization p-shell are presented in [4]):

$$b^{3/2}_{110}=\frac{\kappa}{N}\frac{3\sqrt{2}}{5}\left[d_1q_0\cos(\delta_0-\delta_1)+\sqrt{\frac{7}{10}}d_1q_2\cos(\delta_1-\delta_2)\right.$$

$$+\frac{\sqrt{3}}{\sqrt{35}}d_3q_2\cos(\delta_2-\delta_3)+\frac{2\sqrt{3}}{\sqrt{7}}d_3q_4\cos(\delta_3-\delta_4)\Bigg], \quad (12.9)$$

$$b_{330}^{3/2}=-\frac{\kappa}{N}\frac{3\sqrt{2}}{5}\Bigg[\frac{\sqrt{6}}{7\sqrt{5}}d_1q_2\cos(\delta_1-\delta_2)+\frac{2\sqrt{6}}{7}d_1q_4\cos(\delta_1-\delta_4)$$
$$+\frac{\sqrt{2}}{\sqrt{7}}d_3q_0\cos(\delta_0-\delta_3)+\frac{8}{7\sqrt{5}}d_3q_2\cos(\delta_2-\delta_3)+\frac{2}{7}d_3q_4\cos(\delta_3-\delta_4)\Bigg], \quad (12.10)$$

$$b_{211}^{3/2}=\frac{\kappa}{N}\frac{3}{10}\Bigg[d_1q_0\cos(\delta_0-\delta_1)+\sqrt{\frac{14}{5}}d_1q_2\cos(\delta_1-\delta_2)$$
$$-\frac{3\sqrt{3}}{\sqrt{35}}d_3q_2\cos(\delta_2-\delta_3)-\frac{4}{\sqrt{21}}d_3q_4\cos(\delta_3-\delta_4)\Bigg], \quad (12.11)$$

$$b_{231}^{3/2}=-\frac{\kappa}{N}\frac{1}{5}\Bigg[\frac{9\sqrt{3}}{2\sqrt{35}}d_1q_2\cos(\delta_1-\delta_2)+\frac{6}{21}d_1q_4\cos(\delta_1-\delta_4)$$
$$+d_3q_0\cos(\delta_0-\delta_3)-\frac{2\sqrt{2}}{\sqrt{35}}d_3q_2\cos(\delta_2-\delta_3)-\frac{4\sqrt{2}}{\sqrt{7}}d_3q_4\cos(\delta_3-\delta_4)\Bigg], \quad (12.12)$$

$$b_{111}^{3/2}=-\frac{3i\kappa}{N}\Bigg[\frac{\sqrt{3}}{10}d_1q_0\sin(\delta_0-\delta_1)+\frac{1}{2\sqrt{35}}d_3q_2\sin(\delta_2-\delta_3)$$
$$+\frac{2}{5\sqrt{7}}d_3q_4\sin(\delta_3-\delta_4)\Bigg], \quad (12.13)$$

$$b_{331}^{3/2}=\frac{3i\kappa}{N}\Bigg[\frac{1}{5\sqrt{3}}d_1q_4\sin(\delta_1-\delta_4)+\frac{4}{15\sqrt{7}}d_3q_0\sin(\delta_0-\delta_3)$$
$$+\frac{2\sqrt{2}}{21\sqrt{5}}d_3q_2\sin(\delta_2-\delta_3)+\frac{\sqrt{2}}{105}d_3q_4\sin(\delta_3-\delta_4)\Bigg], \quad (12.14)$$

where $\kappa=\omega/c$ is the photon momentum and $N=d_1^2+d_3^2$.

There is a relation between the coefficients b

$$b_{110}^{3/2}+\sqrt{\frac{7}{3}}b_{330}^{3/2}-2\sqrt{2}b_{211}^{3/2}+2\sqrt{3}b_{231}^{3/2}=0 \quad (12.15)$$

In this case, the coefficients b with $j=5/2$ are expressed in terms of coefficients b with $j=3/2$ by a simple ratio:

$$b_{ijk}^{5/2} = -\frac{2}{3} b_{ijk}^{3/2}. \tag{12.16}$$

As in the case of dipole parameters, substitutions similar to (12.3) are used to calculate the coefficients b in the RPAE:

$$\begin{aligned} &d_l^2 \to |D_l|^2, \\ &d_l q_j \cos(\delta_l - \delta_j) \to \mathrm{Re}\big[D_l Q_j^* \mathrm{e}^{i(\delta_l - \delta_j)}\big], \\ &d_l q_j \sin(\delta_l - \delta_j) \to \mathrm{Im}\big[D_l Q_j^* \mathrm{e}^{i(\delta_l - \delta_j)}\big]. \end{aligned} \tag{12.17}$$

12.2 Numerical Model

12.2.1 *Spin Polarization Parameters in the Dipole Approximation*

As noted in Sect. 12.1.1, parameters $A^j(\omega)$, $\alpha^j(\omega)$, and $\xi^j(\omega)$ can be obtained by expressions as follows:

$$\begin{aligned} |D_{l\pm 1}|^2 &= (\mathrm{Re}D_{l\pm 1})^2 + (\mathrm{Im}D_{l\pm 1})^2, \\ \mathrm{Re}[D_{l+1} D_{l-1}^* e^{i(\delta_{l+1} - \delta_{l-1})}] &= (\mathrm{Re}D_{l+1}\mathrm{Re}D_{l-1} + \mathrm{Im}D_{l+1}\mathrm{Im}D_{l-1}) \cos(\delta_{l+1} - \delta_{l-1}) \\ &\quad + (\mathrm{Re}D_{l+1}\mathrm{Im}D_{l-1} - \mathrm{Im}D_{l+1}\mathrm{Re}D_{l-1}) \sin(\delta_{l+1} - \delta_{l-1}), \\ \mathrm{Im}[D_{l+1} D_{l-1}^* e^{i(\delta_{l+1} - \delta_{l-1})}] &= (\mathrm{Im}D_{l+1}\mathrm{Re}D_{l-1} - \mathrm{Re}D_{l+1}\mathrm{Im}D_{l-1}) \cos(\delta_{l+1} - \delta_{l-1}) \\ &\quad + (\mathrm{Re}D_{l+1}\mathrm{Re}D_{l-1} + \mathrm{Im}D_{l+1}\mathrm{Im}D_{l-1}) \sin(\delta_{l+1} - \delta_{l-1}). \end{aligned} \tag{12.18}$$

We also introduce the notation

$$K = \frac{(-1)^{j-l-1/2}}{(2j+1)(D_{l-1}^2 + D_{l+1}^2)}. \tag{12.19}$$

Then, formulas (12.2) can be rewritten in the following ways:

$$A^j(\omega) = K[l d_{l+1}^2 - (l+1) d_{l-1}^2], \tag{12.20}$$

$$\begin{aligned} \alpha^j(\omega) = \frac{2K}{(2l+1)} &\{l(l+2)d_{l+1}^2 - (l^2-1)d_{l-1}^2 \\ &- 3\sqrt{l(l+1)}\big[(\mathrm{Re}D_{l+1}\mathrm{Re}D_{l-1} + \mathrm{Im}D_{l+1}\mathrm{Im}D_{l-1}) \cos(\delta_{l+1} - \delta_{l-1}) \\ &+ (\mathrm{Re}D_{l+1}\mathrm{Im}D_{l-1} - \mathrm{Im}D_{l+1}\mathrm{Re}D_{l-1}) \sin(\delta_{l+1} - \delta_{l-1})\big]\big\}, \end{aligned} \tag{12.21}$$

$$\xi^j(\omega) = 3K\sqrt{l(l+1)}\big[(\mathrm{Im}D_{l+1}\mathrm{Re}D_{l-1} - \mathrm{Re}D_{l+1}\mathrm{Im}D_{l-1})\cos(\delta_{l+1} - \delta_{l-1}) \\ +(\mathrm{Re}D_{l+1}\mathrm{Re}D_{l-1} + \mathrm{Im}D_{l+1}\mathrm{Im}D_{l-1})\sin(\delta_{l+1} - \delta_{l-1})\big], \qquad (12.22)$$

Formulas (12.20–12.22) are obviously correct for calculating the dipole parameters of spin polarization both in RPAE and in the HF approximation. In the latter case, it should be assumed that $\mathrm{Re}D_j \equiv d_j$, $\mathrm{Im}D_j \equiv 0$.

The ***spinpol*** program makes it possible to calculate the dipole parameters of the spin polarization both in the HF approximation and in RPAE. In this case, matrix elements in the HF and RPAE approximations are preliminarily calculated by the ***phoabs*** program. The values of the Hartree–Fock phases δ_{l+1}, δ_{l-1} are determined in the ***fchf*** program (see Chap. 6).

12.2.2 Spin Polarization Parameters with Allowance for Quadrupole Terms

Let us consider the spin polarization parameters with allowance for the quadrupole terms for the case of photoionization of *d* shells. In order for the calculations to be carried out both in the HF approximation and in the RPAE, we use the substitutions (12.17), writing them down in more detail:

$$\begin{aligned}
&d_l^2 \to |D_l|^2 = (\mathrm{Re}D_l)^2 + (\mathrm{Im}D_l)^2,\\
&d_l q_j \cos(\delta_l - \delta_j) \to \mathrm{Re}\Big[D_l Q_j^* e^{i(\delta_l - \delta_j)}\Big] = (\mathrm{Re}D_l\mathrm{Re}Q_j + \mathrm{Im}D_l\mathrm{Im}Q_j)\cos(\delta_l - \delta_j)\\
&\quad + (\mathrm{Re}D_l\mathrm{Im}Q_j - \mathrm{Im}D_l\mathrm{Re}Q_j)\sin(\delta_l - \delta_j),\\
&d_l q_j \sin(\delta_l - \delta_j) \to \mathrm{Im}\Big[D_l Q_j^* e^{i(\delta_l - \delta_j)}\Big] = (\mathrm{Im}D_l\mathrm{Re}Q_j - \mathrm{Re}D_l\mathrm{Im}Q_j)\cos(\delta_l - \delta_j)\\
&\quad + (\mathrm{Re}D_l\mathrm{Re}Q_j + \mathrm{Im}D_l\mathrm{Im}Q_j)\sin(\delta_l - \delta_j).
\end{aligned} \qquad (12.23)$$

Substituting (12.23) in (12.9)–(12.14), we obtain general formulas for calculating the non-dipole parameters of spin polarization in RPAE and in the HF approximation. In the latter case, the imaginary parts of the dipole and quadrupole matrix elements are equal to zero. Let us additionally define the following parameters characterizing the degree of polarization:

$$P_1^{\pm 1}(\omega) = \frac{-\frac{3}{\sqrt{2}}\left(B_{111}^j - \frac{\sqrt{7}}{2\sqrt{2}}B_{331}^j\right)}{1 + \frac{\beta}{4}}, \qquad (12.24)$$

$$P_2^{\pm 1}(\omega) = \pm\frac{\frac{\sqrt{6}}{2}\left(-B_{211}^j + \sqrt{\frac{3}{2}}B_{231}^j\right)}{1 + \frac{\beta}{4}}. \qquad (12.25)$$

The values of these parameters are sometimes measured experimentally.

The indicated spin polarization parameters that account for the quadrupole terms for the *d*-subshells are calculated in the ***spinpolq*** program. For this purpose, dipole matrix elements are preliminarily found using the ***phoabs*** program [6] and quadrupole matrix elements using the ***gos*** program [7].

References

1. N.A. Cherepkov, Phys. Lett. A**40** (1972)
2. N.A. Cherepkov, JETP **65** (1973)
3. N.A. Cherepkov, J. Phys. B: At. Mol. Opt. Phys. **14** (1981)
4. N.A. Cherepkov, S.K. Semenov, J. Phys. B.: At. Mol. Opt. Phys. **34** (2001)
5. M.Ya. Amusia, N.A. Cherepkov., L.V. Chernysheva, Z. Felfli, A.Z. Msezane, J. Phys. B.: At. Mol. Opt. Phys. **38** (2005)
6. .L.V. Chernysheva, S.K. Semenov, N.A. Cherepkov, *Numerical Calculation of the Partial Cross Sections for Photoionization of Atoms with Allowance for Correlations in Several Transitions* (Ioffe Institute, No. 1620, Leningrad, 1993) (In Russian)
7. L.V. Chernysheva, M.Ya. Amusia, S.I. Sheftel, *Numerical Calculation of Atomic Generalized Oscillator Strength Allowing for Many-Electron Correlations Within Two Transitions* (Ioffe Institute, No. 495, Leningrad, 1975) (In Russian)

Chapter 13
Angular Distribution of Secondary Electrons Arising from the Scattering of Fast Particles by Atoms

In this chapter, we present an approach to calculating the GOS density differential in the angle of the outgoing electron and the angular anisotropy parameters for secondary electrons in the scattering of fast particles for intermediate values of the momentum q transferred from the projectile to the target atom. For small values of q, the same parameters are calculated in the programs described in Chap. 11.

13.1 Theoretical Model

In one-electron approximation, the GOS densities, differential in the emission angle and energy of an electron with a linear momentum **k** ionized from a subshell with the principal quantum number n and angular momentum l, are given by the following formula [LL, 1]:

$$\frac{\mathrm{d}f_{nl}(q,\omega)}{\mathrm{d}\Omega}=\frac{1}{2l+1}\frac{2\omega}{q^2}\sum_m|\langle nlms|\exp(i\mathbf{qr})|\varepsilon\mathbf{ks}\rangle|^2. \tag{13.1}$$

Here $\mathbf{q}=\mathbf{p}-\mathbf{p}'$, $\mathbf{p}'$ and $\mathbf{p}$ are the linear momenta of inelastically scattered fast electron in the initial and final states. Its energies in these states E and E' are connected with their moments by relations $p=\sqrt{2E}$ and $p'=\sqrt{2E'}$; Ω is the solid angle, m is the projection of the angular momentum, and s is the spin of the electron knocked out of the atom. Note that $\omega=E-E'$ and $\varepsilon=\omega-I_{nl}$ are the energy of the knocked-out electron.

The values of the transferred energy ω are limited by the inequality $0\le\omega\le pq$, as opposed to the equality $\omega=cq$ for the case of the photoionization. To be able to consider an incident particle as fast, its velocity must be much higher than the velocity of electrons in the ionizable subshell, i.e., inequality $\sqrt{2E}\gg R^{-1}$, where R

M. Ya. Amusia and L. V. Chernysheva, *Computation of Atomic and Molecular Processes*, Springer Series on Atomic, Optical, and Plasma Physics 117,
https://doi.org/10.1007/978-3-030-85143-9_13

is the subshell radius, must be valid. The linear momentum q transferred to the atom can be considered small if $qR \leq 1$.

Presenting $\exp(i\mathbf{qr})$ as a sum of the products of radial and angular parts and performing analytical integration over the angular variables, we obtain for the GOS in the one-electron HF approximation:

$$g_{nl,kl',L}(q) \equiv \int_0^\infty R_{nl}(r) j_L(qr) R_{kl'}(r) r^2 \mathrm{d}r, \tag{13.2}$$

where $j_L(qr)$ is the spherical Bessel function and $R_{nl(kl')}(r)$ is the radial parts of the electron wave functions in the HF approximation in the initial and final states, respectively.

Consider the angular distribution of the emitted electrons relative to direction of $\mathbf{q}$ [2, 3]. For this, it is convenient to choose the Z-axis coinciding with the direction $\mathbf{q}$, which makes it possible to choose $\theta_\mathbf{q} = \varphi_\mathbf{q} = 0$ in (13.1). Since we are considering the ionization of a certain nl-subshell, for simplicity of notation and taking into account the conservation of the energy of a fast electron in the inelastic scattering process, from which it follows $k = \sqrt{2(\omega - I_{nl})}$, it is convenient to introduce a notation $g_{nl,kl',L}(q) \equiv g_{kl'L}(q)$.

The formulas for the GOS can be generalized to include interelectronic correlations within RPAE. This is achieved by replacing $g_{kl'L'}(q)$ the modulus $\tilde{G}_{kl'L'}(q)$ and scattering phases by $\bar{\delta}_{l'} = \delta_{l'} + \Delta_{l'}$, where the expressions $G_{kl'L'}(q) \equiv \tilde{G}_{kl'L'}(q)\exp(i\Delta_{l'})$ are solutions to the system of RPAE Eq. (10.9) [AM, 2, 3], which, after separating the angular and spin parts, acquires the form similar to (10.10):

$$\begin{aligned} \langle \varepsilon l'|G^L(\omega,q)|nl\rangle &= \langle \varepsilon l'|g^L(qr)|nl\rangle \\ &+ \left(\sum_{\varepsilon''l'' \leq F, \varepsilon'''l''' > F} - \sum_{\varepsilon''l'' F >, \varepsilon'''l''' \leq F} \right) \frac{\langle \varepsilon'''l'''|G^L(\omega,q)|\varepsilon''l''\rangle \langle \varepsilon''l'', \varepsilon l'|U|\varepsilon'''l''', nl\rangle_L}{\omega - \varepsilon_{\varepsilon'''l'''} + \varepsilon_{\varepsilon''l''} + i\eta(1 - 2n_{\varepsilon'''l'''})} \end{aligned} \tag{13.3}$$

Here $\leq F (> F)$, as in the previous chapters, denotes the summation over the occupied (free) atomic levels in the target atom. Summation over free levels includes integration over the continuous spectrum, $n_{\varepsilon l}$—Fermi step function equal to 1 for $nl \leq F$ and 0 for $nl > F$. The L-component of the Coulomb interelectron interaction matrix element is defined as

$$\begin{aligned} \langle \varepsilon''l'', \varepsilon l'|U|\varepsilon'''l''', nl\rangle_L &= \langle \varepsilon''l'', \varepsilon l'|r_<^L/r_>^{L+1}|\varepsilon'''l''', nl\rangle \\ &- \langle \varepsilon''l'', \varepsilon l'|r_<^L/r_>^{L+1}|nl, \varepsilon'''l'''\rangle \end{aligned} \tag{13.4}$$

In the last formula, the notation of the smaller (large) radii $r_<(r_>)$ of the coordinates of the interacting electrons follows from the well-known expansion of the electron–electron Coulomb interaction in terms of Legendre polynomials

$|\mathbf{r}_1 - \mathbf{r}_2|^{-1} = \sum_L (r_<^L / r_>^{L+1}) P_L(\cos\vartheta)$, where ϑ—the angle between the vectors $\mathbf{r}_1$ and $\mathbf{r}_2$.

The following expression within the framework of the RPAE is valid for the GOS density of the nl-subshell $dF_{nl}(q,\omega)/d\Omega$, which is differential with respect to the angle of the outgoing electron:

$$\frac{\mathrm{d}F_{nl}(q,\omega)}{\mathrm{d}\Omega} = \sum_{L'L''} \frac{\mathrm{d}F_{nl}^{L',L''}(q,\omega)}{\mathrm{d}\Omega} = \frac{\omega\pi}{(\omega - I_{nl})q^2} \sum_{L'L''} (2L'+1)(2L''+1) i^{L'-L''}$$
$$\times \sum_{l'=|L'-l|}^{L'+l} \sum_{l''=|L''-l|}^{L''+l} \tilde{G}_{kl'L'}(q)\tilde{G}_{kl''L''}(q) i^{ll''-l'}(2l'+1)(2l''+1)$$
$$e^{i(\bar{\delta}_{l'} - \bar{\delta}_{l''})} \begin{pmatrix} L' \, l \, l' \\ 0\,0\,0 \end{pmatrix} \begin{pmatrix} l'' \, l \, L'' \\ 0\,0\,0 \end{pmatrix}$$
$$\sum_{L=|l'-l''|}^{l'+l''} P_L(\cos\theta)(-1)^{L+l}(2L+1) \begin{pmatrix} l' \, L \, l'' \\ 0\,0\,0 \end{pmatrix} \begin{pmatrix} L \, L' \, L'' \\ 0\,0\,0 \end{pmatrix} \begin{Bmatrix} L \, L' \, L'' \\ l \, l'' \, l' \end{Bmatrix}. \quad (13.5)$$

The partial value of GOS $F_{nl}(q,\omega)$ in RPAE is obtained after integrating the expression (13.5) over $d\Omega$. It leads to the following expression:

$$F_{nl}(q,\omega) = \sum_{L'} F_{nl}^{L'}(q,\omega)$$
$$= \frac{4\omega\pi^2}{(\omega - I_{nl})q^2} \sum_{L'} (2L'+1) \sum_{l'=|L'-l|}^{L'+l} [\tilde{G}_{kl'L'}(q)]^2 (2l'+1) \begin{pmatrix} L' \, l \, l' \\ 0\,0\,0 \end{pmatrix}^2. \quad (13.6)$$

Note that at small q, the dipole contribution to the GOS and $F_{nl}(q,\omega)$ dominates and $(\omega - I_{nl})F_{nl}(q,\omega)$ is simply proportional to the photoionization cross section $\sigma_{nl}(\omega)$ (see Chap. 10 and [AM]). To compare the results obtained using the well-known formulas for photoionization with allowance for low-order non-dipole corrections, we consider such small values of q, for which it is sufficient to take into account the terms with $L', L'' \leq 2$. In this case, the angular distribution of the GOS, described by formula (13.5), can be presented, similarly to the case of photoionization, as (11.4)

$$\frac{\mathrm{d}F_{nl}(q,\omega)}{\mathrm{d}\Omega} = \frac{F_{nl}(q,\omega)}{4\pi}\left\{1 - \frac{\beta_{nl}^{(in)}(\omega,q)}{2} P_2(\cos\theta)\right.$$
$$\left. + q\left[\gamma_{nl}^{(in)}(\omega,q)P_1(\cos\theta) + \eta_{nl}^{(in)}(\omega,q)P_3(\cos\theta) + \varsigma_{nl}^{(in)}(\omega,q)P_4(\cos\theta)\right]\right\}. \quad (13.7)$$

The obvious difference is the q-dependence of the coefficients and the additional term $\varsigma_{nl}^{(in)}(\omega, q)P_4(\cos\theta)$. Even so, the expressions for $\beta_{nl}^{(in)}(\omega, q), \gamma_{nl}^{(in)}(\omega, q)$, $\eta_{nl}^{(in)}(\omega, q)$, and $\varsigma_{nl}^{(in)}(\omega, q)$ through $g_{kl'L'}(q)$ are too complex compared to the relations for $\beta_{nl}(\omega)$, $\gamma_{nl}(\omega)$, and $\eta_{nl}(\omega)$ in the case of photoionization. Therefore, it is more convenient to present the results for s-, p-, and d-subshells separately. We will show further that, while $F_{nl}(q, \omega) \sim \sigma(\omega)$, similar relations are not valid for the anisotropy parameters [2].

Let us start with s-subshells, for which, as follows from formula (13.5), the differential GOS in the above-mentioned $L', L'' \leq 2$ approximation is given by the relations [ACY, (Sect. 3.7)]:

$$\frac{\mathrm{d}F_{n0}(q,\omega)}{\mathrm{d}\Omega} = \sum_{L',L''=0}^{2} \frac{dF_{n0}^{L',L''}(q,\omega)}{d\Omega} = \frac{F_{n0}(q,\omega)}{4\pi}$$
$$\times\{1+\frac{6}{W_0}\tilde{G}_{11}\left[\tilde{G}_{00}\cos(\bar{\delta}_0-\bar{\delta}_1)+2\tilde{G}_{22}\cos(\bar{\delta}_1-\bar{\delta}_2)\right]P_1(\cos\theta)$$
$$+\frac{2}{7W_0}\left[21\tilde{G}_{11}^2+5\tilde{G}_{22}(7\tilde{G}_{00}\cos(\bar{\delta}_0-\bar{\delta}_2)+5\tilde{G}_{22})\right]P_2(\cos\theta)$$
$$+18\tilde{G}_{11}\tilde{G}_{22}\cos(\bar{\delta}_1-\bar{\delta}_2)\,P_3(\cos\theta)+\frac{90}{7}\tilde{G}_{22}^2P_4(\cos\theta)\Big\}$$
$$\equiv\frac{F_{ns}(q,\omega)}{4\pi}\times\Big\{1-\frac{1}{2}\beta_{ns}^{(in)}(q,\omega)P_2(\cos\theta)$$
$$+q\left[\gamma_{ns}^{(in)}(q,\omega)P_1(\cos\theta)+\eta_{ns}^{(in)}(q,\omega)P_3(\cos\theta)\right]\Big\} \tag{13.8}$$

where

$$F_{ns}=\frac{4\pi^2\omega}{(\omega-I_{n0})q^2}W_0;\,W_0=\tilde{G}_{00}^2+3\tilde{G}_{11}^2+5\tilde{G}_{22}^2 \tag{13.9}$$

Expressions for p- and d-subshells are much more complex than expressions for s. On the other hand, the angular anisotropy parameters for these subshells are larger, since the corresponding outer p- and d-subshells are noticeably larger than the s-shells. The $4d$-subshell in Xe deserves special attention. Multielectron effects are particularly important in it due to presence of the well-known dipole Giant resonance.

For differential GOS of p-subshells, $l = 1$, we have the following expression [ACY (Sect. 3.7)]:

$$\frac{\mathrm{d}F_{np}(q,\omega)}{\mathrm{d}\Omega} = \sum_{L',L''=0}^{2} \frac{\mathrm{d}F_{np}^{L',L''}(q,\omega)}{\mathrm{d}\Omega}$$
$$=\frac{F_{np}}{4\pi}\Big\{1+\frac{1}{5W_1}\Big[10\tilde{G}_{01}(2\tilde{G}_{12}-\tilde{G}_{10})\cos(\bar{\delta}_0-\bar{\delta}_1)$$

$$+4\tilde{G}_{21}\Big((5\tilde{G}_{10}-\tilde{G}_{12})\cos(\bar{\delta}_1-\bar{\delta}_2)+9\tilde{G}_{32}\cos(\bar{\delta}_2-\bar{\delta}_3)\Big)\Big\}$$
$$\times\ P_1(\cos\theta)+\frac{2}{7W_1}\Big[7\tilde{G}_{21}(\tilde{G}_{21}-2\tilde{G}_{01}\cos(\bar{\delta}_0-\bar{\delta}_2))$$
$$+7\tilde{G}_{12}(\tilde{G}_{12}-2\tilde{G}_{10})+3\tilde{G}_{32}((7\tilde{G}_{10}-2\tilde{G}_{12})\cos(\bar{\delta}_1-\bar{\delta}_3)+4\tilde{G}_{32})\Big]P_2(\cos\theta)$$
$$-\frac{6}{5W_1}\Big[6\tilde{G}_{21}\tilde{G}_{12}\cos(\bar{\delta}_1-\bar{\delta}_2)+\tilde{G}_{32}(5\tilde{G}_{01}\cos(\bar{\delta}_0-\bar{\delta}_3)-4\tilde{G}_{21}\cos(\bar{\delta}_2-\bar{\delta}_3))\Big]$$
$$P_3(\cos\theta)+\frac{18}{7W_1}\tilde{G}_{32}\Big[\tilde{G}_{32}-4\tilde{G}_{12}\cos(\bar{\delta}_1-\bar{\delta}_3)\Big]P_4(\cos\theta)\Big\}\equiv\frac{F_{np}(q,\omega)}{4\pi}$$
$$\times\ 1-\frac{\beta_{np}^{(in)}(q,\omega)}{2}P_2(\cos\theta)$$
$$+q\big[\gamma_{np}^{(in)}(q,\omega)P_1(\cos\theta)+\eta_{np}^{(in)}(q,\omega)P_3(\cos\theta)+\zeta_{np}^{(in)}(q,\omega)P_4(\cos\theta)\big]\big\},\tag{13.10}$$

where

$$F_{np}=\frac{4\pi^2\omega}{(\omega-I_{n1})q^2}W_1,\tag{13.11}$$

and

$$W_1=\tilde{G}_{10}^2+\tilde{G}_{01}^2+2\Big[\tilde{G}_{21}^2+\tilde{G}_{12}^2\Big]+3\tilde{G}_{32}^2\tag{13.12}$$

For differential GOS of d-subshells, $l=2$, we have the following expression [ACY (Sect. 3.7)]:

$$\frac{\mathrm{d}F_{nd}(q,\omega)}{\mathrm{d}\Omega}=\frac{F_{nd}}{4\pi}\{1+\frac{6}{W_2}\Big[14\tilde{G}_{11}(\tilde{G}_{22}-\tilde{G}_{20})\cos(\bar{\delta}_1-\bar{\delta}_2)$$
$$-\ 14\tilde{G}_{11}\tilde{G}_{02}\cos(\bar{\delta}_0-\bar{\delta}_1)$$
$$+\ 3\tilde{G}_{31}\Big((7\tilde{G}_{20}-2\tilde{G}_{22})\cos(\bar{\delta}_2-\bar{\delta}_3)+12\tilde{G}_{42}\cos(\bar{\delta}_3-\bar{\delta}_4)\Big)\Big]P_1(\cos\theta)$$
$$+\ \frac{2}{245W_2}\Big[1029(\tilde{G}_{11}^2+6\tilde{G}_{31}^2)-18522\tilde{G}_{11}\tilde{G}_{31}\cos(\bar{\delta}_1-\bar{\delta}_3)$$
$$+\ 1225\tilde{G}_{02}(7\tilde{G}_{20}-10\tilde{G}_{22})\cos(\bar{\delta}_0-\bar{\delta}_2)-125\tilde{G}_{22}(98\tilde{G}_{20}+15\tilde{G}_{22})$$
$$+\ 450\tilde{G}_{42}((49\tilde{G}_{20}-20\tilde{G}_{22})\cos(\bar{\delta}_2-\bar{\delta}_4)+25\tilde{G}_{42})\Big]P_2(\cos\theta)$$
$$+\ \frac{18}{W_2}\Big[2\tilde{G}_{11}(\tilde{G}_{22}\cos(\bar{\delta}_1-\bar{\delta}_2)-6\tilde{G}_{42}\cos(\bar{\delta}_1-\bar{\delta}_4))+\tilde{G}_{31}(7\tilde{G}_{02}\cos(\bar{\delta}_0-\bar{\delta}_3)$$
$$-\ 8\tilde{G}_{22}\cos(\bar{\delta}_2-\bar{\delta}_3)+6\tilde{G}_{42}\cos(\bar{\delta}_3-\bar{\delta}_4))\big]P_3(\cos\theta)$$
$$+\ \frac{90}{49W_2}\Big[20\tilde{G}_{22}^2+\tilde{G}_{42}(98\tilde{G}_{02}\cos(\bar{\delta}_0-\bar{\delta}_4)$$

$$- 100\tilde{G}_{22}\cos(\bar{\delta}_2 - \bar{\delta}_4) + 27\tilde{G}_{42})\Big] P_4(\cos\theta)\} \equiv \frac{F_{nd}(q,\omega)}{4\pi}$$

$$\times \left\{1 - \frac{\beta_{nd}^{(in)}(q,\omega)}{2} P_2(\cos\theta)\right.$$

$$\left. + q\left[\gamma_{nd}^{(in)}(q,\omega)P_1(\cos\theta) + \eta_{nd}^{(in)}(q,\omega)P_3(\cos\theta) + \zeta_{nd}^{(in)}(q,\omega)P_4(\cos\theta)\right]\right\}, \tag{13.13}$$

where

$$F_{nd} = \frac{4\pi^2\omega}{35(\omega - I_{n2})q^2} W_2, \tag{13.14}$$

$$W_2 = 35\tilde{G}_{20}^2 + 42\tilde{G}_{11}^2 + 63\tilde{G}_{31}^2 + 35\tilde{G}_{02}^2 + 50\tilde{G}_{22}^2 + 90\tilde{G}_{42}^2. \tag{13.15}$$

The matrix elements G included in (13.14)–(13.15) are obtained from the solution of an equation similar to (13.3), but with a different normalization. Namely, the first term on the right is not g there, but the Bessel function $j_L(qr)$. They differ by a numerical factor. This factor is determined by the considered angular momenta. This refers to the moment transferred during the scattering process, as well as the moments of the electron knocked out of the atom in its initial and final states.

13.2 Numerical Model

13.2.1 Angular Anisotropy Parameters for s-subshells

We start with the dipole angular anisotropy parameter for the s-subshell. Using (13.8), we obtain the following relation

$$\beta_{ns}^{(in)}(q,\omega) = -\frac{4}{7W_0}\left\{21\tilde{G}_{11}^2 + 5\tilde{G}_{22}\left[7\tilde{G}_{00}\cos(\bar{\delta}_0 - \bar{\delta}_2) + 5\tilde{G}_{22}\right]\right\}, \tag{13.16}$$

where $W_0 = \tilde{G}_{00}^2 + 3\tilde{G}_{11}^2 + 5\tilde{G}_{22}^2$. Here, $\tilde{G}_{00}, \tilde{G}_{11}, \tilde{G}_{22}$ are the modules of the GOS matrix elements [2] for the transitions, $s-s(L=0), s-p(L=1)$, and $s-d(L=2)$, respectively, which are determined, as above, by the following relation.

$$G_{lL}(\omega) \equiv \tilde{G}_{lL}(\omega)\exp[i\Delta_{lL}(\varepsilon)], \tag{13.17}$$

After appropriate substitutions, we obtain

$$\beta_{ns}^{(in)}(q,\omega) = -\frac{4}{7}\frac{\left\{21\tilde{G}_{11}^2 + 5\tilde{G}_{22}\left[7\tilde{G}_{00}\cos(\overline{\delta}_0 - \overline{\delta}_2) + 5\tilde{G}_{22}\right]\right\}}{\tilde{G}_{00}^2 + 3\tilde{G}_{11}^2 + 5\tilde{G}_{22}^2} \tag{13.18}$$

The non-dipole angular anisotropy parameters are obtained from the same formula (13.8):

$$\gamma_{ns}^{(in)}(q,\omega) = \frac{6}{q(\tilde{G}_{00}^2 + 3\tilde{G}_{11}^2 + 5\tilde{G}_{22}^2)}\tilde{G}_{11}\left[\tilde{G}_{00}\cos(\overline{\delta}_0 - \overline{\delta}_1) + 2\tilde{G}_{22}\cos(\overline{\delta}_1 - \overline{\delta}_2)\right], \tag{13.19}$$

$$\eta_{ns}^{(in)}(q,\omega) = \frac{18}{q(\tilde{G}_{00}^2 + 3\tilde{G}_{11}^2 + 5\tilde{G}_{22}^2)}\tilde{G}_{11}\tilde{G}_{22}\cos(\overline{\delta}_1 - \overline{\delta}_2). \tag{13.20}$$

One can see that all angular anisotropy parameters are expressed in terms of the matrix elements of the GOS for various transitions [2, 3, AC, ACY]. Matrix elements are calculated in the program ***gos*** using the formula:

$$\begin{aligned}\left\langle kl_2\middle|g_L^r(q)\middle|nl_1\right\rangle &= \sqrt{(2l'+1)(2l+1)}\begin{pmatrix} l\,L\,l' \\ 0\,0\,0\end{pmatrix}\int P_{kl'}\tilde{j}_L(qr)P_{nl}dr \\ &\equiv c_L\int P_{kl'}j_L(qr)P_{nl}dr\end{aligned} \tag{13.21}$$

Here $\tilde{j}_L(qr)$ is the spherical Bessel function divided by q^L, and $c_L = \sqrt{(2l+1)(2l'+1)}\begin{pmatrix} l\,L\,l' \\ 0\,0\,0\end{pmatrix}\frac{1}{q^L}$.

Let us consider the transitions that should be taken into account to determine the angular anisotropy parameters of the s-subshells:

1. For the monopole $ns - \varepsilon s$ transition ($L - 0$), we determine the coefficient $c_0 = 1$. The modulus of the matrix element GOS for each energy ω is found from the expression $\tilde{G}_{00} = \sqrt{(\text{Re}G_{00})^2 + (\text{Im}G_{00})^2}$. Its phase is equal to $\overline{\delta}_0 = \delta_0^{HF} + \Delta_{00}$, while the phase correction is calculated by the formula $\Delta_{00} = \text{arctg}(\text{Im}G_{00}/\text{Re}G_{00})$.
2. For the dipole ($L = 1$) transition $ns - \varepsilon p$, we have coefficient $c_1 = -1/q$. We find the modulus of the matrix element from the expression $\tilde{G}_{11} = \sqrt{(\text{Re}G_{11}/c_1)^2 + (\text{Im}G_{11}/c_1)^2}$. The phase is equal to $\overline{\delta}_1 = \delta_1^{HF} + \Delta_{11}$, while correction to it is equal to $\Delta_1 = \text{arctg}(\text{Im}G_{11}/ReG_{11})$. The division by c_1 is the result of different normalizations, as discussed at the end of Sect. 13.1.
3. For the quadrupole ($L = 2$) transition $ns - \varepsilon d$, we obtain the coefficient $c_2 = 1/q^2$. The modulus of the matrix element is found using the expression $\tilde{G}_{22} = q^2\sqrt{(\text{Re}G_{22})^2 + (\text{Im}G_{22})^2}$, and the phase is equal to $\overline{\delta}_2 = \delta_2^{HF} + \Delta_{22}$, with a correction to it $\Delta_{22} = \text{arctg}(\text{Im}G_{22}/\text{Re}G_{22})$.

In the above formulas, $\mathrm{Re}G_{00}$, $\mathrm{Re}G_{11}$, $\mathrm{Re}G_{22}$ are the real, and $\mathrm{Im}G_{00}$, $\mathrm{Im}G_{11}$, $\mathrm{Im}G_{22}$ are the imaginary parts of the GOS matrix elements for the indicated transitions, δ_0^{HF}, δ_1^{HF}, δ_2^{HF} are the HF phases, and Δ_{00}, Δ_{11}, Δ_{22} are the corrections to the HF phases.

The wave functions and HF phases are calculated in the ***scfhf***, ***fchf*** programs, the matrix elements of the GOS (real and imaginary parts)—in the ***gos*** program, and, finally, the dipole and non-dipole parameters of the angular anisotropy and the differential in the angle of the outgoing electron density of the GOS (13.8)—using ***difgos0***.

13.2.2 Angular Anisotropy Parameters for p-subshells

The dipole angular anisotropy parameter is obtained from formulas (13.10) and (13.11) by the appropriate substitutions

$$\begin{aligned}\beta_{np}^{(in)}(q,\omega) = & -\frac{4}{7}\frac{1}{\tilde{G}_{10}^2+\tilde{G}_{01}^2+2\tilde{G}_{21}^2+2\tilde{G}_{12}^2+3\tilde{G}_{32}^2}\\ & \times\Big\{7\tilde{G}_{21}[\tilde{G}_{21}-2\tilde{G}_{01}\cos(\bar{\delta}_0-\bar{\delta}_2)]\\ & +7\tilde{G}_{12}(\tilde{G}_{12}-2\tilde{G}_{10})\\ & +3\tilde{G}_{32}\Big[(7\tilde{G}_{10}-2\tilde{G}_{12})\cos(\bar{\delta}_1-\bar{\delta}_3)+4\tilde{G}_{32}\Big]\Big\}\end{aligned} \tag{13.22}$$

The non-dipole angular anisotropy parameters are also found by appropriate substitutions into (13.10), (13.11):

$$\begin{aligned}\gamma_{np}^{(in)}(q,\omega) = & \frac{1}{5q}\frac{1}{\tilde{G}_{10}^2+\tilde{G}_{01}^2+2(\tilde{G}_{21}^2+\tilde{G}_{12}^2)+3\tilde{G}_{32}^2}\\ & \Big\{10\tilde{G}_{01}(2\tilde{G}_{12}-\tilde{G}_{10})\cos(\bar{\delta}_0-\bar{\delta}_1)\\ & +4\tilde{G}_{21}\Big[(5\tilde{G}_{10}-\tilde{G}_{12})\cos(\bar{\delta}_1-\bar{\delta}_2)+9\overline{G}_{32}\cos(\bar{\delta}_2-\bar{\delta}_3)\Big]\Big\},\end{aligned} \tag{13.23}$$

$$\begin{aligned}\eta_{np}^{(in)}(q,\omega) = & \frac{6}{5q}\frac{1}{\tilde{G}_{10}^2+\tilde{G}_{01}^2+2(\tilde{G}_{21}^2+\tilde{G}_{12}^2)+3\tilde{G}_{32}^2}\\ & \times\Big\{6\tilde{G}_{21}\tilde{G}_{12}\cos(\bar{\delta}_1-\bar{\delta}_2)+\tilde{G}_{32}\Big[5\tilde{G}_{01}\cos(\bar{\delta}_0-\bar{\delta}_3)-4\tilde{G}_{21}\cos(\bar{\delta}_2-\bar{\delta}_3)\Big]\Big\},\end{aligned} \tag{13.24}$$

$$\zeta_{np}^{(in)}(q,\omega) = \frac{18}{7q}\frac{\tilde{G}_{32}\Big[\tilde{G}_{32}-4\tilde{G}_{12}\cos(\bar{\delta}_1-\bar{\delta}_3)\Big]}{\tilde{G}_{10}^2+\tilde{G}_{01}^2+2(\tilde{G}_{21}^2+\tilde{G}_{12}^2)+3\tilde{G}_{32}^2}. \tag{13.25}$$

Let us consider the transitions that should be taken into account for determining the parameters of the angular anisotropy of *p*-subshells:

1. For the monopole transition $np - \varepsilon p$, we determine the coefficient $c_0 = -1.73205$. The modulus of the GOS matrix element for each ω energy is determined through its real and imaginary parts using the relation $\tilde{G}_{10} = \sqrt{(ReG_{10})^2 + (\mathrm{Im}G_{10})^2}/c_0$. The phase is equal to $\overline{\delta}_1 = \delta_1^{HF} + \Delta_{10}$, and the correction to it is determined using the formula $\Delta_{10} = \mathrm{arctg}(\mathrm{Im}G_{10}/\mathrm{Re}G_{10})$.
2. For dipole transition $np - \varepsilon s$, we have $c_1 = -1/q$. The module of the matrix element is found from the expression $\tilde{G}_{01} = \sqrt{(\mathrm{Re}G_{01})^2 + (\mathrm{Im}G_{01})^2}/c_1$. The phase is $\overline{\delta}_{01} = \delta_0^{HF} + \Delta_{01}$, and the correction to it is determined using the formula $\Delta_{01} = \mathrm{arctg}(\mathrm{Im}G_{01}/\mathrm{Re}G_{01})$.
3. For the dipole transition $np - \varepsilon d$, we have $c_1 = 1.4142/q$. We find the modulus of the matrix element using the relation $\tilde{G}_{21} = \sqrt{(\mathrm{Re}G_{21})^2 + (\mathrm{Im}G_{21})^2}/c_1$. The phase is $\overline{\delta}_{21} = \delta_2^{HF} + \Delta_{21}$, and the correction to it is $\Delta_{21} = \mathrm{arctg}(\mathrm{Im}G_{21}/\mathrm{Re}G_{21})$.
4. For the quadrupole transition $np - \varepsilon p$, we have $c_2 = 1.0954/q^2$. The module of the matrix element is found from the expression $\tilde{G}_{12} = \sqrt{(\mathrm{Re}G_{12})^2 + (\mathrm{Im}G_{12})^2}/c_2$. The phase is $\overline{\delta}_{12} = \delta_1^{HF} + \Delta_{12}$, and the correction to it is $\Delta_{12} = \mathrm{arctg}(\mathrm{Im}G_{12}/\mathrm{Re}G_{12})$. The phases $\overline{\delta}_{12}$ and $\overline{\delta}_{10}$ are close to each other. Therefore, for qualitative evaluation of $\overline{\delta}_1$, any of them can be selected.
5. For the quadrupole transition $np - \varepsilon f$, we have $c_2 = (-1.3416)/q^2$. The module of the matrix element is found from the expression $\tilde{G}_{32} = \sqrt{(\mathrm{Re}G_{32})^2 + (\mathrm{Im}G_{32})^2}/c_2$. The phase is $\overline{\delta}_{32} = \delta_3^{HF} + \Delta_{32}$, and the correction to the phase is $\Delta_{32} = \mathrm{arctg}(\mathrm{Im}G_{32}/\mathrm{Re}G_{32})$.

The wave functions and HF phases are calculated using the ***fchf*** program, the matrix elements of the GOS (their real and imaginary parts)—by the ***gos*** program, and, finally, the dipole and non-dipole angular anisotropy parameters and the density of GOS, differential in angle of the outgoing electron (13.10), are found by the ***difgos1***.

13.2.3 Angular Anisotropy Parameters for d-subshells

The dipole angular anisotropy parameter is obtained from comparing two parts of the formulas (13.12):

$$\beta_{nd}^{(in)}(q,\omega) = -\frac{4}{245W_2}\Big\{1029(\tilde{G}_{11}^2 + 6\tilde{G}_{31}^2) - 18522\tilde{G}_{11}\tilde{G}_{31}\cos(\overline{\delta}_1 - \overline{\delta}_3)$$
$$+ 1225\tilde{G}_{02}(7\tilde{G}_{20} - 10\tilde{G}_{22})\cos(\overline{\delta}_0 - \overline{\delta}_2) - 125\tilde{G}_{22}(98\tilde{G}_{20} + 15\tilde{G}_{22})$$
$$+450\tilde{G}_{42}\Big[(49\tilde{G}_{20} - 20\tilde{G}_{22})\cos(\overline{\delta}_2 - \overline{\delta}_4) + 25\tilde{G}_{42}\Big]\Big\}, \qquad (13.26)$$

Here $W_2 = 35\tilde{G}_{20}^2 + 42\tilde{G}_{11}^2 + 63\tilde{G}_{31}^2 + 35\tilde{G}_{02}^2 + 50\tilde{G}_{22}^2 + 90\tilde{G}_{42}^2$.
After appropriate substitutions, we obtain

$$\beta_{nd}^{(in)}(q,\omega) = -\frac{4}{245(35\tilde{G}_{20}^2 + 42\tilde{G}_{11}^2 + 63\tilde{G}_{31}^2 + 35\tilde{G}_{02}^2 + 50\tilde{G}_{22}^2 + 90\tilde{G}_{42}^2)} \Big[1029(\tilde{G}_{11}^2 + 6\tilde{G}_{31}^2) - 18522\tilde{G}_{11}\tilde{G}_{31}\cos(\bar{\delta}_1 - \bar{\delta}_3) + 1225\tilde{G}_{02}(7\tilde{G}_{20} - 10\tilde{G}_{22})\cos(\bar{\delta}_0 - \bar{\delta}_2) - 125\tilde{G}_{22}(98\tilde{G}_{20} + 15\tilde{G}_{22}) + 450\tilde{G}_{42}((49\tilde{G}_{20} - 20\tilde{G}_{22})\cos(\bar{\delta}_2 - \bar{\delta}_4) + 25\tilde{G}_{42})\Big]. \quad (13.27)$$

The non-dipole angular anisotropy parameters are obtained in a similar way from the same formulas:

$$\gamma_{nd}^{(in)}(q,\omega) = \frac{6}{q}\frac{1}{35\tilde{G}_{20}^2 + 42\tilde{G}_{11}^2 + 63\tilde{G}_{31}^2 + 35\tilde{G}_{02}^2 + 50\tilde{G}_{22}^2 + 90\tilde{G}_{42}^2} \times \Big\{14\tilde{G}_{11}(\tilde{G}_{22} - \tilde{G}_{20})\cos(\bar{\delta}_1 - \bar{\delta}_2) - 14\tilde{G}_{11}\tilde{G}_{02}\cos(\bar{\delta}_0 - \bar{\delta}_1) + 3\tilde{G}_{31}\Big[(7\tilde{G}_{20} - 2\tilde{G}_{22})\cos(\bar{\delta}_2 - \bar{\delta}_3) + 12\tilde{G}_{42}\cos(\bar{\delta}_3 - \bar{\delta}_4)\Big]\Big\}, \quad (13.28)$$

$$\eta_{nd}^{(in)}(q,\omega) = \frac{18}{q}\frac{1}{35\tilde{G}_{20}^2 + 42\tilde{G}_{11}^2 + 63\tilde{G}_{31}^2 + 35\tilde{G}_{02}^2 + 50\tilde{G}_{22}^2 + 90\tilde{G}_{42}^2} \Big\{2\tilde{G}_{11}[\tilde{G}_{22}\cos(\bar{\delta}_1 - \bar{\delta}_2) - 6\tilde{G}_{42}\cos(\bar{\delta}_1 - \bar{\delta}_4)] + \tilde{G}_{31}[7\tilde{G}_{02}\cos(\bar{\delta}_0 - \bar{\delta}_3) - 8\tilde{G}_{22}\cos(\bar{\delta}_2 - \bar{\delta}_3) + 6\tilde{G}_{42}\cos(\bar{\delta}_3 - \bar{\delta}_4)]\Big\}, \quad (13.29)$$

$$\zeta_{nd}^{(in)}(q,\omega) = \frac{90}{49q}\frac{1}{35\tilde{G}_{20}^2 + 42\tilde{G}_{11}^2 + 63\tilde{G}_{31}^2 + 35\tilde{G}_{02}^2 + 50\tilde{G}_{22}^2 + 90\tilde{G}_{42}^2} \times \Big\{20\tilde{G}_{22}^2 + \tilde{G}_{42}[98\tilde{G}_{02}\cos(\bar{\delta}_0 - \bar{\delta}_4) - 100\tilde{G}_{22}\cos(\bar{\delta}_2 - \bar{\delta}_4) + 27\tilde{G}_{42}]\Big\}. \quad (13.30)$$

Let us consider the transitions that should be accounted for to determine the angular anisotropy parameters for d-subshells:

1. For the monopole transition $nd - \varepsilon d$, we find the coefficient $c_0 = -2.2361$. The modulus of the matrix element for each energy ω is determined through the real and imaginary parts of GOS using the formula $\tilde{G}_{20} = \sqrt{(\mathrm{Re}G_{20})^2 + (\mathrm{Im}G_{20})^2}/c_0$. The phase is equal to $\bar{\delta}_{20} = \delta_2^{HF} + \Delta_{20}$, while the correction to it is determined using the formula $\Delta_{20} = \mathrm{arctg}(\mathrm{Im}G_{20}/\mathrm{Re}G_{20})$.

2. For the dipole transition $nd - \varepsilon p$, the coefficient is $c_{11} = -1.4142/q$. The modulus of the matrix element for each energy ω is determined through the real and imaginary parts of the GOS by the formula $\tilde{G}_{11} = \sqrt{(\mathrm{Re}G_{11})^2 + (\mathrm{Im}G_{11})^2}/c_{11}$. The phase is equal to $\overline{\delta}_1 = \delta_1^{HF} + \Delta_1$, while the correction to it is determined using the formula $\Delta_{11} = \mathrm{arctg}(\mathrm{Im}G_{11}/\mathrm{Re}G_{11})$.
3. For the dipole transition $nd - \varepsilon f$, the coefficient is $c_{31} = -1.73205/q$. The modulus of the matrix element for each energy ω is determined through the real and imaginary parts by the formula $\tilde{G}_{31} = \sqrt{(\mathrm{Re}G_{31})^2 + (\mathrm{Im}G_{31})^2}/c_{31}$. The phase is equal to $\overline{\delta}_3 = \delta_3^{HF} + \Delta_{31}$, while $\Delta_{31} = \mathrm{arctg}(\mathrm{Im}G_{31}/\mathrm{Re}G_{31})$ is a correction to it.
4. For the quadrupole transition $nd - \varepsilon s$, the coefficient is $c_{02} = 1/q^2$. The module of the matrix element for each energy ω is found by the formula $\tilde{G}_{02} = \sqrt{(\mathrm{Re}G_{02})^2 + (\mathrm{Im}G_{02})^2}/c_{02}$. The phase is equal to $\overline{\delta}_0 = \delta_0^{HF} + \Delta_{02}$, while the correction to it is $\Delta_{02} = \mathrm{arctg}(\mathrm{Im}G_{02}/\mathrm{Re}G_{02})$.
5. For the quadrupole transition $nd - \varepsilon d$, the coefficient is $c_{22} = (-1.1952)/q^2$. The modulus of the matrix element for each energy ω is found by the formula $\tilde{G}_{22} = \sqrt{(\mathrm{Re}G_{22})^2 + (\mathrm{Im}G_{22})^2}/c_{22}$. The phase is equal to $\overline{\delta}_2 = \delta_2^{HF} + \Delta_{22}$, and $\Delta_{22} = \mathrm{arctg}(\mathrm{Im}G_{22}/\mathrm{Re}G_{22})$ is a correction to it. The phases $\overline{\delta}_{22}$ and $\overline{\delta}_{20}$ are close to each other, so any of them could be chosen in determination of $\overline{\delta}_2$.
6. For the quadrupole transition $nd - \varepsilon h$, the coefficient is $c_{42} = (-1.6036)/q^2$. The modulus of the matrix element for each energy ω is determined by the formula $\tilde{G}_{42} = \sqrt{(\mathrm{Re}G_{42})^2 + (\mathrm{Im}G_{42})^2}/c_{42}$. The phase is equal to $\overline{\delta}_4 = \delta_4^{HF} + \Delta_{42}$, while the correction to it is $\Delta_{42} = \mathrm{arctg}(\mathrm{Im}G_{42}/\mathrm{Re}G_{42})$.

Wave functions and HF phases are calculated in the ***scfhf*** and ***fchf*** programs. The real and imaginary parts of the matrix elements of the GOS are obtained using the ***gos*** program [4]. The dipole and non-dipole angular anisotropy parameters and the GOS density differential in angle of the emitted electron (13.12) are calculated using the ***difgos2*** program.

References

1. M. Inokuti, Inelastic collisions of fast charged particles with atoms and molecules—The Bethe Theory revisited. Rev. Mod. Phys. **43** (1971)
2. M.Ya. Amusia, L.V. Chernysheva, E.Z. Liverts, Phys. Rev. A**85**, 4, 042722 (2012)
3. M.Ya. Amusia, L.V. Chernysheva, V.G. Yarzhemsky, *Handbook of Theoretical Atomic Physics, Data for Photon Absorption, Electron Scattering, Vacancies Decay* (Springer, Heidelberg, New York, Dordrecht, London, 2012)
4. L.V. Chernysheva, M.Ya. Amusia, S.I. Sheftel, *Numerical Calculation of Atomic Generalized Oscillator Strength with Wave Functions in the Hartree-Fock Approximation and Taking into Account Many-Electron Correlations in One Transition* (Ioffe Institute, No. 493, Leningrad, 1973) (In Russian)

Chapter 14
Scattering of Particles of Low and Medium Energies by Atoms

In this chapter, a connection is established between the self-energy part of the single-particle Green's function and the phase shift of the incident particle in elastic scattering, as well as the shift in the energy of the atomic electron caused by correlation effects. Formulas for calculating the phase shifts of inelastic scattering are presented.

Formulas are also presented that allow one to take into account the scattering phases of an electron and the change in its wave function under the action of the dynamic polarization of the target atom by an incident particle.

The scattering of a positron and a $\mu^{\pm}-$meson by an atom and the capture of a $\mu^{-}-$meson by an atom are considered.

14.1 Theoretical Model

14.1.1 The Self-energy Part of the One-Particle Green's Function

It is convenient to divide the irreducible self-energy part of the single-particle Green's function Σ into two parts: the Hartree–Fock part and the one determined by the correlation interaction, so that one has:

$$\Sigma = \Sigma^{\text{HF}} + \Sigma^{\text{cor}} \tag{14.1}$$

The self-energy part in the Hartree–Fock approximation Σ^{HF} is the self-consistent field (see Chap. 4), with which the atom acts on the scattered electron. We will assume that the wave function of the incident electron in the Hartree–Fock approximation and the scattering phase δ_l^{HF} are known. With this approach, the exchange of an

M. Ya. Amusia and L. V. Chernysheva, *Computation of Atomic and Molecular Processes*, Springer Series on Atomic, Optical, and Plasma Physics 117,
https://doi.org/10.1007/978-3-030-85143-9_14

incident electron with atomic electrons is taken into account already in the zero approximation.

In the lowest order in Σ^{cor}, the phase shift of the scattering of the electron wave with the angular momentum l in comparison with its Hartree–Fock value is [LL]

$$\Delta\delta_{l_1}(E) = -\pi\langle\phi_{\nu_1}|\Sigma^{cor}|\phi_{\nu_1}\rangle \tag{14.2}$$

where ϕ_{ν_1} is the Hartree–Fock wave function of the incident electron (1.8), E_1 and l_1 are its energy and angular momentum, respectively.

Within the framework of so-called simplified RPAE or SRPAE, the diagonal matrix element Σ^{cor} can be diagrammatically presented in the following form:

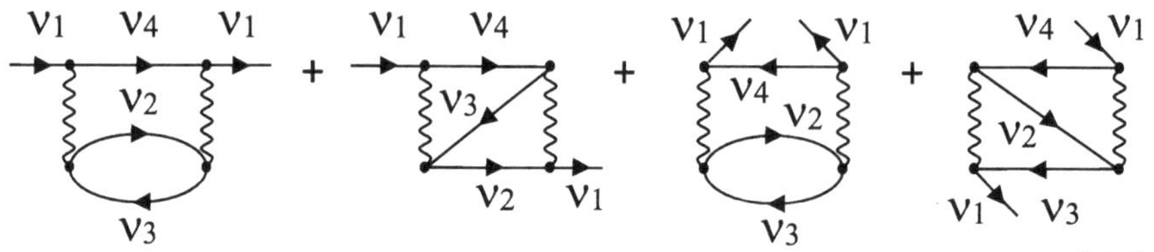

(14.3)

In accord with the correspondence rules (see, e.g., [ACY] or [AICC]), this matrix element is written in the form (see notations in Chap. 1):

$$\langle\nu_1|\Sigma^{cor}|\nu_1\rangle = \left(\sum_{\nu_4\nu_2>F,\nu_3\leq F} + \sum_{\nu_4\nu_2\leq F,\nu_3>F}\right)\frac{\langle\nu_1\nu_3|V|\nu_4\nu_2\rangle\langle\nu_4\nu_2|U|\nu_1\nu_3\rangle}{E_1 - E_4 - E_2 + E_3 + i\eta(1-2n_{\nu_4})} \tag{14.4}$$

Knowledge of the diagonal matrix element Σ^{cor} allows, along with elastic scattering of electrons by atoms, to describe also other atomic processes and characteristics. The imaginary part of the matrix element Σ^{cor} determines the imaginary part of the scattering phase shift, which is associated with inelastic processes, namely the excitation of an atom or its ionization by an incident particle.

When the energy of the incident electron E_1 exceeds the threshold of inelastic processes, the imaginary addition to the phase μ_l that follows from formula (14.4), after integration over angular and summation over spin variables, is determined by the expression

$$\mu_{l_1} = \pi^2\sum_L\sum_{\nu_4\nu_2>F,\nu_3\leq F}\frac{\langle\nu_1\nu_3\|V_L\|\nu_4\nu_2\rangle\langle\nu_4\nu_2\|U_L\|\nu_1\nu_3\rangle}{(2L+1)(2l_1+1)}\delta(E_1 - E_4 - E_2 + E_3) \tag{14.5}$$

The total cross section for inelastic scattering of electrons by an atom is

$$\sigma_r = \frac{2\pi}{p^2}\sum_{l=0}^{\infty}(2l+1)\left[1 - |\exp(2i\delta_l)|^2\right] \tag{14.6}$$

Here k is the momentum of the incident electron. The scattering phases δ_l are complex:

$$\delta_l = \lambda_l + i\mu_l \tag{14.7}$$

Substituting (14.7) into (14.6), we obtain the following expression for the inelastic scattering cross section:

$$\sigma_r = \frac{2\pi}{p^2}\sum_{l=0}^{\infty}(2l+1)\left[1 - \exp(-4\mu_l)\right] \tag{14.8}$$

For $\mu_l \ll 1$

$$\sigma_r = \frac{2\pi}{p^2}\sum_{l=0}^{\infty}(2l+1)\mu_l \tag{14.9}$$

The differential in angle θ cross section of the elastically scattered electron by an atom can be presented in the following way [LL]:

$$\frac{d\sigma}{d\Omega} = |f(\theta)|^2, \tag{14.10}$$

where

$$f(\theta) = \frac{1}{2ip}\sum_{l=0}^{\infty}(2\ell+1)\left(e^{2i\delta_l} - 1\right)P_l(\cos\theta) \tag{14.11}$$

Substituting (14.7) into (14.10) and (14.11), we obtain the following expression for the differential scattering cross section

$$\begin{aligned}\frac{d\sigma}{d\Omega} = \frac{1}{p^2}\sum_{l_1,l_2=0}^{\infty} & (2\ell_1+1)(2\ell_2+1)e^{-\mu_2-\mu_1}P_{l_2}(\cos\theta)P_{l_1}(\cos\theta) \\ & \times\left[\cos\left(\lambda_{l_1}-\lambda_{l_2}\right)\left(\sin\lambda_{l_1}\sin\lambda_{l_2}\mathrm{ch}\mu_{l_1}\mathrm{ch}\mu_{l_2} + \cos\lambda_{l_1}\cos\lambda_{l_2}\mathrm{sh}\mu_{l_1}\mathrm{sh}\mu_{l_2}\right)\right. \\ & \left. - \sin\left(\lambda_{l_1}-\lambda_{l_2}\right)\left(\cos\lambda_{l_1}\sin\lambda_{l_2}\mathrm{sh}\mu_{l_1}\mathrm{ch}\mu_{l_2} - \sin\lambda_{l_1}\cos\lambda_{l_2}\mathrm{ch}\mu_{l_1}\mathrm{sh}\mu_{l_2}\right)\right]\end{aligned} \tag{14.12}$$

In concrete calculations, the angular momentum sums are limited to four terms with $l = 0 \div 3$.

Through the matrix element Σ^{cor} calculated between the wave functions of the occupied states, the energy shift of the hole level (in comparison with its Hartree–Fock value) due to the correlation interaction of atomic electrons is expressed.

The one-particle Green's function of a hole can be written as follows [1]:

$$G(E) = \frac{1}{E - E^{\mathrm{HF}} - \Sigma^{\mathrm{cor}}(E)}, \tag{14.13}$$

where E^{HF} is the level energy in the Hartree–Fock approximation.

The energy of the hole level ν_1 can be determined by finding the pole of the Green's function $G(E)$, i.e., by equating to zero the denominator in expression (14.14). So, the exact one-particle energy is the solution to the equation $\tilde{E}_{\nu_1} = E_{\nu_1} + \Sigma^{\mathrm{cor}}_{\nu_1}\left(\tilde{E}_{\nu_1}\right)$.

Near the pole, Green's function (14.14) can be represented in the form

$$G(E) = \frac{F_{\nu_1}}{E - \tilde{E}_{\nu_1}}, \tag{14.14}$$

where

$$F_{\nu_1} = \frac{1}{1 - \partial\Sigma^{\mathrm{cor}}_{\nu_1}(E)/\partial E|_{E\to\tilde{E}_{\nu_1}}}; \tag{14.15}$$

$$0 \le F_{\nu_1} \le 1 \tag{14.16}$$

is called the spectroscopic factor of level ν_1. It characterizes the fraction of a pure Hartree–Fock state ν_1 in the complex conglomerate of different excitations that this state becomes under the influence of electron–electron interaction (for more details, see [AM]).

The self-energy part of the hole $\Sigma^{\mathrm{cor}}_{\nu_1}(E)$ is a complex function. The real part of the matrix element $\Sigma^{\mathrm{cor}}_{\nu_1}(E)$ is equal to the shift of the level energy ν_1 in comparison with its Hartree–Fock value. The matrix $\Sigma^{\mathrm{cor}}_{\nu_1}(E)$ is diagonal in angular momentum. So, it. does not violate the spherical symmetry of the atomic field. In what follows, we will assume that:

1. The distance between different subshells is large, so that the matrix Σ^{cor} has a quasi-diagonal structure and in principal quantum numbers, i.e., for any j one has

$$\left\langle \nu_1 \middle| \Sigma^{\mathrm{cor}}(E) \middle| \nu_1 \right\rangle \gg \left\langle \nu_{j\neq 1} \middle| \Sigma^{\mathrm{cor}}(E) \middle| \nu_1 \right\rangle \tag{14.17}$$

2. The Hartree–Fock value of the level energy E_1^{HF} is close to the true value:

$$\Sigma^{\mathrm{cor}}(E_1) \approx \Sigma^{\mathrm{cor}}\left(E_1^{\mathrm{HF}}\right) \tag{14.18}$$

and the spectroscopic factor is close to 1.

Taking these approximations into account, the energy shift of the level ν_1 in comparison with its Hartree–Fock value is determined by the formula:

$$E_1 - E_1^{\mathrm{HF}} = \Delta E_1 = \mathrm{Re}\langle \nu_1 | \Sigma_{\nu_1}^{\mathrm{cor}}(E_1^{\mathrm{HF}}) | \nu_1 \rangle \tag{14.19}$$

In the framework of SRPAE, ΔE_1 is equal to:

$$\Delta E_1 = \sum_L \frac{1}{(2L+1)(2l_1+1)} \left(\sum_{\nu_4\nu_2 > F, \nu_3 \le F} + \sum_{\nu_4\nu_2 \le F, \nu_3 > F} \right) \frac{\langle \nu_1 \nu_3 \| V_L \| \nu_4 \nu_2 \rangle \langle \nu_4 \nu_2 \| U_L \| \nu_1 \nu_3 \rangle}{E_1 - E_4 - E_2 + E_3 + i\eta(1 - 2n_{\nu_4})} \tag{14.20}$$

The imaginary part of the matrix element Σ^{cor} taken between the wave functions of the occupied states determines the half-width of the hole level ν_1 relative to Auger decay:

$$\Gamma_1 = 2\mathrm{Im}\langle \nu_1 | \Sigma^{\mathrm{cor}} | \nu_1 \rangle \tag{14.21}$$

In SRPAE, Γ_1 equals

$$\Gamma_1 = 2\pi \sum_L \sum_{\nu_4\nu_2 \le F, \nu_3 > F} \frac{\langle \nu_1 \nu_3 \| V_L \| \nu_4 \nu_2 \rangle \langle \nu_4 \nu_3 \| U_L \| \nu_1 \nu_3 \rangle}{(2L+1)(2l_1+1)} \delta(E_1 - E_4 - E_2 + E_3) \tag{14.22}$$

14.1.2 *Scattering Phases and Wave Functions with Exact Account of the Self-Energy Part*

When studying the scattering of electrons by atoms, the wave function of the incident electron ϕ^{N+1} is determined in the field of the frozen core, i.e., the ground state of a neutral atom. The asymptotic of the radial wave function in the continuous spectrum has the form (compare with (6.14)):

$$P_{\mathrm{El}}^{N+1}(r)\big|_{r\to\infty} \sim \frac{1}{\sqrt{\pi p}} sin\left(pr - \frac{\pi l}{2} + \delta_l^{\mathrm{HF}}(E) \right) \tag{14.23}$$

where $p = \sqrt{E}$, E in Ry, is the electron momentum, δ_l^{HF} is the scattering phase in the HF approximation. The exact radial wave function of the scattered electron satisfies the following equation:

$$P_{\mathrm{El}}^{\Sigma}(r) = P_{\mathrm{El}}^{N+1}(r) - \frac{1}{\hat{H}_l^{\mathrm{HF}} - E} \int \Sigma_l^{\mathrm{cor}}\left(r, r', E\right) P_{\mathrm{El}}^{\Sigma}\left(r'\right) dr' \tag{14.24}$$

with the asymptotic

$$P_{El}^{\Sigma}(r) \sim \frac{1}{\sqrt{\pi p}}\left[\sin\left(pr - \frac{\pi l}{2} + \delta_l^{\mathrm{HF}}(E)\right) + \tan \Delta\delta_l \cdot \cos\left(pr - \frac{\pi l}{2} + \delta_l^{\mathrm{HF}}(E)\right)\right]. \tag{14.25}$$

where $\Delta\delta_l$ is the correction to the Hartree–Fock phase due to the interaction Σ^{cor}:

$$\delta_l(E) = \delta_l^{\mathrm{HF}}(E) + \Delta\delta_l(E). \tag{14.26}$$

Here $\Sigma_l^{\mathrm{cor}}(r, r', E)$ is the so-called irreducible self-energy part of the one-particle Green's function (see Sect. 2.18), $\hat{H}_l^{\mathrm{HF}}$ is the radial part of the Hartree–Fock operator, whose eigenfunctions P_{El}^{N+1} are determined by the HF equation:

$$\left(\hat{H}_l^{\mathrm{HF}} - E\right) P_{\mathrm{El}}^{N+1}(r) = 0. \tag{14.27}$$

From the asymptotic (14.25) and (14.24), we can obtain the expression

$$\mathrm{tg}\Delta\delta_l(E) = -\pi\left\langle El\left|\overline{\Sigma}(E)\right|El\right\rangle, \tag{14.28}$$

where $\overline{\Sigma}$ is the so-called reducible self-energy part, which is the solution to Eq. (2.13):

$$\left\langle El\left|\overline{\Sigma}(\omega)\right|E'l\right\rangle = \left\langle El\left|\Sigma^{\mathrm{cor}}(\omega)\right|E'l\right\rangle + P\int \left\langle El\left|\Sigma^{\mathrm{cor}}(\omega)\right|E''l\right\rangle\left\langle E''l\left|\overline{\Sigma}(\omega)\right|E'l\right\rangle \frac{dE''}{\omega - E''}. \tag{14.29}$$

Here, $\mathrm{P}\int$ denotes the principal value of the integral $\int$. Integration over E'' includes summation over discrete states, including occupied (hole) levels. The matrix elements Σ^{cor} in (14.29) are calculated using the Hartree–Fock wave functions:

$$\left\langle El\left|\Sigma^{\mathrm{cor}}(\omega)\right|E'l\right\rangle = \int \Sigma_l^{\mathrm{cor}}\left(r, r', \omega\right) P_{\mathrm{El}}^{*N+1}(r) P_{E'l}^{N+1}\left(r'\right) dr dr' \tag{14.30}$$

Equation (14.29) can be diagrammatically represented in the following way:

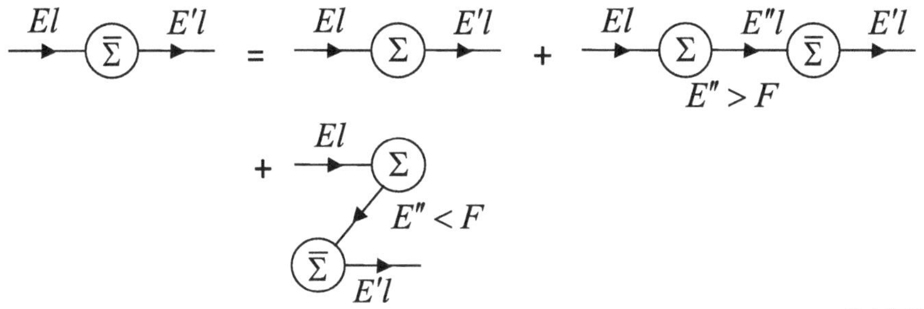

(14.31)

Equation (14.29) and its diagrammatic form (14.31) are often called the Dyson equation.

When considering the scattering of an electron by an ion, due to the interference of electron waves scattered by the potentials Σ^{cor} and the Coulomb one, Eq. (14.29) should be solved, even if the matrix elements $\langle El|\Sigma^{\text{cor}}(\omega)|E'l\rangle$ are small. In this case, $\langle El|$ determines the electron wave function in the ion field.

14.1.3 Phases, Scattering Cross Sections, and Wave Functions in the Frame of the SRPAE

Thus, the problem is reduced to the calculation of the self-energy part Σ^{cor} (14.30) and the solution of Eq. (14.29). To find the self-energy part (14.30), one can use RPAE, which has proven itself in other calculations as a quite accurate approach. However, due to complexity of using this method in the scattering problem, we will consider a simpler option—SRPAE.

In the first order of the perturbation theory for the Coulomb interaction of an incident electron with atomic electrons, for Σ there are only Hartree–Fock diagrams [AM, AICC]. Since we consider the Hartree–Fock approximation as the initial approach, the HF contribution has already been taken into account by the choice as wave functions the $P^{N+1}(r)$ functions. In the second order of perturbation theory, for Σ^{cor} there are four diagrams (14.3), the contribution of which must be accounted for. All these diagrams correspond to the lowest RPAE order, thus forming SRPAE. Analytically, their contribution is described by the following expression:

$$\langle El|\Sigma^{\text{cor}}(\omega)|E'l\rangle = \left(\sum_{\substack{\nu_4\nu_2>F,\\ \nu_3\leq F,L}} - \sum_{\substack{\nu_4\nu_2\leq F,\\ \nu_3>F,L}} \right) \frac{1}{(2L+1)(2l+1)} \frac{\langle El,\nu_3\|V_L\|\nu_4\nu_2\rangle\langle\nu_4\nu_2\|U_L\|E'l,\nu_3\rangle}{\omega - E_4 - E_2 + E_3 + i\eta\left(1-2n_{\nu_4}\right)} \tag{14.32}$$

After the matrix elements are calculated, Eq. (14.29) can be solved to obtain the reducible self-energy part $\overline{\Sigma}(E)$, through which the corrections to the scattering phase are expressed (14.28):

$$\Delta\delta_l(E) = \mathrm{arctg}\left[-\pi\left\langle El\middle|\overline{\Sigma}(E)\middle|El\right\rangle\right] \tag{14.33}$$

At energies exceeding the atomic ionization threshold, the expression for the matrix element defined by (14.32) becomes complex, and with it the solution of Eq. (14.29) and the correction to the phase (14.28) acquire an imaginary part.

Let us present the phase in accord with (14.7) as

$$\delta_l(E) = \lambda_l(E) + i\mu_l(E), \tag{14.34}$$

where

$$\lambda_l(E) = \delta_l^{\mathrm{HF}}(E) + \mathrm{Re}\Delta\delta_l(E), \tag{14.35}$$

and

$$\mu_l(E) = \mathrm{Im}\Delta\delta_l(E). \tag{14.36}$$

Knowledge of the phase (14.34) allows one to find the cross sections of both elastic

$$\sigma_{el} = \frac{4\pi}{p^2}\sum_{l=0}^{\infty}(2l+1)(\cosh 2\mu_l - \cos 2\lambda_l)\exp(-2\mu_l), \tag{14.37}$$

and inelastic

$$\sigma_{in} = \frac{2\pi}{p^2}\sum_{l=0}^{\infty}(2l+1)\left[1 - \exp(-4\mu_l)\right] \tag{14.38}$$

electron–atom scattering [2, LL].

In addition to the cross sections for elastic and inelastic scattering, knowing $\overline{\Sigma}$ allows one to find the wave function of the scattered electron in RPAE. Indeed, Eq. (14.24) can be written in terms of the reducible self-energy part (14.29) as follows:

$$P_{\mathrm{El}}^{\Sigma}(r) = P_{\mathrm{El}}^{N+1}(r) + P\int \frac{P_{E'l}^{N+1}(r)\left\langle E'l\middle|\overline{\Sigma}(E)\middle|El\right\rangle}{E - E'}dE'. \tag{14.39}$$

The matrix element $\overline{\Sigma}$ between the Hartree–Fock hole states determines the shift of the hole energy in comparison with its Hartree–Fock value. Expression (14.29) for $\overline{\Sigma}$ differs from expression (14.4) for Σ^{cor} in that it contains corrections of higher order in Σ^{cor}, determined by (14.24).

The integral equation, e.g., (14.39), would be reduced to a set of one-dimensional algebraic equations if instead of a continuous spectrum E' we consider a number of discrete values ε_i equidistant in momentum, i.e., $p_i = p_0 + i\Delta p (i = 1, 2, \ldots)$; $p_0^2 = E_0$ (energy expressed in Ry and p in atomic units). Solving the integral equation is equivalent to diagonalizing the following matrix:

$$M_{ij} = \varepsilon_i \delta_{ij} + A_i \langle \varepsilon_i l | \Sigma(E_0) | \varepsilon_j l \rangle A_j \tag{14.40}$$

Here A_i are the coefficients of numerical integration. Usually, $\Sigma(E_0)$ only weakly depends on E_0 at $E_0 \to 0$, and it can be neglected. With this assumption, the solution is especially easy to find. Otherwise, the dependence on energy E_0 should be fixed, and E_0 would be determined only after the whole diagonalization procedure. Knowing the eigenvalues (14.40), the coefficients $C_l(\varepsilon)$ that are results of M_{ij} diagonalization become determined, as well as the wave functions. If the lowest eigenvalue (14.40) turns out to be negative, this means that a negative ion has been formed in this partial wave—by an N-electron atom and an incident electron $N + 1$.

14.1.4 Scattering of Slow Positrons by Atoms

For a positron, as well as a positively charged $\mu-$meson and other charged particles, $\overline{\Sigma}$ is obtained from Σ^{cor} using (14.29), while the matrix element Σ^{cor} is given in SRPAE, without exchange of incident particles with atomic electrons. Of course, when calculating Σ^{cor} and solving the Hartree equation for an incident particle, one should account for their real mass and charge.

The study of "positron–atom" and "positron–molecule" scattering is a very active area of research (see [3] and references therein). It is interesting to compare this process with electron collisions on the same objects. The effect of the "projectile–target" interaction is strong at low collision energies. Therefore, here we are concentrating on this area of energies.

In both cases, namely for electron and positron scattering, the cross section is determined by the electrostatic and polarization potentials, by which the target acts upon the projectile. However, for incident electrons, the exchange with target electrons is important. For positrons, exchange does not exist at all. Therefore, at first glance, the process of scattering of a positron looks simpler than that of electrons. In addition, the positron–atom static potential, being generally repulsive, forces positrons to move in the target region preferably far from nucleus, where the potential is small, thus weakening atomic potential total action. This is opposite to the situation in electron–atom (molecule) collision. The contrast looks even stronger when we consider that, as is usually believed, the polarization potential for both particles, an electron and a positron, is attractive.

Thus, for electrons, the total potential must be stronger than for positrons. At first glance, therefore, it seems that any simple approach that is good for electrons should be applicable, at least with equal accuracy, for positrons. For example, the use of the second-order approximation in calculating the polarization interaction discussed in Sects. 14.1.1 and 14.1.3 [see (14.4) and (14.25)], should lead to success when considering the scattering of a positron by an atom. In this case, only the first of the diagrams (14.3) will make a nonzero contribution due to the absence of exchange. One would expect this approach to be at least as good at describing the scattering of a positron by an atom as it is describing scattering of an electron by an atom. It turned out, however, that this point of view is incorrect [4].

The reason of this mistake in argumentation is the following. The positron, which is "pushed out" by the potential of the atom from the target, can strongly interact with a virtually excited electron, which is also outside the target. They form a kind of connected state, which can be called "virtual positronium". This is a very important mechanism that strongly influences the scattering cross section [4–8].

Diagrammatically, this mechanism is taken into account in a modified second-order diagram, which describes the polarization interaction of the positron as follows

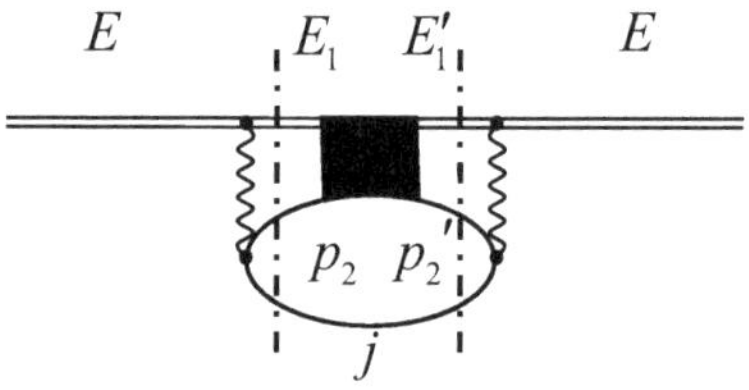

(14.41)

Here, the shaded square denotes the interaction of a positron with a virtually excited electron, including the formation of "virtual positronium", which has to be accounted for the framework of the many-body perturbation theory—MBPT. The positron itself is represented by a double line.

In the lowest order in the interaction between a positron and an atomic electron, the analytical expression for the polarization interaction in the scattering of a positron is given by the following expression

$$\left\langle E\ell \left\| \hat{\Sigma}^{l}(E) \right\| E\ell \right\rangle = \frac{1}{(2l+1)} \sum_{\substack{j\leq F,L \\ \varepsilon_2 l_2 > F}} \int_0^\infty \frac{\left|\left\langle E\ell, \varepsilon_j l_j \| V_L \| E_1 \ell_1, \varepsilon_2 \ell_2 \right\rangle\right|^2}{(2L+1)\left(E - E_1 - \varepsilon_2 + \varepsilon_j + i\delta\right)} dE_1 d\varepsilon_2 \tag{14.42}$$

Here the second term with U_L in (14.4) is replaced by the term with V_L, since there is no exchange for the positron.

It turned out that the interaction between the positron and the virtually excited atomic electron is extremely important. Here we describe a method that significantly improves the agreement with experiment and provides a simple qualitative explanation for the large difference in positron scattering cross sections for noble gases and their neighbors, alkali atoms. The method was first proposed in [4] and further developed in [5–7]. Its idea is to take into account the entire complex set of diagrams included in (14.41) simply by introducing the binding energy of a real *positronium* Ps, I_{Ps} into the energy denominator (14.42):

$$\left\langle E\ell\left\|\hat{\Sigma}^{l}_{\mathrm{Ps}}(E)\right\|E\ell\right\rangle = \frac{1}{(2l+1)} \sum_{\substack{j\le F, L\\ \varepsilon_2 l_2 > F}} \int_0^{\infty} \frac{\left|\left\langle E\ell,\varepsilon_j l_j\left\|V_L\right\|E_1\ell_1,\varepsilon_2\ell_2\right\rangle\right|^2}{(2L+1)\left(E-E_1-\varepsilon_2+\varepsilon_j+I_{Ps}+i\delta\right)} dE_1 d\varepsilon_2. \tag{14.43}$$

Since the introduction of this approach in [4], a number of more complex calculations have been performed (e.g., [8, 9]), which use the idea of the formation of "virtual positronium", but replace the positron–electron wave function $|E_1\ell_1, \varepsilon_2\ell_2>$ in (14.43) with the wave function of the free positronium, which is not affected by the field of the residual ion.

The elastic $\sigma^{el+}(E)$, inelastic $\sigma^{in+}(E)$, and total $\sigma^{\mathrm{tot}+}(E)$ scattering cross sections for a positron with energy E, expressed in terms of the partial scattering phases $\delta_\ell(E)$, are given by formulas (14.37), (14.38).

14.1.5 Scattering and Capture of μ−Mesons by Atoms

The wave functions of a μ−meson colliding with an atom are found by solving Eq. (1.17), with substitution of the charge e_μ and mass m_μ of this particle as Ze_1 and m_1 accordingly. The scattering cross section is determined by formulas (14.5)–(14.8), where $p = \sqrt{m_\mu E_\mu}$ (E_μ in Ry). The presence of $m_\mu \gg 1$ in the expression for p leads to the fact that in specific calculations, a much larger number of phases δ_l must be taken into account.

Correlation corrections in the second order of perturbation theory can be depicted by the first diagram (14.3). Analytically, the corresponding expression is:

$$\left\langle \nu_1\left|\Sigma^{\mathrm{cor}}_{\mu}\right|\nu_1\right\rangle = \int_{\nu_{2\mu}} \sum_{\nu_4>F,\nu_3\le F} \left|\left\langle \nu_{1\mu}\nu_3\left|V_\mu\right|\nu_{2\mu}\nu_4\right\rangle\right|^2 (E_1-E_4-E_2+E_3+i\eta)^{-1}. \tag{14.44}$$

This expression differs from (14.4) by the elimination of the exchange between the meson and atomic electrons and the replacement of the electron–electron potential

V by the electron–meson V^{μ} one. The corresponding correction to the scattering phase is given by formula (14.28).

When considering the scattering of a positive meson, like a positron (see Sect. 14.1.4), it is necessary to include the interaction of the incident particle with a virtually excited target electron, which can lead to the formation of a *quasi-mesonium* bound state $e\mu^{+}$. The problem is similar to that discussed above in connection with the scattering of a positron by an atom. The simplest solution is to use the obviously modified relation (14.43):

$$\left\langle E_{\mu}\ell\left\| \hat{\Sigma}_{\mu}^{l}(E_{\mu})\right\| E_{\mu}\ell\right\rangle = \frac{1}{(2l+1)}\sum_{\substack{j\le F,L\\ E_2l_2>F}}\int_{0}^{\infty}\frac{\left|\left\langle E_{\mu}\ell,\, E_j l_j\left\| V_L^{\mu}\right\| E_{1\mu}\ell_1,\, E_2\ell_2\right\rangle\right|^2}{(2L+1)\left(E_{\mu}-E_{1\mu}+E_j-E_2+I_{e\mu^+}+i\delta\right)}dE_{1\mu}dE_2. \tag{14.45}$$

Here $I_{e\mu^+}$ is an ionization potential of $e\mu^{+}$ that is very close to the ionization potential of hydrogen atom.

The inelastic scattering of a positive meson by an atom, leading to ionization, is described in accordance with diagram (1.42), where the final state of the meson $\nu_{2\mu}$ is in the continuous spectrum. The corresponding cross section is expressed via the imaginary part of the scattering phase (14.7) using formula (14.8), where $p^2 = m_{\mu}E_{\mu}$. The imaginary part of the phase is found using the relations (14.59)–(14.61) presented below, obtained from the matrix elements of the $\mu-$ meson self-energy part, given by formula (14.45).

At $E_{\mu} > (E_j - E_2)$, the meson energy is high enough to create an electron–hole pair, while the meson remains unbound. In this case, the denominator in (14.45) can be equal to zero, which leads to the appearance of a nonzero imaginary part, which takes into account inelastic processes.

When a negative meson is scattered, another possibility arises. Namely, it can be captured to one of the levels in the field of the atom, and the excess energy is transferred to the electron. This process is called Auger meson capture and is described by diagram (1.42).

Analytically, the cross section for Auger meson capture by an atom is determined by the expression

$$\sigma_{\mu}^{\mathrm{Aug}}(E_1) = \frac{8\pi^3}{m_{\mu}E_1}\sum_{Ll_1}\sum_{\nu_3>F,\nu_4\le F}\frac{1}{(2L+1)}\left|\left\langle \nu_{2\mu}\nu_3\left\| V_L^{\mu}\right\|\nu_{1\mu}\nu_4\right\rangle\right|^2, \tag{14.46}$$

where E_1 is the initial energy of the meson ($E_1 > 0$), l_1 is its angular momentum, $\left\langle \nu_{2\mu}\nu_3\left\| V_L^{\mu}\right\|\nu_{1\mu}\nu_4\right\rangle$ is the reduced Coulomb matrix element corresponding to diagram (1.42), $E_3 = E_{1\mu} + E_4 - E_{2\mu}$; the index μ marks the states and energies of the

$\mu-$meson. It is understood that the meson is captured at a discrete level, and one of the atomic electrons appears in the continuous spectrum.

The captured meson can start moving from one level to another, its transition energy leads to the emission of additional atomic electrons. The probability of the Auger transition of a meson from one discrete state ν_1 to another ν_2 is given by an expression written like (14.46) in the following form:

$$W_{E_{2\mu}}^{\mu \text{Aug}} = 2\pi \sum_{\substack{\nu_{2'} L, \nu_3 > F, \\ \nu_4 \leq F}} \frac{1}{2L+1} \left| \langle \nu_{2'\mu} \nu_3 \| V_L \| \nu_{2\mu} \nu_4 \rangle \right|^2, \tag{14.47}$$

where $E_{2\mu}$ is the meson energy at the discrete level ($E_{2\mu} < 0$) occupied after capture, the cross section of which is described by formula (14.46).

14.2 Numerical Model

14.2.1 *Scattering Cross Section and Phases in the Framework of the SRPAE*

The shift of the partial scattering phases from their Hartree–Fock values in SRPAE is equal to:

$$\Delta\delta_{l_1}(E_1) = -\pi \sum \frac{1}{(2L+1)(2l_1+1)} \left(\sum_{\nu_4 \nu_2 > F, \nu_3 \leq F} + \sum_{\nu_4 \nu_2 \leq F, \nu_3 > F} \right)$$
$$\frac{\langle \nu_1 \nu_3 \| V_L \| \nu_4 \nu_2 \rangle \langle \nu_4 \nu_2 \| U_L \| \nu_1 \nu_3 \rangle}{E_1 - E_4 - E_2 + E_3 + i\eta(1 - 2n_{\nu_4})} \tag{14.48}$$

where $\langle \nu_1 \nu_3 \| V_L \| \nu_4 \nu_2 \rangle$ are the reduced matrix elements (9.33). The combined matrix element U, which includes the contribution of direct and exchange terms, is defined by formula (9.34). To find the principal value of the integral in the vicinity of the pole, we use the procedure described in Sect. 9.2.4.

The ***scat*** program calculates the diagonal matrix element Σ^{cor} (14.4) or the phase shift of the scattering for a fixed value L (see (14.5))

$$\Delta\delta_{Ll_1}(E_1) = -\pi \frac{1}{(2L+1)(2l_1+1)} \left(\sum_{\substack{\nu_4 \nu_2 > F, \\ \nu_3 \leq F}} + \sum_{\substack{\nu_4 \nu_2 \leq F, \\ \nu_3 > F}} \right)$$
$$\frac{\langle \nu_1 \nu_3 \| V_L \| \nu_4 \nu_2 \rangle \langle \nu_4 \nu_2 \| U_L \| \nu_1 \nu_3 \rangle}{E_1 - E_4 - E_2 + E_3 + i\eta(1 - 2n_{\nu_4})}. \tag{14.49}$$

Wave functions are selected in the following way. The incident electron ν_1 is described by the wave function ϕ^{N+1}, since it moves in the field of a neutral atom. By the appropriate choice of the wave functions of virtual electrons, one can account for the contribution of many-electron correlations in different ways, namely

- By choosing $\phi^{N}_{\nu_2}$ and $\phi^{N+1}_{\nu_4}$ thus taking into account part of RPAE correlations in the interaction between an electron ν_2 and a hole ν_3, while the electron ν_4 moves in the field of a neutral atom;
- If functions of virtual states ν_4 and ν_2 are both ϕ^{N+1}, then this means complete neglect of the interaction of virtual electrons with a vacancy ν_3.
- The choice $\phi^{N}_{\nu_4}$ and $\phi^{N}_{\nu_2}$ implies that the screening of the vacancy field ν_3.is neglected for both electrons ν_4 and ν_2.

Through the diagonal matrix element for fixed E_1, l_1, L, the imaginary correction to the phase (14.36) and the cross section for elastic and inelastic scattering in the SRPAE are calculated ((14.37), (14.38)).

14.2.2 Scattering Phases with Account for Polarization Corrections

As usual in numerical calculations, the integral over the continuous spectrum is replaced in formula (14.29) by summation, assuming the upper limit of integration to be finite. The energy value at which the integration can be interrupted with a stable result, independent of the upper limit, is usually several ionization potentials of the outer shell. This energy region makes the main contribution to Σ^{cor}. As a rule, in the Hartree–Fock approximation, there are no bound states of a negative ion. Therefore, the sum over discrete states in formula (14.29) contains only the sum over hole states with a given orbital angular momentum l, if any.

When calculating the principal value of the integral in the vicinity of the pole, the procedure described in Sect. 9.2.4 is applied. In accordance with formula (14.28), we are interested in the solution of Eq. (14.29) at $\omega = E = E'$, in other words, the position of the pole in (14.29) always coincides with the points where the value of the integrand is known. Therefore, in formulas (9.69) and (9.72) it is necessary to find the limit $x_\lambda \to x$. Since we accept the number of discrete points at an interval to be $N = 3$, the limit for even value i gives:

$$Q_i = 0, \quad Q_{i\pm 1} = C_{i\pm 1} + \frac{C_i}{2} \tag{14.50}$$

An odd value i corresponds to the junction of two segments, so calculations according to (9.72) must be performed on two adjacent segments

$$Q_i = 0, \qquad Q_{i\pm1} = C_{i\pm1} + \frac{2}{3}C_i,$$
$$Q_{i\pm2} = C_{i\pm2} - \frac{1}{6}C_i \tag{14.51}$$

If there is no pole, then $Q_i = C_i$. For discrete states $Q_i = 1$.

When solving (14.29), we proceed, as usual, to integration over the momentum $p = \sqrt{E}$, since the wave functions are calculated in the ***fchf*** program with a constant step in p. As a result, (14.29) takes the form:

$$\langle El|\overline{\Sigma}(\omega)|E'l\rangle = \langle El|\Sigma^{\text{cor}}(\omega)|E'l\rangle + \sum_i Q_i D_i \langle El|\Sigma^{\text{cor}}(\omega)|E_i l\rangle\langle E_i l|\overline{\Sigma}(\omega)|E'l\rangle, \tag{14.52}$$

where

$$D_i = \frac{2p_i}{\omega - p_i^2} \tag{14.53}$$

for the values i corresponding to the states of the continuous spectrum and

$$D_i = \frac{1}{\omega - E_i} \tag{14.54}$$

for the values i corresponding to the states of the discrete spectrum.

The quantities $\langle El|\Sigma(\omega)|E'l\rangle$ form a $t \times t$ matrix, where $t = t_1 + t_2$ is the number of states, with t_1 belonging to the discrete spectrum, and t_2—to continuous spectrum selected for integration. For energies E exceeding the ionization potential of the atom, the values $\Sigma^{\text{cor}}(\omega)$ become complex: $\Sigma^{\text{cor}}(\omega) = \Sigma'(\omega) + i\Sigma''(\omega)$. Separation of the real and imaginary parts in (14.52) leads to the following set of equations:

$$\begin{pmatrix} \overline{\Sigma}' \\ \overline{\Sigma}'' \end{pmatrix} = \begin{pmatrix} \Sigma' \\ \Sigma'' \end{pmatrix} + \begin{pmatrix} A & -B \\ B & A \end{pmatrix} \begin{pmatrix} \overline{\Sigma}' \\ \overline{\Sigma}'' \end{pmatrix} \tag{14.55}$$

where

$$A_{jj}(\omega) = \langle E_i l|\Sigma'(\omega)|E_{jl}\rangle Q_j D_j$$
$$B_{jj}(\omega) = \langle E_i l|\Sigma''(\omega)|E_{jl}\rangle Q_j D_j \tag{14.56}$$

So, the symbolic solution of (14.55) is

$$\begin{pmatrix} \overline{\Sigma}' \\ \overline{\Sigma}'' \end{pmatrix} = \begin{pmatrix} 1-A & B \\ -B & 1-A \end{pmatrix}^{-1} \begin{pmatrix} \Sigma' \\ \Sigma'' \end{pmatrix} \tag{14.57}$$

Thus, the solution of Eq. (14.55) is reduced to inversion of $2t \times 2t$ matrix. The equation is consistently solved for all points $\omega = E_i$ in the continuous spectrum except the first and the last, where the pole coincides with the integration limits and the formulas that we use become inapplicable [10].

At each value ω, a phase correction is calculated according to Eq. (14.33):

$$\mathrm{Re}\Delta\delta_l(E_i) = \begin{cases} 1/2\arctan\left(2a/\left(1-a^2-b^2\right)\right), & |a|<1, \\ (\pi a)/(2|a|), & |a|=1, \\ 1/2\arctan\left(2a/\left(1-a^2-b^2\right)\right)+(\pi a)/(2|a|), & a|>1 \end{cases} \quad (14.58)$$

$$\mathrm{Im}\Delta\delta_l(E_i) = \frac{1}{2}ln\frac{(1+b)^2+a^2}{\sqrt{\left(1-a^2-b^2\right)^2+4a^2}}, \quad (14.59)$$

$$a = -\pi\left\langle E_i l\middle|\Sigma'(E_i)\middle|E_i l\right\rangle, \quad (14.60)$$

$$b = -\pi\left\langle E_i l\middle|\Sigma''(E_i)\middle|E_i l\right\rangle. \quad (14.61)$$

The phase δ_l^{HF} in the Hartree–Fock approximation is calculated in the ***fchf*** program simultaneously with the wave function $P_{\mathrm{El}}^{N+1}(r)$. And the ***scat*** program finds the real and imaginary parts of the phases. The ***scat*** program can also be used to calculate the scattering amplitudes of electrons by an ion, or charged particles by an atom or ion. Then, the one-particle wave function for the incident particles is taken as P^{N+1}, while $\overline{\Sigma}$ is calculated in accordance to (14.29).

The wave function of a scattered electron can also be found by solving the differential equation that is obtained from Eq. (14.24) by multiplying it by the difference $\left(\hat{H}_l^{\mathrm{HF}} - E_0\right)$:

$$\left(\hat{H}_l^{\mathrm{HF}} - E_0\right)P_{E_0 l}^{\Sigma}(r) + \int\limits_0^\infty \Sigma_l^{\mathrm{cor}}\left(r, r'E_0\right)P_{E_0 l}^{\Sigma}\left(r'\right)dr' = 0 \quad (14.62)$$

Explicit evaluation of $\Sigma_l^{\mathrm{cor}}\left(r, r', E_0\right)$ can be avoided if one deals with matrix elements $\left\langle E_1 l|\Sigma_l^{\mathrm{cor}}\left(r, r', E_0\right)|E_2 l\right\rangle$. To do this, we will expand $P_{El}^{\Sigma}(r)$ into a series of one-electron wave functions in the Hartree–Fock approximation, [see. (4.3)] for occupied states, and $P_{\mathrm{El}}^{N+1}(r)$ for unoccupied states and states of a continuous spectrum.

Let us represent $P_{\mathrm{El}}^{\Sigma}(r)$ in the following way:

$$P_{E_0 l}^{\Sigma}(r) = \int\limits_{\varepsilon} C_l(\varepsilon)P_{\varepsilon l}(r). \quad (14.63)$$

For occupied states $P_{\varepsilon l}(r) \rightarrow P_{nl}(r)$ [see. (4.3)], $n, l \leq F$, and for unoccupied states $P_{\varepsilon l}(r) \rightarrow P_{\varepsilon l}^{N+1}(r)$ [see (6.5)].

The coefficients $C_l(\varepsilon)$ can be found using the equation obtained from (14.62) by the transformation (14.63), subject to orthogonality condition (6.11) (see Chap. 6):

$$E_0 C_l(E) = EC_l(E) + \int\limits_{E'} < El|\Sigma(E_0)|E'l > C_l(E') \tag{14.64}$$

14.2.3 Scattering of Slow Positrons by Atoms

The first step of the approach described above is the use of the HF approximation for the target atom and the frozen HF core for an incident positron, naturally without exchange between the positron and atomic electrons. Thus, to calculate the HF phase shift $\delta_\ell^{HF}(E)$ of a positron, it is necessary to solve the following equation for the positron wave function $\phi_{e+}(\vec{r})$, in which the interaction of the positron with the atom does not include the exchange terms

$$\left[-\frac{\Delta}{2} + \frac{Z}{r} - \int \frac{d\vec{r}\rho(r')}{\left|\vec{r} - \vec{r}'\right|}\right]\phi_{e+}(\vec{r}) = E\phi_{e+}(\vec{r}). \tag{14.65}$$

Here $\rho(r)$ is the density of electrons in an atom. The asymptotic in r of the radial part $P_{E\ell}^{e+}(r)$ of the function $\phi_{e+}(\vec{r})$ determines the phase shift.

The second step involves the inclusion of the polarization interaction. The formation of virtual positronium is taken into account by changing the energy of the intermediate state, shifting it by the value of the free positronium binding energy 0.5Ry. Assuming that the polarization interaction $\hat{\sum}_{Ps}$ is weak, it is possible to find the correction $\Delta\delta_\ell(E)$ to the HF partial scattering phase of the positron, which appears due to action of $\hat{\sum}_{Ps}$, using the following formula:

$$\Delta\delta_\ell(E) = -\pi\left\langle E\ell\left\|\sum_{Ps}^{l}(E)\right\|E\ell\right\rangle, \tag{14.66}$$

where $\hat{\sum}_{(Ps)}^{l}(E)$ is the ℓ-component of $\hat{\sum}_{(Ps)}$ determined from the formula (14.43).

The scattering cross section, that accounts for polarization interaction, is determined by Eqs. (14.37) and (14.38), where $\delta_l^{HF}(E)$ is replaced by phases $\delta_\ell(E) = \delta_\ell^{HF}(E) + \Delta\delta_\ell(E)$ for E, to $E < I$, where I is the ionization potential of the target atom. For higher positron energies, the positron scattering phases given by (14.43) are complex, and their imaginary part $\mathrm{Im}\Delta\delta_\ell(E) \equiv \mu_l(E)$ determines the cross section for inelastic positron–atom scattering. The elastic scattering cross section for $E > I$ is described by (14.37).

So, the ***scat*** program [11, 12], depending on the specification of the initial data, can give phases, elastic and inelastic scattering cross sections for electrons and positrons in the HF and SRPAE approximations and with account of polarization corrections (see (14.24)). The differential in scattering angle cross section for incident electrons or positrons is calculated by the ***anglscat*** program that uses formula (14.12).

14.2.4 Cross Sections of μ–Mesons Scattering and Capture

To calculate the cross section and the probability of μ–meson scattering and capture, knowledge of the reduced matrix elements $\langle \nu_2\nu_3 \| V_L \| \nu_1\nu_4 \rangle$ and the proper self-energy part of the one-particle Green's function of the meson (14.45) is required.

The ***meson*** program allows one to calculate depending on the energy value E_1, the cross section σ_μ^{Aug} of Auger capture of a meson by an atom $(E_{1\mu} > 0)$ or the probability W_μ^{Aug} of a meson Auger transition $(E_{2\mu} < 0)$.

For this, first of all, it is necessary to calculate the ground state wave functions of the atom in the Hartree–Fock approximation using the ***scfhf*** program. The wave function of the initial state of the meson in the continuous spectrum $\nu_{1\mu} \equiv (E_{1\mu} l_1)$ is of the form [LL]

$$\phi^+_{\nu_\mu}(x) = \sqrt{\frac{\pi}{2pm_\mu}} \sum_{l=0}^{\infty} 4\pi i^l \exp(i\delta_l) \frac{1}{r} P^\mu_{\nu_1}(r) Y_{lm}(\vartheta, \phi) Y^*_{lm}(\vartheta_1, \phi_1) \tag{14.67}$$

where δ_l is the phase of the wave function. The radial part $P^\mu_{\nu_{1\mu}}(r)$ is the solution of the Hartree equation for the meson in the "frozen" field of the ground state of the atom (program ***fchf***), normalized to the energy δ-function. The asymptotic of the function $P^\mu_{\nu_1}(r)$ is

$$P^\mu_{\nu_1}(r) \to \sqrt{\frac{2m_\mu}{\pi p_1}} \sin\left(p_1 r + \frac{Q}{p_1}\ln(2p_1 r) - \frac{\pi l_1}{2} + \delta_{l_1}\right) \tag{14.68}$$

where Q is the core charge (for a neutral atom $Q = 0$). The wave function of the discrete final state of the meson $\nu_2 \equiv (n_2 l_2)$ has the following form:

$$\phi_{\nu_2}(x) = \frac{1}{r} P_{\nu_2}(r) Y_{\text{lm}}(\vartheta, \phi) \tag{14.69}$$

Its radial part is calculated in the field of the "frozen" Hartree–Fock core with a hole ν_4, which means the solution of the Hartree equation without self-consistency with the wave functions of the ground state. It is assumed that the number of electrons in the ν_4-subshell is one less than in the ground state.

The wave function ϕ_{ν_3} of the emitted Auger electron $\nu_3 \equiv (E_3 l_3)$ is calculated in the field of the frozen Hartree–Fock core with a hole ν_4. To this should be added the field produced by the meson in the state ν_2, if it is discrete. $P_{\nu_2}(r)$ is a radial part of ϕ_{ν_3}. As in the case of the meson, it is normalized to the delta energy function.

The wave functions of the emitted Auger electron are determined at the energy satisfying the conservation law:

$$E_3 = E_\mu - E_{1\mu} + E_4 \tag{14.70}$$

It is well known [LL] that for the application of perturbation theory, the complete set of one-particle wave functions must be used as a basis. The above-mentioned functions of the initial and final states of the meson and the electron refer to different Hamiltonians. This choice of wave functions in the calculations is equivalent to the summation of an infinite series of diagrams, where each takes into account an additional term of the Coulomb interaction of an Auger electron ν_3 with a vacancy ν_4 or a meson $\nu_{2\mu}$.

References

1. V.M. Galitsky, in: *Applications of the Methods of Quantum Field Theory to Many-Body Problems*, ed. by A.I. Alekseeva. (Nauka, Moscow, 1963) (In Russian)
2. D.R. Bates, A. Fundaminsky A., H.S.W. Massey, Trans. Roy. Soc. A **243** (London, 1950)
3. Gribakin, Ludlow, J. Phys. Rev. A **70**, 032720 (2004)
4. M.Ya. Amusia, N.A. Cherepkov, L.V. Chernysheva, S.G. Shapiro, J. Phys. B **9** (1976)
5. E.F. Varracchio, J. Phys. B: At. Mol. Opt. Phys. **23** (1990)
6. E.F. Varracchio, L.A. Parcell, J. Phys. B: At. Mol. Opt. Phys. **25** (1992)
7. M.Ya. Amusia, N.A. Cherepkov, L.V. Chernysheva, JETP **124**, 1 (2003)
8. G.F. Gribakin and W.A.J. King, Phys. B: At. Mol. Opt. Phys. **27** (1994)
9. V.A. Dzuba, V.V. Flambaum, G.F. Gribakin, W.A. King, J. Phys. B: At. Mol. Opt. Phys. **29** (1996)
10. C. Bloch, Proc. Scuola Internazionale di Fisica 'E. Fermi', Course XXXVI (Academic Press, New York, London, 1966)
11. L.V. Chernysheva, M.Ya. Amusia, S.G. Shapiro, *Numerical Calculation of the Diagonal Matrix Element of the Self-Energy Part of the One-Particle Green's Function in a Simplified Random Phase Approximation with Exchange* (Ioffe Institute, No. 504, Leningrad, 1976) (In Russian)
12. L.V. Chernysheva, M.Ya. Amusia, N.A. Cherepkov, *Numerical Calculation of scattering phases in the simplified random phase approximation with exchange* (Ioffe Institute, No. 663, Leningrad, 1980) (In Russian)

Chapter 15
Non-radiative or Auger Vacancy Decay

This chapter provides formulas for calculating the Auger widths of atomic levels, both in the one-electron approximation and with the inclusion of corrections in this process in the framework of RPAE and perturbation theory.

A method is presented to calculate the probability of double Auger decay, i.e., a process in which two vacancies decay with the transfer of their energy to one escaping electron. Formulas are given for the amplitude of the process and the Coulomb matrix elements.

15.1 Theoretical Model

15.1.1 One-Electron Auger Decay

The width Γ_{ν}^{A} of the Auger decay of the level $\nu_1 \equiv (n, l, m, \sigma)$ to the final state of an atom with two vacancies ν_2,ν_3 and an ejected electron ν_4 in the lowest order of the perturbation theory in the interelectron interaction V is determined by the square of the modulus of the amplitudes depicted by the diagrams [cf. (1.36)]

M. Ya. Amusia and L. V. Chernysheva, *Computation of Atomic and Molecular Processes*, Springer Series on Atomic, Optical, and Plasma Physics 117,
https://doi.org/10.1007/978-3-030-85143-9_15

(15.1)

After summation over the quantum numbers $\nu_2,\ \nu_3,\ \nu_4$ of the final state, consistent with the energy conservation law $E_1 = E_2 + E_3 + E_4$, we obtain the total Auger width of the level ν_1. The width of the level in the atomic system of units is equal to the probability of Auger decay W per unit time. In higher orders in the electron–electron interaction, corrections appear to the Auger decay amplitude shown in (15.1). Within the RPAE frame, correlations mean taking into account, along with the amplitude (15.1), the following processes:

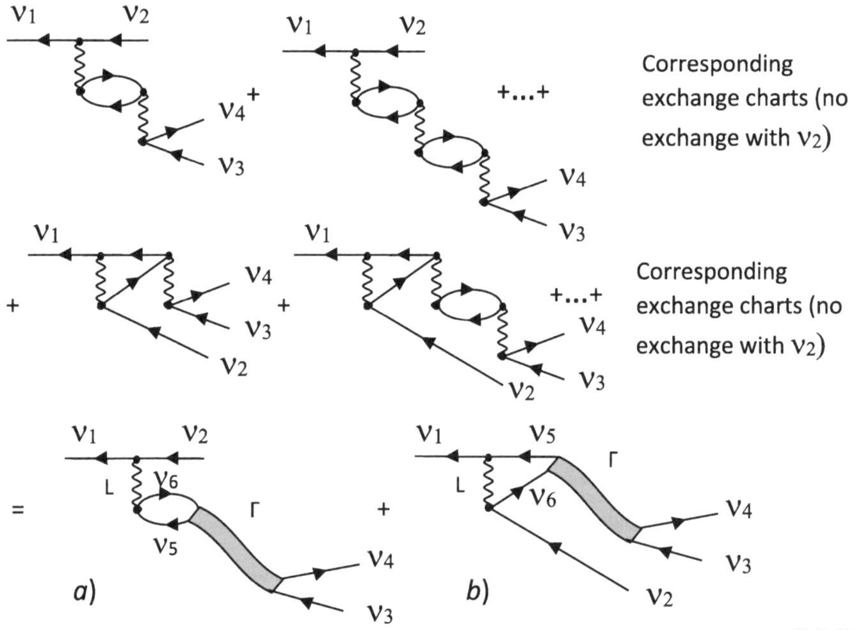

(15.2)

The direct matrix element of Auger decay W_{dir} in the RPAE is the sum of matrix elements (15.1a) and (15.2a) and is written analytically in the form:

$$\langle \nu_1 \nu_4 | W_{\mathrm{dir}} | \nu_2 \nu_3 \rangle = \langle \nu_1 \nu_4 | V | \nu_2 \nu_3 \rangle +$$

$$\left(\sum_{\nu_5\le F,\nu_6>F} - \sum_{\nu_5>F,\nu_6\le F}\right)\frac{\langle\nu_1\nu_6|V|\nu_2\nu_5\rangle\langle\nu_5\nu_4|\Gamma(\omega)|\nu_6\nu_3\rangle}{\omega - E_6 + E_5 + i\eta(1-2n_{\nu_6})}, \tag{15.3}$$

where $\Gamma(\omega)$ is the effective interaction in the RPAE (1.58). However, the matrix element of Auger decay can be found by performing the intermediate calculation of the effective interaction. The following equation is obtained as a result of an iterative procedure, based on the diagrams depicted in the form (15.2):

$$\langle\nu_1\nu_4|W_{\text{dir}}|\nu_2\nu_3\rangle = \langle\nu_1\nu_4|V|\nu_2\nu_3\rangle + \left(\sum_{\nu_5\le F,\nu_6>F} - \sum_{\nu_5>F,\nu_6\le F}\right)$$
$$\frac{\langle\nu_1\nu_6|W_{\text{dir}}|\nu_2\nu_5\rangle\langle\nu_5\nu_4|U|\nu_6\nu_3\rangle}{\omega - E_6 + E_5 + i\eta(1-2n_{\nu_6})}. \tag{15.4}$$

A similar equation is obtained from the exchange matrix element and is obtained from the sums of diagrams (15.1b) and (15.2b). It is a combination of these equations that is used to determine the Auger decay matrix element with correlations.

The choice of basis wave functions for calculating matrix elements is ambiguous. From the point of view of many-body theory, the Hartree–Fock functions obtained in the field of a neutral atom should be used to describe the ν_1. ν_2, ν_3 hole and ν_4 electronic states. Physically, this choice is equivalent to the assumption that Auger decay proceeds quickly: The time interval between the appearance of a hole ν_1 and its decay into a state ν_2, ν_3, ν_4 is too short for the rearrangement of atomic electron shells. Therefore, shells can be described by the wave functions of a neutral atom, in accordance with the concept of the "frozen" core (see Sect. 2.3). However, for the removed electron it is convenient to use the wave functions ϕ^N and $\phi^{N(LS)}$ (see Sect. 8.1.1), which describe the motion in the field of a single ion. These functions account for the diagonal in the hole state and directed "time forward" RPAE diagrams,

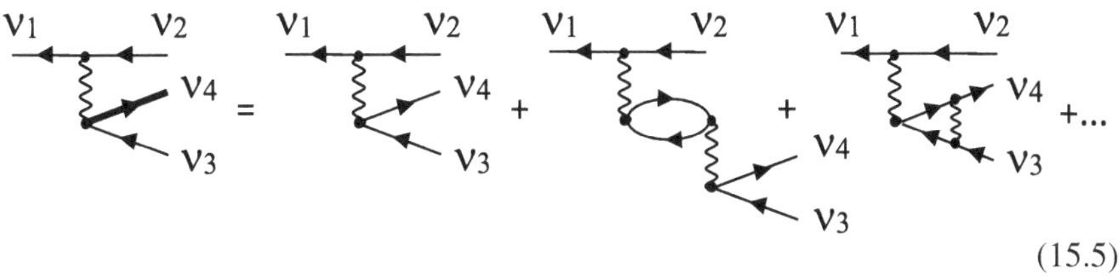

(15.5)

In (15.5), the thick line describes the removed electron in the field of the ion, and the thin lines describe holes and electrons in the field of a neutral atom.

The appropriate choice of wave functions also allows one to take into account the fact that the ejected electron ν_4 moves in the field of an atom with two holes ν_2 and ν_3, rearranged as a result of their formation, and not in the field with one hole.

For this, one must first solve the Hartree–Fock equations of the ground state of an atom with holes ν_2 and ν_3. Then, the self-consistent field of such an ion determines the wave functions of the electron ν_4. From the many-body theory point of view, this allows one to take into account, in addition to those shown in Figs. (15.1) and (15.2), a number of other diagrams. The simplest of them are

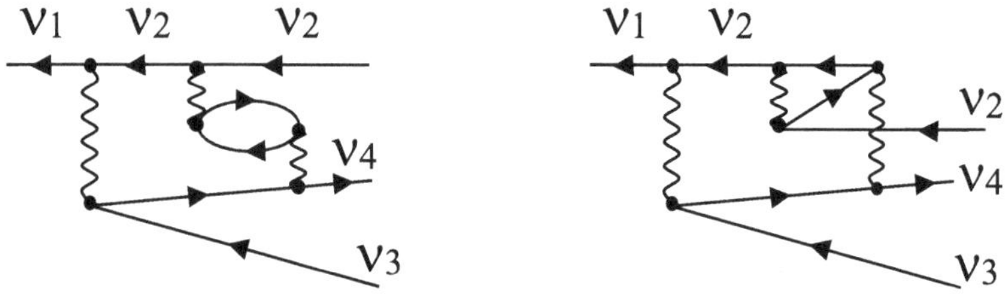

(15.6)

These diagrams describe rearrangement, i.e., the reaction of an atom to the appearance of a hole ν_2, and, consequently, alteration of the field in which the Auger electron moves. In this case, the initial wave functions of the hole states ν_1, ν_2, and ν_3 themselves are determined in the field of a neutral atom.

Solving the corresponding Hartree–Fock equations for an atom with a two-hole state, one can also take into account that the ejected electron moves in the field of the residual ion, which is rearranged due to the presence of two holes.

In turn, the wave functions of holes ν_1, ν_2, and ν_3 can be determined in the field of the rearranged ion. In this case, they are usually considered as belonging to different bases. In this approach, however, it is difficult to indicate which higher order diagrams of the perturbation theory in interelectron interaction are taken into account.

Multielectron correlations are taken into account in two transitions $\nu_2 \to \nu_4$ and $\nu_3 \to \nu_4$, including the interaction of these transitions with each other, and the quantum numbers of hole states ν_2 and ν_3 may coincide. In this case, the sum over hole states $i \leq F$ includes no more than two terms, ν_2 and ν_3.

In principle, correlations that go beyond RPAE may also be important. First of all, they lead to a change in the transition energy due to a change in the final state energy caused by the interaction between holes, which is illustrated by examples of such diagrams:

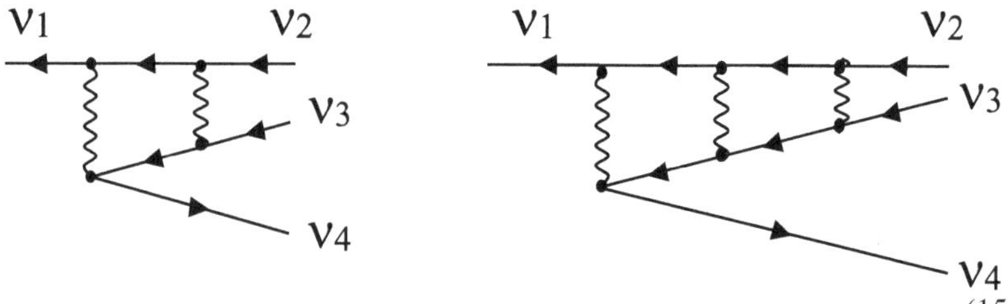

(15.7)

The same type of diagram describes the mixing of final state configurations, which sometimes also make a noticeable contribution [1] To account for the possible hole levels energy shift in comparison with their Hartree–Fock values, the program provides for the possibility of using both Hartree–Fock and other energy levels, for example, obtained experimentally.

15.1.2 Double Auger Decay

Double Auger decay is a process of transition of an atom from an initial state with two vacancies ν_1 and ν_2 to a final state with three vacancies ν_4, ν_5,ν_6 and an electron ν_3 in continuous spectrum state. The probability of such a process $\Gamma^A_{\nu_1\nu_2}$ is determined by the equation

$$\Gamma^A_{\nu_1\nu_2} = 2\pi \left|M_{\nu_1\nu_2}\right|^2 = 2\pi \sum_{\nu_4\nu_5\nu_6\le F, l_3} \left|M_{\nu_1\nu_2,\nu_4\nu_5\nu_6\varepsilon_3 l_3}\right|^2 \quad (15.8)$$

Here the summation is over all hole states permitted by the laws of energy conservation of hole states and over all admissible values of the angular momentum l_3. The matrix element $M_{\nu_1\nu_2,\nu_4\nu_5\nu_6\varepsilon_3 l_3}$ is related to the transition operator $\hat{M}$ by the obvious relation

$$M_{\nu_1\nu_2,\nu_4\nu_5\nu_6\varepsilon_3 l_3} = \left\langle \phi_{\nu_1\nu_2} \middle| \hat{M} \middle| \phi_{\nu_4\nu_5\nu_6\varepsilon_3 l_3} \right\rangle \quad (15.9)$$

where $\hat{M}$ is the operator of the double Auger decay.

The decay probability has the dimension of energy and is expressed in Ry. The Many-body perturbation theory (MBPT) [2, 3] is usually applicable to the consideration of this process, since atomic shells are well separated in energy and the Coulomb interaction between electrons of different shells is relatively weak.

In the lowest order of the perturbation theory in the electron–electron interaction, the following diagrams contribute to the amplitude of the process:

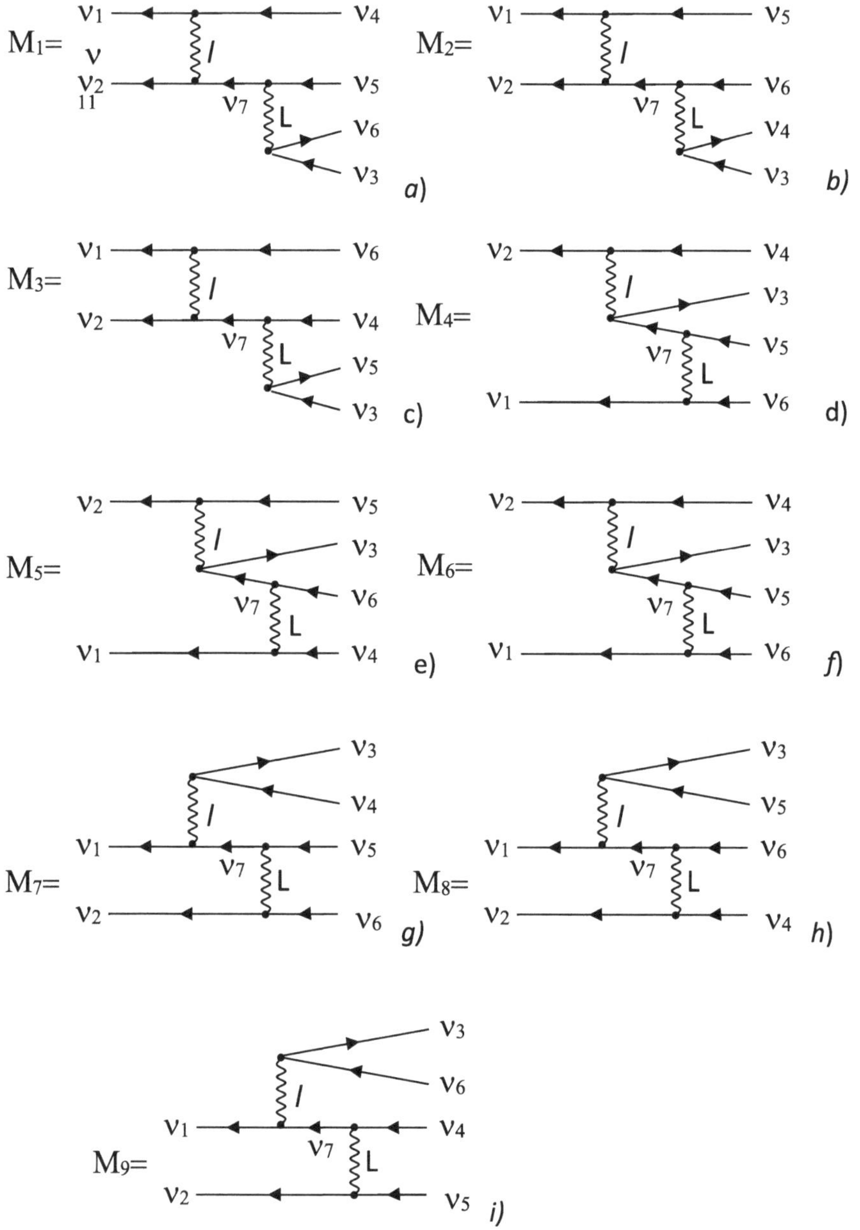

(15.10)

The corresponding analytical expressions are given below:

$$M_1 = \sum_{\nu_7} \langle \nu_7 \nu_3 | U | \nu_5 \nu_6 \rangle \langle \nu_1 \nu_2 | U | \nu_4 \nu_7 \rangle \left[E^{-1}_{\nu_1 \nu_2 \nu_4 \nu_7} \right], \tag{15.11a}$$

$$M_2 = \sum_{\nu_7} \langle \nu_7 \nu_3 | U | \nu_6 \nu_4 \rangle \langle \nu_1 \nu_2 | U | \nu_5 \nu_7 \rangle \left[E^{-1}_{\nu_1 \nu_2 \nu_5 \nu_7} \right], \tag{15.11b}$$

$$M_3 = \sum_{\nu_7} \langle \nu_7 \nu_3 | U | \nu_4 \nu_5 \rangle \langle \nu_1 \nu_2 | U | \nu_6 \nu_7 \rangle \left[E^{-1}_{\nu_1 \nu_2 \nu_6 \nu_7} \right], \tag{15.11c}$$

$$M_4 = \sum_{\nu_7} \langle \nu_7 \nu_1 | U | \nu_5 \nu_6 \rangle \langle \nu_2 \nu_3 | U | \nu_4 \nu_7 \rangle \left[E^{-1}_{\nu_2 \nu_3 \nu_4 \nu_7} \right], \tag{15.11d}$$

$$M_5 = \sum_{\nu_7} \langle \nu_7 \nu_1 | U | \nu_6 \nu_4 \rangle \langle \nu_2 \nu_3 | U | \nu_5 \nu_7 \rangle \left[E^{-1}_{\nu_2 \nu_3 \nu_5 \nu_7} \right], \tag{15.11e}$$

$$M_6 = \sum_{\nu_7} \langle \nu_7 \nu_1 | U | \nu_4 \nu_5 \rangle \langle \nu_2 \nu_3 | U | \nu_6 \nu_7 \rangle \left[E^{-1}_{\nu_2 \nu_3 \nu_6 \nu_7} \right], \tag{15.11f}$$

$$M_7 = \sum_{\nu_7} \langle \nu_7 \nu_2 | U | \nu_5 \nu_6 \rangle \langle \nu_3 \nu_1 | U | \nu_4 \nu_7 \rangle \left[E^{-1}_{\nu_3 \nu_1 \nu_4 \nu_7} \right], \tag{15.11g}$$

$$M_8 = \sum_{\nu_7} \langle \nu_7 \nu_2 | U | \nu_6 \nu_4 \rangle \langle \nu_3 \nu_1 | U | \nu_5 \nu_7 \rangle \left[E^{-1}_{\nu_3 \nu_1 \nu_5 \nu_7} \right], \tag{15.11h}$$

$$M_9 = \sum_{\nu_7} \langle \nu_7 \nu_2 | U | \nu_4 \nu_5 \rangle \langle \nu_3 \nu_1 | U | \nu_6 \nu_7 \rangle \left[E^{-1}_{\nu_3 \nu_1 \nu_6 \nu_7} \right], \tag{15.11i}$$

Summation is performed over both vacancy ($\nu_7 \leq F$) and electron ($\nu_7 > F$) intermediate one-particle states, including integration over states of the continuous spectrum.

Note that

$$\begin{aligned} \langle \nu_1 \nu_3 | U | \nu_2 \nu_4 \rangle &= \langle \nu_1 \nu_3 | V | \nu_2 \nu_4 \rangle - \langle \nu_1 \nu_3 | U | \nu_4 \nu_2 \rangle, \\ E_{\nu_1 \nu_2 \nu_3 \nu_4} &= E_1 + E_2 - E_3 - E_4 \end{aligned} \tag{15.12}$$

where E_i are the one-electron energies, their Hartree–Fock values in the "frozen" core, or values that account for rearrangement due to the presence of one, two, or three vacancies. In principle, the experimental values of the energies of the initial, intermediate, and final states of the ion can also be used in calculations.

After integrating over angle variables and summation over spin projections, the expression for the partial amplitudes of the M_j in LS-coupling has the form:

$$\begin{aligned} M_1 = \sum_{L,l,l_7} (-1)^{L_1+L_2+l_1+l_6+L} [L_2][L_3] \begin{Bmatrix} L_1 & l_4 & l_7 \\ l & l_2 & l_1 \end{Bmatrix} \begin{Bmatrix} L_1 & L_2 & L \\ l_6 & l_3 & L_3 \end{Bmatrix} \begin{Bmatrix} L_1 & L_2 & L \\ l_5 & l_7 & l_4 \end{Bmatrix} \times \\ \delta_{L_1 L_4} \delta_{M_{L_1} M_{L_4}} \left[A \cdot V^L_{\nu_7 \nu_3 \nu_5 \nu_6} W^l_{\nu_1 \nu_2 \nu_4 \nu_7} + (-1)^{-S_1} \cdot B \cdot V^L_{\nu_7 \nu_3 \nu_5 \nu_6} W^l_{\nu_1 \nu_2 \nu_4 \nu_7} + \right. \\ \left. (-1)^{-S_1} \cdot A \cdot W^L_{\nu_7 \nu_3 \nu_5 \nu_6} V^l_{\nu_1 \nu_2 \nu_4 \nu_7} + B \cdot W^L_{\nu_7 \nu_3 \nu_5 \nu_6} W^l_{\nu_1 \nu_2 \nu_4 \nu_7} \right] \left(E^{-1}_{\nu_7 \nu_3 \nu_5 \nu_6} - E^{-1}_{\nu_1 \nu_2 \nu_4 \nu_7} \right), \end{aligned} \tag{15.13}$$

$$M_2 = \sum_{L,l,l_7} (-1)^{L_1+l_2+l_3+L}[L_2][L_3]\begin{Bmatrix} L_1 & l_5 & l_7 \\ l & l_2 & l_1 \end{Bmatrix}\begin{Bmatrix} L_1 & l_5 & l_7 \\ L_3 & L_2 & l_6 \\ l_3 & l_4 & L \end{Bmatrix}\delta_{L_1L_4}\delta_{M_{L_1}M_{L_4}}\times$$
$$\big[(-1)^{-S_1-S_2}\cdot B\cdot V^L_{\nu_7\nu_3\nu_6\nu_4}V^l_{\nu_1\nu_2\nu_5\nu_7} + (-1)^{-S_1}\cdot A\cdot V^L_{\nu_7\nu_3\nu_6\nu_4}W^l_{\nu_1\nu_2\nu_5\nu_7} +$$
$$(-1)^{-S_2}\cdot B\cdot W^L_{\nu_7\nu_3\nu_6\nu_4}V^l_{\nu_1\nu_2\nu_5\nu_7} + A\cdot W^L_{\nu_7\nu_3\nu_6\nu_4}V^l_{\nu_1\nu_2\nu_5\nu_7}\big]\big(E^{-1}_{\nu_7\nu_3\nu_6\nu_4} - E^{-1}_{\nu_1\nu_2\nu_5\nu_7}\big), \tag{15.14}$$

$$M_3 = \sum_{L,l,l_7} (-1)^{L_3+l_2+l_3+l_5}[L_2][L_3]\begin{Bmatrix} L_1 & l_6 & l_7 \\ l & l_2 & l_1 \end{Bmatrix}\begin{Bmatrix} l_3 & l_7 & L_2 \\ l_6 & L_3 & L_1 \end{Bmatrix}\begin{Bmatrix} l_3 & l_7 & L_2 \\ l_4 & l_5 & L \end{Bmatrix}\times$$
$$\delta_{L_1L_4}\delta_{ML_1ML_4}B\big[V^L_{\nu_7\nu_3\nu_4\nu_5}V^l_{\nu_1\nu_2\nu_6\nu_7} + (-1)^{-S_2}V^L_{\nu_7\nu_3\nu_4\nu_5}W^l_{\nu_1\nu_2\nu_6\nu_7} +$$
$$(-1)^{-S_1}W^L_{\nu_7\nu_3\nu_4\nu_5}V^l_{\nu_1\nu_2\nu_6\nu_7} + (-1)^{-S_1-S_2}W^L_{\nu_7\nu_3\nu_4\nu_5}W^l_{\nu_1\nu_2\nu_6\nu_7}\big]$$
$$\big(E^{-1}_{\nu_7\nu_3\nu_4\nu_5} - E^{-1}_{\nu_1\nu_2\nu_6\nu_7}\big), \tag{15.15}$$

$$M_4 = \sum_{L,l,l_7,x} (-1)^{l_4+l_6+l}[L_2][L_3]\,[x]^2\begin{Bmatrix} l_1 & l_7 & x \\ l_5 & l_6 & L \end{Bmatrix}\begin{Bmatrix} L_3 & l_4 & x \\ l_5 & l_6 & L_2 \end{Bmatrix}\begin{Bmatrix} L_1 & l_2 & l_1 \\ l_3 & l & l_7 \\ L_3 & l_4 & x \end{Bmatrix}\times$$
$$\delta_{L_1L_4}\delta_{M_{L_1}M_{L_4}}\big[B\cdot V^L_{\nu_7\nu_1\nu_5\nu_6}V^l_{\nu_2\nu_3\nu_4\nu_7} + (-1)^{-S_1}A\cdot V^L_{\nu_7\nu_1\nu_5\nu_6}W^l_{\nu_2\nu_3\nu_4\nu_7} +$$
$$(-1)^{-S_2}\cdot B\cdot W^L_{\nu_7\nu_1\nu_5\nu_6}V^l_{\nu_2\nu_3\nu_4\nu_7} + (-1)^{-S_1-S_2}\cdot B\cdot W^L_{\nu_7\nu_1\nu_5\nu_6}W^l_{\nu_2\nu_3\nu_4\nu_7}\big]$$
$$\big(E^{-1}_{\nu_7\nu_1\nu_5\nu_6} - E^{-1}_{\nu_2\nu_3\nu_4\nu_7}\big), \tag{15.16}$$

$$M_5 = \sum_{L,l,l_7,x} (-1)^{L_2+l_1+l_7+l+x}[L_2][L_3]\begin{Bmatrix} l_6 & l_4 & x \\ l_5 & L_3 & L_2 \end{Bmatrix}\begin{Bmatrix} l_1 & l_7 & x \\ l_6 & l_4 & L \end{Bmatrix}\begin{Bmatrix} L_1 & l_2 & l_1 \\ l_3 & l & l_7 \\ L_3 & l_5 & x \end{Bmatrix}\times$$
$$\delta_{L_1L_4}\delta_{M_{L_1}M_{L_4}}\big[A\cdot V^L_{\nu_7\nu_1\nu_6\nu_4}V^l_{\nu_2\nu_3\nu_5\nu_7} + (-1)^{-S_2}\cdot B\cdot V^L_{\nu_7\nu_1\nu_6\nu_4}W^l_{\nu_2\nu_3\nu_5\nu_7} +$$
$$(-1)^{-S_1}\cdot B\cdot W^L_{\nu_7\nu_1\nu_6\nu_4}V^l_{\nu_2\nu_3\nu_5\nu_7} + B\cdot W^L_{\nu_7\nu_1\nu_6\nu_4}W^l_{\nu_2\nu_3\nu_5\nu_7}\big]\big(E^{-1}_{\nu_7\nu_1\nu_6\nu_4} - E^{-1}_{\nu_2\nu_3\nu_5\nu_7}\big), \tag{15.17}$$

$$M_6 = \sum_{L,l,l_7} (-1)^{L_2+l_4+l_5+l}[L_2][L_3]\begin{Bmatrix} l_1 & l_7 & L_2 \\ l_4 & l_5 & L \end{Bmatrix}\begin{Bmatrix} L_1 & l_2 & l_1 \\ l_3 & l & l_7 \\ L_3 & l_6 & L_2 \end{Bmatrix}\times$$
$$\delta_{L_1L_4}\delta_{M_{L_1}M_{L_4}}\big[(-1)^{-S_1-S_2}\cdot B\cdot V^L_{\nu_7\nu_1\nu_4\nu_5}V^l_{\nu_2\nu_3\nu_6\nu_7} +$$
$$(-1)^{-S_1}\cdot B\cdot V^L_{\nu_7\nu_1\nu_4\nu_5}W^l_{\nu_2\nu_3\nu_6\nu_7} + (-1)^{-S_1}\cdot A\cdot W^L_{\nu_7\nu_1\nu_4\nu_5}V^l_{\nu_2\nu_3\nu_6\nu_7} +$$
$$A\cdot W^L_{\nu_7\nu_1\nu_4\nu_5}W^l_{\nu_2\nu_3\nu_6\nu_7}\big]\big(E^{-1}_{\nu_7\nu_1\nu_4\nu_5} - E^{-1}_{\nu_2\nu_3\nu_6\nu_7}\big), \tag{15.18}$$

$$M_7 = \sum_{L,l,l_7,x} (-1)^{L_1+l_3+l_5+l}[L_2][L_3]\,[x]^2\begin{Bmatrix} l_4 & l_2 & x \\ l_5 & L & l_7 \\ L_2 & l_6 & L_3 \end{Bmatrix}\begin{Bmatrix} l_4 & l_2 & x \\ l & l_1 & l_7 \\ l_3 & L_1 & L_3 \end{Bmatrix}\times$$

$$\delta_{L_1L_4}\delta_{M_{L_1}M_{L_4}}\big[(-1)^{-S_1-S_2}\cdot B\cdot V^L_{\nu_7\nu_2\nu_5\nu_6}V^l_{\nu_3\nu_1\nu_4\nu_7}+$$
$$(-1)^{-S_2}\cdot B\cdot V^L_{\nu_7\nu_2\nu_5\nu_6}W^l_{\nu_3\nu_1\nu_4\nu_7}+(-1)^{-S_1}\cdot B\cdot W^L_{\nu_7\nu_2\nu_5\nu_6}V^l_{\nu_3\nu_1\nu_4\nu_7}+$$
$$A\cdot W^L_{\nu_7\nu_2\nu_5\nu_6}W^l_{\nu_3\nu_1\nu_4\nu_7}\big]\big(E^{-1}_{\nu_7\nu_2\nu_5\nu_6}-E^{-1}_{\nu_3\nu_1\nu_4\nu_7}\big), \tag{15.19}$$

$$M_8=\sum_{L,l,l_7,x}(-1)^{L_2+l_2+l_3+l_5+l_7}[L_2][L_3]\begin{Bmatrix} l_5 & l_2 & x\\ L & L_2 & l_4\end{Bmatrix}\begin{Bmatrix} l_3 & l_7 & x\\ L & L_2 & l_6\end{Bmatrix}\begin{Bmatrix} l_3 & l & l_5\\ L_1 & l_1 & l_2\\ L_3 & l_7 & x\end{Bmatrix}\times$$
$$\delta_{L_1L_4}\delta_{M_{L_1}M_{L_4}}\big[B\cdot V^L_{\nu_7\nu_2\nu_6\nu_4}V^l_{\nu_3\nu_1\nu_5\nu_7}\,(-1)^{-S_1}\,B\cdot V^L_{\nu_7\nu_2\nu_6\nu_4}W^l_{\nu_3\nu_1\nu_5\nu_7}+$$
$$(-1)^{-S_1}\cdot A\cdot W^L_{\nu_7\nu_2\nu_6\nu_4}V^l_{\nu_3\nu_1\nu_5\nu_7}+(-1)^{-S_1-S_2}\cdot B\cdot W^L_{\nu_7\nu_2\nu_6\nu_4}W^l_{\nu_3\nu_1\nu_5\nu_7}\big]$$
$$\big(E^{-1}_{\nu_7\nu_2\nu_6\nu_4}-E^{-1}_{\nu_3\nu_1\nu_5\nu_7}\big), \tag{15.20}$$

$$M_9=\sum_{L,l,l_7}(-1)^{L_1+l_2+l_3+l_5+l}[L_2][L_3]\begin{Bmatrix} l_2 & l_7 & L_2\\ l_4 & l_5 & L\end{Bmatrix}\begin{Bmatrix} L_2 & l & L_1\\ l_3 & L_3 & l_6\end{Bmatrix}\begin{Bmatrix} l_1 & l_2 & L_1\\ L_2 & l & l_7\end{Bmatrix}\times$$
$$\delta_{L_1L_4}\delta_{M_{L_1}M_{L_4}}\big[A\cdot V^L_{\nu_7\nu_2\nu_4\nu_5}V^l_{\nu_3\nu_1\nu_6\nu_7}+(-1)^{-S_1}\cdot A\cdot V^L_{\nu_7\nu_2\nu_4\nu_5}W^l_{\nu_3\nu_1\nu_6\nu_7}+$$
$$(-1)^{-S_2}\cdot B\cdot W^L_{\nu_7\nu_2\nu_4\nu_5}V^l_{\nu_3\nu_1\nu_6\nu_7}+B\cdot W^L_{\nu_7\nu_2\nu_4\nu_5}W^l_{\nu_3\nu_1\nu_6\nu_7}\big]\big(E^{-1}_{\nu_7\nu_2\nu_4\nu_5}-E^{-1}_{\nu_3\nu_1\nu_6\nu_7}\big). \tag{15.21}$$

In Expressions (15.13)–(15.21) $\begin{Bmatrix} l_1 & l_2 & l_3\\ l_4 & l_5 & l_6\end{Bmatrix}$ are the Wigner $6j$-coefficients, and $\begin{Bmatrix} l_1 & l_2 & l_3\\ l_4 & l_5 & l_6\\ l_7 & l_8 & l_9\end{Bmatrix}$—Wigner $9j$-coefficients [4]; $[L]=(2L+1)^{1/2}$, A и B are spin factors defined as follows:

$$A=(-1)^{S_1+S_3-1/2}[S_1]^{-1}[S_3]\{\,1/2\ 1/2\ S_1\,\}\{S_1\ S_2\ 1/2\}\delta_{S_1S_2}\delta_{S_1S_4}\delta_{M_{S_1}M_{S_4}} \tag{15.22}$$

$$B=(-1)^{S_1+S_2+S_3-1/2}[S_2][S_3]\begin{Bmatrix} 1/2 & 1/2 & S_1\\ 1/2 & S_3 & S_2\end{Bmatrix}\delta_{S_1S_4}\delta_{M_{S_1}M_{S_4}} \tag{15.23}$$

Here L_1S_1, L_4S_4, L_2S_2, L_3S_3 are terms of the initial, final, and two intermediate states, respectively; l and L are the angular moments transferred in the matrix elements of the Coulomb interaction. Direct $V^{\lambda}_{\nu_1\nu_2\nu_3\nu_4}$ and exchange $V^{\lambda}_{\nu_1\nu_4\nu_3\nu_2}$ reduced Coulomb matrix elements are determined by the following formula, which coincides with (9.32):

$$V^{\lambda}_{\nu_1\nu_2\nu_3\nu_4}=\langle\nu_1\nu_3\|V^{\lambda}\|\nu_2\nu_4\rangle=[l_1][l_2][l_3][l_4]\begin{pmatrix} l_1 & \lambda & l_2\\ 0 & 0 & 0\end{pmatrix}\begin{pmatrix} l_3 & \lambda & l_4\\ 0 & 0 & 0\end{pmatrix}R_{\lambda}(\nu_1\nu_2\nu_3\nu_4) \tag{15.24}$$

$$V^{\lambda}_{\nu_1\nu_4\nu_3\nu_2} = [\lambda]\sum_{\lambda'}(-1)^{\lambda+\lambda'}\begin{Bmatrix} l_1 & \lambda & l_2 \\ l_3 & \lambda' & l_4 \end{Bmatrix} V^{\lambda'}_{\nu_1\nu_4\nu_3\nu_2} \tag{15.25}$$

$$R_{\lambda}(\nu_1\nu_2\nu_3\nu_4) = \int_0^{\infty} dr\, P_{\nu_1}(r)P_{\nu_2}(r)\int_0^{\infty} dr' \frac{r_<^{\lambda}}{r_>^{\lambda+1}} P_{\nu_3}(r')P_{\nu_4}(r') \tag{15.26}$$

where $P_{\nu}(r)$ are the one-electron radial Hartree–Fock wave functions, $\nu \equiv nl$ $(\nu \leq F)$, or $\nu \equiv [n(\varepsilon)l]$, $(\nu > F)$.

15.2 Numerical Model

15.2.1 Probability of One-Electron Auger Decay

The probability of atomic level decay is inversely proportional to its width. The width Γ_1^A of the level ν_1 relative to the decay to the final state $\nu_2,\ \nu_4,\ \nu_3$ in the lowest order of perturbation theory (15.1) is described by the formula

$$\Gamma^A_{\nu_1,\nu_2\nu_3\nu_4} = \frac{4\pi}{(2l_1+1)(2L+1)}\left[|\langle\nu_1\nu_4\|V_L\|\nu_2\nu_3\rangle|^2 + |\langle\nu_1\nu_4\|V_L\|\nu_3\nu_2\rangle|^2 - \right.$$
$$\left. - \langle\nu_1\nu_4\|V_L\|\nu_2\nu_3\rangle\langle\nu_1\nu_4\|V_L\|\nu_3\nu_2\rangle\right],\ \nu \equiv (nl). \tag{15.27}$$

Here $\langle\nu_1\nu_4\|V_L\|\nu_2\nu_3\rangle$ are the reduced Coulomb matrix elements obtained from the usual ones after integration over the angular variables and separation of the factors containing the projections of the angular momenta. Note that the width of the level, like all energy values in our computing codes, is measured in Ry. The reduced Coulomb matrix elements are calculated using the formulas given in Chap. 9.

In RPAE, the matrix elements of Auger decay are obtained for each transition from an equation similar to (9.53):

$$W_{\alpha\delta}(\omega) = V_{\alpha\delta} + S_{\beta}U_{\alpha\beta}\Pi_{\beta}(\omega)W_{\beta\delta}(\omega) \tag{15.28}$$

Here ω is the Auger decay energy equal to the energy difference $E_2 - E_1$; $V_{\alpha\delta}$—Coulomb matrix elements of the "three holes—one particle" type (15.1); $U_{\alpha\beta}$ is the matrix of Coulomb elements that takes into account the interaction of two transitions $\alpha \equiv (\nu_1\nu_2)$, $\delta \equiv (\nu_3\nu_4)$, $\beta \equiv (\nu_5\nu_6)$ (9.37); S_{β} denotes summation and integration over all β.

Linear complex Eq. (15.28) is solved similarly to (9.73). The level width in the RPAE is determined at energy

$$\omega_1 = E_4 - E_3, \tag{15.29}$$

close to the Auger decay energy ω. For this, in the ***auger*** program, when calculating the wave functions ν_4, the step over p is chosen so that the energy ω_1 at the selected point is close enough to ω, i.e. $E_4 = E_2 - E_1 + E_3$.

To calculate the Auger width of the atomic level in RPAE, only on-shell matrix elements are used $\langle \nu_1 \nu_4 \| W(\omega) \| \nu_2 \nu_3 \rangle$, $\langle \nu_1 \nu_6 \| W(\omega) \| \nu_2 \nu_5 \rangle$. The width of the level in RPAE is calculated by Formulas (15.8), (15.9), into which matrix elements in the RPAE are substituted instead of the matrix elements in the lowest order.

15.2.2 The Probability of Double Auger Decay

To calculate the reduced matrix elements in (15.24), (15.25), it is convenient to pass from the two-dimensional integral to the one-dimensional integral (9.36), which is calculated by the Simpson formula with account of alteration of the variable (4.42).

When calculating the energy integrals in (15.13)–(15.21), as in the study of the photoelectric effect and electron scattering, one cannot directly use the formulas for approximate integration, since the upper limit of integration is equal to infinity, and in addition, the expressions under the integral in some cases can have a pole.

As in the other cases mentioned, the infinite upper limit of integration is replaced by a sufficiently large finite one, and its value is determined by the rate of decrease of the Coulomb matrix elements. Equations (9.69–9.72) are used to calculate integrals with a pole in the expression under the integration sign.

Partial amplitudes M_j, $j = 1, 2, \ldots, 9$ are calculated using Formulas (15.13)–(15.21) with predefined angular and spin factors for fixed values L, l, l_7. The total amplitude is the sum of all nine partials. The probability is calculated using Formula (15.8).

The radial wave functions of the ground state of an atom or ion are calculated using the *scfhf* program. The wave functions of the excited state can be determined in a self-consistent field (***scfhf*** program) or in the field of the frozen core of an atom or ion (***fchf*** program).

By choosing the field in which the wave functions are calculated, we can account for some infinite sequences of higher-order diagrams [1]. As noted in Sect. 8.1.1, using in the calculations the wave functions found in the field of an atom with one vacancy, formed as a result of ionization of the ith electron, is equivalent to the summation of an infinite series of “time forward” diagrams’ diagonal in hole states. An even wider class of diagrams, where a hole interacts with an electron without changing its energy, and only the projections of the spin and angular momentum change, is achieved if the total moment and spin of the electron–hole pair are fixed. Such functions are referred to as $P_i^{N(LS)}(r)$. The functions $P_i^N(r)$ and $P_i^{N(LS)}(r)$, which in the general case are not orthogonal, can be orthogonalized to the functions of the ground state by including the off-diagonal parameters in the Hartree–Fock equations [5].

The rearrangement of shells due to excitation of an atom can be taken into account by solving the Hartree–Fock equation for the excited atomic state, where this state is determined by the electronic configuration [6].

The ***auger*** program [7] allows one to calculate the Auger widths of atomic levels taking into account many-electron correlations in n-transitions.

The ***dauger*** program [8] allows calculating the amplitudes, and thus the probabilities, of other atomic processes, such as, for example, the Auger effect, in which a state with one vacancy decays by emitting two electrons. In this process, triple ions are formed. The ***dauger*** program permits also to consider the decay of autoionization states with emission of two electrons, decay of a two-hole state with excitation of an electron into discrete level, as well as the decay of one vacancy with the simultaneous excitation and emission of electrons.

References

1. W.N. Asaad, Nucl. Phys. **66** (1965)
2. N.H. March, W.H. Young, S. Sampanthar, *The Many-Body Problem in Quantum Mechanics* (Cambridge University Press, London, 1967)
3. A.A. Abrikosov, L.P. Gor'kov, I.E. Dzyaloshinsky, *Methods of Quantum Field Theory in Statistical Physics* (Dover Publications Inc., New York, 1963)
4. D.A. Varshalovich, A.N. Moskalev, V.K. Khersonskii, *Quantum Theory of Angular Momentum* (World Scientist, Singapore, 1988).
5. L.V. Chernysheva, N.A. Cherepkov, V. Radojevič, Frozen-core Hartree-Fock program for atomic discrete and continuum states. Comput. Phys. Commun. **18**, 87 (1979)
6. T.A. Carlson, M.O. Krause, Phys. Rev. A **137** (1965)
7. L.V. Chernysheva, M.Ya. Amusia, V.K. Ivanov, S.G. Shapiro, *Numerical Calculations of the Width of Atomic Levels Relative to Auger Decay with Allowance for Multielectron Correlations in Two Transitions* (Ioffe Institute, No. 536, Leningrad, 1977) (In Russian)
8. L.V. Chernysheva, M.Ya. Amusia, V.A. Kilin, I.S. Li, *Numerical Calculations of the Probability of Double Auger Decay* (Ioffe Institute, No. 944, Leningrad, 1985) (In Russian)

Chapter 16
One-Photon Decay of Single- and Double-Hole States

This chapter describes the one-photon decay of single- and double-hole states of atoms. One-photon decay is a process in which the initial state—one or two vacancies—decays with emission of a single photon. Formulas are given for calculating the amplitudes and probabilities of these processes.

16.1 Theoretical Model

16.1.1 One-Photon Decay of States with a Single Vacancy

When an electron is removed from an atomic shell, excluding the outermost one, two types of vacancy decay are possible: radiative and Auger. Chapter 15 is devoted to Auger decay. Here, we will discuss the decay by photon emission. In the case of radiative decay, the filling of a vacancy by an electron of the outer shell is accompanied by the emission of a photon with energy ω. The probability of a radiative transition from a closed subshell with angular momentum l_f and the number of electrons N, $l_{\nu_f}^N$ to a vacancy state ν_1^{-1} is given by the following expression [AICC]:

$$W_{\nu_1 \to \nu_f} = \frac{4\omega}{3c^3[l_{\nu_1}]} \left| \left\langle \nu_1 l_{\nu_1} \left\| d^{(r,\nabla)}(\omega) \right\| \nu_f l_{\nu_f} \right\rangle \right|^2, \tag{16.1}$$

where the dipole matrix elements $d^{r,\nabla}(\omega)$ are defined by formulas (9.17), (9.18).

In atomic units, the probability of a radiative transition W is simply inversely proportional to the linewidth Γ, $W = 1/\Gamma$ of the vacancy (see, e.g., [1]). In the lowest order of the perturbation theory, i.e., neglecting the vacancy rearrangement, the total linewidth, which is inversely proportional to the lifetime of the level ν_1, is represented as the sum of two quantities, the Auger Γ^A and radiative Γ^γ widths:

M. Ya. Amusia and L. V. Chernysheva, *Computation of Atomic and Molecular Processes*, Springer Series on Atomic, Optical, and Plasma Physics 117,
https://doi.org/10.1007/978-3-030-85143-9_16

$$\Gamma_{\nu_1} = \sum_{jk} \Gamma^{A}_{\nu_1 \to \nu_f \nu_k} + \sum_{f} \Gamma^{\gamma}_{\nu_1 \to \nu_f} \tag{16.2}$$

In the first term, the summation includes all energetically possible two-hole states $j^{-1}k^{-1}$, and in the second, it goes over all outer, with respect to ν_1-subshells. If we neglect the usually small contribution of non-dipole transitions to the linewidth, then the second sum is limited by the outer shells with $l_{\nu_f} = l_{\nu_1} \pm 1$.

Radiative decay can be included, like the Auger decay in the self-energy part of the Green's function [2]. The contribution to the self-energy part of the Green's function of the radiation filling of ν_1-vacancy by ν_f electron from the subshell $l^N_{\nu_f}$ is given by the following diagram (16.3):

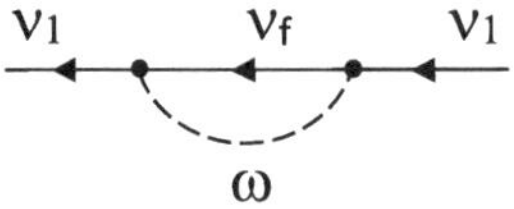

(16.3)

Consider the transition from a state $l^{-1}_{\nu_1} l^N_{\nu_f}(LS)(L'S')$ with a vacancy in the inner subshell ν_1 to the final state $l^{N-1}_{\nu_f}(L_1 S_1)$. Integration over angular and summation over spin variables gives the following formula for the contribution of the radiative transition to the width of the ν_1 level:

$$\Gamma^{\gamma}_{\nu_1 \to \nu_f} = \frac{4\omega}{3c^3}[LS]N_{\nu_f} \begin{Bmatrix} L' & L_1 & 1 \\ l_{\nu_f} & l_{\nu_1} & L \end{Bmatrix}^2 \left(G^{LS}_{L_1 S_1}\right)^2 \left|\langle \nu_1 l_1 \| d^{(r,\nabla)}(\omega) \| \nu_f l_{\nu_f} \rangle\right|^2 \tag{16.4}$$

Accounting for the interaction of a vacancy with other electrons beyond the HF approximation mixes it with other states, so that in the experimentally observed structure the initial vacancy is represented with a statistical weight equal to the spectroscopic factor $F_{\nu_1} = \left[1 - \partial \Sigma^{\text{cor}}_{\nu_1}(E)/\partial E|_{E \to \tilde{E}_{\nu_1}}\right]^{-1}$ (14.15) [AM, ACY]. Thus, the following formula is valid for the radiative linewidth, which differs by the factor F_{ν_1} from that given, for example, in [AICC]

$$\Gamma_{\nu_1 \nu_f} = F_{\nu_f} \frac{4\omega}{3c^3 [l_{\nu_1}]} \left|\langle \nu_1 l_{\nu_1} \| d^{(r,\nabla)}(\omega) \| \nu_f l_{\nu_f} \rangle\right|^2 \tag{16.5}$$

The contributions of radiative transitions to the linewidth are significant for the inner shells of heavy atoms. Note that for inner and even many intermediate shells, because of the large or significant distance between them in energy, it is sufficient to include the corrections to the self-energy part only in the second order in the electron–electron interaction. We consider the imaginary part $\Sigma^{\text{cor}}_{\nu_1}(E)$ to be small in comparison with the real one, which is true for most atomic subshells. This allows us to write $\partial \Sigma^{\text{cor}}_{\nu_1}(E)/\partial E|_{E \to \tilde{E}_{\nu_1}}$ in (14.15) as

$$\left.\frac{\partial\Sigma_{\nu_1}^{\text{cor}}(E)}{\partial E}\right|_{E\to\tilde{E}_{\nu_1}} \simeq \left.\frac{\partial\text{Re}\Sigma_{\nu_1}^{\text{cor}}(E)}{\partial E}\right|_{E\to\tilde{E}_{\nu_1}} = \frac{\partial}{\partial E}\left[\sum_L \frac{1}{(2L+1)(2l_1+1)}\right.$$
$$\left.\left.\left(\sum_{\nu_4\nu_2\le F} P\int_{\nu_3>F} + \sum_{\nu_3\le F} P\int_{\nu_4\nu_2>F}\right) \times \frac{\langle\nu_1\nu_3\|V_L\|\nu_4\nu_2\rangle\langle\nu_4\nu_2\|U_L\|\nu_1\nu_3\rangle}{E-E_4-E_2+E_3}\right]\right|_{E\to\tilde{E}_{\nu_1}}, \tag{16.6}$$

where

$$\tilde{E}_{\nu_1} = E_{\nu_1} + \text{Re}\Sigma_{\nu_1}^{\text{cor}}\left(E_{\nu_1}\right). \tag{16.7}$$

and P is the symbol of the principal value. These relations are obtained using (14.15) and (14.20).

Concrete calculations show that F_{ν_1} is usually close to 1 [3], although there are some cases of impressive deviations from 1 [4, AM].

Electron correlations of RPAE type are taken into account by replacing in (16.5) operators $d^{(r,\nabla)}(\omega)$ from (9.17) to (9.18) by $D^{(r,\nabla)}(\omega)$ from (9.67) to (9.68). The influence of correlations can be quite strong, up to making Γ_{kf} close to zero due to destructive interference of the direct decay channel and that going through the action of virtual excitation of other electrons upon the decay channel [1, 5, AICC].

16.1.2 One-Photon Decay of Double-Vacancy States

The one-photon decay of a double-vacancy state is a correlative decay of a two-vacancy state by emitting of a single photon with energy ω and polarization $\vec{e}$ as a result of filling the initial vacancies ν_1 and ν_2 with electrons ν_3 and ν_4 from other subshells. The probability of such a process in the framework of the dipole approximation is determined as [6]

$$\Gamma_{\nu_1\nu_2}^{\gamma} = \frac{\alpha^3\omega^3}{(2L+1)(2S+1)} \sum_{M_L M_{L'} M_S M_{S'}} |M|^2. \tag{16.8}$$

where $M = \left\langle\Psi_f(L'S'M_{L'}M_{S'})|(\mathbf{De})|\Psi_i(LSM_LM_S)\right\rangle$ is the amplitude of the radiative transition from the initial state $\Psi_i(LSM_LM_S)$ with ν_1 and ν_2 to the final state $\Psi_f\left(L'S'M_{L'}M_{S'}\right)$ with vacancies ν_3 and ν_4; LS and $L'S'$ are the orbital and spin moments of these states, and $M_LM_S,\ M_{L'}M_{S'}$ are their projections; $D = \sum_i d_i$ is the operator of the atomic dipole moment. The probability (16.8) has the dimension of energy (Ry).

When calculating the decay amplitude, perturbation theory can usually be applied, since atomic subshells in most of the cases are quite far from each other. Because of this, the matrix elements of the Coulomb interaction between electrons of different subshells are relatively small. We use many-body perturbation theory (MBPT) [7, AM].

In the lowest order of the perturbation theory in the residual electron–electron interaction, in the basis of the Hartree–Fock one-electron wave functions, the amplitude of one-photon decay is determined by the following diagrams:

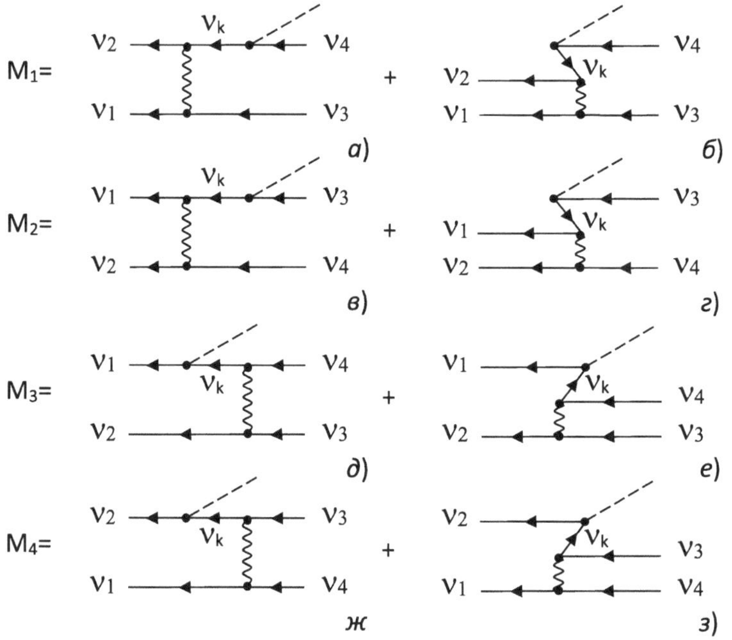

(16.9)

where the left and right arrows denote, respectively, one-hole (vacancy) and one-electron states, the dashed line denotes the emission of an electromagnetic field quantum, and the wavy line is the Coulomb interaction.

The corresponding analytical expressions are:

$$M = \sum_{i=1}^{4} M_i \tag{16.10}$$

$$M_1 = \sum_{\Pi} B \sum_{\nu_k} \langle \nu_1 \nu_2 | U | \nu_3 \nu_k \rangle \langle \nu_k | (\mathbf{er}) | \nu_4 \rangle E^{-1}_{\nu_1 \nu_2 \nu_3 \nu_k}, \tag{16.11}$$

$$M_2 = \sum_{\Pi} B \sum_{\nu_k} \langle \nu_1 \nu_2 | U | \nu_k \nu_4 \rangle \langle \nu_k | (\mathbf{er}) | \nu_3 \rangle E^{-1}_{\nu_1 \nu_2 \nu_k \nu_4}, \tag{16.12}$$

$$M_3 = \sum_{\Pi} B \sum_{\nu_k} \langle \nu_2 \nu_k | U | \nu_3 \nu_4 \rangle \langle \nu_1 | (\mathbf{er}) | \nu_k \rangle E^{-1}_{\nu_1 \nu_2 \nu_3 \nu_4}, \tag{16.13}$$

$$M_4 = \sum_{\Pi} B \sum_{\nu_k} \langle \nu_1 \nu_2 | U | \nu_3 \nu_k \rangle \langle \nu_2 | (\mathbf{er}) | \nu_k \rangle E^{-1}_{\nu_1 \nu_k \nu_4 \nu_3}. \tag{16.14}$$

In Eqs. (16.11)–(16.14), $E_{klmn} = E_k + E_l - E_m - E_n$, E_ν are the Hartree–Fock single-particle energies, while the matrix elements of the interparticle interaction are determined according to (15.12).

The summation in (16.11)–(16.14) is performed over hole ($\nu_k \le F$) and discrete excited states ($\nu_k > F$), including integration over states of the continuous spectrum.

The amplitudes (16.11)–(16.14) include the summation $\sum\limits_{\Pi} \equiv \sum\limits_{m_1 m_2 m_3 m_4 \mu_1 \mu_2 \mu_3 \mu_4}$ over the projections of the orbital and spin moments of the initial $\nu_1 \nu_2$ and final vacancies $\nu_3 \nu_4$, as well as a factor B defined by following relation

$$B = (-1)^{l_1 - l_2 + l_3 - l_4 + M_L + M_{L'} + M_S + M_{S'} + \sum\limits_{i=1}^{4} (l_i + s_i - m_i - \mu_i)} \begin{pmatrix} l_1 & l_2 & L \\ m_1 & m_2 & -M_L \end{pmatrix} \times$$
$$\begin{pmatrix} 1/2 & 1/2 & S \\ \mu_1 & \mu_2 & -M_S \end{pmatrix} \begin{pmatrix} l_3 & l_4 & L' \\ m_3 & m_4 & -M_{L'} \end{pmatrix} \begin{pmatrix} 1/2 & 1/2 & S' \\ \mu_3 & \mu_4 & -M_{S'} \end{pmatrix}, \tag{16.15}$$

where $\begin{pmatrix} l_1 & l_2 & l_3 \\ m_1 & m_2 & m_3 \end{pmatrix}$ is the Wigner $3j$-symbols. The factor B and the summation over projections is caused by the binding of the angular momenta of holes $\nu_1 \nu_2 (\nu_3 \nu_4)$ in a certain term $LS(L'S')$ [8].

After integration over the angular and summation over the spin variables and projections of the orbital and spin moments of holes, the partial amplitudes M_i can be represented as

$$M_i = A \sum_{l l_k} \Big[R_i^{(d)}(l, l_k) Y_i^{(d)}(l, l_k) + R_i^{(e)}(l, l_k) Y_i^{(e)}(l, l_k) \Big], \tag{16.16}$$

where $R_i^{(d)}(l, l_k)$ and $R_i^{(e)}(l, l_k)$ are defined as follows:

$$R_1^{(d)}(l, l_k) = \sum_{n_k \le F} \big\langle \nu_1 \nu_2 \| V^{(l)} \| \nu_3 \nu_k \big\rangle \langle \nu_k \| d \| \nu_4 \rangle E^{-1}_{\nu_1 \nu_2 \nu_3 \nu_k},$$
$$R_1^{(e)}(l, l_k) = \sum_{n_k > F} \big\langle \nu_1 \nu_2 \| V^{(l)} \| \nu_3 \nu_k \big\rangle \langle \nu_k \| d \| \nu_4 \rangle E^{-1}_{\nu_1 \nu_2 \nu_3 \nu_k}. \tag{16.17}$$

$$R_2^{(d)}(l,l_k) = \sum_{n_k \le F} \left\langle \nu_1\nu_2 \| V^{(l)} \| \nu_k\nu_4 \right\rangle \langle \nu_k \| d \| \nu_3 \rangle E^{-1}_{\nu_1\nu_2\nu_k\nu_4},$$
$$R_2^{(e)}(l,l_k) = \sum_{n_k > F} \left\langle \nu_1\nu_2 \| V^{(l)} \| \nu_k\nu_4 \right\rangle \langle \nu_k \| d \| \nu_3 \rangle E^{-1}_{\nu_1\nu_2\nu_k\nu_4}. \tag{16.18}$$

$$R_3^{(d)}(l,l_k) = \sum_{n_k \le F} \left\langle \nu_2\nu_k \| V^{(l)} \| \nu_3\nu_4 \right\rangle \langle \nu_1 \| d \| \nu_k \rangle E^{-1}_{\nu_2\nu_k\nu_3\nu_4},$$
$$R_3^{(e)}(l,l_k) = \sum_{n_k > F} \left\langle \nu_2\nu_k \| V^{(l)} \| \nu_3\nu_k \right\rangle \langle \nu_1 \| d \| \nu_k \rangle E^{-1}_{\nu_2\nu_k\nu_3\nu_4}. \tag{16.19}$$

$$R_4^{(d)}(l,l_k) = \sum_{n_k \le F} \left\langle \nu_1\nu_k \| V^{(l)} \| \nu_4\nu_3 \right\rangle \langle \nu_2 \| d \| \nu_k \rangle E^{-1}_{\nu_1\nu_k\nu_4\nu_3},$$
$$R_4^{(e)}(l,l_k) = \sum_{n_k > F} \left\langle \nu_1\nu_k \| V^{(l)} \| \nu_4\nu_3 \right\rangle \langle \nu_2 \| d \| \nu_k \rangle E^{-1}_{\nu_1\nu_k\nu_4\nu_3}. \tag{16.20}$$

Here, the superscripts (d) and (e) correspond to direct and exchange matrix elements, respectively.

The products of the angular and spin factors are denoted as $Y_i^{(d)}(l,l_k)$ and the formulas for them are given below:

$$\begin{aligned}
Y_1^{(d)}(l,l_k) &= (-1)^{l_1+L'} \begin{Bmatrix} L & 1 & L' \\ l_4 & l_3 & l_k \end{Bmatrix} \begin{Bmatrix} L & l_k & l_3 \\ l & l_1 & l_2 \end{Bmatrix}, \\
Y_1^{(e)}(l,l_k) &= (-1)^S\, Y_1^{(d)}(l,l_k), \\
Y_2^{(d)}(l,l_k) &= (-1)^{l_1+l_3+l_4+L} \begin{Bmatrix} L & 1 & L' \\ l_4 & l_3 & l_k \end{Bmatrix} \begin{Bmatrix} L & l_k & l_4 \\ l & l_2 & l_1 \end{Bmatrix} \\
Y_2^{(e)}(l,l_k) &= (-1)^S\, Y_2^{(d)}(l,l_k), \\
Y_3^{(e)}(l,l_k) &= (-1)^{l_2+l_3+l_k} \begin{Bmatrix} L & 1 & L' \\ l_k & l_2 & l \end{Bmatrix} \begin{Bmatrix} l_k & l_2 & L' \\ l_3 & l_4 & l \end{Bmatrix}, \\
Y_3^{(d)}(l,l_k) &= (-1)^S\, Y_3^{(e)}(l,l_k), \\
Y_4^{(e)}(l,l_k) &= (-1)^{L+L'+l_3+1} \begin{Bmatrix} L & 1 & L' \\ l_k & l_1 & l_2 \end{Bmatrix} \begin{Bmatrix} l_k & l_1 & L' \\ l_4 & l_3 & l \end{Bmatrix}, \\
Y_4^{(d)}(l,l_k) &= (-1)^S\, Y_4^{(e)}(l,l_k)
\end{aligned} \tag{16.21}$$

The factor A is

$$A = N_i N_f \sum_q e_q \begin{pmatrix} L & 1 & L' \\ M_L & q & -M_{L'} \end{pmatrix} [L][L'] \tag{16.22}$$

Here, e_q is the spherical component of the polarization vector **e**, N_i and N_f are the normalization factors of the initial and final states $N_i = 1$ $\left(N_f = 1\right)$, if holes ν_1, ν_2 (ν_3, ν_4) are not equivalent, $1/\sqrt{2}$ otherwise. Recall that $\begin{Bmatrix} a & b & c \\ d & e & f \end{Bmatrix}$ is the Wigner $6j$-coefficient [IS].

16.2 Numerical Model

16.2.1 The Probability of One-Photon Decay of Single-Vacancy States

The probability of one-photon radiative decay of a vacancy is calculated using formula (16.5). The central point is the calculation of the dipole matrix element that is done in the same way as for the oscillator strength, as described in Sect. 9.2.1. The transition energy ω is determined by the relation

$$\omega_{\nu_1\nu_2} = \tilde{E}_{\nu_2} - \tilde{E}_{\nu_1} = E_{\nu_2} - E_{\nu_1} + \mathrm{Re}\Sigma^{\mathrm{cor}}_{\nu_2}\left(E_{\nu_1}\right) - \mathrm{Re}\Sigma^{\mathrm{cor}}_{\nu_1}\left(E_{\nu_1}\right) \approx E_{\nu_2} - E_{\nu_1}. \tag{16.23}$$

In most cases, the approximate value $\omega_{\nu_1\nu_2} \approx E_{\nu_2} - E_{\nu_1}$ accurately enough presents the actual decay energy. The calculation of the self-energy part is carried out according to the formula (14.20).

If we consider those one-particle levels and take into account only those corrections to them, for which the denominator $E_1 - E_4 - E_2 + E_3$ in (14.20) does not vanish, differentiation with respect to E and finding the spectroscopic factor F is simple. It is necessary to use the formula following from (16.6) for this case:

$$\frac{\partial \Sigma^{\mathrm{cor}}_{\nu_1}(E)}{\partial E|_{E\to\tilde{E}_{\nu_1}}} \simeq -\left[\sum_L \frac{1}{(2L+1)(2l_1+1)}\right.$$
$$\left.\left(\sum_{\nu_4\nu_2\le F}\int_{\nu_3>F} + \sum_{\nu_3\le F}\int_{\nu_4\nu_2>F}\right)\frac{\langle\nu_1\nu_3\|V_L\|\nu_4\nu_2\rangle\langle\nu_4\nu_2\|U_L\|\nu_1\nu_3\rangle}{\left[\tilde{E}_{\nu_1} - E_4 - E_2 + E_3\right]^2}\right] \tag{16.24}$$

This formula is inapplicable if the energy denominator vanishes—direct calculation using (16.24) leads to divergence. To find the factor F, one must have in mind that when differentiating the principal value of an integral, one must also differentiate over the limits of integration. Exactly presenting, it is

$$\frac{\partial}{\partial y} P \int_0^\infty \frac{f(x)}{x-y} dx = \lim_{\varepsilon \to 0} \left[\int_0^{y-\varepsilon} \frac{f(x)}{(x-y)^2} dx + \int_{y+\varepsilon}^\infty \frac{f(x)}{(x-y)^2} dx + \frac{2f(x)}{\varepsilon} \right] \quad (16.25)$$

However, instead of using the computationally delicate formula (16.25), it is easier to calculate the self-energy part $\Sigma^{\mathrm{cor}}_{\nu_1}(E)$, determined by (14.20), at 5–7 points near $\tilde{E}_{\nu_1}$ with subsequent numerical differentiation of the resulting function over E. When calculating the decay probability, the transition energy can also be taken from experiment.

16.2.2 The Probability of One-Photon Decay of Double-Vacancy States

The width of the one-photon decay of the double-vacancy state is determined by formula (16.8). The transition amplitude itself is determined by expressions (16.11)–(16.14). After separating the angular and spin parts, the matrix elements (16.17)–16.20) are calculated. In (16.17)–(16.20), the direct reduced Coulomb matrix elements are denoted as $\langle \nu_1 \nu_2 \| V^{(\lambda)} \| \nu_3 \nu_4 \rangle$, while the exchange matrix elements are denoted as $\langle \nu_1 \nu_3 \| V^{(\lambda)} \| \nu_2 \nu_4 \rangle$ (9.34). The quantities $\langle \nu_2 \| d \| \nu_k \rangle$ in formulas calculating the decay probability are the reduced dipole matrix elements (9.20)–(9.21).

The ***phodec*** program at first determines all the required reduced Coulomb and dipole matrix elements. The radial parts $R_i^{(d)}$, $R_i^{(e)}$ and angular factors $Y_i^{(d)}$, $Y_i^{(e)}$ are then calculated. After that, finally, the partial amplitudes $M_i (i = 1, 2, 3, 4)$ are determined by formulas (16.11)–(16.14) at a fixed value of the transmitted angular momenta l and given initial ν_1, ν_2, final ν_3, ν_4, and intermediate ν_k states.

The calculation of the reduced Coulomb and dipole matrix elements and integration over the states of the continuous spectrum in $R_i^{(d)}$, $R_i^{(e)}$, including the integration of expressions containing a pole, is described in Sect. 9.2.

The total amplitude M of the joint decay of two vacancies with the emission of one photon is equal to the sum of four partial amplitudes $M_i (i = 1, 2, 3, 4)$. The probability of one-photon decay of a double-vacancy state is found by formula (16.8), where the summation and averaging over the projections of the orbital moments of the initial and final states, give the factor 1/3, and the summation and averaging over spin projection lead to $(2S + 1)$.

To determine the probability of one-photon decay of a double-vacancy state, in addition to the process amplitude, knowledge of the energy of the emitted photon is necessary. In the zeroth approximation, it is equal to the difference between the one-electron energies of the final ν_3, ν_4 and initial ν_1, ν_2 vacancies $\omega^{(0)} = E_{\nu_3} + E_{\nu_4} - E_{\nu_1} - E_{\nu_2}$. However, calculations by this formula can sometimes be insufficiently accurate for identification experimentally observed spectral lines. In such cases, the interaction between vacancies in both the initial and final states should be taken into account. Since the perturbation theory should also be applicable to energy calculations, the first order of the perturbation theory in the electron–electron interaction is a fairly good approximation for ω:

$$\omega = \omega^{(0)} + \langle \nu_1 \nu_2 | U | \nu_1 \nu_2 \rangle - \langle \nu_3 \nu_4 | U | \nu_3 \nu_4 \rangle \tag{16.26}$$

In *LS*-coupling approximation, it transforms into the formulas

$$\omega = \omega^{(0)} + (-1)^L \sum_\lambda \begin{Bmatrix} l_1 & l_2 & L \\ l_2 & l_1 & \lambda \end{Bmatrix} \left[\langle \nu_1 \nu_2 \| V^{(\lambda)} \| \nu_1 \nu_2 \rangle + (-1)^S \langle \nu_1 \nu_2 \| V^{(\lambda)} \| \nu_1 \nu_2 \rangle \right] +$$
$$(-1)^{L'+1} \sum_\lambda \begin{Bmatrix} l_3 & l_4 & L' \\ l_4 & l_3 & \lambda \end{Bmatrix} \left[\langle \nu_3 \nu_4 \| V^{(\lambda)} \| \nu_3 \nu_4 \rangle + (-1)^S \langle \nu_3 \nu_4 \| V^{(\lambda)} \| \nu_3 \nu_4 \rangle \right]. \tag{16.27}$$

This formula permits to determine ω quite accurately. Improvement of the result can be achieved either by taking into account higher orders within the MBPT frame or by using accurate enough experimental values.

In general, the ***phodec*** program gives the partial amplitudes of one-photon decay and the total probability of one-photon decay.

References

1. M.Ya. Amusia, N.A. Cherepkov, Case Studies in Atomic Physics, vol. 5 (1975), p. 2
2. M. Ohno, G.Z. Wendin, Phys. D **5** (1987)
3. M.H. Chen, F.P. Larkins, B. Crasemann, At. Data Nucl. Data Tables **45** (1990)
4. M.Ya. Amusia, L.V. Chernysheva, V.G. Yarzhemsky, *Handbook of Theoretical Atomic Physics, Data for Photon Absorption, Electron Scattering, Vacancies Decay* (Springer, Heidelberg, New York, Dordrecht, London, 2012)
5. M.Ya. Amusia, M.P. Kazachkov, N.A. Cherepkov, L.V. Chernysheva, Phys. Lett. A **39**, 2 (1972)
6. M.Ya. Amusia, V.A. Kilin, A.N. Kolesnikova, I.S. Lee, Sov. J. Tech. Phys. Lett. **10** (1984) (In Russian)
7. A.A. Abrikosov, L.P. Gor'kov, I.E. Dzyaloshinsky, *Methods of Quantum Field Theory in Statistical Physics* (Dover Publications Inc., New York, 1963)
8. B.R. Judd, *Second Quantization and Atomic Spectroscopy* (The John Hopkins Press, Baltimore, 1967)

Chapter 17
Bremsstrahlung Spectra in High and Intermediate Energy Collisions

This chapter describes a method for calculating the bremsstrahlung of incident high and intermediate energy electrons scattered by an atom, taking into account the dynamic polarization of the target atom by the incident particle in the collision process. Formulas are given for calculating the form factor, bremsstrahlung cross section, and matrix elements included in these expressions.

17.1 Theoretical Model

17.1.1 High Energy Particles

In one-particle Born approximation, the bremsstrahlung spectrum of a fast incident particle is determined by the formula [BLP]:

$$\frac{d\sigma}{d\omega} = \frac{32z^2}{3c^3 p_1^2 \omega m_1^2} \int\limits_{p_1-\sqrt{p_1^2-m_1\omega}}^{2p_1} z^4 \left[Z - F(q)\right]^2 \frac{dq}{q} \tag{17.1}$$

Here ω is the energy of the emitted photon; $q = |\,\mathbf{p}_1 - \mathbf{p}_2\,|$ is the impulse transferred to the target atom, the minimum value of which is determined by the conservation energy law, i.e., in the selected system of units, where the energy is in Rydberg, by the expression $p_1^2 = p_2^2 + m_1\omega$, z and m_1—are the charge and mass of the incident particle. At large values of the linear momentum p_1 of the incoming and p_2 outgoing particles, the upper limit for the transferred momentum can be taken equal to $2p_1$,Z is the charge of the nucleus.

The form factor of an atom $F(q)$ is, by definition, the matrix element of an operator $\exp(i\mathbf{qr})$ [see. (13.2)]:

M. Ya. Amusia and L. V. Chernysheva, *Computation of Atomic and Molecular Processes*, Springer Series on Atomic, Optical, and Plasma Physics 117,
https://doi.org/10.1007/978-3-030-85143-9_17

$$F(q) = \langle 0|M(q)|0\rangle = \sum_{\nu,l} \frac{(-1)^l N_\nu}{\sqrt{2l+1}} \langle \nu \| M_0(q) \| \nu \rangle, \tag{17.2}$$

where N_ν is the number of electrons in ν-subshell with angular momentum l, $|\nu\rangle$ are the radial wave functions of the ground state in the Hartree–Fock approximation, and $M_0(q)$ is the monopole component of the operator $\exp(i\mathbf{qr})$, $M_0(qr) \sim j_0(qr)$. Summation is performed over all occupied atomic subshells. Relation (17.2) is applied by us in cases when each atomic shell is either closed or semi-closed.

The bremsstrahlung process, which we now describe, can be represented by the following diagrams [1]:

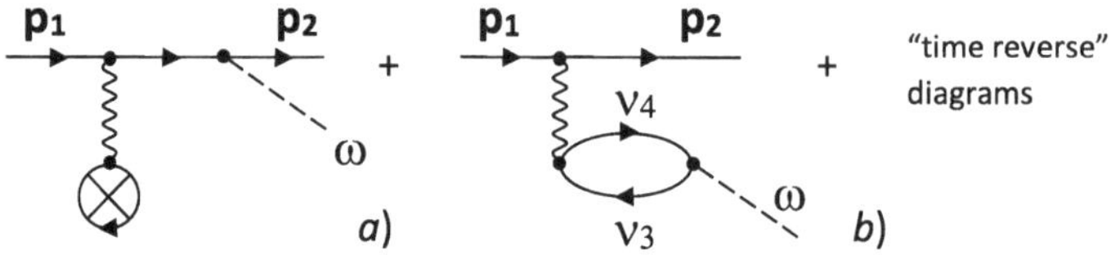

(17.3)

Here p_1, p_2 are linear moments of a fast incident and scattered electron described by a plane wave.

The calculation of the bremsstrahlung spectrum (17.1) corresponds to the diagram (17.3a), and the crossed circle depicts the atomic form factor. By including the influence of the dynamic polarizability of the atom (17.3b), the spectral distribution of bremsstrahlung $d\sigma/d\omega$ takes the form:

$$\frac{d\sigma}{d\omega} = \frac{32}{3c^3 p_1^2 \omega m_1^2} \int_{p_1 - \sqrt{p_1^2 - m_1\omega}}^{2p_1} |a_1[Z - F(q)] + a_2\Pi(\omega, q)|^2 \frac{dq}{q} \tag{17.4}$$

Here $\Pi(\omega, q)$ is the polarization amplitude equal to the sum of diagrams (17.3b) and the same diagram with "time revers." It is expressed through dipole matrix elements $\langle \nu_3 \| d \| \nu_4 \rangle$, $\langle \nu_3 \| M_1(qr) \| \nu_4 \rangle$, $l_4 = l_3 \pm 1$ [AC]:

$$\Pi(\omega, q) = \omega^2 \sum_{\nu_3 \le F} \int_{\nu_4 > F} \frac{E_4 - E_3}{\omega^2 - (E_4 - E_3)^2 + i\eta} \langle \nu_3 \| d \| \nu_4 \rangle \langle \nu_4 \| M_1(qr) \| \nu_3 \rangle \tag{17.5}$$

Integration over the variable ν_4 implies also, as usual, summation over discrete spectrum states. The parameters a_1 and a_2 are determined by the ratios $a_1 = z^2$, $a_2 = zm_1$. The choice of their different values determines the spectral distributions for different types of incident particles.

The values $a_1 = a_2 = 1$ imply accounting for the atomic polarizability in the bremsstrahlung spectra with electrons as projectile. To calculate the spectrum of the positron bremsstrahlung, the coefficients $a_1 = -a_2 = 1$ must be taken. In the case of scattering of heavy particles, for example, $\mu-$ mesons or protons, the coefficients

$a_1 = 0$ and $a_2 = m_1$ should be used, since the direct amplitude (the first term in (17.4)) is suppressed in comparison with the polarization amplitude due to the presence of the factor $1/m_1 \ll 1$.

When calculating the polarization amplitude within the framework of one transition, only, one atomic subshell ν_3 is taken into account in the sum (17.5), the moment l_4 is fixed, and the integration is performed over ν_4. Expression (17.5) was obtained under the condition that the interaction of electrons with electromagnetic field is described in the "length form." A similar expression can be obtained in the "velocity form."

In terms of diagrams, the bremsstrahlung process in RPAE is represented as follows:

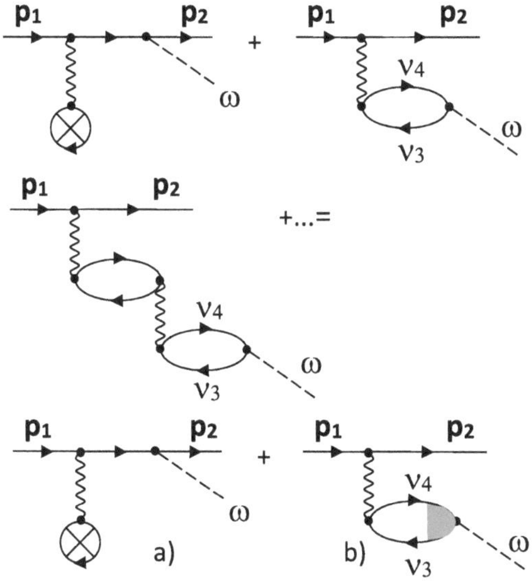

(17.6)

It is implied that terms with permutation of moments of electron interaction and photon emission as compared to (17.6a and b) are accounted for in (17.6).

Corrections to the polarization amplitude arising in higher orders with respect to the electron–electron interaction are taken into account by replacing $\langle \nu_3 \| d \| \nu_4 \rangle$ with dipole matrix elements in RPAE (1.57). As a result, the polarization amplitude corresponding to diagrams (17.6a) and (17.6b) is expressed through the reduced matrix element of the effective correlation interaction $D(\omega)$:

$$\Pi(\omega, q) = \omega^2 \sum_{\nu_3 \le F} \int_{\nu_4 > F} \frac{E_4 - E_3}{\omega^2 - (E_4 - E_3)^2 + i\eta} \langle \nu_3 \| D \| \nu_4 \rangle \langle \nu_4 \| M_1(qr) \| \nu_3 \rangle \quad (17.7)$$

The angular distribution of bremsstrahlung photons is determined as follows [1, 2]:

$$\frac{d^2\sigma}{d\omega d\vartheta} = \frac{1}{4\pi}\frac{d\sigma}{d\omega}\left[1 - \gamma(\omega)P_2(\cos\vartheta)\right]. \tag{17.8}$$

Here ϑ is the angle between the incident electron momentum and the direction of photon emission.

The angular anisotropy parameter $\gamma(\omega)$ can be represented as [3]

$$\gamma(\omega) = \left(\frac{d\sigma}{d\omega}\right)^{-1} \frac{16}{3c^3 p_1^2 \omega m_1^2} \times \int\limits_{p_1-\sqrt{p_1^2-m_1\omega}}^{2p_1} |a_1[Z - F(q)] + a_2 m_1 \Pi(\omega, q)|^2 \left[3\left(\frac{m_1\omega}{p_1 q} + \frac{q}{p_1}\right)^2 - 1\right]\frac{dq}{q}. \tag{17.9}$$

The angle ϑ is connected with ω and the transferred linear momentum q by the following relation

$$2P_2(\cos\vartheta) = 3\left(\frac{m_1\omega}{p_1 q} + \frac{q}{p_1}\right)^2 - 1,$$

17.1.2 Intermediate Energy Particles

In the region of intermediate and low energies, the bremsstrahlung is described by the following diagrams:

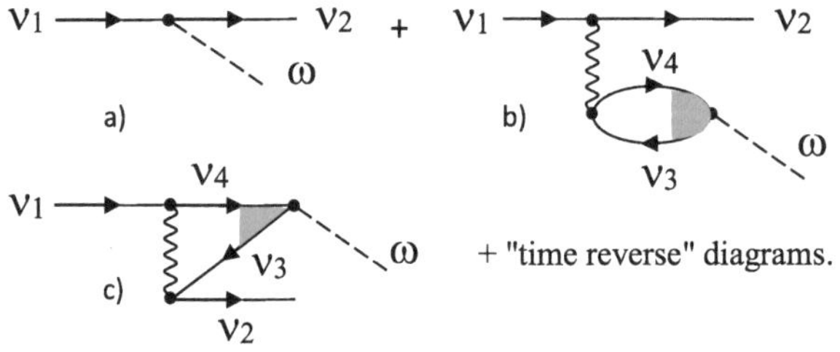

(17.10)

We assume that both the scattered electron (ν_1, ν_2) and the excited atomic electron ν_4 are in the self-consistent Hartree–Fock field. Particle states ν_i are determined by energy and orbital angular momentum $\nu_i = (E_i, l_i)$. Diagram (17.10a) represents the "ordinary" bremsstrahlung, while diagrams (17.10b) and (17.10c) and corresponding "time reverse" diagrams depicts "atomic" or "polarization" bremsstrahlung. Diagram (17.10b) stands for the direct bremsstrahlung amplitude, while diagram (17.10c) gives the exchange one. If the projectile is not an electron, there are no exchange diagrams.

Let us represent the wave function of a scattered particle as an expansion in partial waves [LL]:

$$\psi_{\mathbf{p}}^{\pm}(\mathbf{r}) = 4\pi\sqrt{\frac{\pi m}{p}}\sum_{l,m} i^l \exp(\pm i\delta_l(p))Y_{lm}^{*}(\mathbf{p}/p)Y_{lm}(\mathbf{r}/r)\frac{P_{pl}(r)}{r} \qquad (17.11)$$

The negative sign in Formula (17.11) corresponds to the incident particle, while the positive sign is related to the scattered particle. The radial parts in (17.11) $P_{pl}(r)$ are solutions of the radial Hartree–Fock equation (or the Hartree equation, if the incident particle is not an electron) in the field of the "frozen" core. They are normalized for $r \to \infty$ in accordance with (6.14).

The bremsstrahlung cross section in the frequency range $\omega > I$ can be calculated using the amplitudes represented by diagrams (17.10) and the wave functions ν_1, ν_2 defined in (17.11). The absolute cross section is the sum of partial cross sections:

$$\omega\frac{d\sigma}{d\omega} = \sum_{l_1=0}^{\infty}\omega\frac{d\sigma_{l_1}}{d\omega} \qquad (17.12)$$

where the spectra $d\sigma_{l_1}/d\omega$ are defined as follows:

$$\frac{d\sigma_{l_1}}{d\omega} = z^2\frac{32\pi}{3c^3}\frac{\omega^3}{p_1^2}\sum_{l_2=l_1\pm 1}^{\infty}\left|F_{l_2 l_1}(\omega)\right|^2 \qquad (17.13)$$

In (17.13), z is the charge of the projectile, $F_{l_1 l_2}$ is the total amplitude of bremsstrahlung, including both ordinary and polarization terms:

$$F_{l_1 l_2}(\omega) = F_{l_1 l_2}^{\mathrm{ord}}(\omega) + F_{l_1 l_2}^{\mathrm{pol}}(\omega) \qquad (17.14)$$

Here

$$F_{l_1 l_2}^{\mathrm{ord}}(\omega) = \sqrt{(2l_1+1)(2l_2+1)}\begin{pmatrix} l_2 & 1 & l_1 \\ 0 & 0 & 0 \end{pmatrix}\int_0^{\infty} r dr\, P_{\nu_1}(r)P_{\nu_2}(r) \qquad (17.15)$$

is the amplitude of the ordinary bremsstrahlung.

To calculate the amplitude of the polarization bremsstrahlung for the frequency ω near the ionization threshold I_{ν_3} of the chosen multielectron shell ν_3, the contributions of this shell itself $F_{l_1 l_2}^{\nu_3}(\omega)$ and the subshells external to it $F_{l_1 l_2}^{\mathrm{outer}}(\omega)$ must be taken into account. The contribution of the inner shells can be neglected, since their polarizabilities at $\omega \geq I_{\nu_3}$ are small. Thus, the amplitude of the polarization bremsstrahlung is presented as the sum of two terms:

$$F^{\text{pol}}_{l_1 l_2}(\omega) = F^{\nu_3}_{l_1 l_2}(\omega) + F^{\text{outer}}_{l_1 l_2}(\omega) \tag{17.16}$$

Here the term $F^{\nu_1}_{l_4 l_3}(\omega)$ is defined as follows:

$$F^{\nu_3}_{l_1 l_2}(\omega) = -\frac{1}{3}\int\limits_{\nu_4 > F}\left[\frac{\langle\nu_3\nu_1|U|\nu_4\nu_2\rangle}{\omega + E_3 - E_4 + i\eta} - \frac{\langle\nu_3\nu_1|U|\nu_2\nu_4\rangle}{\omega - E_3 + E_4}\right] D_{\nu_4\nu_3}(\omega) \tag{17.17}$$

where $l_4 = l_3 \pm 1$, U are the reduced dipole Coulomb matrix elements with direct and exchange terms (9.34) in it. For example,

$$U_{\nu_3\nu_1\nu_4\nu_2} = 2V_{\nu_3\nu_1\nu_4\nu_2} - aV_{\nu_3\nu_1\nu_2\nu_4} \tag{17.18}$$

Factor a in (17.18) is equal to 1 if the incident particle is an electron, and zero otherwise.

In RPAE, the reduced effective matrix elements $D_{\nu_2\nu_1}(\omega)$ are determined from (9.56), and (9.17–9.18) gives the dipole matrix elements $d_{\nu_2\nu_1}$.

The term $F^{\text{outer}}_{l_1 l_2}(\omega)$ in (17.16) takes into account the contribution of virtual electron excitations of the outer subshells to the amplitude of the polarization bremsstrahlung. It is calculated under the assumption that the energy ω is considerably higher than the outer shell typical excitation energy. For these energies ω, RPAE corrections for the outer shell can be neglected. The expression for $F^{\text{outer}}_{l_4 l_3}(\omega)$ includes direct and exchange terms, respectively, F and $\tilde{F}$:

$$F^{\text{outer}}_{l_1 l_2}(\omega) = F_{l_1 l_2}(\omega) + a\,\tilde{F}_{l_1 l_2}(\omega). \tag{17.19}$$

Here

$$\tilde{F}_{l_1 l_2}(\omega) = \frac{\sqrt{(2l_1+1)(2l_2+1)}}{\omega}\begin{pmatrix} l_1 & 1 & l_2 \\ 0 & 0 & 0 \end{pmatrix}$$
$$\sum_{\nu_0, L} N_{\nu_0} \int\limits_0^\infty\int\limits_0^\infty dr_1 dr_2 \frac{r_<^L}{r_>^{L+1}} P_{\nu_0}(r_1) r_1 P_{\nu_0}(r_2)$$
$$\times\left[\begin{pmatrix} l_4 & L & l_0 \\ 0 & 0 & 0 \end{pmatrix}^2 P_{\nu_1}(r_2)P_{\nu_2}(r_1) - \begin{pmatrix} l_3 & L & l_0 \\ 0 & 0 & 0 \end{pmatrix}^2 P_{\nu_1}(r_1)P_{\nu_2}(r_2)\right], \tag{17.20}$$

$$F_{l_1 l_2}(\omega) = \frac{\sqrt{(2l_1+1)(2l_2+1)}}{\omega}\begin{pmatrix} l_1 & 1 & l_2 \\ 0 & 0 & 0 \end{pmatrix}$$
$$\sum_{\nu_0} N_{\nu_0} \int\limits_0^\infty dr\, P_{\nu_1}(r)P_{\nu_2}(r) \int\limits_0^r dr_1 P^2_{\nu_0}(r_1). \tag{17.21}$$

In (17.20) and (17.21), the summation is performed over all outer electron shells with quantum numbers ν_0. Here N_0 is the number of electrons on the subshell ν_0; $P_{\nu_0}(r)$ is the electron wave function at the level ν_0. The factor a in (17.19) is the same as in (17.18).

17.2 Numerical Model

17.2.1 Bremsstrahlung Spectrum for High Energy Particles

Dipole matrix elements $\langle \nu_3 \| d \| \nu_4 \rangle$ are equal to

$$\langle \nu_3 \| d \| \nu_4 \rangle = \sqrt{(2l_3+1)(2l_4+1)} \begin{pmatrix} l_3 & 1 & l_4 \\ 0 & 0 & 0 \end{pmatrix} \int_0^\infty P_{\nu_3}(r) P_{\nu_4}(r) r dr, \tag{17.22}$$

where

$$\sqrt{(2l_3+1)(2l_4+1)} \begin{pmatrix} l_3 & 1 & l_4 \\ 0 & 0 & 0 \end{pmatrix} = (-1)^{\frac{l_3+l_4+1}{2}} \sqrt{\max(l_3, l_4)} \tag{17.23}$$

Matrix elements (17.22) are calculated as described in Sect. 9.2.1.

The radial wave functions $P_{\nu_3}(r)$, $P_{\nu_4}(r)$ and corresponding energies E_3, E_4 are calculated in the Hartree–Fock approximation using ***scfhf*** program for the ground state $|\nu_3\rangle$ and using ***fchf*** program for excited atomic states $|\nu_4\rangle$.

The effective interaction dipole matrix elements $\langle \nu_3 \| D(\omega) \| \nu_4 \rangle$ (1.56) are calculated as described in Sect. 9.2.3.

The matrix elements $\langle \nu_3 \| M_1(qr) \| \nu_4 \rangle$ of dipole transitions $\nu_3 \to \nu_4 (l_4 = l_3 \pm 1)$ are determined by the formula:

$$\langle \nu_3 \| M_1(qr) \| \nu_4 \rangle = \frac{1}{q} \sqrt{(2l_3+1)(2l_4+1)} \begin{pmatrix} l_3 & 1 & l_4 \\ 0 & 0 & 0 \end{pmatrix} \int P_{\nu_3}(r) P_{\nu_4}(r) j_1(qr) dr \tag{17.24}$$

in the same way as it was done in Sect. 10.2.1.

The monopole transition matrix element used in calculating the form factor $F(q)$ (17.2) is

$$\langle \nu_3 \| M_0(q) \| \nu_3 \rangle = (2l_3+1) \begin{pmatrix} l_3 & 0 & l_3 \\ 0 & 0 & 0 \end{pmatrix} \int_0^\infty P_{\nu_3}^2(r) j_0(qr) dr \tag{17.25}$$

In the case of a spherically symmetric atom, matrix elements with $l > 0$ do not contribute to the form factor.

The polarization amplitude is determined for all possible virtual excitation energy values $\omega_{43} = E_4 - E_3$, where E_4 are the continuous spectrum energy states. Pole at points $E_1 - E_2 - E_4 + E_3 = 0$ is treated in the same way as in 9.2.4.

The integrand in (17.4) changes rapidly at small values of q. Therefore, when integrating, we choose a logarithmic scale with respect to q, so that

$$\Delta \ln q = \frac{\ln q_{\max} - \ln q_{\min}}{k - 1} = \frac{\ln\left[2p_1/\left(p_1 - \sqrt{p_1^2 - m_1\omega}\right)\right]}{k - 1} \tag{17.26}$$

where k is the number of points of integration for the transferred linear moment q, which varies in the range $[p_1 - \left(p_1^2 - m_1\omega\right)^{1/2}, 2p_1]$.

The lower limit of integration over lnq in (17.4) depends on ω. Therefore, it can coincide with one of the integration points or be shifted from it to the left or right. If the lower limit $\ln[p_1 - \left(p_1^2 - m_1\omega\right)^{1/2}]$ is shifted relative to $\ln q$, then the integration starts at the nearest point $\ln q_0$. In this case, $\ln q_0 \neq \ln[p_1 - \left(p_1^2 - m_1\omega\right)^{1/2}]$, and this shift is taken into consideration by introducing the following correction ζ to (17.4):

$$\zeta = \frac{32}{3c^3 p^2\omega}\left[\ln\left(p_1 - \sqrt{p_1^2 - m_1\omega}\right) - \ln q_0\right][a_1(Z - F(q)) + a_2\Pi(\omega, q_0)]^2 \tag{17.27}$$

Integration over lnq is performed using the Simpson formula [BZ].

After changing the variable, the spectrum becomes described by the following expression:

$$\frac{d\sigma(\omega)}{d\omega} = \frac{32}{3c^3 p^2\omega m_1^2}\int\limits_{\ln q_{\min}}^{\ln q_{\max}} |a_1(Z - F(q)) + a_2 m_1\Pi(\omega, \ln q)|^2\, d(\ln q) \tag{17.28}$$

When calculating the polarization amplitude using ***brem*** program [2], a certain hole state is fixed and $\Pi(\omega, q)$ is determined under the condition that ω exceeds the shell ionization threshold. The summation is performed over discrete electron–hole excitations, and for the continuous energy spectrum of a particle above the ionization threshold, it is replaced by integration over E. The integrand corresponding to the continuous spectrum has a pole at energy $E = \omega + E_3$. Therefore, the polarization amplitude is complex:

$$\Pi(\omega, q) = \mathrm{Re}\Pi(\omega, q) + i\mathrm{Im}\Pi(\omega, q)$$

where

$$\mathrm{Re}\Pi(\omega, q) = 2\omega^2 P \int\limits_{\nu_4} \frac{(E_4 - E_3)\langle\nu_3\|d\|\nu_4\rangle\langle\nu_4\|M_1(q)\|\nu_3\rangle}{\omega^2 - (E_4 - E_3)^2}, \tag{17.29}$$

$$\mathrm{Im}\Pi(\omega, q) = -\pi\omega^2\langle\nu_3\|d\|E_3 + \omega\rangle\langle\omega + E_3\|M_1(q)\|\nu_3\rangle \tag{17.30}$$

The integral in (17.29) is represented by its principal value, which is marked by the letter P. The integral also implies summation over discrete excited states. To calculate the integral in expression (17.29), the generalized Simpson formula is used. The procedure for determining the principal value of the integral (17.29) is described in Subsection 9.2.4. The matrix elements included in (17.30) are at the points of the pole when $E_4 = \omega + E_3$.

In the ***breml*** program [2], the polarization amplitude is calculated under the condition that ω does not exceed the ν_3-shell ionization threshold, $\Pi(\omega, q)$ is real in this region and is determined as follows:

$$\Pi(\omega, q) = 2\omega^2 \int\limits_{\nu_3 \leq F < \nu_4} \frac{E_4 - E_3}{\omega^2 - (E_4 - E_3)^2}\langle\nu_3\|d\|\nu_4\rangle\langle\nu_4\|M_1(q)\|\nu_3\rangle \tag{17.31}$$

where the integration sign also implies summation over discrete states.

The expressions for the cross section (17.28) and the angular anisotropy parameter (17.9) remain unchanged. The photon energy below the threshold is chosen arbitrarily. The initial value ω_0, step along ω and the number of integration points, at which $\sigma(\omega)$ and $\gamma(\omega)$ are determined.

The polarization amplitude in the ***bremco*** program [2] (17.7) is calculated under the condition that ω exceeds the ionization threshold of the subshell in which the hole ν_3 is considered. The parameter $\gamma(\omega)$ is determined by the values $F(q)$ (17.2), $d\sigma/d\omega$ (17.4), and $\Pi(\omega, q)$ (17.7).

17.2.2 Bremsstrahlung Spectrum for Intermediate Energy Particles

17.2.2.1 Matrix Element of Conventional Bremsstrahlung

The reduced dipole matrix element in (17.15) includes the integral of two continuous spectrum wave functions:

$$I_{12} = \int\limits_0^\infty r P_{\nu_1}(r) P_{\nu_2}(r) dr \tag{17.32}$$

The value I_{12} is calculated in the following way. The region of integration over r is divided into two segments—[0, R] and [R, ∞]. Then the integral (17.32) can be represented as the sum of two integrals:

$$I_{12} = \int_0^R r dr P_{\nu_1}(r) P_{\nu_2}(r) + \int_R^\infty r dr P_{\nu_1}(r) P_{\nu_2}(r) \equiv I_{12}^{(<)} + I_{12}^{(>)} \tag{17.33}$$

The choice of R is determined by the condition $R >> R_A$, R_A is the atomic radius. In the region $r \le R_A$ the wave functions P_{ν_4} and P_{ν_3} are obtained from the Hartree–Fock equation in the field of the "frozen" core and $I_{12}^{(<)}$ is calculated by Simpson's formula. In the region $r > R$, the influence of the atomic field on the wave functions can be neglected. Then P_{ν_4} и P_{ν_3} can be found in the form of a superposition of functions describing radial free motion:

$$P_\nu(r) = A_l(p)\sqrt{\frac{\pi p r}{2}}\left[I_{l+1/2}(pr)\cos\delta_l(p) - Y_{l+1/2}(pr)\sin\delta_l(p)\right]. \tag{17.34}$$

Here $I_{l+1/2}(pr)$ and $Y_{l+1/2}(pr)$ are the Bessel and Neumann functions, respectively. The normalization factor $A_l(p)$ and phase $\delta_l(p)$ are determined by combining (17.34) with a numerical solution in the area $r \le R$. After substituting (17.34) into $I_{12}^{(>)}$ from (17.33), the following integral appears:

$$C_{l_1 l_2} \equiv \int_R^\infty r^2 dr \cdot W_{l_2+1/2}(p_2 r, \delta_{l_2}) W_{l_1+1/2}(p_1 r, \delta_{l_1}) \tag{17.35}$$

Here $W_j(\kappa r, \delta_l)$ is a linear combination of cylindrical functions:

$$W_j(pr, \delta_l) \equiv I_j(pr)\cos\delta_l(p) - Y_j(pr)\sin\delta_l(p) \tag{17.36}$$

In (17.35), the angular moments l_1 and l_2 are connected by the equality $l_2 = l_1 \pm 1$. To be specific, consider the case $l_2 = l_1 + 1$. We apply the well-known recurrence relations between cylindrical functions and their linear combinations similar to (17.36):

$$z W_{j+1}(z, \delta_l) = 2j W_j(z, \delta_l) - z W_{j-1}(z, \delta_l). \tag{17.37}$$

Using this formula, relation (17.35) is reduced to the following equality

$$C_{l_1+1, l_1}(p_2, p_1) = 2\sum_{n=0}^{n_{\max}}\left[\frac{2l_1 + 1 - 4n}{2p_2} S_{l_1+1/2-2n} - \frac{2l_1 - 1 - 4n}{2p_1} S_{l_1-1/2-2n}\right] + C_{\mathrm{res}}. \tag{17.38}$$

Here

$$n_{\max} = \begin{cases} (l_1/2) - 1 \text{ if } l_1 \text{ is even or } l_1 = 0 \\ (l_1 - 1)/2 \text{ if } l_1 \text{ is odd} \end{cases} \tag{17.39}$$

$$C_{\text{res}} = \begin{cases} \pi\sqrt{p_2 p_1} \int\limits_R^\infty dr\, r^2 W_{3/2}(p_2 r, \delta_2) W_{1/2}(p_1 r, \delta_1), \text{ if } l_1 \text{ is even or } l_1 = 0 \\ \pi\sqrt{p_2 p_1} \int\limits_R^\infty dr r^2 W_{1/2}(p_2 r, \delta_2) W_{3/2}(p_1 r, \delta_1), \text{ if } l_1 \text{ is odd} \end{cases} \tag{17.40}$$

while S_j is the integral is defined as

$$S_j = \int\limits_R^\infty dr \cdot r W_j(p_2 r, \delta_2) W_j(p_1 r, \delta_1), \tag{17.41}$$

which can be represented explicitly as [3]:

$$S_j = \frac{\kappa_3 R}{\kappa_4^2 - \kappa_3^2}\left[W_{j+1}(\kappa_3 R, \delta_3) W_j(\kappa_4 R, \delta_4) - W_j(\kappa_3 R, \delta_3) W_{j+1}(\kappa_4 R, \delta_4)\right] \tag{17.42}$$

Using Formulas (17.35)–(17.42), we obtain the final expression for $I_{12}^{(>)}$:

$$\begin{aligned} I_{12}^{(>)} = {} & \frac{A_{l_2}(p_2) A_{l_1}(p_1)}{2} \pi\sqrt{p_2 p_1} \\ & \times \left[2 \sum_{n=0}^{n_{\max}} \left\{ \frac{2l_1 + 1 - 4n}{2} \left[W_{l_1+3/2-2n}(p_2 R, \delta_2) \cdot W_{l_1+1/2-2n}(p_1 R, \delta_1) \right.\right.\right. \\ & \left. - \frac{p_1}{p_2} W_{l_4+3/2-2n}(p_1 R, \delta_1) \cdot W_{l_1+1/2-2n}(p_2 R, \delta_2) \right] \\ & + \frac{2l_1 + 1 - 4n}{2} \left[W_{l_1+1/2-2n}(p_1 R, \delta_1) \cdot W_{l_1-1/2-2n}(p_2 R, \delta_2) \right. \\ & \left.\left.\left. - \frac{p_2}{p_1} W_{l_4+1/2-2n}(p_2 R, \delta_2) \cdot W_{l_1-1/2-2n}(p_1 R, \delta_1) \right] \right\} \frac{R}{p_1^2 - p_2^2} + C_{\text{res}} \right] \end{aligned} \tag{17.43}$$

The explicit form C_{res} looks like this:

$$C_{res} = \frac{\sin[(p_2 + p_1)R + \delta_2 + \delta_1]}{(p_2 + p_1)^2} - \frac{\sin[(p_2 - p_1)R + \delta_2 - \delta_1]}{(p_2 - p_1)^2} + \left\{ \frac{\sin[(p_2 + p_1)R + \delta_2 + \delta_1]}{p_2 + p_1} - \frac{\sin[(p_2 - p_1)R + \delta_2 - \delta_1]}{p_2 - p_1} \right\} \frac{1}{p_2} - R\left\{ \frac{\cos[(p_2 + p_1)R + \delta_2 + \delta_1]}{p_2 + p_1} - \frac{\cos[(p_2 - p_1)R + \delta_2 - \delta_1]}{p_2 - p_1} \right\} \quad (17.44)$$

for even l_1 and

$$C_{\mathrm{res}} = \frac{\sin[(p_2 - p_1)R + \delta_2 - \delta_1]}{(p_2 - p_1)^2} - \frac{\sin[(p_2 + p_1)R + \delta_2 + \delta_1]}{(p_2 - p_1)^2} + R\left\{ \frac{\cos[(p_2 + p_1)R + \delta_2 + \delta_1]}{p_2 + p_1} - \frac{\cos[(p_2 - p_1)R + \delta_2 - \delta_1]}{p_2 - p_1} \right\} \quad (17.45)$$

for odd l_1.

In (17.38)–(17.45), the following notations are used: $\delta_2 \equiv \delta_{l_2}$, $\delta_1 \equiv \delta_{l_1}$. Equations 17.38 and (17.43)–(17.45) imply that $l_2 = l_1 + 1$. If $l_2 = l_1 - 1$, then all the corresponding formulas are obtained by replacing the indices 2 by 1 and vice versa.

In the ***bsiea*** program [4], along with the calculated integral $I_{12}^{(<)}$, the matrix element of the direct bremsstrahlung (17.7) is determined.

17.2.2.2 Matrix Element of Polarization Bremsstrahlung

This matrix element is determined by Eqs. (17.16)–(17.21). First, consider the algorithm for calculating the amplitude $F_{l_1 l_2}^{\nu_3}(\omega)$ in (17.17). The matrix element $F_{l_1 l_2}^{\nu_3}(\omega)$ is determined under the condition that ω exceeds the ν_3-subshell ionization threshold, i.e., $\omega > I_{\nu_3}$. Therefore, the real and imaginary parts of (17.17) are equal to, respectively:

$$\mathrm{Re} F_{l_1 l_2}^{\nu_3}(\omega) = -\frac{1}{3}\left[2P \int\limits_{\nu_4 > F} \frac{\omega_{43} U_{\nu_3\nu_1,\nu_4\nu_2} \mathrm{Re} D_{\nu_4\nu_3}(\omega)}{\omega_{43}^2 - \omega^2} - \pi U_{\nu_3\nu_1,\nu_4\nu_2} \mathrm{Im} D_{\nu_4\nu_3}(\omega)|_{\omega_{43} = \omega} \right], \quad (17.46)$$

$$\operatorname{Im} F^{\nu_4}_{l_1 l_2}(\omega) = -\frac{1}{3}\left[2 \int\limits_{\nu_4 > F} \frac{\omega_{43} U_{\nu_3\nu_1,\nu_4\nu_2} \operatorname{Im} D_{\nu_4\nu_3}(\omega)}{\omega_{43}^2 - \omega^2} + \pi U_{\nu_3\nu_1,\nu_4\nu_2} \operatorname{Re} D_{\nu_4\nu_3}(\omega)|_{\omega_{43}=\omega} \right] \tag{17.47}$$

Here $\omega_{43} \equiv E_4 - E_3$. The summation over the orbital angular momenta of the intermediate states in (17.46) and (17.47) is limited to terms with $l_4 = l_3 \pm 1$. Integration over the continuous spectrum is performed numerically with a step Δp_4. The complete set of included excited states is equal to k, where k_1 is the number of discrete states and $(k - k_1)$ is the number of states in the continuous spectrum. The values $k,\ k_1$ and Δp_4 have to be chosen in such a way that the transitions $(E_3, l_3) \rightarrow (E_4, l_3 \pm 1)$ would account for the main part of the oscillator strength of the main transition of the corresponding subshell.

The amplitudes (17.46) and (17.47) are determined for the frequencies

$$\omega = E_4 + I_{\nu_3}, \qquad E_4 > 0 \tag{17.48}$$

Condition (17.48) makes the calculation of the principal values of integrals much easier (see Sect. 9.2.4).

For numerical integration, it is convenient to separate singular and non-singular terms in (17.46) and (17.47). The former is found using the procedure described in Subsection 9.2.4. The latter have no singularities, and the integrals of these terms over the states of the continuous spectrum are calculated simply by the Simpson formula. This leads to the condition that the number of states in the continuous spectrum $(k - k_1)$ must be odd. The Coulomb matrix elements $\langle \nu_3\nu_1|U|\nu_4\nu_2\rangle$ and $\langle \nu_3\nu_2|U|\nu_4\nu_1\rangle$ in (17.46) and (17.47) are defined in Subsection 9.2.2.

The ***bsiea*** program provides the ability to calculate the amplitude $F^{\nu_3}_{l_1 l_2}(\omega)$ using the one-particle operator of interaction between atomic electrons and the electromagnetic field. In this case, $\operatorname{Im} D_{\nu_2\nu_1}(\omega)$ in (17.46) and (17.47) is equal to zero, and $\operatorname{Re} D_{\nu_2\nu_1}(\omega)$ is replaced by a matrix element $d_{\nu_2\nu_1}$.

Now consider the calculation of two terms $F^{\text{outer}}_{l_4 l_3}$ in (17.19). Let us take $F_{l_4 l_3}$ (17.21) as an example. After the usual change of the variable r to ρ (4.42), (17.21) [see. also (17.23)] takes the form

$$F_{l_4 l_3}(\omega)() = \frac{(-1)^{\max\{l_3, l_4\}}}{\omega^2} \sqrt{\max\{l_4, l_3\}} \int\limits_0^{\rho_{\max}} d\rho P_{\nu_4}(\rho) P_{\nu_3}(\rho) \left[\alpha r(\rho) + \beta\right]^{-2}$$
$$\times \int\limits_0^{\rho} d\rho_1 r^2(\rho_1) \left[\alpha r(\rho) + \beta\right]^{-2} \sum_{\nu_0} N_{\nu_0} P^2_{\nu_0}(\rho_1) \tag{17.49}$$

Integration over ρ and ρ_1 is done with a constant step h. The internal integral over ρ_1 is calculated for all discrete values ρ using the integration formulas for two or three points. Further, the outer integral is calculated using the Simpson formula. The calculation $\tilde{F}_{l_4 l_3}$ is performed in a similar way.

17.2.2.3 Bremsstrahlung Amplitudes on the Mass Shell

The methods for finding the matrix elements described above are formal procedures that do not take into consideration the relationship between the linear momenta p_1 and p_2, which follows from the conservation energy law (recall that energy is in Ry):

$$p_1^2 = p_2^2 + \omega \tag{17.50}$$

When this condition is fulfilled, the amplitude of the bremsstrahlung is "on the mass shell," which makes it possible to describe a real physical process. However, a certain complexity comes from this equation. As noted, the determination of the principal values of the integrals in (17.46) and (17.47) is carried out under condition (17.48).

Choosing suitable values of ω saves counting time. Note, however, that in the program ***fchf,*** which we use to calculate the wave functions of excited states of an atomic electron (p_{ν_3}) and scattered particles (p_{ν_1} and p_{ν_2}), these functions are calculated using the grid of momentum, not energy. Therefore, it is impossible for a given value of the linear momentum p_1 of the incident particle and the ionization potential I_{ν_1} to construct two sequences of energies $E_1 = p_1^2$ и $E_2 = p_2^2$, satisfying simultaneously the following two conditions:

(a) Sets p_{ν_2} and p_{ν_3} are chosen to be equidistant in the linear momentum scale;
(b) For fixed values of p_4 and I_{ν_3}, and for each frequency ω, defined as $\omega = E_4 + I_{\nu_3}$, there is a value $E_2 = p_1^2 - \omega$ belonging to a set of functions P_{ν_3}, which corresponds to zero of the energy denominator in the matrix element depicted by a diagram in (17.10).

To avoid this complication and calculate the ordinary and polarization bremsstrahlung amplitudes with account of energy conservation, the ***bsiea*** program uses the following procedure:

(1) The set $(k - k_1)$ of continuous spectrum excited states wave functions p_{ν_2} is determined equidistant in the linear momentum scale. Thus, the sets $\{ E_2^{(i)} \}$ and $\{ \omega^{(i)} = E_2^{(i)} - I_{\nu_1}, i = 1, \ldots, k - k_1 \}$ are obtained.
(2) For the energy E_1 of an incident particle, we find a set of f-wave functions $P_{\nu_2}^{(j)}(r)$ of a scattered particle, equidistant along the momentum scale. Thus, we get a set $\{ P_2^{(j)},\ j = 1, \ldots, f \}$. Now we order this set according to the magnitude of the momentum of its members, i.e., $p_2^{(1)} < \ldots < p_2^{(f)}$. The minimum value $p_2^{(1)}$, momentum step size Δp_2, and numbers f must be chosen so that for each frequency ω, the following condition is fulfilled:

$$p_1^{(1)} < \sqrt{E_1 - \omega^{(i)}} < p_1^{(f)} \tag{17.51}$$

For the data E_1, $E_3^{(i)} - E_4 - \omega^{(i)}$ exists, obviously, the energy of the scattered particle provided that a photon with a frequency $\omega^{(i)}$ is emitted.

(3) For each value of the final linear momentum of the scattered particle $p_3^{(j)}$, we calculate the amplitudes $F_{l_4 l_3}^{v_1}$, $F_{l_4 l_3}^{\text{outer}}$, and $F_{l_4 l_3}^{pol}$ according to the algorithms described above. As a result, we get sets f of matrix elements.

(4) Calculate energy values $E_2^{(i)} = E_1 - \omega^{(i)}$ for all frequencies $\omega^{(i)}$. By means of quadratic interpolation, the amplitudes of the ordinary and polarization bremsstrahlung radiation are obtained in the ***bsiea*** program from the sets obtained in (3) for physical values of the scattered particle energy $E_2^{(i)}$.

References

1. M.Ya. Amusia, Atomic Bremsstrahlung. Phys. Rep. **162**(5), 249–335 (1988)
2. L.V. Chernysheva, N.B. Avdonina, M.Ya. Amusia, M.Y. Kuchiev, Numerical calculation of fast-electron induced atomic bremsstrahlung. (Ioffe Institute, No. 865, Leningrad, 1983) (In Russian)
3. A.P. Prudnikov, Y.A. Brychkov, O.I. Marychev, *Integrals and Series*; *Special Functions*, vol 1–3, (Gordon and Breach Science Publishers, New York 1986)
4. L.V. Chernysheva, N.B. Avdonina, M.Ya. Amusia, A.V. Korol, Numerical calculation of the spectrum of bremsstrahlung induced by intermediate energy particles (Ioffe Institute, No. 1314, Leningrad, 1989) (In Russian)

Chapter 18
Photoabsorption Cross Section with Account for Inelastic Photoelectron Scattering

This chapter describes a method for calculating the cross section of a process in which a photoelectron on its way from an atom collides inelastically with the residual ion, thus changing its own energy, angular momentum, and spin projection. Formulas are given for calculating the photoabsorption dipole matrix elements and the photoionization cross section taking this effect into account.

18.1 Theoretical Model

Within the framework of RPAE, it is assumed that the photoelectron moves in the field of a frozen core with a vacancy formed due to photoionization. However, in fact, the field acting upon the photoelectron is not identical to the Hartree–Fock field due to the virtual excitations of the electrons of the residual ion, that is, the polarization of the ion by the photoelectron. This field variation is taken into consideration by introducing the self-energy operator $\hat{\Sigma}$ (see Chap. 14). In the language of diagrams, the photoionization amplitude with allowance for $\hat{\Sigma}$ is represented by the following sequence:

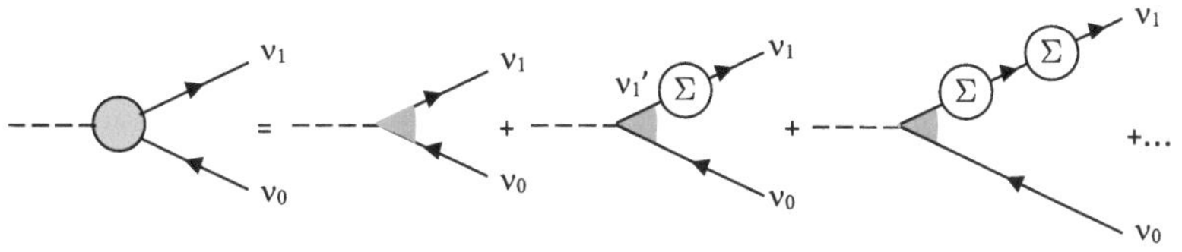

(18.1)

As in the case of elastic scattering of an electron by an atom (see Chap. 14), $\hat{\Sigma}$ is an addition to the static atomic potential acting on an electron.

Let us describe the equation for the photoelectron wave function in integral form. We assume that the electron moves in a centrally symmetric field, so that it is easy and

M. Ya. Amusia and L. V. Chernysheva, *Computation of Atomic and Molecular Processes*, Springer Series on Atomic, Optical, and Plasma Physics 117,
https://doi.org/10.1007/978-3-030-85143-9_18

convenient to separate the angular and spin parts of its wave function. Therefore, the integral equation for the radial components of the wave function has a form similar to (14.24):

$$P_{El}^{\Sigma}(r) = P_{El}(r) - \int_0^\infty \int_0^\infty G_l^0\left(r, r', E\right)\Sigma_l^{\rm cor}\left(r', r'', E\right)P_{El}^{\Sigma}\left(r''\right)dr'dr''. \quad (18.2)$$

Here $P_{El}(r)$ is the radial part of the wave function in the HF approximation, $G_l^0\left(r, r', E\right)$ is the radial part of the Green's function in the zero approximation, i.e., in the HF field, $\Sigma_l^{\rm cor}\left(r', r'', E\right)$ is the irreducible self-energy part of the one-particle Green's function. The explicit form of $G_l^0\left(r, r', E\right)$ is given by the formula

$$G_l^0\left(r, r', E\right) = \int_{E'} \frac{P_{E'l}(r)P_{E'l}\left(r'\right)}{E - E' + i\delta}. \quad (18.3)$$

Integration over the continuous spectrum E' here also implies summation over discrete single-particle energy levels.

The equation for the radial part of Green's function, known as the Dyson equation, is:

$$G_l^{\Sigma}\left(r, r', E\right) = G_l^0\left(r, r', E\right) - \int_0^\infty \int_0^\infty G_l^0\left(r, r'', E\right)\Sigma_l^{\rm cor}\left(r'', r''', E\right)G_l^{\Sigma}\left(r''', r', E\right)dr''dr''' \quad (18.4)$$

After replacing the wave function of the final state in the matrix element of the dipole operator in RPAE by P_{El}^{Σ}, (18.2) leads to the following equations:

$$\left\langle il_i\middle|D^{\Sigma}(\omega)\middle|El\right\rangle = \langle il_i|D(\omega)|El\rangle + \int_{E'} \frac{\langle il_i|D(\omega)|E'l\rangle\left\langle E'l\middle|\tilde{\Sigma}(\omega)\middle|El\right\rangle}{E - E' + i\delta},$$

$$\left\langle E'l\middle|\tilde{\Sigma}(\omega)\middle|El\right\rangle = \left\langle E'l\middle|\Sigma^{\rm cor}(\omega)\middle|El\right\rangle + \int_{E''} \frac{\langle E'l|\Sigma^{\rm cor}(\omega)|E''l\rangle\left\langle E''l\middle|\tilde{\Sigma}(\omega)\middle|El\right\rangle}{E - E' + i\delta}. \quad (18.5)$$

Here $\tilde{\Sigma}^l(\omega)$ is the reducible self-energy part of the one-electron Green's function, $El,\ E'l,\ il_i$ are states of $\nu_1,\ \nu_1'$-photoelectron, and ν_0-vacancy, respectively.

Note that for calculating the photoionization cross section, Eq. (18.5) is more convenient, which leads to complex values of D^{Σ} и Σ. They differ from (14.24) to (14.29) by imaginary additions $i\delta$ in the denominator.

Since the photoelectron in the case of atomic ionization moves in the Coulomb field of the ion, the asymptotic of the radial wave function instead of (14.17) is determined by the expression

$$P_{El}^{\Sigma}(r) \sim \frac{1}{\sqrt{\pi p}} \sin\left(pr - \frac{\pi l}{2} + \delta_l^{\text{HF}}(E) + \Delta\delta_l(E) + \frac{1}{p} \ln 2pr \right) \qquad (18.6)$$

with an additional Coulomb phase $(1/p)\ln 2pr$, where $p = \sqrt{E}$, and $\Delta\delta_l(E)$ is the correction to the scattering phase due to action of the polarization potential $\langle El \sum^{cor}(E) E'l \rangle$.

Through formulas (18.2) and (18.5), the additional phase shift $\Delta\delta_l(E)$ can be associated with the diagonal matrix element $\left\langle El \middle| \tilde{\Sigma}(E) \middle| E'l \right\rangle$. Note that (14.29) is another definition of a potential, similar to $\langle El | \overline{\Sigma}(E) | El \rangle$ that facilitates the calculation of the scattering phases, but not the photoionization cross section. The procedure for solving Eq. (18.5) is not identical to the definition $\langle El | \overline{\Sigma}(E) | El \rangle$ given in (14.29). The matrix element $\left\langle El \middle| \tilde{\Sigma}(E) \middle| E'l \right\rangle$ is related to the phase $\Delta\delta_l(E)$ as follows:

$$\exp[i\Delta\delta_l(E)] \sin[\Delta\delta_l(E)] = -\pi \left\langle El \middle| \tilde{\Sigma}(E) \middle| El \right\rangle \qquad (18.7)$$

This formula is not identical to (14.28).

The one-electron photoionization cross section that accounts for the scattering of the photoelectron by the residual ion is expressed in terms of the dipole matrix element $\langle il_i | D^{\Sigma}(\omega) | El \rangle$ according to the following formula:

$$\sigma_i^+ = \frac{4\pi^2 \alpha a_0^2 \omega N_1}{3(2l_i + 1)} \left[\left| \langle il_i | D^{\Sigma}(\omega) | E, l+1 \rangle \right|^2 + \left| \langle il_i | D^{\Sigma}(\omega) | E, l-1 \rangle \right|^2 \right] \qquad (18.8)$$

where α is the fine structure constant, a_0 is the Bohr radius, l_i is the orbital moment of the ionized shell, N_1 is the number of electrons in it.

Knowledge of all dipole matrix elements $\langle i | D^{\Sigma}(\omega) | E \rangle$, i.e., not only those that lie on the energy shell, i.e., meet the condition $E = \omega + E_i$, allows the total photoabsorption cross section to be calculated. According to the optical theorem (see, e.g., [LL] or [1]), photoabsorption is proportional to the imaginary part of the photon elastic scattering amplitude by an atom:

$$\sigma_{\text{tot}} = (4\pi/\omega) \text{Im} f_{\text{el}}(\omega) \qquad (18.9)$$

The photon elastic scattering amplitude is described by the following diagrammatical equation:

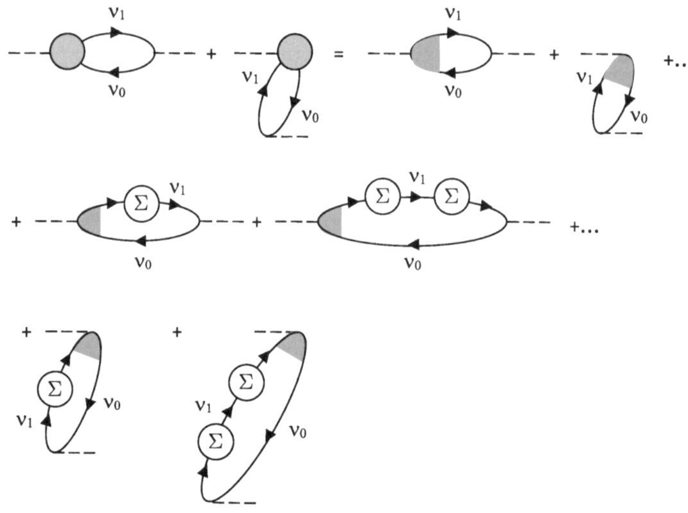

(18.10)

Neglecting $\Sigma^{cor}(\omega)$ for the photoelectron, the amplitude of elastic scattering of a photon can be described by the first two terms on the right-hand side of (18.10). Then the cross section related to the amplitude by Eq. (18.9) will be identical to the photoionization cross section in RPAE.

The optical theorem is based on the so-called unitarity relation. The latter require that the shape of the diagram sequence (18.10) does not change with the permutation of the left and right vertexes in diagrams similar to those shown on the left side of (18.10). This requirement is obviously fulfilled within RPAE, which clearly follows from the graphical definition of the photon scattering amplitude in RPAE (18.11):

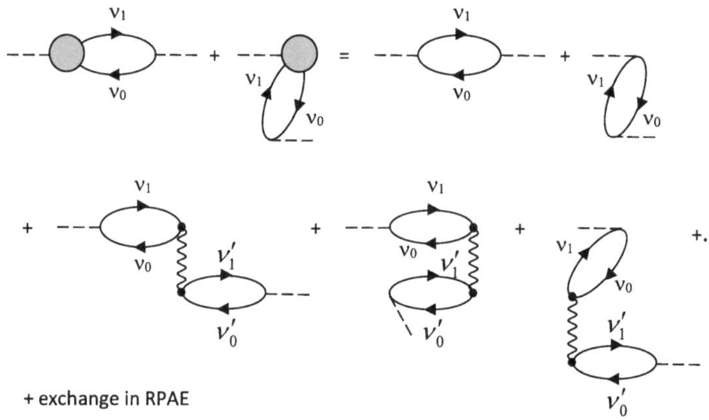

(18.11)

The graphic Eq. (18.10) is asymmetric with respect to the permutation of the right and left vertexes in its left side. Therefore, (18.10) is only an approximate amplitude of elastic photon scattering. To obtain the exact amplitude, Eq. (18.10) must be changed by introducing self-energy corrections to the photoabsorption diagrams in the RPAE (18.11).

To do this, the internal lines should be changed in such a way that they described one-electron states no longer in the HF approximation, but including also the corrections shown in figure (18.1) and accounted for by Eq. (18.2). The larger the error in the description of photoabsorption in the framework of the RPAE, the larger is the contribution of corrections of the type (18.1) to the photoionization amplitude.

Consider the ionization of the intermediate or inner shell. The absorption of a photon and the subsequent scattering of a photoelectron in this case are separated in space. The distance scale is determined by the size of the ionizable shell. Therefore, the entire ionization process can actually be divided into two stages. First, the electrons of the intermediate or inner subshell absorb a photon, and one of them leaves the subshell, subsequently being scattered by the outer electrons of the atom. As a result of this scattering, another atomic electron can be excited or knocked out. Within this approximation, the total photoionization cross section is determined by the result of the calculation in the RPAE (see Subsection 9.2.3). In fact, it was found in the studied Xe and Ba atoms, that the corrections to the total cross section were small, since the data calculated without them in RPAE were in good agreement with the experimental results.

Including the contribution of Σ, the cross section has the following form:

$$\sigma_{l\to l'}^{\rm tot} = -\frac{4\pi\alpha a_0^2\omega N_1}{3(2l_i+1)}{\rm Im}\left[\int\limits_{E'}\frac{\langle il_i\|d\|E'l'\rangle\langle E'l'\|D^{\Sigma}(\omega)\|il_i\rangle}{E-E'+i\delta}\right] \tag{18.12}$$

where $< il_i\|d\|E'l' >$ is the dipole matrix element in the Hartree–Fock approximation, $l' = l_i + 1$ or $l' = l_i - 1$ (cf. (9.6) and discussion there).

The amplitude of elastic scattering of a photon of frequency ω by an atom is proportional to its dynamic dipole polarizability $\alpha(\omega)$ [1].

In the Hartree–Fock approximation, $\alpha(\omega)$ is the sum of the polarizability of each atomic shell i:

$$\alpha_i(\omega) = -\frac{2N_1a_0^3}{3(2l_i+1)}\int\limits_{E'}\left\{\frac{(I_i+E')}{\omega^2-(I_i+E')^2+i\delta}\right.$$
$$\left.\left[\left|\langle n_il_i\|d\|E',l_i+1\rangle\right|^2+\left|\langle n_il_i\|d\|E',l_i-1\rangle\right|^2\right]\right\} \tag{18.13}$$

In RPAE, which is used here, and including also the self-energy part of the photoelectron, we have for the contribution to the dipole polarizability of one transition $l_i \to l'$ the following relation:

$$\alpha_{l_i \to l'}(\omega) = -\frac{N_1 a_0^3}{3(2l+1)} \int\limits_{E'} \left[\frac{\langle n_i l_i \| d \| E'l' \rangle \langle E'l'_i \| D^{\Sigma}(\omega) \| n_i l_i \rangle}{\omega - (I_i + E') + i\delta} + \frac{\langle n_i l_i \| D^{\Sigma}(\omega) \| E'l' \rangle \langle E'l' \| d \| n_i l_i \rangle}{\omega + (I_i + E') + i\delta} \right],$$

$$\alpha(\omega) = \sum_{i, l' = l \pm 1} \alpha_{l_i \to l'}(\omega). \tag{18.14}$$

Note that due to the unitarity relation, the real and imaginary parts of the dipole polarizability are related as follows:

$$\mathrm{Re}\alpha(\omega) = \frac{2}{\pi} P \int \frac{\omega' \mathrm{Im}\alpha(\omega')}{\omega'^2 - \omega^2} d\omega', \tag{18.15}$$

where P is the symbol of the principal value.

Equation (18.15) allows to control whether the unitarity relation holds in the approximation used, and, accordingly, whether (18.12) is correct in this case. The adequacy of the used approximation for individual atoms is verified by the agreement reached between the results of calculations in the *length* and *velocity* forms.

18.1.1 Numerical Model

18.1.1.1 General Consideration

To find $\hat{\Sigma}^{kop}$ in the second order of the perturbation theory, it is necessary to account for the contribution of four diagrams:

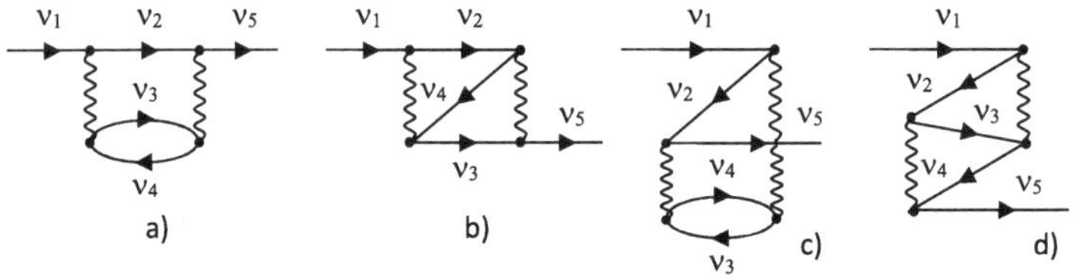

(18.16)

Equation (14.4) is the analytical expression for the contribution of these diagrams. The calculation is performed in the ***scat*** program [2].

Phosep program [3] accounts for multiple transitions; several different intermediate states describing excitation with different initial ν_4 and different final states ν_3 (18.16).

Unlike Chap. 14, which deals with the scattering of an electron by an atom, we now consider a photoelectron undergoing scattering by a positive ion. Therefore, among the states ν_1, ν_5 (18.16), discrete excited levels should also be included.

However, if, instead, we consider negative ion photoionization with account for inelastic photoelectron scattering, there is no discrete excited levels among the states ν_1, ν_5 (18.16).

The result essentially depends on the choice of wave functions in $\hat{\Sigma}$ (18.16), just as in the final and intermediate states of diagrams (18.1). As noted in Chap. 6, using the wave functions of excited states calculated in the field of the "frozen" core of a singly charged ion as the wave function of a particle is equivalent to summing up an infinite series of time-forward RPAE diagrams where the hole state remains the same that is exemplified in (18.17):

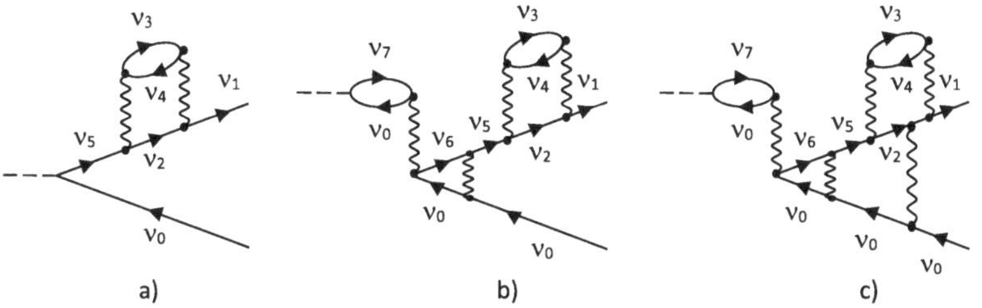

(18.17)

All sets of excited states ($\nu_1, \nu_2, \nu_3. \nu_5$ and so on) are calculated independently of each other. Summation is performed over all of these states, except for ν_0.

In such a way, different diagrams in higher orders of perturbation theory can be included. For example, by using the function $P^{N(LS)}$ obtained from the Hartree–Fock Eq. (6.27) in the field of a frozen core with a hole in the shell ν_0 (see Subsection 8.1.1) takes into account an infinite series of RPAE diagrams diagonal in the hole ν_0, like (18.17a, b) (see Sect. 8.1 and (1.50b)). It is difficult to include the action of a vacancy ν_4 upon a photoelectron in an intermediate state ν_2. However, this can be reasonably neglected, since the electron ν_3 screens the ν_4 hole field, preventing its action upon photoelectron ν_2. States ν_2 are usually described by the same wave functions $P^{N(LS)}$ as states ν_1, ν_5. Thus, diagrams similar to (18.17b) are taken into account.

The choice of the function describing the state ν_3 can either include or ignore the formation of a hole ν_0 as a result of photon absorption.

Neglecting the influence of the field of an internal vacancy on a vacancy ν_4 in the outer shell and on an electron ν_3 excited from it is equivalent to the assumption that the field of a hole ν_0 is screened by photoelectron ν_2. As in the case of scattering, this approach can be called the simplified random phase approximation with exchange, or SRPAE (see Sect. 1.7).

Alternatively, as the field in which the photoelectron moves one can consider the self-consistent field of the ion formed as a result of photoionization of the initial atom which becomes rearranged earlier than the photoelectron leaves the atom. This approach is equivalent to the rearrangement approximation (GRPAE, see Sect. 1.8). In this case, the interaction of a hole ν_4 and an electron ν_3 with a hole ν_0 is also taken into account:

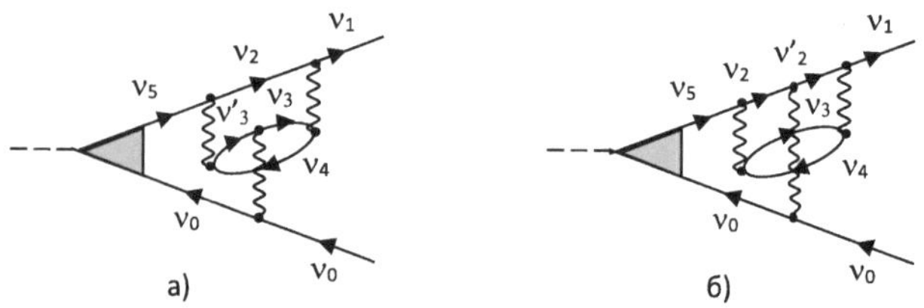

(18.18)

The slower the photoelectron, the more accurate is this approximation. Therefore, it turns out to be valid near the ν_0-subshell threshold. As a rule, the rearrangement in calculations of the functions included in the self-energy part must be taken into account for the same range of photon energies, as in the case of GRPAE. The thresholds of ν_0- and ν_4-subshells are either taken from experiment or calculated in the limit of a completely rearranged ion.

18.1.1.2 Irreducible Self-energy Part and Effective Dipole Matrix Elements

Equation (18.5) for $\tilde{\Sigma}$ is solved by replacing its integral over the continuous spectrum with a sum over discrete energy levels (see Sect. 9.2). As a result, Eq. (18.5) takes the following form:

$$\begin{aligned}\left\langle E''l\middle|\tilde{\Sigma}(\varepsilon)\middle|E'l\right\rangle &= \langle E''l|\Sigma(\varepsilon)|E'l\rangle \\ &+ \sum_i Q_i D_i \langle E''l|\Sigma(\varepsilon)|E_i l\rangle \left\langle E_i l\middle|\tilde{\Sigma}(\varepsilon)\middle|E'l\right\rangle \\ &- i\pi \langle E''l|\Sigma(\varepsilon)|\varepsilon l\rangle \left\langle \varepsilon l\middle|\tilde{\Sigma}(\varepsilon)\middle|E'l\right\rangle \end{aligned} \tag{18.19}$$

This equation differs from (14.52) by the presence of the imaginary term on its right side. In (18.19) the factor D_i is defined by the same expression as in Chap. 14 (cf. (14.53)):

$$D_i = \frac{2p_i}{\varepsilon - p_i^2} \tag{18.20}$$

for i, corresponding to the continuous spectrum states, and

$$D_i = \frac{1}{\varepsilon - E_i} \tag{18.21}$$

for i, corresponding to discrete states; p_i is the linear momentum of the photoelectron, Q_i are the coefficients in the Simpson method.

The quantities $\langle El|\Sigma(\varepsilon)|E'l'\rangle$ form the $(k \times k)$ matrix, with $k = k_1 + k_2$, where k_1 is the number of discrete states and k_2 is the number of states of the continuous spectrum. For ε exceeding the ionization threshold of the outer atomic subshell, the values $\Sigma(\varepsilon)$ are complex: $\Sigma(\varepsilon) = \Sigma'(\varepsilon) + i\Sigma''(\varepsilon)$. After separating the real and imaginary parts in (18.19), we obtain the following system of equations, which in matrix form can be represented similarly to (14.55):

$$\begin{pmatrix} \tilde{\Sigma}' \\ \tilde{\Sigma}'' \end{pmatrix} = \begin{pmatrix} \Sigma' \\ \Sigma'' \end{pmatrix} + \begin{pmatrix} A & -B \\ B & A \end{pmatrix} \begin{pmatrix} \tilde{\Sigma}' \\ \tilde{\Sigma}'' \end{pmatrix} \tag{18.22}$$

However, in contrast to case (14.56), in this equation A and B include imaginary parts:

$$\begin{aligned} A_{ij}(\varepsilon) &= \langle E_i l|\Sigma'(\varepsilon)|E_j l\rangle Q_j D_j + \pi\delta_{j\varepsilon}\langle E_i l|\Sigma'(\varepsilon)|E_j l\rangle \\ B_{ij}(\varepsilon) &= \langle E_i l|\Sigma''(\varepsilon)|E_j l\rangle Q_j D_j - \pi\delta_{j\varepsilon}\langle E_i l|\Sigma''(\varepsilon)|E_j l\rangle \end{aligned} \tag{18.23}$$

Equation (18.22) can be solved, permitting to obtain $\tilde{\Sigma}'(\varepsilon)$ и $\tilde{\Sigma}''(\varepsilon)$:

$$\begin{pmatrix} \tilde{\Sigma}' \\ \tilde{\Sigma}'' \end{pmatrix} = \begin{pmatrix} 1-A & B \\ -B & 1-A \end{pmatrix}^{-1} \begin{pmatrix} \Sigma' \\ \Sigma'' \end{pmatrix} \tag{18.24}$$

Thus, the solution of Eq. (18.5) is reduced to the inversion of the $2k \times 2k$. The equation is solved sequentially for all values $\varepsilon = E_i$ in the continuous spectrum, except for the first and the last ones, where the pole coincides with the integration limits and the numerical integration formulas that we use are not applicable. For each value ε, the phase correction is calculated in accordance with (18.7):

$$\mathrm{Re}\Delta\delta_l = -\frac{1}{2}\arcsin\left(\frac{\pi a}{\sqrt{A}}\right), \quad \mathrm{Im}\Delta\delta_l = -\frac{1}{4}\ln\left(4A^2\right) \tag{18.25}$$

where

$$\begin{aligned} A &= \pi^2 a^2 + \left(\pi b + \frac{1}{2}\right)^2 \\ a &= \left\langle E_i l\left|\tilde{\Sigma}'(E_i)\right|E_i l\right\rangle, \; b = \left\langle E_i l\left|\tilde{\Sigma}''(E_i)\right|E_i l\right\rangle \end{aligned} \tag{18.26}$$

The effective dipole matrix elements are calculated according to Eq. (18.5) with the precalculated dipole matrix elements $D_{i\varepsilon'}(\varepsilon)$ in RPAE (for a detailed description of the procedure for calculating the dipole matrix elements $D_{i\varepsilon'}(\varepsilon)$ in the RPAE "on the energy shell," see Subsection 9.2.4). The same procedure is actually used for the calculation of RPAE dipole matrix elements $D_{i\varepsilon'}(\varepsilon)$ "off the energy shell." Integration over the states of the continuous spectrum in (18.5) is performed by replacing the continuous variable with a discrete one. Further, the standard procedure according to the Simpson method is applied, the same as for integrals in Eqs. (18.12) and (18.15).

So, the ***phosep*** program [3, AC] allows to calculate the dipole matrix elements of photoabsorption and photoionization cross sections accounting for the photoelectron scattering with other atomic electrons on its way off the atom, as well as single photoionization of intermediate atomic shells with account of the self-energy part.

References

1. V.B. Berestetskii, E.M. Lifshits, L.P. Pitaevskii, *Quantum Electrodynamics* (Butterworth-Heinemann, Oxford, 1982)
2. L.V. Chernysheva, M.Ya. Amusia, D. Davidovich, N.A. Cherepkov, Numerical calculation of scattering phases in simplified version of the random phase approximation with exchange (Ioffe Institute, No. 663, Leningrad, 1980) (In Russian)
3. L.V. Chernysheva, M.Ya. Amusia, K.L Tsemekhman, Numerical calculation of the cross sections for photoabsorption and single photoionization of intermediate atomic shells with account of the photoelectron self-energy part (Ioffe Institute, No. 1492, Leningrad, 1990) (In Russian)

Chapter 19
Ionization and Excitation of an Atom by Electron Impact

This chapter shows how, using the wave functions of the ground and excited single-electron states, as well as the wave functions of an incident and inelastically scattered electron, one can calculate the cross section for ionization and excitation of an atom by electron impact in the Born approximation with plane and distorted waves. In the latter case, the effect of the atomic field is accounted for within the Hartree–Fock approximation. It is shown how to calculate the cross section for the excitation of triplet levels by electrons.

19.1 Theoretical Model

19.1.1 Ionization of an Atom by Electron Impact

The process of ionization of an atom by electron impact in Born approximation with account for exchange is depicted by the following diagrams:

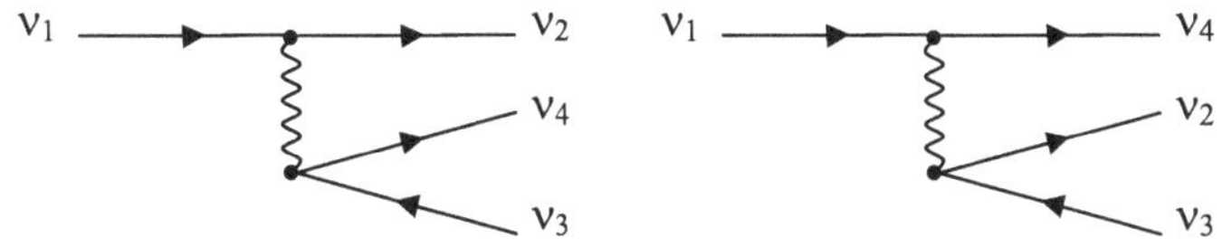

(19.1)

The triple-differential ionization cross section, which fixes the energy of one of the electrons and the angle of emission of both ν_2, ν_4 electrons in the distorted waves Born approximation (DWBA) is:

$$\frac{d^3\sigma}{d\Omega_2 d\,\Omega_4\, dE_4} = \frac{p_4 p_2}{2(2\pi)^5 p_1} \sum_{\sigma_1\sigma_2\sigma_3\sigma_4} \left| \left\langle \nu_4^{(-)} \nu_2^{(-)} \right| U \left| \nu_1^{(+)} \nu_3 \right\rangle \right|^2 \qquad (19.2)$$

M. Ya. Amusia and L. V. Chernysheva, *Computation of Atomic and Molecular Processes*, Springer Series on Atomic, Optical, and Plasma Physics 117,
https://doi.org/10.1007/978-3-030-85143-9_19

Here, $\Omega_{2,4}, E_4 = p_4^2, p_2$, respectively, are the emission angle, energy and momentum ($E_1 = p_1^2$) of the incident ν_1, scattered ν_4, knocked-out ν_2 and incident ν_4 electrons in the field of an ion with a vacancy ν_3.

Wave functions $\left|\nu^{(\pm)}\right\rangle \equiv \phi_\nu^{(\pm)}(x)$ are represented as a superposition of products of radial, angular, and spin wave functions. The radial wave function of the electron in the continuous spectrum $P^{(\pm)}(r)$ is normalized, like (6.15), to the delta function of the linear momentum. Its asymptotic for $r \to \infty$ is represented as a combination of a plane wave and converging (+) or diverging (−) spherical waves. The Coulomb interaction U in (19.2) includes the direct and exchange parts, which corresponds to inclusion of both singlet ($S = 0$) and triplet ($S = 1$) scattering. Summation over projections of particles spins in the final state and averaging in the initial state correspond to scattering and registration of unpolarized electrons, i.e., without fixing the orientation of the spin.

One-electron wave functions can be calculated as $\phi^N \phi^{N+1}$, $\phi^{N(LS)}$ or accounting for rearrangement processes. Functions $\phi_{\nu_4}^{(+)}\phi_{\nu_1}^{(-)}$, $\phi_{\nu_2}^{(-)}$ are able to include atomic field effect (see Chaps. 6 and 8). The wave function of the incident $\phi_{\nu_4}^{(+)}$ is a solution of the Hartree–Fock Eqs. (6.2), (6.3). It is calculated in the field of a neutral atom $\phi_{\nu_4}^{N+1}$.

The wave function of the knocked-out electron $\phi_{\nu_2}^{(-)}$ is calculated in the field of a single ion with a vacancy $\nu_3 - \phi_{\nu_3}^N$. The wave function of a scattered electron $\phi_{\nu_1}^{(-)}$ can be calculated both in the field of a neutral atom ($\phi_{\nu_1}^{N+1}$ for a fast electron) and in the field of an ion ($\phi_{\nu_1}^{N(LS)}$ for a slow electron), depending on whether screening of the ion field by knocked-out electron seems to be significant.

Excitation of an atom by electron impact in the Born approximation is described by the same diagrams as its ionization (19.1). However, in this case, the state of the atomic electron after scattering ν_2 belongs to the discrete spectrum.

The differential cross section for the excitation of an atom by an electron impact in the DWBA is

$$\frac{d\sigma}{d\Omega_4} = \frac{1}{2(2\pi)^2}\frac{p_4}{p_1}\sum_{\sigma_1\sigma_2\sigma_3\sigma_4}\left|\left\langle \nu_1^{(-)}\nu_2|U|\nu_4^{(+)}\nu_3\right\rangle\right|^2 \tag{19.3}$$

If both the incident and scattered electrons are fast, then the action of the atomic field on them can be neglected. In this case, the wave functions $\phi_{\nu_4}^{(+)}$, $\phi_{\nu_1}^{(-)}$ become plane waves $\exp(i\mathbf{kr})$. The triple-differential ionization cross section of an atom in the plane wave Born approximation (PWBA) is (see Chap. 10):

$$\frac{d^3\sigma}{d\Omega_2\, d\Omega_4\, dE_2} = \frac{4p_2}{q^4 p_1}\left|\left\langle \nu_4^{(-)}|\exp(i\mathbf{qr})|\nu_3\right\rangle\right|^2 \tag{19.4}$$

Here, $\mathbf{q} = \mathbf{p}_2 - \mathbf{p}_1$ is the linear momentum transferred to the atom in the process of inelastic scattering. In PWBA, the exchange between the knocked-out and scattered electrons is neglected and only singlet scattering takes place.

We emphasize that in (19.4), in contrast to expression for the GOS (10.8), there is a matrix element in which the state of the outgoing electron is specified not by energy and angular momentum, but by linear momentum, i.e., by energy and direction of electron emission.

The differential cross section for excitation of an atom in PWBA is also determined by formula (19.4); however, $\phi_{\nu_2}^{(-)}$ is replaced by the one-electron wave function of excited state ϕ_{ν_2}, which is always characterized by principal quantum number and angular momentum. In this case, (19.4) reduces to (10.2).

19.1.2 *Ionization of an Atom at the Threshold of Triplet Autoionizing Level Excitation with Allowance for "Post-collision Interaction"*

By autoionization decay, we mean the decay of a discrete excited state, the energy of which exceeds the ionization potential.

In the lowest order in interelectron interaction, excitation of the triplet autoionization state with its subsequent decay is described by the following diagram:

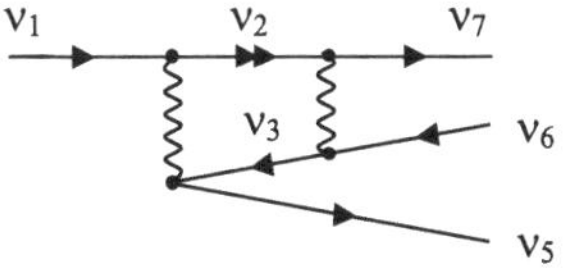

(19.5)

Inclusion of *post-collision interaction* (PCI) into consideration implies taking into account the interaction between the slow electron ν_5 and atomic field, which alters due to autoionization decay, and is carried out by summing the following infinite series of diagrams:

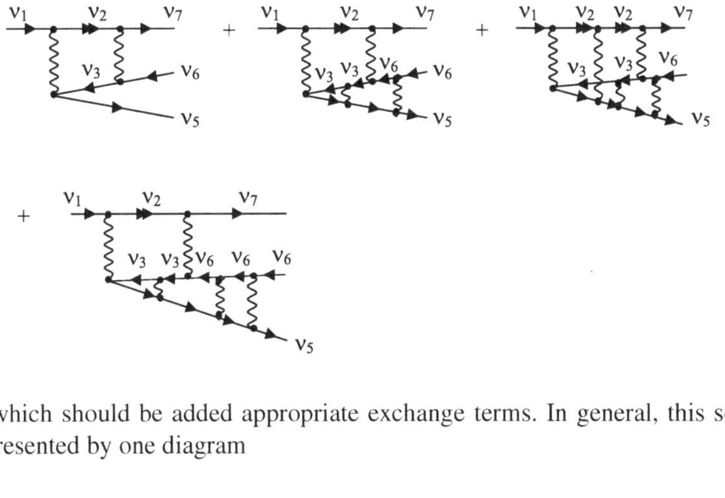

(19.6)

to which should be added appropriate exchange terms. In general, this set can be represented by one diagram

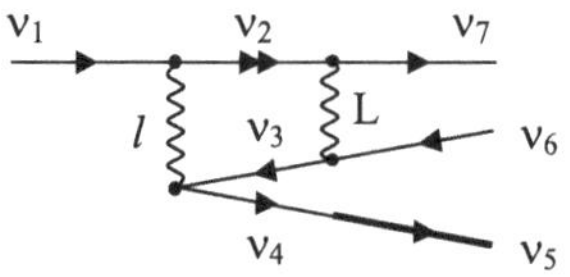

(19.7)

where the thick electron line ν_5 corresponds to an electron moving in the field of an ion with vacancy ν_6, and not ν_3, while the thin line ν_4 depicts an electron in the field of an atom excited into an autoionization state $\nu_3^{-1}\nu_2$. Their junction corresponds to the overlap integral of the corresponding wave functions $\langle \nu_4 | \tilde{\nu}_5 \rangle$.

19.1.3 Excitation of a Triplet Level by Electrons

In the first order of DWBA, the triplet level excitation of an atom by electrons in diagrammatic language is represented as follows:

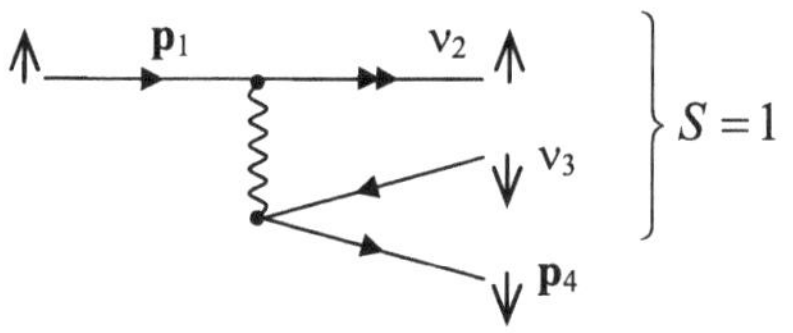

(19.8)

Here, the incident electron linear momentum is $\mathbf{p}_1$, and $\mathbf{p}_4$ is the linear momentum of the knocked-out electron; the arrows $\uparrow\downarrow$ indicate the directions of their spins. Since the diagram considers triplet level excitation in first order of approximation, the exchange diagram does not contribute to (19.8).

The differential excitation cross section, the amplitude of which is represented by diagram (19.8), is

$$\frac{d\sigma}{dE_4\, d\Omega_4} = 2\pi p_4 |\langle \mathbf{p}_1\, \nu_3 | V | \nu_2\, \mathbf{p}_4 \rangle|^2. \tag{19.9}$$

Here, Ω_4 is the solid emission angle of the emitted particle, $E_4 = p_4^2$, $\langle \mathbf{p}_1\, \nu_3 | V | \nu_2\, \mathbf{p}_4 \rangle$ is the Coulomb matrix element between the one-electron wave functions of the continuous spectrum and discrete one-electron states. The level energy satisfies energy conservation law:

$$E_1 - E_4 - E_2 + E_3 = 0 \tag{19.10}$$

The radial wave function $P_{\nu_2,\nu_3}(r)$ is normalized in energy scale, and $\phi^{(+)}_{\mathbf{p}_1}$ and $\phi^{(-)}_{\mathbf{p}_4}$ are normalized in the linear momentum $\mathbf{p}$ scale.

In the second order of DWBA, there are corrections to the process described by the following diagram:

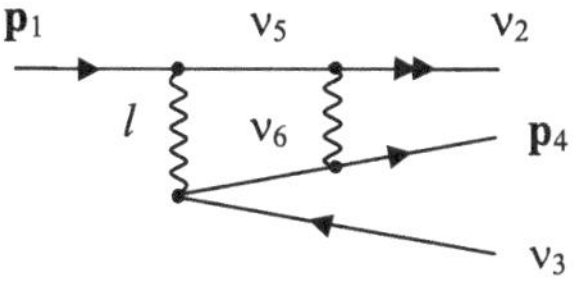

(19.11)

This process accounts for the contribution of the off-diagonal matrix element of the interaction between the scattered and knocked-out electrons [1]. The amplitude of triplet level excitation in the second order of perturbation theory $\langle \mathbf{p}_1\, \nu_3 | V | \nu_2\, \mathbf{p}_4 \rangle_{\mathrm{II}}$ is determined by the sum of corresponding matrix elements. The cross section is given by expression (19.9) with matrix elements

$$\langle \mathbf{p}_1\, \nu_3 | V | \nu_2\, \mathbf{p}_4 \rangle_{\mathrm{II}} = \langle \mathbf{p}_1\, \nu_3 | V | \nu_2 \mathbf{p}_4 \rangle + \int\limits_{\mathbf{p}_5, \mathbf{p}_6} \frac{\langle \mathbf{p}_1\, \nu_3 | V |\, \mathbf{p}_5\, \mathbf{p}_6 \rangle \langle \mathbf{p}_5\, \mathbf{p}_6 | V | \nu_2 \mathbf{p}_4 \rangle}{E_1 - E_5 - E_6 + E_3 + i\eta} \tag{19.12}$$

instead of just the first term $\langle \mathbf{p}_1\, \nu_3 | V | \nu_2\, \mathbf{p}_4 \rangle$. Expanding the continuous spectrum electrons wave functions into partial waves, carrying out subsequent integration over angular variables and summation over the projections of the angular momentum and spin (19.9), one obtains

$$\frac{d^2\sigma}{dE_4 d\Omega_4} = \frac{4\pi}{p_1^2} \sum_{m_3 m_2} \Bigg| \sum_{l_1 l_4 l, m_4 m} (-1)^{m_3+m+l_1+l_4}\, e^{-i(\delta_{l_1}+\delta_{l_4})} \times \sqrt{2l_1+1} \begin{pmatrix} l_1 & l_2 & l \\ 0 & m_2 & -m \end{pmatrix} \begin{pmatrix} l & l_4 & l_1 \\ -m & m_4 & -m_1 \end{pmatrix} \langle \nu_1 \nu_3 \| V \| \nu_2 \nu_4 \rangle_{\mathrm{II}} Y^*_{l_4 m_4}(\Omega_4) \Bigg|^2, \tag{19.13}$$

where δ_{l_1}, δ_{l_4} are elastic scattering phases of electrons in the field of an atom or ion, respectively. The matrix element $\langle \nu_1 \nu_3 \| V \| \nu_2 \nu_4 \rangle_{\mathrm{II}}$ is defined as follows:

$$\begin{aligned} \langle \nu_1 \nu_3 \| V \| \nu_2 \nu_4 \rangle_{\mathrm{II}} = {} & \langle \nu_1 \nu_3 \| V \| \nu_2 \nu_4 \rangle \\ & - \sum_{l_5 l_6 l' L} (-1)^{l_4+l_5+l_6+l+l_3} (2l+1) \begin{Bmatrix} l_4 & l & l_3 \\ L & l_6 & l' \end{Bmatrix} \begin{Bmatrix} l & l_4 & l_1 \\ L & l_5 & l' \end{Bmatrix} \\ & \int\limits_{\nu_5 \nu_6} \frac{\langle \nu_1 \nu_3 \| V_L \| \nu_5 \nu_6 \rangle \langle \nu_5 \nu_6 \| V_{l'} \| \nu_2 \nu_4 \rangle}{E_1 - E_5 - E_6 + E_3 + i\eta}. \end{aligned} \tag{19.14}$$

This relation accounts for the triplet excitation processes in the first (first term) and second (subsequent terms) order in DWBA corresponding to diagrams (19.8) and (19.11). In terms of diagrams, expression (19.14) corresponds to the sum of the diagrams

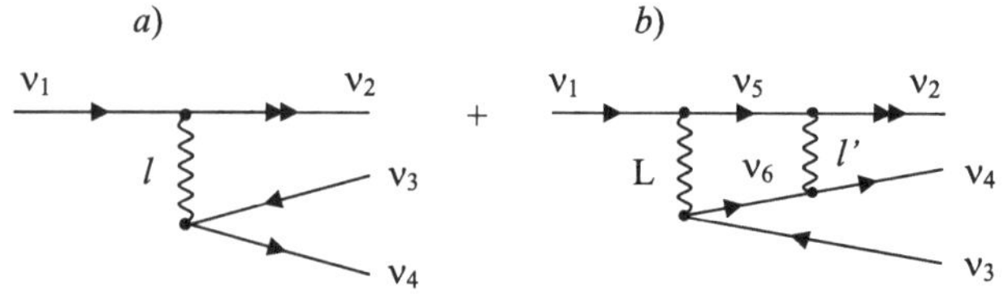

(19.15)

which differ from (19.8) and (19.11) by replacing momenta with a set of quantum numbers $nl(El)$.

19.2 Numerical Model

19.2.1 Ionization of an Atom by Electron Impact

To calculate the cross section for ionization of an atom by an electron impact in DWBA, the wave function $\phi_\nu^\pm(r)$ is expanded into a series in partial waves:

$$\phi_\nu^{(\pm)}(r) = 4\pi\sqrt{\frac{\pi}{p}}\sum_{l,m} i^l e^{\pm i\delta_l}\frac{P_\nu(r)}{r}Y_{lm}^*\left(\frac{\mathbf{p}}{p}\right)Y_{lm}\left(\frac{\mathbf{r}}{r}\right) \tag{19.16}$$

Here, δ_l is the scattering phase, $P_\nu(r)$ is the radial function (6.5)–(6.9) with asymptotic at infinity, determined according to (6.14):

$$P_\nu(r) \to \frac{1}{\sqrt{\pi p}}\sin\left(pr - \frac{\pi}{2}l + \delta_l\right) \tag{19.17}$$

The number of terms in expansion (19.16), which must be included to ensure the desired accuracy, can be estimated as follows. The formula for the cross section (19.2) includes the overlap integral of wave functions $\phi_{\nu_2}^{(-)}$ and ϕ_{ν_3}. The function ϕ_{ν_3} is concentrated near the nucleus at a distance of the order of I_3^{-1}, where I_3 is the ν_3-shell ionization potential. Therefore, in expansion (19.16), it is sufficient to represent the behavior of function $\phi_{\nu_2}^{(-)}$ up to these distances, i.e., confine ourselves to partial components with quantum numbers $l_2 \leq l_{2_{\max}} = p_2/I_3$. Similarly, we can estimate $l_{4_{\max}}$ and $l_{1_{\max}}$. The Coulomb electron–electron interaction is expanded in a series of terms of partial angular momentum of excitation [LL]:

$$\frac{1}{|\mathbf{r}_1 - \mathbf{r}_2|} = \sum_{l=0}^{\infty}\sum_{m=-l}^{l}\frac{4\pi}{(2l+1)}Y_{lm}^*\left(\frac{\mathbf{r}_1}{r_1}\right)Y_{lm}\left(\frac{\mathbf{r}_2}{r_2}\right)\frac{r_<^l}{r_>^{l+1}}, \tag{19.18}$$

where $r_{<(>)}^l$ is the smaller (larger) of the radii $r_{1(2)}$.

Let us choose the direction of the incident electron beam as the axis z and plane of scattering xz as passing through the vectors $\mathbf{p}_4$ and $\mathbf{p}_1$. Then,

$$Y_{l_4 m_4}\left(\frac{\mathbf{p}_4}{p_4}\right) = \begin{cases} \sqrt{(4\pi)/(2l_4+1)}, & m_4 = 0, \\ 0, & m_4 \neq 0. \end{cases} \tag{19.19}$$

Substituting expansions (19.16) and (19.18) into (19.2) and performing integration over angles and summation over electron spin projections, we obtain the expression for the ionization cross section:

$$\frac{d\sigma^3}{d\Omega_2 d\Omega_4 dE_4} = \frac{2\pi^3}{E_1} \sum_{m_3} \left(\left| T_{m_3}^{(3)} \right|^2 + 3 \left| T_{m_3}^{(1)} \right|^2 \right). \tag{19.20}$$

Here, the triplet $T_{m_3}^{(3)}$ and singlet $T_{m_3}^{(1)}$ amplitudes are equal, respectively, to

$$\begin{aligned} T_{m_3}^{(3,1)} = &\sum_{l_4 l_1 l l_2} \sqrt{2l_4+1}\, i^{(-l_1+l_4-l_2)} e^{i(\delta_{l_4}+\delta_{l_1}+\delta_{l_2})} (-1)^{m_2} \\ &\times \begin{pmatrix} l_4 & l & l_1 \\ -m_4 & m & 0 \end{pmatrix} \begin{pmatrix} l_2 & l & l_5 \\ -m_2 & m & m_3 \end{pmatrix} Y_{l_1 m_1}\left(\frac{p_4}{p_4}\right) Y_{l_2 m_2}\left(\frac{p_2}{p_2}\right) U_{\nu_1 \nu_4 \nu_2 \nu_3}^{(3,1)}, \end{aligned} \tag{19.21}$$

where

$$U_{\nu_1 \nu_4 \nu_2 \nu_3}^{(3,1)} = \langle \nu_1 \nu_2 \| V_l \| \nu_4 \nu_3 \rangle - i^{3(1)-1} (2l+1) \sum_{l'} (-1)^{l+l'} \begin{Bmatrix} l_1 & l & l_4 \\ l_4 & l' & l_3 \end{Bmatrix} \langle \nu_1 \nu_2 \| V_l \| \nu_3 \nu_4 \rangle \tag{19.22}$$

is, respectively, the sum or difference of the direct and exchange reduced Coulomb matrix elements (9.32)–(9.34). In the program ***impact*** (see Sect. 2.16) $3j$- and $6j$-symbols are calculated for large l. As a result, in usual notation of $3j$- and $6j$-symbols it is required to determine factorials of large numbers, which overflow the computer bit grid. Calculation of $3j$- and $6j$-symbols through the binomial coefficients [2] permits to expand (19.16) up to $l_{\max} = 30$.

The spherical functions in (19.16) are calculated by the following formula:

$$Y_{lm}(\vartheta, \phi) = (-1)^{(m+|m|)/2} \sqrt{4\pi}\, e^{im\phi}\, \tilde{P}_l^{|m|}(\cos\vartheta) = e^{im\phi} Y_{lm}(\vartheta, 0) \tag{19.23}$$

Here, $\tilde{P}_l^{(m)}(\cos\vartheta)$ is the normalized Legendre polynomial [3]. Complex factors in (19.21) are combined as follows:

$$\begin{aligned} &(-1)^{m_2}\, i^{(-l_1+l_4-l_2)} \exp\big(i(\delta_{l_4}+\delta_{l_1}+\delta_{l_2})\big) \exp(i m_2 \phi_2) = \exp\big(i \Phi_{l_4 l_1 l_2 l_{m_2}}\big), \\ &\Phi_{l_4 l_1 l_2 l_{m_2}} = \delta_{l_4} + \delta_{l_1} + \delta_{l_2} + \frac{\pi}{2}(-l_1+l_4-l_2) + m_2(\phi_2+\pi) \end{aligned} \tag{19.24}$$

As a result, the amplitudes $T_m^{(3,1)}$ are calculated by formulas:

$$\begin{aligned}
\mathrm{Re}T_{m_3}^{(3,1)} &= \sum_{l_4 l_1 l l_2, m_1=m, m_2} \sqrt{2l_4+1}\begin{pmatrix} l_1 & l & l_4 \\ -m_1 & m & 0 \end{pmatrix} \\
&\times \begin{pmatrix} l_2 & l & l_3 \\ -m_2 & -m & m_3 \end{pmatrix} zY_{l_1 m_1}(\vartheta_1, 0) Y_{l_2 m_2}(\vartheta_2, 0) U^{(3,1)}_{\nu_1 \nu_4 \nu_2 \nu_3} \cos \Phi_{l_4 l_1 l_2 l_{m_2}}, \\
\mathrm{Im}T_{m_3}^{(3,1)} &= \sum_{l_4 l_1 l l_2, m_1=m, m_2} \sqrt{2l_4+1}\begin{pmatrix} l_1 & l & l_4 \\ -m_1 & m & 0 \end{pmatrix} \\
&\times \begin{pmatrix} l_2 & l & l_3 \\ -m_2 & m & m_3 \end{pmatrix} Y_{l_1 m_1}(\vartheta_1, 0) Y_{l_2 m_2}(\vartheta_2, 0) U^{(3,1)}_{\nu_1 \nu_4 \nu_2 \nu_3} \sin \Phi_{l_4 l_1 l_2 l_{m_2}}
\end{aligned} \tag{19.25}$$

In ***impact2*** program [4], the cross section of electron impact excitation in the DWBA is calculated by the formula

$$\frac{d\sigma}{d\Omega_4} = \frac{2\pi^3}{E_1} \sum_{m_3 m_2} \left(\left| T^{(3)}_{m_3 m_2} \right|^2 + 3 \left| T^{(1)}_{m_3 m_2} \right|^2 \right) \tag{19.26}$$

Here

$$\begin{aligned}
\mathrm{Re}T_{m_3 m_2}^{(3,1)} &= \sum_{l_4 l l_1, m=m_3-m_2, m_1=m} \begin{pmatrix} l_1 & l & l_4 \\ -m_1 & m & 0 \end{pmatrix} \begin{pmatrix} l_2 & l & l_3 \\ -m_2 & -m & m_3 \end{pmatrix} \\
& Y_{l_1 m_1}(\vartheta_1, 0) U^{(3,1)}_{\nu_1 \nu_4 \nu_2 \nu_3} \cos \Phi_{l_4 l_1}, \\
\mathrm{Im}T_{m_3 m_2}^{(3,1)} &= \sum_{l_4 l l_1, m=m_3-m_2, m_1=m} \begin{pmatrix} l_1 & l & l_4 \\ -m_1 & m & 0 \end{pmatrix} \begin{pmatrix} l_2 & l & l_3 \\ -m_2 & -m & m_3 \end{pmatrix} \\
& Y_{l_1 m_1}(\vartheta_1, 0) U^{(3,1)}_{\nu_1 \nu_4 \nu_2 \nu_3} \sin \Phi_{l_4 l_1}, \\
\Phi_{l_4 l_1} &= \frac{\pi}{2}(-l_1 + l_4) + \delta_{l_1} + \delta_{l_4}
\end{aligned} \tag{19.27}$$

To calculate the triple-differential ionization cross section of an atom in PWBA in ***impact1*** program [4], we choose the axis z along the transferred momentum $\mathbf{q}$ direction. Then, similar to (19.19), we have

$$Y^*_{lm}\left(\frac{\mathbf{q}}{q}\right) = \begin{cases} \sqrt{(2l+1)/(4\pi)}, & m_4 = 0, \\ 0, & m_4 \neq 0. \end{cases} \tag{19.28}$$

Substituting expansion (19.16) into (19.4) with subsequent integration over angles, it is obtained:

$$\frac{d^3\sigma}{d\Omega_2\, d\Omega_4 dE_4} = \frac{4p_4}{q^4 p_1}\sum_{m_3}\left|\sum_{l_2 l,m_2}(-1)^{m_2}\, i^{l-l_2}e^{i\delta_{l_2}}\begin{pmatrix} l_3 & l & l_2 \\ -m_3 & 0 & m_2\end{pmatrix} Y_{l_2 m_2}\left(\frac{\mathbf{p}_2}{p_2}\right)\left\langle \nu_2\left\| M_l^{(r)}(q,\omega)\right\|\nu_3\right\rangle\right|^2. \tag{19.29}$$

Here, in the HF approximation, matrix element $\langle \nu_2 \| G(q,\omega) \| \nu_3\rangle$ is simplified transforming into the following expression

$$\left\langle \nu_2\left\| M_l^{(r)}(q,\omega)\right\|\nu_3\right\rangle \to \sqrt{(2l_2+1)(2l_3+1)}\begin{pmatrix} l_2 & l & l_3 \\ 0 & 0 & 0\end{pmatrix}\int_0^\infty P_{\nu_2}(r)P_{\nu_3}(r)j_l(qr)dr \tag{19.30}$$

We represent the spherical function $Y_{l_2 m_2}(\vartheta_2\phi_2)$ as follows:

$$Y_{l_2 m_2}(\vartheta_2\phi_2) = \exp(im_2\phi_2)Y_{l_2 m_2}(\vartheta_2, 0) \tag{19.31}$$

Due to condition for summation of angular momenta projections, in the sum over m_2 in (19.29), only the term with $m_3 = m_2$ is nonzero. The factor $(-1)^{m_2}\exp(im_3\phi_2)$ can be taken out of the internal sum sign and omitted, as being equal by modulus to one. Therefore, the cross section is independent on angle ϕ_2 at all that reflects the well-known property of axial symmetry of PWBA results relative to the vector **q**.

Let us denote $i^{l-l_2}\exp\left(i\delta_{l_2}\right)$ as $\exp\left(i\Phi_{ll_2}\right)$, where $\Phi_{ll_2} = \delta_{l_2} + \pi(l - l_2)$. Then, the cross section (19.29) becomes equal to

$$\frac{d^3\sigma}{d\Omega_4\, d\Omega_2\, dE_4} = \frac{4p_4}{q^4 p_1}\sum_{m_3}\left|T_{m_3}\right|^2, \tag{19.32}$$

where

$$\begin{aligned}
\mathrm{Re}T_{m_3} &= \sum_{ll_2}(2l+1)\begin{pmatrix} l_3 & l & l_2 \\ -m_3 & 0 & m_3\end{pmatrix}\left\langle \nu_2\left\| M_l^{(r)}(q)\right\|\nu_3\right\rangle Y_{l_2 m_3}\left(\frac{p_2}{p_2}\right)\cos\Phi_{ll_2},\\
\mathrm{Im}T_{m_3} &= \sum_{ll_2}(2l+1)\begin{pmatrix} l_3 & l & l_2 \\ -m_3 & 0 & m_3\end{pmatrix}\left\langle \nu_2\left\| M_l^{(r)}(q)\right\|\nu_3\right\rangle Y_{l_2 m_3}\left(\frac{p_2}{p_2}\right)\sin\Phi_{ll_2}
\end{aligned} \tag{19.33}$$

Calculating of the reduced matrix elements $\left\langle \nu_2 \| M_l^{(r)}(q) \| \nu_3\right\rangle$ in (19.30) is similar to that described in Sect. 10.2 for GOS. The excitation cross section of an atom in PWBA is calculated using the ***impact2*** program [5].

19.2.2 Ionization Cross Section of an Atom by Electrons at the Threshold of Triplet Autoionizing Level Excitation with Allowance for "Post-collision Interaction"

After integration over angular variables and summation over spins, the differential cross section of inelastic process corresponding to diagram (19.7) acquires the form:

$$\frac{d\sigma}{dE_3}=\frac{4\pi^3}{E_1^2}\sum_{l_1 l_7 l_5 l L l' L' \tilde{l}}(-1)^{L-l+L'-l'}\left(2\tilde{l}+1\right)\begin{Bmatrix} l_1 & \tilde{l} & l_7\\ L & l_2 & l\end{Bmatrix}\begin{Bmatrix} l_1 & \tilde{l} & l_7\\ L' & l_2 & l'\end{Bmatrix}$$
$$\times\begin{Bmatrix} l_5 & \tilde{l} & l_6\\ L & l_3 & l\end{Bmatrix}\begin{Bmatrix} l_5 & \tilde{l} & l_6\\ L' & l_3 & l'\end{Bmatrix}$$
$$\times\int_{\nu_4}\frac{\langle\nu_2\nu_4\|V_l\|\nu_1\nu_3\rangle\langle\nu_7\nu_3\|V_L\|\nu_2\nu_6\rangle\langle\tilde{\nu}_5\|\nu_4\rangle}{E_1-E_2+E_3-E_4+i\Gamma/2}$$
$$\int_{\nu_4}\frac{\langle\nu_2\nu_4\|V_{l'}\|\nu_1\nu_3\rangle\langle\nu_7\nu_3\|V_{L'}\|\nu_2\nu_6\rangle\langle\tilde{\nu}_5\|\nu_4\rangle}{E_1-E_2+E_3-E_4+i\Gamma/2},\tag{19.34}$$

where the reduced Coulomb matrix elements $\langle\nu_1\nu_2\|V_l\|\nu_3\nu_4\rangle$ are determined by Eq. (9.33), p_1 is the linear momentum of the incident electron; $lLl'L',\tilde{l}$ are the angular momenta transferred in the Coulomb interaction matrix elements. The energies of fast E_1, E_7 and slow E_5 electrons are related by the conservation law $E_1-E_7-E_5+E_6=0$, where E_6 is the energy of a vacancy in the final state. The width of the autoionization state is denoted by Γ. Integration over the energy of an intermediate electron with energy E_4 also implies the summation over the excited discrete states. The overlap integrals are given by formula

$$\langle\tilde{\nu}_5\|\nu_4\rangle=\int_0^\infty\tilde{P}_{\nu_5}(r)P_{\nu_4}(r)dr.\tag{19.35}$$

Here, $\tilde{P}_{\nu_5}(r)$ is the radial one-electron HF wave function in the field of an atom with a vacancy ν_6 and $P_{\nu_4}(r)$ is the same function, but calculated in the field of autoionization excitation $\nu_2\nu_3$.

When calculating the matrix elements in (19.34), Hartree–Fock wave functions are used, calculated in different fields (the ***fchf*** program, see Chaps. 2 and 8). For vacancies ν_3, ν_6 and incident electron ν_1, eigenfunctions of the Hartree–Fock operator of a neutral atom are applied. In case of an excited state of the target atom, ν_1 are functions ϕ^{N+1} in the field of the frozen core. The wave functions of fast autoionization electron ν_7 and slow electron ν_5 are found as functions $\phi^{N(LS)}$ in the field of a single ion with a vacancy ν_6. The wave functions of the discrete excited state ν_2 should rather be found with account of rearrangement, i.e., including the self-consistency of $\nu_2\nu_3$ configuration. Finally, the wave functions of slow electrons in intermediate state ν_4 are defined as the Hartree–Fock wave functions ϕ^{N+1} in the field of an excited atom in the state with a configuration $\nu_2\nu_3$.

When an s-electron is excited into an autoionization state ($l_3 = 0$), expression (19.34) is greatly simplified and takes the form:

$$\frac{d\sigma}{dE_5} = \sum_{l_1 l_7 l_5 l L} \frac{d\sigma_{l_1 l_7 l_5 l L}}{dE_5}, \tag{19.36}$$

where

$$\frac{d\sigma_{l_1 l_7 l_5 l L}}{dE_5} = \frac{4\pi^3}{E_1} \frac{|\langle \nu_7 \nu_3 \| V_L \| \nu_2 \nu_6 \rangle|^2}{(2l_2+1)(2l+1)(2L+1)} \left| \int_{\nu_4} \frac{\langle \nu_2 \nu_4 \| V_l \| \nu_1 \nu_3 \rangle \langle \tilde{\nu}_5 \| \nu_4 \rangle}{E_1 - E_2 + E_3 - E_4 + i\Gamma/2} \right|^2 \tag{19.37}$$

Here, the integral sign also implies summation over the states of the continuous spectrum.

The ***autpci*** program [5] calculates contributions to the cross section of each term for fixed values $l_1 l_7 l_5 l$, L. Estimation of the number of partial waves l_1 in each case is carried out individually. For each partial wave l_1, $(d\sigma_{l_1 l_7 l_5 l L})/(dE_5)$ (19.37) are calculated with different values $l_7 l_5 l$ L.

When calculating the matrix element $\langle \nu_7 \nu_3 \| V_L \| \nu_2 \nu_6 \rangle$, its weak dependence on energy E_7 is assumed. Therefore, in (19.37), this matrix element can be considered as a constant value. This matrix element is computed preliminary in the ***gos*** program (See Sect. 2.9) and given as an input parameter to the ***autpci*** program (Sect. 2.17).

Autoionizing decay channel is determined by the states of both vacancy and emitted electron. Several decay channels are possible, for example:

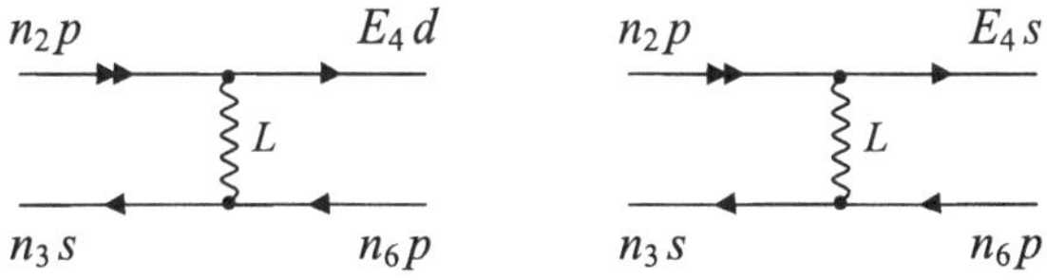

(19.38)

The ***autpci*** program accounts for possible presence of two decay channels. For each matrix element, the contribution to (19.37) is calculated for given values l_7 and L. The overlap integrals $\langle \tilde{\nu}_5 \| \nu_4 \rangle$ in (19.37) are calculated in accordance with (19.35). The integral at E_4 (ν_4) in (19.34) is transformed into a sum of two terms corresponding to real and imaginary parts ReInt and ImInt.

If integration is performed over momentum $p_4 = \sqrt{E_4}$, the following equation is obtained:

$$\mathrm{ReInt} + \mathrm{ImInt} = 2 \int_{p_4} \frac{p_4 \langle \nu_2 \nu_4 \| V_l \| \nu_1 \nu_3 \rangle \langle \tilde{\nu}_5 \| \nu_4 \rangle \left(E_1 - E_2 + E_3 - p_4^2\right)}{\left(E_1 - E_2 + E_3 - p_4^2\right)^2 + \Gamma^2/4}$$
$$- i\Gamma \int_{p_4} \frac{p_4 \langle \nu_2 \nu_4 \| V_l \| \nu_1 \nu_3 \rangle \langle \tilde{\nu}_5 \| \nu_4 \rangle}{\left(E_1 - E_2 + E_3 - p_4^2\right)^2 + \Gamma^2/4}. \tag{19.39}$$

The autoionization state width Γ is taken either from experiment or calculated in advance using ***gos*** program. Each integral in (19.39) is calculated using Simpson formula [BZ].

The electron inelastic scattering cross sections with excitation of an autoionization level can also be calculated in the case when a slow scattered electron as a result of post-collision interaction (PCI) is captured to a discrete level in the field of a single-charged ion. Then, for an electron ν_5, the wave function refers to a discrete spectrum instead of continuous.

In the case when the electron is excited into an autoionization state with $l_3 \neq 0$, i.e., not from the s-shell, the scattering cross section must be calculated in accordance with (19.37). The ***autpci*** program is also used for this purpose. In this case, in relation (19.37), one should substitute values of integrals over E_4, the real and imaginary parts of which are calculated for each value E_5 and E_1 at fixed angular momenta.

19.2.3 Excitation of a Triplet Level by Electrons

Cross section calculation of (19.13) for the process shown in figure (19.15) is reduced, first of all, to calculating matrix elements $\langle \nu_1 \nu_3 \| V_l \| \nu_2 \nu_4 \rangle_{\mathrm{II}}$ for each fixed state $\nu_1,\ \nu_3,\ \nu_2,\ \nu_4$.

In the case when a triplet level $ns \to n's$ ($l_3 = l_2 = 0$) is excited, the coefficient in front of integral in (19.14) is simplified, and the matrix element has the form

$$\langle \nu_1 \nu_3 \| V_l \| \nu_2 \nu_4 \rangle_{\mathrm{II}} = \langle \nu_1 \nu_3 \| V_l \| \nu_2 \nu_4 \rangle + \sum_{l_5 l_6} \frac{(-1)^{l_5 + l_6}}{\sqrt{(2l_5 + 1)(2l_6 + 1)}}$$
$$\times \int_{\nu_5 \nu_6} \frac{\langle \nu_1 \nu_3 \| V_{l_6} \| \nu_5 \nu_6 \rangle \langle \nu_5 \nu_6 \| V_{l_5} \| \nu_2 \nu_4 \rangle}{E_1 - E_5 - E_6 + E_3 + i\eta} = [\langle \nu_1 \nu_3 \| V \| \nu_2 \nu_4 \rangle$$
$$+ \sum_{l_5 l_6} \frac{(-1)^{l_5 + l_6}}{\sqrt{(2l_5 + 1)(2l_6 + 1)}} \times P \int_{\nu_5 \nu_6} \frac{\langle \nu_1 \nu_3 \| V_{l_6} \| \nu_5 \nu_6 \rangle \langle \nu_5 \nu_6 \| V_{l_5} \| \nu_2 \nu_4 \rangle}{E_1 - E_5 - E_6 + E_3} \Bigg]$$
$$- i\pi \sum_{l_5 l_6} \frac{(-1)^{l_5 + l_6}}{\sqrt{(2l_5 + 1)(2l_6 + 1)}} \int_{\nu_5 \nu_6} \langle \nu_1 \nu_3 \| V_{l_6} \| \nu_5 \nu_6 \rangle$$
$$\times \langle \nu_5 \nu_6 \| V_{l_5} \| \nu_2 \nu_4 \rangle \delta(E_1 - E_5 - E_6 + E_5) = \mathrm{Re} V_{\mathrm{II}, l \nu_1} + i \mathrm{Im} V_{\mathrm{II}, l \nu_1}. \tag{19.40}$$

where

$$\mathrm{Re}V_{\mathrm{II},l\nu} \equiv \langle \nu_1\nu_3 \| V_l \| \nu_2\nu_4\rangle - \sum_{l_5 l_6} \mathrm{Re}\tilde{V}_{\mathrm{II},\nu_5\nu_6\nu_1} \tag{19.41}$$

$$V_{\mathrm{II},l\nu_1} = \sum_{l_5 l_6} V_{\mathrm{II},\nu_5\nu_6\nu_1} \tag{19.42}$$

The ***intripl*** program allows to calculate contributions to real and imaginary parts (19.40) of individual terms corresponding to certain angular momenta of intermediate states l_5 and l_6 for each value l_1. In other words, the ***intripl*** program calculates real and imaginary parts of the second-order partial amplitude, represented by diagram (19.15b) with $L = l_6,\ l' = l_5$, namely

$$\mathrm{Re}\tilde{V}_{\mathrm{II},\nu_5\nu_6\nu_1} = \frac{(-1)^{l_5+l_6}}{\sqrt{(2l_5+1)(2l_6+1)}} P \int_{\nu_5\nu_6} \frac{\langle \nu_1\nu_3 \| V_{l_6} \| \nu_5\nu_6\rangle\langle \nu_5\nu_6 \| V_{l_5} \| \nu_2\nu_4\rangle}{E_1 - E_5 - E_6 + E_3}, \tag{19.43}$$

$$\mathrm{Im}V_{\mathrm{II},\nu_5\nu_6\nu_1} = -\pi \int_{\nu_5} \frac{\langle \nu_1\nu_3 \| V_{l_6} \| \nu_5\nu_6\rangle\langle \nu_5\nu_6 \| V_{l_5} \| \nu_2\nu_4\rangle \delta(E_1 - E_5 - E_6 + E_3)}{(-1)^{l_5+l_6}\sqrt{(2l_5+1)(2l_6+1)}}, \tag{19.44}$$

and the first-order amplitude $\langle \nu_1\nu_3 \| V_l \| \nu_2\nu_4\rangle$ also.

Second-order amplitudes are determined not only for $ns \to n's$, but also for other triplet excitations. Note that the integrals in (19.43) and (19.44) must be multiplied by numerical factors a_1 for $\mathrm{Re}V_{\mathrm{II},\nu_5\nu_6\nu_1}$ and a_2 for $\mathrm{Im}V_{\mathrm{II},\nu_5\nu_6\nu_1}$:

$$a_1 = (2l+1)\sqrt{(2l_5+1)(2l_6+1)} \begin{Bmatrix} l_4 & l & l_3 \\ L & l_6 & l' \end{Bmatrix} \begin{Bmatrix} l & l_4 & l_1 \\ L & l_5 & l' \end{Bmatrix} \tag{19.45}$$

$$a_2 = (2l+1)\sqrt{(2l_5+1)(2l_6+1)} \begin{Bmatrix} l_4 & l & l_3 \\ L & l_6 & l' \end{Bmatrix} \begin{Bmatrix} l & l_4 & l_1 \\ L & l_5 & l' \end{Bmatrix} \tag{19.46}$$

The choice of wave functions for calculating the matrix elements in (19.43), (19.44) is not unambiguous. From the point of view of the distorted wave method, it is advisable to have in mind that each channel has its own "distorting" HF-potential, and in different states the electron "feels" different HF-fields. So, in a state ν_1, the field of a neutral atom acts on it, and its wave function is, of course, ϕ^{N+1}. The ν_4 electron moves in the field of an atom excited into triplet state, and its wave function is determined by the configuration of corresponding self-consistent field. The wave function of an intermediate electron with quantum numbers ν_6 is calculated in a similar way, i.e., HF function of an electron in $(\nu_5\nu_3)^3(L)$ field. Electron wave functions in states ν_2 and ν_5 are calculated as the Hartree–Fock functions of triplet states in the field of an atom with a vacancy ν_3: $\phi^{N(LS)}_{\nu_3}$.

As noted in Chap. 8, using the wave functions of the excited states found in the frozen field of a single ion as the wave functions of particles is equivalent to summing an infinite sequence of time-forward RPAE diagrams where the hole state does not change. For example, graph (19.8) is equivalent to the following infinite sequence

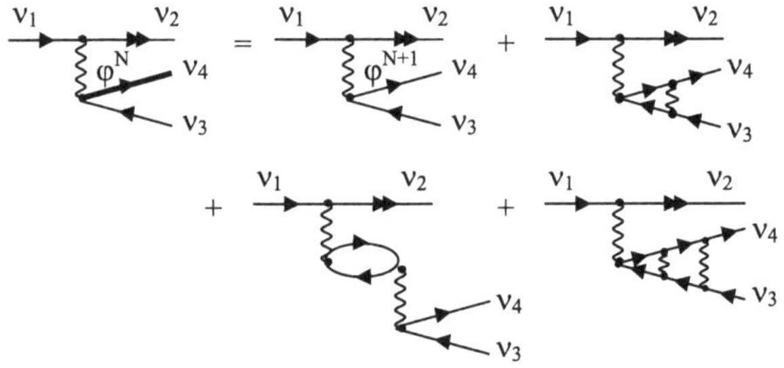

(19.47)

if, when calculating it, the wave function $\phi_{\nu_4}^N$ found in the field of a frozen core with a hole in the state ν_3 is used as the wave function of the particle ν_4. The diagonal in the state of the field diagrams is accounted for by ϕ^N for electrons in states $\nu_2,\ \nu_5,\ \nu_6$. The calculation of the double integral in (19.40) is carried out using the following relations:

$$\int\limits_{\nu_5\nu_6} \frac{\langle \nu_1\nu_3 \| V_{l_6} \| \nu_5\nu_6 \rangle \langle \nu_5\nu_6 \| V_{l_5} \| \nu_2\nu_4 \rangle}{E_1 - E_5 - E_6 + E_3 + i\eta} = \int\limits_{\nu_5} \left\{ P \int\limits_{\nu_6} \left[\frac{\langle \nu_1\nu_3 \| V_{l_6} \| \nu_5\nu_6 \rangle \langle \nu_5\nu_6 \| V_{l_5} \| \nu_2\nu_4 \rangle}{E_1 - E_5 - E_6 + E_3} \right.\right.$$
$$\left.\left. -\ i\pi \langle \nu_1\nu_3 \| V_{l_6} \| \nu_5\nu_6 \rangle \langle \nu_5\nu_6 \| V_{l_5} \| \nu_2\nu_4 \rangle \delta(E_1 - E_5 - E_6 + E_5)) \right] \right\} \tag{19.48}$$

When calculating the principal value of the integral in (19.48), in pole vicinity the procedure described in Sect. 9.2.4 is applied. In calculation of (19.48), as usual, a transition is made to a variable—to linear momentum $p_6 = \sqrt{E_6}$. The integration over E_6 in (19.48) is reduced to summation:

$$\begin{aligned} &\sum_j \left[\langle \nu_1\nu_3 \| V_{l_6} \| \nu_5\nu_6 \rangle \langle \nu_5\nu_6 \| V_{l_5} \| \nu_2\nu_4 \rangle Q_j D_j \right. \\ &\left. -i\pi \langle \nu_1\nu_3 \| V_{l_6} \| \nu_5\nu_6 \rangle \langle \nu_5\nu_6 \| V_{l_5} \| \nu_2\nu_4 \rangle \delta(E_1 - E_5 - E_6 + E_5) \right] \end{aligned} \tag{19.49}$$

where

$$D_j = \frac{2p_j}{E_1 - E_5 + E_3 - p_j^2} \tag{19.50}$$

When summing over discrete excited states $Q_j = 1$, and

$$D_j = \frac{1}{E_1 - E_5 + E_3 - E_j} \tag{19.51}$$

where E_j is the energy of the discrete excited state ν_6.

To calculate the Coulomb matrix elements included in the imaginary part of expression (19.44) between states whose energies satisfy the conservation law, we use wave functions of particles in the state ν_6 with the energy closest to the value $E_6 = E_1 - E_5 + E_3$.

Integration over dE_5 in real and imaginary parts of relation (19.48), including summation over discrete states, is performed by Simpson formula [BZ].

The ***intripl*** program [6] provides the computation of matrix elements in the second order of the DWBA method for any triplet transition. In this case, the values $a_1 \mathrm{Re} V_{II,\nu_3\nu_4\nu_1}$ and $a_2 \mathrm{Im} V_{II,\nu_3\nu_4\nu_1}$ contribute to the real and imaginary parts of $\langle \nu_1\nu_3 \| V_l \| \nu_2\nu_4 \rangle_{\mathrm{II}}$, respectively.

Knowledge of $\mathrm{Re} V_{\mathrm{II},\nu_3\nu_4\nu_1}$ and $\mathrm{ImRe} V_{\mathrm{II},\nu_3\nu_4\nu_1}$, in accordance with (19.41) and (19.42), makes it possible to calculate the total cross section for the excitation of the triplet level:

$$\frac{d\sigma}{dE_4} = \frac{4\pi^3}{p_1^2} \sum_{l_1,l_4} \frac{1}{(2l_1+1)} \left[\left(\mathrm{Re} V_{II,l\nu_1} \right)^2 + \left(\mathrm{Im} V_{II,l\nu_1} \right)^2 \right] \tag{19.52}$$

In the case of $s \to s$ transition, the expression for the cross section (19.13) is simplified and reduced to the form

$$\begin{aligned} \frac{d^2\sigma}{dE_4 d\Omega_4} &= \frac{\pi^2}{p_1^2} \sum_{l,l'} \Big\{ \cos\left[\delta_l(p_1) + \delta_l(p_4) - \delta_{l'}(p_1) - \delta_{l'}(p_4)\right] \\ &\times \left[\mathrm{Re} V_{II,l\nu_1} \mathrm{Re} V_{II,l'\nu_1} + \mathrm{Im} V_{II,l\nu_1} \mathrm{Im} V_{II,l'\nu_1} \right] \\ &+ \sin\left[\delta_l(p_1) + \delta_l(p_4) - \delta_{l'}(p_1) - \delta_{l'}(p_4)\right] \\ &\times \left[\mathrm{Im} V_{II,l\nu_1} \mathrm{Re} V_{II,l'\nu_1} - \mathrm{Re} V_{II,l\nu_1} \mathrm{Im} V_{II,l'\nu_1} \right] \Big\} P_l(\cos\vartheta_4) P_{l'}(\cos\vartheta_4), \end{aligned} \tag{19.53}$$

where $P_l(\cos\vartheta)$ is Legendre polynomial of lth order that depends on emitted electron angle ϑ. The number of partial waves $\max(l, l')$ is usually taken to be about 10. The individual terms in the sums (19.41) and (19.42) are calculated in accordance with the definition (19.43) and (19.44).

In ***twosec*** program [6], the cross section $d^2\sigma/dE_4 d\Omega_4$ is calculated by the formula (19.53). The values $\mathrm{Re} V_{\mathrm{II},l\nu_1}$ and $\mathrm{Im} V_{\mathrm{II},l\nu_1}$ in (19.52) are preliminarily found using the ***intripl*** program. The number of partial waves l_1 is determined by the incident electron energy and is estimated separately in each case. The Legendre polynomials are calculated using the standard procedure. For excitation $s \to s$, the sum over $l_1,\ l_4,\ l$ in (19.52) is reduced only to the sum over l_1 (since $l_1 = l_4 = l$).

References

1. H. Bransden, K.H. Winters, J. Phys. B: Atom. Mol. Phys. **8** (1975)
2. D.A. Varshalovich, A.N. Moskalev, V.K. Khersonskii, *Quantum Theory of Angular Momentum* (World Scientist, Singapore, 1988)
3. M. Abramovitz, I.A. Stegun (eds) *Handbook of Mathematical Functions with Formulas, Graphs and Mathematical Tables* (NBS, AMS-55, 1964)
4. L.V. Chernysheva, M.Ya. Amusia, A.S. Kheifets, Numerical calculation of the differential cross sections for excitation and ionization of an atom by electron impact (Ioffe Institute, No. 1531, Leningrad, 1990) (In Russian)
5. S.A. Sheinerman, M.Ya. Amusia, M.Y. Kuchiev, Numerical calculation of the cross section of atomic ionization near the thresholds of inelastic processes (Post-collision interaction) (Ioffe Institute, No. 777, Leningrad, 1982) (In Russian)
6. L.V. Chernysheva, M.Ya. Amusia, S.A. Sheinerman, Numerical calculation of the cross section of the electron-induced triplet excitation in the second order of the distorted-wave approximation (Ioffe Institute, No. 776, Leningrad, 1982) (In Russian)

Chapter 20
Photoionization of Endohedrals and Vacancies Decay in Them

This chapter discusses endohedrals photoionization cross section and corresponding dipole and non-dipole angular distributions of photoelectrons. Fullerene, inside which atom A is located, is considered approximately as a static potential of zero or finite thickness. In addition, dipole and quadrupole fullerene polarization under the action of incoming electromagnetic radiation is accounted for.

Calculations can be performed for any spherically symmetric one- and two-layer endohedrals $A@C_N$, $A@C_{N1}@C_{N2}$. It is assumed that within these endohedrals, atoms A are located at fullerenes center. Concrete calculations were carried out for $N = 60$, $N_1 = 60$, and $N_2 = 240$.

20.1 Theoretical Model

20.1.1 General Remarks

In Chaps. 20 and 21, we consider photoionization cross section of endohedrals—atoms A, stuffed inside a fullerene shell, and fast electron scattering by them, as well as the decay of vacancies formed in these processes in atom A. Photoionization of endohedrals is a relatively new area of atomic physics, attracting more and more attention (see, for example, [1]). Fullerene is a shell formed by a significant number $N \gg 1$ of carbon atoms C. The most widely considered fullerene is C_{60}—a structure close to a sphere, consisting of 60 C atoms.

In principle, a high-precision ab initio quantum–mechanical calculation of processes involving an endohedral is much more difficult than that for an isolated individual atom. Therefore, we will replace the real fullerene with a spherically symmetric static potential well with zero or finite, but small in comparison with the fullerene radius R, width. The internal degrees of freedom will be reduced to

M. Ya. Amusia and L. V. Chernysheva, *Computation of Atomic and Molecular Processes*, Springer Series on Atomic, Optical, and Plasma Physics 117,
https://doi.org/10.1007/978-3-030-85143-9_20

the ability of fullerene to be polarized under the action of the incoming photon or electron. Thus, details of individual electronic structure of the fullerene shell are completely neglected.

Within the framework of the described approximations, endohedrals can be considered simply as "big atoms" [2], using approaches and formulas given above in Chaps. 9–11, when considering photoionization and inelastic fast electrons scattering.

Peculiarities of endohedral structure, namely its large radius R in comparison with the radius r_A of an atom A, $R \gg r_A$, and the small thickness b of the fullerene shell, $b \ll R$, permit to significantly simplify the problem. Cross sections for the case of a fullerene shell, which is described by zero thickness potential, can be represented as products of purely atomic cross sections, reflection factors F, and polarization factors G. In this case, it is not necessary to solve the RPAE equations explicitly for endohedral as a whole. It is enough only to find one-electron functions accounting for the action of fullerene shell potential and find alteration in the photon field due to action of fullerene shell. This is what this chapter is about.

20.1.2 An Atom Inside a Fullerene Shell

In the case of zero thickness, the fullerene shell potential acting upon the emitted electron is

$$V(r) = -V_0\,\delta(r - R) \tag{20.1}$$

To find the reflection factor F, due to validity of inequality $R \gg r_A$, it is sufficient to restrict ourselves to a simplified one-electron equation in comparison with (6.2), in which exchange with atom A is neglected. As a result, instead of (6.5) we have

$$\left(-\frac{\mathrm{d}^2}{\mathrm{d}r^2} + U(r) - V_0\delta(r - R) + \frac{l(l+1)}{r^2}\right)P_{pl}(r) = EP_{pl}(r). \tag{20.2}$$

We are looking for a solution to the Schrödinger Eq. (20.2) for $E = p^2 > 0$ (energy is as usual in Ry) in the following form:

$$P_{pl}(r) = \begin{cases} F_l(p)u_{pl}(r)\ (r \le R) \\ u_{pl}(r)\cos\delta_l - \upsilon_{pl}(r)\sin\delta_l\ (r > R) \end{cases}. \tag{20.3}$$

Here $u_{pl}(r)$ and $\upsilon_{pl}(r)$ are, respectively, regular and irregular solutions of the Schrödinger equation at $r = 0$ for an atom placed inside a fullerene shell. Potential $U(r)$ is the self-consistent field created by atomic nucleus together with the rest of the atomic electrons, in which the ionized electron moves.

Wave function $R_{pl}(r)$ must be continuous at $r = R$. This gives

$$F_l(p)u_{Pl}(R) = u_{pl}(R)\cos\delta_l - \upsilon_{pl}(R)\sin\delta_l \tag{20.4}$$

On the other hand, the second derivative of the wave function has a discontinuity at this point, while the first must be continuous. Thus, integrating (20.2) near the potential well surface, we obtain

$$\frac{1}{2}\int\limits_{R-\varepsilon}^{R+\varepsilon} P''_{pl}(r)\,\mathrm{d}r\big|_{\varepsilon\to 0} = \lim_{\varepsilon\to 0}\frac{1}{2}\left[P'_{pl}(R+\varepsilon) - P'_{pl}(R-\varepsilon)\right] = -V_0 P_{pl}(R) \tag{20.5}$$

Substituting relation (20.3) into Eq. (20.5), we obtain the second equation relating the phase δ and amplitudes $F_l(p)$

$$-2V_0 F_l(p)u_{pl}(R) = u'_{pl}(R)\cos\delta_l - \upsilon'_{pl}(R)\sin\delta_l - F_l(p)u'_{pl}(R) \tag{20.6}$$

Let us rewrite last equation as follows:

$$F_l(p)u'_{pl}(R) = u'_{pl}(R)\cos\delta_l - \upsilon'_{pl}(R)\sin\delta_l + 2V_0 F_l(p)u_{pl}(R) \tag{20.7}$$

Now we multiply Eqs. (20.4) and. (20.7) by $u'_{pl}(R)$ and $u_{pl}(R)$, respectively. Then, if to subtract one of the resulting equations from the other, one gets

$$2V_0 F_l(p)u^2_{pl}(R) = sin\delta_l\left[u_{pl}(R)\upsilon'_{pl}(R) - \upsilon_{pl}(R)u'_{pl}(R)\right] \tag{20.8}$$

Inserting here the expression for Wronskian

$$u_{pl}(r)\upsilon'_{pl}(r) - \upsilon_{pl}(r)u'_{pl}(r) = p \tag{20.9}$$

we get

$$F_l(p) = \frac{p\sin\delta_l}{2V_0 u^2_{pl}(R)}. \tag{20.10}$$

In turn, substitution of relation (20.10) into the right-hand side of (20.4) gives

$$\mathrm{tg}\delta_l = \frac{u^2_{pl}(R)}{u_{pl}(R)\upsilon_{pl}(R) + \frac{p}{2V_0}}. \tag{20.11}$$

The squared amplitude final expression takes the following form:

$$|F_l(p)|^2 = \frac{p^2}{\left[p + 2V_0 u_{pl}(R)\upsilon_{pl}(R)\right]^2 + 4V_0^2 u^2_{pl}(R)} \tag{20.12}$$

20.1.2.1 An Atom in a Double Fullerene Shell

For two infinitely thin spherical layers (two shells), we have instead of (20.2) the following Schrödinger equations:

$$\frac{1}{2}\left[P''_{kl}-\frac{l(l+1)}{r^2}P_{kl}\right]+[V_1\delta(r-R_1)+V_2\delta(r-R_2)-U(r)+E]P_{pl}=0, \tag{20.13}$$

with solutions

$$P_{pl}(r)=\begin{cases}F_l(p)u_{pl}(r)\,(r\le R_1)\\ C_1u_{pl}(r)+C_2\upsilon_{pl}(r)(R_1<r\le R_2)\\ u_{pl}(r)\cos\delta_l-\upsilon_{pl}(r)\sin\delta_l\,(R_2<r)\end{cases} \tag{20.14}$$

Let us introduce the following notation simplifying outlook of the formulas:

$$\begin{aligned}&F_l(p)\equiv F,\\ &u_{pl}(R_1)\equiv u_1,\ u_{pl}(R_2)\equiv u_2,\\ &\upsilon_{pl}(R_1)\equiv\upsilon_1,\ \upsilon_{pl}(R_2)\equiv\upsilon_2.\end{aligned} \tag{20.15}$$

Then the continuity condition for wave function (20.14) at the point $r=R_1$ takes the following form:

$$Fu_1=C_1u_1+C_2\upsilon_1 \tag{20.16}$$

Integration of Eq. (20.13) near $r=R_1$ gives

$$Fu'_1=C_1u'_1+C_2\upsilon'_1+2V_1Fu_1 \tag{20.17}$$

Multiplying relations (20.16) and (20.17) by u'_1 and u_1, respectively, and then subtracting the first equation from the second, one obtains

$$C_2(u_1\upsilon'_1-\upsilon_1u'_1)=-2V_1Fu_1^2. \tag{20.18}$$

Substituting (20.9) into (20.18), it is obtained

$$C_2p=-2V_1Fu_1^2. \tag{20.19}$$

Eliminating F from (20.16) and (20.19), the first equation connecting the coefficients C_1 и C_2 is obtained

$$2V_1u_1^2C_1=-(p+2V_1u_1\upsilon_1)C_2. \tag{20.20}$$

Now consider the point $r = R_1$ outside outer spherical layer. From the continuity condition for the wave function (20.14) at this point, one has

$$(C_1 u_2 + C_2 \upsilon_2) = u_2 \cos\delta_l - \upsilon_2 \sin\delta_l \tag{20.21}$$

Integrating Eq. (20.13) near $r = R_2$ gives

$$-2V_2(C_1 u_2 + C_2 \upsilon_2) = u_2' \cos\delta_l - \upsilon_2' \sin\delta_l - C_1 u_2' - C_2 \upsilon_2' \tag{20.22}$$

Multiplying (20.21) and (20.22) by u_2' andu_2, respectively, and then subtracting the first equation from the second, one can get

$$\sin\delta_l = -C_2 + \frac{2V_2 u_2(C_1 u_2 + C_2 \upsilon_2)}{p}. \tag{20.23}$$

Multiplying (20.21) and (20.22) by υ_1' and v_2, respectively, and then, subtracting the first equation from the second, it is obtained

$$\cos\delta_l = C_1 + \frac{2V_2 \upsilon_2(C_1 u_2 + C_2 \upsilon_2)}{p}. \tag{20.24}$$

In deriving (20.23) and (20.24), we have used relation (20.21) for the Wronskian.

With the help of well-known relations for trigonometric functions, the second equation connecting coefficients C_1 and C_2 is obtained

$$[pC_1 + 2V_2\upsilon_2(C_1 u_2 + C_2\upsilon_2)]^2 + [pC_2 - 2V_2 u_2(C_1 u_2 + C_2 \upsilon_2)]^2 = p^2. \tag{20.25}$$

Finally, by solving the system of Eqs. (20.20) and (20.25), it is obtained for coefficients C_1 and C_2

$$C_1 = \frac{p(p + 2V_1 u_1 \upsilon_1)}{\sqrt{\zeta}}, \tag{20.26}$$

$$C_2 = -\frac{2pV_1 u_1^2}{\sqrt{\zeta}}, \tag{20.27}$$

where

$$\begin{aligned}\zeta = {} & p^4 + 4p^3(V_1 u_1 \upsilon_1 + V_2 u_2 \upsilon_2) + 16V_1^2 V_2^2 u_1^2 (u_2 \upsilon_1 - u_1 \upsilon_2)^2 (u_2^2 + \upsilon_2^2) - \\ & 16pV_1 V_2 u_1 (u_1 \upsilon_2 - u_2 \upsilon_1)\left[V_1 u_1 (u_1 u_2 + \upsilon_1 \upsilon_2) + V_2 u_2 (u_2^2 + \upsilon_2^2)\right] + \\ & 4p^2\left\{V_1^2 u_1^2 (u_1^2 + \upsilon_1^2) + V_2^2 u_2^2 (u_2^2 + \upsilon_2^2) + 2V_1 V_2 u_1\left[2u_2 \upsilon_1 \upsilon_2 + \upsilon_1 (u_2^2 - \upsilon_2^2)\right]\right\}.\end{aligned} \tag{20.28}$$

Inserting expressions (20.26) and (20.27) for C_1 and C_2, respectively, into (20.23), we obtain

$$\sin\delta_l = \frac{2}{\sqrt{\zeta}}\left\{pV_2u_2^2 + V_1u_1[pu_1 + 2V_2u_2(u_2\upsilon_1 - u_1\upsilon_2)]\right\}. \tag{20.29}$$

Removing the coefficient C_2 from (20.19) and (20.27), we obtain the following simple expression for the amplitude:

$$F \equiv F_l(p) = \frac{p^2}{\sqrt{\zeta}} = \frac{p\sin\delta_l}{2\left[V_1u_1^2 + V_2u_2^2 - 2V_1V_2u_1u_2(u_1\upsilon_2 - u_2\upsilon_1)/p\right]}, \tag{20.30}$$

where

$$\mathrm{tg}\delta_l(p) = \frac{u_1^2 + u_2V_2[u_2/V_1 + 2u_1V_1(u_2\upsilon_1 - u_1\upsilon_2)/p]}{u_1\upsilon_1 + p/2V_1 + u_2\upsilon_2V_2/V_1 - 2u_1\upsilon_2V_2(u_2\upsilon_1 - u_1\upsilon_2)/p}. \tag{20.31}$$

It is easy to check that, setting in (20.30) and (20.31) $R_1 = R_2$, we obtain from them expressions (20.10) and (20.11) for $F_l(p)$ and $\mathrm{tg}\delta_l$, respectively, with replacements $R \to R_1 = R_2$, $R \to R_2$, $V_0 \to (V_1 + V_2)$.

20.1.2.2 Accounting for Fullerene Potential in RPAE

It was shown above that zero thickness potential leads to appearance in one-electron radial wave function, including that in HF, the reflection factor $F_l(p)$ that depends on the electron linear momentum. This is correct for all radial one-electron wave functions included in RPAE equations.

As a result, instead of Eq. (9.1) for the reduced dipole matrix elements, a correspondingly modified equation should be solved

$$\langle\nu_1\|D(\omega)\|\nu_2\rangle = \langle\nu_1\|d\|\nu_2\rangle + \left(\sum_{\nu_3\le F,\nu_4>F} - \sum_{\nu_3>F,\nu_4\le F}\right) \frac{\langle\nu_4\|D(\omega)\|\nu_3\rangle\left|F_{\nu_4}\right|^2\left|F_{\nu_3}\right|^2\langle\nu_3\nu_1\|U_1\|\nu_4\nu_2\rangle}{3\left[\omega - E_{\nu_4} + E_{\nu_3} + i\delta(1 - 2n_{\nu_4})\right]} \tag{20.32}$$

As a rule, in calculations it is possible to neglect the influence of the fullerene potential upon the wave functions of electrons in occupied atomic states. In this case, instead of the product of the squares of the modulus $\left|F_{\nu_4}\right|^2\left|F_{\nu_3}\right|^2$, Eq. (20.32) includes a term $\left|F_{\nu_{4,3}}\right|^2$ for excited states or states of a continuous spectrum $\nu_{4(3)}$ only.

To describe the fullerene shell more accurate, one should use a phenomenological potential not of zero, but of finite thickness, or even of more complex form *V(r)*. The influence of such a fullerene potential on the electron of atom *A* is not reduced to appearance of an additional factor $F_l(k)$. In this case, instead of (20.2), one should solve numerically the equation

$$\left(-\frac{d^2}{dr^2} + U(r) + V(r) + \frac{l(l+1)}{r^2}\right) P_{pl}(r) = E P_{pl}(r) \tag{20.33}$$

Solving this equation, one can find one-electron atom A functions for both occupied and excited states and the continuous spectrum. With their help, it is possible to calculate all the characteristics of endohedrals and processes with their participation using the same formulas and corresponding programs similar to those used for isolated atoms and described in a number of previous chapters.

20.1.2.3 Polarization Factor *G*

The electron fullerene shell effectively changes the intensity of the incident photon beam, which penetrates it and reaches the inner atom A. This effect can be included in the same way as it is done in Chap. 9, when studying the effect of one atomic electron shell on another. Indeed, from a physical point of view, the situation is the same, and only an additional fullerene shell appears along with atomic shells.

A significant simplification is, however, that the interaction between fullerene and atom A electrons is relatively weak. Therefore, it is sufficient to take it into account in the first order of the many-body perturbation theory. This is a consequence of the fact that inequality $R \gg r_A$ is assumed to be valid. As a result, we have only a dipole–dipole contribution for the main term of interaction between atomic and fullerene electrons:

$$V_{AC} \approx \mathbf{r}_A \cdot \mathbf{r}_C / R^3 \tag{20.34}$$

Here $\mathbf{r}_A$ and $\mathbf{r}_C$ are the radius vectors of atomic A and fullerene C_N electrons, respectively. The relation $R \gg r_A$ is valid for atoms, whose ionization potential is much higher than that of fullerene.

In the language of diagram theory, the amplitude of photoabsorption $A@C_N$ is similar to (1.51), where the wavy line denotes the interaction (20.34).

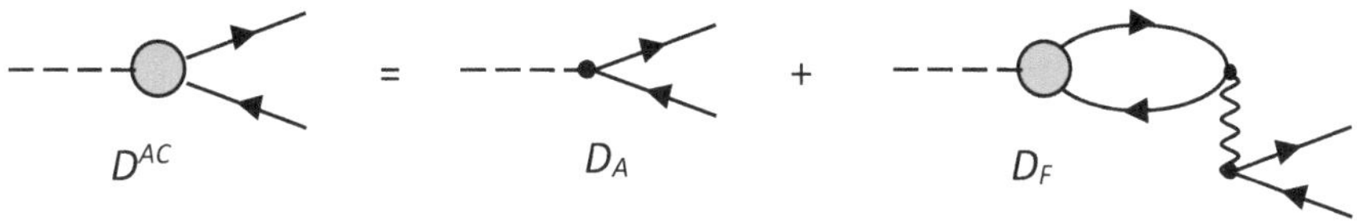

(20.35)

Here D^{AC}, D_A, and D_F are photoionization amplitudes of $A@C_N$, isolated atom A, and fullerene C_N, respectively. Note that so far, the most frequently studied example is $C_N \equiv C_{60}$, although other fullerenes, for example C_{80} also become objects of research. Analytically, we have for (20.35)

$$D^{AC}(\omega) \cong D_A(\omega)\left(1 - \frac{\alpha_C(\omega)}{R^3}\right) \equiv G_C^d(\omega) D_A(\omega), \tag{20.36}$$

where $\alpha_C(\omega)$ is the dipole polarizability of the fullerene shell. Within RPAE frame, D_A is a solution of (9.53). But from the point of view of (20.36), it is inessential, in which approximation is used to describe an isolated atom.

Thus, accounting for polarization of the fullerene shell leads to factor $G_C(\omega)$ called polarization factor.

The combined action of fullerene static potential, described by factor $F_l(p)$ from (20.12), which takes into account reflection of the photoelectron from atom A by fullerene shell, and the factor $G_C(\omega)$ from (20.36), which describes the modification of the incident photon beam, acting upon atom A, due to polarization of the fullerene C_N, leads to the following relation [3]:

$$D_{nl,pl'}^{AC(G)}(\omega) = F_{l'}(p)\left[1 - \frac{\alpha(\omega)}{R^3}\right] D_{nl,pl'}^A(\omega) \equiv F_{l'}(p) G_C^d(\omega) D_{nl,pl'}^A(\omega) \tag{20.37}$$

Here, the superscript AC denotes calculated values that take into account the reflection of the photoelectron by the fullerene static field and the effect of dynamic polarization of fullerene upon the process.

Note that (20.37) can be used with experimental photoionization cross sections $\sigma_{nl,kl'}(\omega)$ of an isolated atom and dipole polarizability $\alpha_C(\omega)$ of fullerene and with factors $F_{l'}(p)$ calculated within the framework of a better model than the bubble potential presented above (20.1).

The factor $G_C^d(\omega)$ in formula (20.37) accounts for the change in the flux of dipole photons. Quadrupole matrix element $Q_A(\omega)$ is modified in the same way as dipole ones by inclusion of diagrams similar to (20.35), leading to the following expression:

$$Q^{AC(G)}(\omega) \cong Q_A(\omega)\left(1 - \frac{\alpha_C^{q(\omega)}}{4R^5}\right) \equiv G_C^q(\omega) Q_A(\omega), \tag{20.38}$$

where $\alpha_C^q(\omega)$ is the fullerene quadrupole polarizability.

Note that factor $F_l(p)$ as a function of ω has strong resonances, which are reflected in $\sigma_{nl,kl'}^{AC}(\omega)$ and are called *confinement resonances* (see [2] and references therein). The dipole polarization factor $G_C^d(\omega)$ has a maximum in the vicinity of giant dipole resonance of the fullerene photoabsorption cross section.

20.1.2.4 Radiative Decay Probability

Let us consider the effect of the C_N fullerene shell on the decay of vacancies in atom *A*. It is natural to distinguish decays of vacancies in endohedral atoms by the degree of participation of fullerene electrons. Namely a vacancy in an endohedral atom can be filled by an external electron of the same atom or by one collectivized electron of fullerene, forming what could be called *atomic and fullerene decays*, respectively. If the atomic vacancy is internal, it can be filled with one of the non-collectivized internal electrons of the *C* atoms that form the fullerene. This process is called *molecular* decay [4].

If an atomic vacancy decays by emission of a photon, the fullerene shell can absorb this photon, become virtually excited, and then emit a photon of the same energy. Thus, vacancy decays along two channels. The first is recombination with the emission of a photon from an isolated atom A. The second is a two-step process through the excitation of electrons in the fullerene shell. The amplitudes of these channels add up in a similar way as they do in isolated atoms (see Chaps. 15 and 16), where the decay probability can change from increasing to almost complete suppressing. The same can happen in Auger decay: The interaction between electrons, inner and outer, can change due to virtual excitations of the C_N electrons, which results in the replacement of the pure Coulomb interaction V by the effective interaction $\Gamma(\omega)$ (1.56), regardless of the transferred energy. The difference between V and $\Gamma(\omega)$ is determined by the intra- and intershell effects.

The endohedral radiative decay amplitude can be represented by diagrams similar to that describing the radiative decay of an isolated atom (see [AICC], Sect. 8.1).

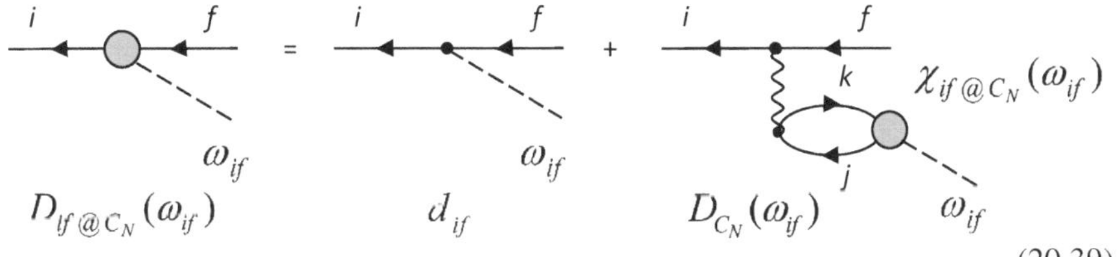

(20.39)

A substantial simplification of this expression is possible, as in the case of photoionization. Let us use the fact that relations $R \gg r_A$ and $R \gg b$, where b is the fullerene shell thickness, are valid. For the amplitude of radiative decay, this leads to a ratio similar to (20.36). As a result, the probability of this decay for an atomic transition $i \to f$ is expressed in terms of the same parameters as in the case of photoabsorption of an endohedral atom:

$$\Gamma^{AC}_{\gamma,if} = \Gamma^{A}_{\gamma,if}\left|1 - \frac{\alpha_C(\omega_{if})}{R^3}\right|^2 \equiv |G_C(\omega)|^2\Gamma^{A}_{\gamma,if} \tag{20.40}$$

Here $\Gamma^{AC(A)}_{\gamma,if}$ is the radiation width of the endohedral (atom) transition $i \to f$ and ω_{if} is the transition energy.

20.1.2.5 Auger Decay Probability

Auger decay can be represented, similarly to (20.39), as the emission of a "photon" in the transition $i \to f$ followed by its absorption and ionization of C_N shell (see [AICC], Sect. 8.1).

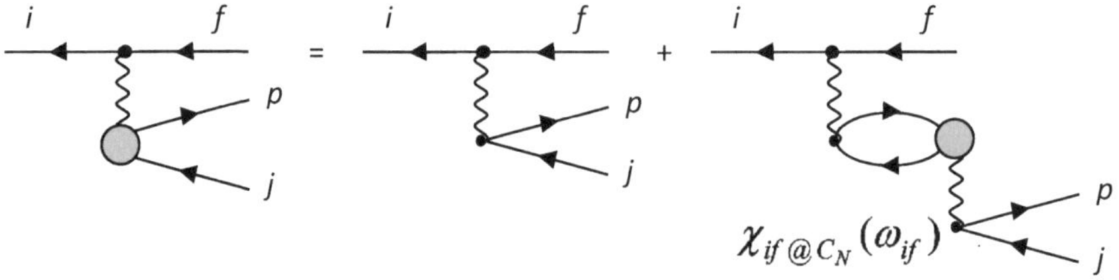

(20.41)

The subvalent s-subshells decay of endohedral noble gas atoms is of particular interest. Auger decay of these vacancies in atoms is forbidden by the energy conservation law. However, Auger decay of these vacancies with emission of fullerene electron p and formation of a hole j in it is allowed in endohedral.

Let us estimate the probability of this process. We can neglect the first term on the right-hand side of (20.41) as electrons of atom and fullerene are relatively very far from each other and therefore are weakly interacting. Then, with the help of (20.34), the probability of Auger decay $\Gamma^{C}_{Au,if}$, determined by the transition $i \to f$, is expressed via the radiative width $\Gamma^{A}_{\gamma,if}$ of the atomic transition $i \to f$ and photoionization cross section $\sigma_c(\omega_{if})$ of C_N electron shell by a photon with energy ω_{if}:

$$\Gamma^{C}_{Au,if} = \Gamma^{A}_{\gamma,if}\frac{3}{8\pi}\left(\frac{c}{\omega_{if}}\right)^4\frac{\sigma_c(\omega_{if})}{R^6} \equiv \Gamma^{A}_{\gamma,if}\eta_{AR}(\omega) \tag{20.42}$$

Estimations show that all quantities except $\Gamma^{A}_{\gamma,if}$ are very large—of the order of $10^4 \div 10^6$. Thus, the decay probability of subvalent vacancies in endohedrals increases by $10^4 \div 10^5$ times, approaching values corresponding to normal Auger decay widths in atom A in the absence of prohibition of the process by energy conservation law.

Using (20.40) and (20.42), one can easily calculate so-called fluorescence yield $J_R(\omega_{if})$, i.e., the ratio of the radiation width to the total width

$$J_R(\omega_{if}) = \frac{8\pi}{3}\left(\frac{\omega_{if}}{c}\right)^4\frac{R^6}{\sigma_c(\omega_{if})}\left|1 - \frac{\alpha_c(\omega_{if})}{R^3}\right|^2 \tag{20.43}$$

Virtual excitations of C_N electrons alter the interactions between atomic electrons involved in the transition from level f to level i, under the influence of electrons of the C_N fullerene. However, it can be demonstrated that this alteration is small [5].

Note that if an electron with energy $E \leq 70$ eV is emitted in the Auger decay of an endohedral atom vacancy, the probability of this process can modify significantly under the influence of the interaction of this electron with static potential of the fullerene shell. As a result, the probability of decay with the removal of an electron with angular momentum l acquires the factor $|F_l(p)|^2$, where $F_l(p)$ is given by (20.30).

20.1.3 Numerical Model

20.1.3.1 Reflection Factor F

For an endohedral atom, reflection factor $F_l(p)$ in photoionization amplitude is introduced, which is given by the following expression:

$$F_l(\omega) = \cos \Delta_l(\omega)\left[1 - \mathrm{tg}\Delta_l(\omega)\frac{\upsilon_{pl}(R)}{u_{pl}(R)}\right] \tag{20.44}$$

where $\Delta_l(p)$ is the additional phase shift due to action of the fullerene shell potential, $u_{pl}(r)$ and $\upsilon_{pl}(r)$ are regular and irregular solutions of the atomic Hartree–Fock equations for a photoelectron with linear momentum $p = \sqrt{E}$, where ε is the photoelectron energy related to the photon energy ω by the relation $\varepsilon = \omega - I_A$, while I_{A} is the ionization potential of atom A.

Factor F is determined for each value ω.

The phase shift is determined, according to (20.11), by the following formula:

$$\mathrm{tg}\Delta_l(\omega) = \frac{u_{pl}^2(R)}{u_{pl}(R)\upsilon_{pl}(R) + \frac{p}{2V_0}} \tag{20.45}$$

Potential well depth is determined so that in potential (20.1), there exists a single bound state with an energy equal to the binding energy I_{C^-} of an electron and neutral fullerene C_N. There exists a formula that connects parameters $\beta = \sqrt{I_{C^-}}$ and R with V_0 (see, for example, [6])

$$V_0 = \frac{\beta}{2}[1 + \cot g(\beta R)] \tag{20.46}$$

The reflection factor [7] is defined differently, depending on whether the wave function belongs to a discrete or continuous spectrum. For the discrete spectrum, we assume that the atomic and endohedral functions coincide. A different situation occurs with the wave functions of the continuous spectrum. For $r < R$, the wave function of the atomic electron is proportional to the atomic one. Outside the fullerene shell, at $r > R$, the wave function becomes a linear combination of regular and irregular solutions of the Schrödinger equation [8].

Regular and irregular solutions of the wave equation are connected to each other by relation.

$$\upsilon_{pl}(r) = k u_{pl}(r) \int \frac{dr}{[u_{pl}(r)]^2}, \tag{20.47}$$

allowing to use only a regular solution when calculating the reflection factor F. To calculate the value of this integral at the point R, we approximate the regular solution in the vicinity of this point by a square trinomial $u_{pl}(r) = a_{pl}r^2 + b_{pl}r + c_{pl}$.

Integral (20.47) takes the form

$$\upsilon_{pl}(r) = p[a_{pl}r^2 + b_{pl}r + c_{pl}] \int \frac{dr}{[a_{pl}r^2 + b_{pl}r + c_{pl}]^2} \tag{20.48}$$

The antiderivative for this integral exists and is given in the tables.

The coefficients a_{pl}, b_{pl}, and c_{pl} are determined by three points at which the wave function is calculated. The regular solution is calculated in an array of points r. For a given value of the shell radius R, our points closest to R are determined in the array of points r, namely at points: $r_{1,2} < R$ и $r_{3,4} > R$.

At all these points, the values of the regular solution are known. For the first three points, we have following system of equations

$$\begin{aligned} u_{pl}(r_1) &= a_{pl}r_1^2 + b_{pl}r_1 + c_{pl} \\ u_{pl}(r_2) &= a_{pl}r_2^2 + b_{pl}r_2 + c_{pl} \\ u_{pl}(r_3) &= a_{pl}r_3^2 + b_{pl}r_3 + c_{pl} \end{aligned} \tag{20.49}$$

This system of equations is solved, and the coefficients of the trinomial are found. The procedure is repeated with the next three points

$$\begin{aligned} u_{pl}(r_2) &= a_{pl}r_2^2 + b_{pl}r_2 + c_{pl} \\ u_{pl}(r_3) &= a_{pl}r_3^2 + b_{pl}r_3 + c_{pl} \\ u_{pl}(r_4) &= a_{pl}r_4^2 + b_{pl}r_4 + c_{pl} \end{aligned} \tag{20.50}$$

Thus, one obtains two sets of coefficients a_{kl}, b_{kl}, and c_{kl}. When calculating the integral (20.47), the arithmetic mean values of these coefficients are taken. As a result, irregular solution of the equation is given by an analytical formula, into which it is substituted $r = R$. The regular solution at the point R is

$$u_{pl}(R) = a_{pl}R^2 + b_{pl}R + c_{pl} \tag{20.51}$$

while the irregular solution is

$$\upsilon_{pl}(R) = -\frac{2a_{pl}R + b_{pl} + 2a_{pl}u(R)\tilde{a}_{plR}}{b_{pl} - 4a_{pl}c_{pl}}p \tag{20.52}$$

Here

$$\tilde{a}_{plR} = 2\frac{1}{\sqrt{4a_{pl}c_{pl} - b_{pl}}}\text{arctg}\frac{(2a_{pl}R + b_{pl})}{\sqrt{4a_{pl}c_{pl} - b_{pl}}},\ if\ 4a_{pl}c_{pl} > b_{pl} \tag{20.53}$$

$$\tilde{a}_{plR} = -\frac{1}{2a_{pl}R + b_{pl}}\ if\ 4a_{pl}c_{pl} = b_{pl} \tag{20.54}$$

$$\tilde{a}_{plR} = \frac{1}{\sqrt{b_{pl} - 4a_{pl}c_{pl}}}\ln\left(\left|\frac{2a_{pl}R + b_{pl} - \sqrt{b_{pl} - 4a_{pl}c_{pl}}}{2a_{pl}R + b_{pl} + \sqrt{b_{pl} - 4a_{pl}c_{pl}}}\right|\right)\ if\ 4a_{pl}c_{pl} < b_{pl} \tag{20.55}$$

and the wave function, after change of variables, is redefined using the expression.

$$\upsilon_{f1}(i) = \upsilon_f(i)\sqrt{\frac{r_1(i)}{(\alpha r_1(i) + \beta)}}\sqrt{\pi k}\ i = 1, N \tag{20.56}$$

where $\upsilon_f(i)$ is the original wave function.

Thus, for each wave function at each point along ω, we obtain the phase shift

$$\Delta_l(\omega) = \arccos\sqrt{\frac{u_{pl}(R)\upsilon_{pl}(R) + k/2V_0}{u_{pl}(R)\upsilon_{pl}(R) + u_p^2(R) + p/2V_0}} \tag{20.57}$$

and reflection factor (20.44).

20.1.3.2 Polarization Factor *G*

The dipole polarization factor $G_C(\omega)$ is determined using relation that follows from (20.37)

$$G_C(\omega) = \left[1 - \frac{\alpha_C(\omega)}{R^3}\right] \tag{20.58}$$

According to (20.58), $G_C(\omega)$ is expressed in terms of the dipole dynamic polarizability of fullerene $\alpha_C(\omega)$. It is determined by oscillator strength $g_{\nu\nu}$ and the dipole photoabsorption cross section of fullerene $\sigma(\omega)$ using relation [ACY, Chap. 1].

$$\alpha_C(\omega) = \sum_{j \le F, k > F} \frac{g_{jk}}{\omega_{jk}^2 - \omega^2} + \frac{c}{2\pi^2} \int_{I_C}^{\infty} \frac{\sigma(\omega')d\omega'}{\omega'^2 - \omega^2 + i\omega\eta}, \eta \to +0 \ \eta \to 0 \quad (20.59)$$

Here I_C is fullerene ionization potential.

Calculating $\alpha_C(\omega)$ from first principles is difficult. It requires knowledge of the fullerene electron wave functions, which, however, can be obtained only using different simplifying approaches, such as, for example, RPAE. However, in some cases, the fullerene photoabsorption total cross section $\sigma_C(\omega)$ is known. So, instead of calculating, the experimentally determined $\sigma_C(\omega)$ can be used, substituting it in (20.59). We prefer to employ experimental cross section $\sigma_C(\omega)$ that is directly proportional to the imaginary part of the dipole polarizability [BLP]

$$\text{Im}\alpha_C(\omega) = c\sigma_C(\omega)/4\pi\omega \quad (20.60)$$

For fullerene, it is also reasonable to neglect, outside poles in (20.59), the contribution of discrete oscillator strengths.

Dependence of $\sigma_C(\omega)$ upon ω is characterized by a powerful maximum called the giant resonance. For C_{60}, for example, it is located at about 22 eV. Experiments usually do not directly give absolute values of $\sigma_C(\omega)$. These can, however, be estimated reliably using a normalization procedure based on the sum rule:

$$(c/2\pi^2) \int \sigma_C(\omega)\text{d}\omega = N \quad (20.61)$$

Here N is the number of collectivized electrons. For C_{60}, it is equal to $N = 240$, which means four collectivized electrons from each carbon atom.

The dispersion relation for the real part $\text{Re}\alpha_C(\omega)$ has the form [BLP]

$$\text{Re}\alpha_C(\omega) = \frac{c}{2\pi^2} \int_{I_C}^{\infty} \frac{\sigma_C(\omega')\text{d}\omega'}{\omega'^2 - \omega^2} \quad (20.62)$$

Determination of fullerene quadrupole polarizability $\alpha_c^q(\omega)$ and the corresponding quadrupole polarization factor $G_c^q(\omega)$ could proceed in a similar way, but is hampered by lack of experimental data on the quadrupole photoabsorption cross section, which is very small.

20.1.3.3 Photoionization Cross Section and Angular Anisotropy Parameters

If potential (20.1) is accounted for in the lowest order of the perturbation theory in V_0, the endohedral atom photoionization cross section has the following form:

$$\sigma_{ij}(\omega) = |F_{l'}(\omega)|^2 \sigma_{ij}^A(\omega) \tag{20.63}$$

Here $\sigma_{ij}^A(\omega)$ is photoionization cross section of atom A due to $i \to j$ transition, which is obtained from (9.2) with an amplitude that is the solution of Eq. (9.55).

If potential (20.1) is taken into account within the framework of RPAE, then formula (20.63) is valid for the cross section, where $\sigma_{ij}^A(\omega)$ is also determined by (9.2), but using amplitude $\langle \nu_1 \| D(\omega) \| \nu_2 \rangle$, which is the solution of (20.32).

Taking the polarization factor into account adds the factor $|G_C(\omega)|^2$ (20.58) to (20.62).

The photoionization differential cross section $d\sigma_i/d\Omega$ of i-subshell, with inclusion of non-dipole corrections for unpolarized light, can be represented in the following form (11.4):

$$\frac{d\sigma_{nl}(\omega)}{d\Omega} = \frac{\sigma_{nl}(\omega)}{4\pi} \left\{ \begin{array}{l} 1 - \dfrac{\beta_{nl}(\omega)}{2} P_2(\cos\theta) \\ + \dfrac{\omega}{c}\left[\gamma_{nl}(\omega) P_1(\cos\theta) + \eta_{nl}(\omega) P_3(\cos\theta)\right] \end{array} \right\} \tag{20.64}$$

where $\beta_{nl}(\omega)$ is the dipole angular anisotropy parameter, $P_i(\cos\theta)$ with $i = 1$, 2, 3 are the Legendre polynomials, and $\gamma_{nl}(\omega)$ and $\eta_{nl}(\omega)$ are the non-dipole angular anisotropy parameters. In one-electron approximation, $\gamma_{nl}(\omega)$ and $\eta_{nl}(\omega)$ are expressed in terms of dipole $l_i \to l' = l_i \pm 1$ and quadrupole $l_i \to l' = l_i \pm 2, 0$ matrix elements, respectively.

In one-electron HF approximation, the dipole angular anisotropy parameter is expressed via reduced dipole matrix elements (9.20) and (9.21)

$$\left\langle nl \left\| \hat{d} \right\| \varepsilon, l \pm 1 \right\rangle = (-1)^{l_>} \sqrt{l_>} \int_0^\infty P_{nl}(r) r P_{\varepsilon l \pm 1}(r) dr \equiv d_{l \pm 1} \tag{20.65}$$

($l_> = l + 1$ for transition to state $l + 1$ and $l_> = l$ for transition to state $l - 1$) by the following relation

$$\beta_{nl}(\omega) = \frac{1}{(2l+1)\left[\left(\tilde{F}_{l-1} d_{l-1}\right)^2 + \left(\tilde{F}_{l+1} d_{l+1}\right)^2\right]} \left[\begin{array}{l} (l-1)\left(\tilde{F}_{l-1} d_{l-1}\right)^2 + (l+2)\left(\tilde{F}_{l+1} d_{l+1}\right)^2 \\ -6\sqrt{l(l+1)}\left(\tilde{F}_{l-1} d_{l-1}\right)\left(\tilde{F}_{l+1} d_{l+1}\right) \cos\left(\tilde{\delta}_{l-1} - \tilde{\delta}_{l+1}\right) \end{array} \right]. \tag{20.66}$$

Here the reflection factor is presented as $F_{l'} \equiv F_{l'}(\varepsilon) = \tilde{F}_{l'}(\varepsilon) \exp[i\Delta_{l'}(\varepsilon)]$ where $\tilde{F}_{l'}(\varepsilon)$ is real; $\tilde{\delta}_{l'}(\varepsilon) = \delta_{l'}(\varepsilon) + \Delta_{l'}(\varepsilon)$, $\delta_{l'}(\varepsilon)$ is the photoelectron phase, ε is its energy, $\varepsilon = \omega - I_{nl}$, and I_{nl} is the i-shell ionization potential. The parameter $\beta_{nl}(\omega)$, including the multi-electron correlations in RPAE (or GRPAE), is expressed through the dipole matrix elements $\left\langle \varepsilon, l\pm 1 \middle\| \hat{D}(\omega) \middle\| nl \right\rangle \equiv D_{l\pm 1}$, which in RPAE become complex quantities. Therefore, expression (20.66) becomes more complicated

$$\beta_{nl}^{AC}(\varepsilon) = \frac{1}{(2l+1)\left[\left(\tilde{F}_{l-1}\tilde{D}_{l-1}\right)^2 + \left(\tilde{F}_{l+1}\tilde{D}_{l+1}\right)^2\right]} \begin{array}{l} \times[(l-1)(\tilde{F}_{l-1}\tilde{D}_{l-1})^2 + (l+2)(\tilde{F}_{l+1}\tilde{D}_{l+1})^2 \\ +6\sqrt{l(l+1)}\tilde{F}_{l-1}\tilde{F}_{l+1}\tilde{D}_{l-1}\tilde{D}_{l+1}\cos(\tilde{\tilde{\delta}}_{l+1} - \tilde{\tilde{\delta}}_{l-1})] \end{array}, \quad (20.67)$$

where $D_{l'} \equiv D_{l'}(\varepsilon) = \tilde{D}_{l'}(\varepsilon)\exp[i\xi_{l'}(\varepsilon)]$, $\tilde{D}_{l'}(\varepsilon)$ is purely really; $\tilde{\tilde{\delta}}_{l'}(\varepsilon) = \tilde{\delta}_{l'}(\varepsilon) + \xi_{l'}(\varepsilon)$. Dipole angular anisotropy parameter is independent upon polarization factor $G_C(\omega)$.

Non-dipole angular anisotrop parameters in RPAE are calculated using formulas

$$\gamma_{nl}^{AC}(\omega) = \frac{6G_C^q(\omega)}{5G_C^d(\omega)\left[(l+1)\left(\tilde{F}_{l+1}\tilde{D}_{l+1}\right)^2 + l\left(\tilde{F}_{l-1}\tilde{D}_{l-1}\right)^2\right]} \times\left\{\frac{(l+1)\tilde{F}_{l+1}}{2l+3}\left[\begin{array}{l} 3(l+2)\tilde{F}_{l+2}\tilde{Q}_{l+2\uparrow\downarrow}\tilde{D}_{l+1}\cos\left(\tilde{\tilde{\delta}}_{l+2} - \tilde{\tilde{\delta}}_{l+1}\right) \\ -l\tilde{F}_l\tilde{Q}_l\tilde{D}_{l+1}\cos\left(\tilde{\tilde{\delta}}_{l+2} - \tilde{\tilde{\delta}}_{l+1}\right) \end{array}\right] - \frac{l\tilde{F}_{l-1}}{2l+1}\left[\begin{array}{l} 3(l-1)\tilde{F}_{l-2}\tilde{Q}_{l-2}\tilde{D}_{l-1}\cos\left(\tilde{\tilde{\delta}}_{l-2} - \tilde{\tilde{\delta}}_{l-1}\right) \\ -(l+1)\tilde{F}_l\tilde{Q}_l\tilde{D}_{l-1}\cos\left(\tilde{\tilde{\delta}}_l - \tilde{\tilde{\delta}}_{l-1}\right) \end{array}\right]\right\}, \quad (20.68)$$

$$\begin{aligned} \eta_{nl}^{AC}(\omega) = {} & \frac{6G_C^q(\omega)}{5G_C^d\left[(l+1)\left(\tilde{F}_{l+1}\tilde{D}_{l+1}\right)^2 + l\left(\tilde{F}_{l-1}\tilde{D}_{l-1}\right)^2\right]} \times \\ & \left\{\frac{(l+1)(l+2)}{(2l+1)(2l+3)}\tilde{F}_{l+2}\tilde{Q}_{l+2}\left[5l\tilde{F}_{l-1}\tilde{D}_{l-1}\cos\left(\tilde{\tilde{\delta}}_{l+2} - \tilde{\tilde{\delta}}_{l-1}\right) - \right.\right. \\ & \left.(l+3)\tilde{F}_{l+1}\tilde{D}_{l+1}\cos\left(\tilde{\tilde{\delta}}_{l+2} - \tilde{\tilde{\delta}}_{l-1}\right)\right] - \frac{(l-1)l}{(2l+1)(2l+1)}\tilde{F}_{l-2}\tilde{Q}_{l-2}\times \\ & \left[5(l+1)\tilde{F}_{l+1}\tilde{D}_{l+1}\cos\left(\tilde{\tilde{\delta}}_{l-2} - \tilde{\tilde{\delta}}_{l+1}\right) - (l-2)\tilde{F}_{l-1}\tilde{D}_{l-1}\cos\left(\tilde{\tilde{\delta}}_{l-2} - \tilde{\tilde{\delta}}_{l-1}\right)\right] + \\ & \left. 2\frac{l(l+1)\tilde{F}_l\tilde{Q}_l}{(2l-1)(2l+3)}\left[(l+2)\tilde{F}_{l+1}\tilde{D}_{l+1}\cos\left(\tilde{\tilde{\delta}}_l - \tilde{\tilde{\delta}}_{l+1}\right) - (l-1)\tilde{F}_{l-1}\tilde{D}_{l-1}\cos\left(\tilde{\tilde{\delta}}_l - \tilde{\tilde{\delta}}_{l-1}\right)\right]\right\} \end{aligned} \quad (20.69)$$

Here, the quadrupole matrix element $\tilde{Q}_{l'}$ is determined by relations (11.10) and (11.18), $Q_{l'}(\omega) = \tilde{Q}_{l'}(\omega)\exp(i\xi_{l'})$ and other notations are the same as in (20.67).

The same parameters can be calculated if A is an atom with semi-closed subshells. To obtain the values $\sigma^{AC(r)}_{nl,kl'\uparrow,\downarrow}(\omega)$, $\beta^{AC(r)}_{nl\uparrow\downarrow}(\omega)$, $\gamma^{AC(r)}_{nl\uparrow\downarrow}(\omega)$, and $\eta^{AC(r)}_{nl\uparrow\downarrow}(\omega)$, one should use formulas (20.63) and (20.66)–(20.69), replacing in them $\tilde{D}_{l'}(\omega), \tilde{Q}_{l''}(\omega)$, $\tilde{\tilde{\delta}}_{l'''}$, and $F_{l'}(k)$ by either $\tilde{D}_{l'\uparrow}(\omega), \tilde{Q}_{l''\uparrow}(\omega)$, $\tilde{\tilde{\delta}}_{l'''\uparrow}$, and $F_{l'\uparrow}(\omega)$ for photoelectrons with spin "up" ($\uparrow$), or by $\tilde{D}_{l'\downarrow}(\omega), \tilde{Q}_{l''\downarrow}(\omega)$, $\tilde{\tilde{\delta}}_{l'''\downarrow}$, and $F_{l'\downarrow}(\omega)$ for photoelectrons with spin "down" ($\downarrow$) (see paragraph 9.2.3).

If fullerene is described by a potential of finite thickness instead of (20.1), the expressions for cross section and angular anisotropy parameters, both dipole and non-dipole, are of the same form as those given in Chaps. 9 and 11 for an atom, and the polarization factor is also included, as in the case of the above-considered bubble potential of zero thickness (20.1). One-electron wave functions states are calculated as atomic in the HF approximation, but with an additional potential that describes the fullerene shell.

References

1. R.A. Phaneuf, A.L.D. Kilcoyne, N.B. Aryal, et.al., Phys. Rev. A **88**, 053402 (2013)
2. M.Ya. Amusia, Chem. Phys. **414**, 168 (2013)
3. M.Ya. Amusia, A.S. Baltenkov, Phys. Lett. A **360** (2006)
4. V. Averbukh, I.B. Müller, L.S. Cederbaum, Phys. Rev. Lett., **93**, 263002 (2004)
5. M.Ya. Amusia, A.S. Baltenkov, Phys. Rev. A **73**, 063206 (2006)
6. M.Ya. Amusia, A.S. Baltenkov, B.G. Krakov. Phys. Let. A **243** (1998)
7. M.Ya. Amusia, A.S. Baltenkov, U. Becker, Phys. Rev. A **62**, 012701 (2000)
8. J. Luberek, G. Wendin, Chem. Phys. **248**(3–4), 147 (1996)

Chapter 21
Inelastic Scattering of Fast Electrons by Endohedrals

In this chapter, we consider the cross section for scattering of fast electrons by endohedrals [1, 2]. This cross section is expressed in terms of generalized oscillator strength (GOS). Formulas are given for calculating the angular distributions of secondary electrons emitted from the endohedral during its ionization. The general approach is similar to that developed in Chaps. 10 and 13 for isolated atoms.

21.1 Theoretical Model

21.1.1 Endohedral with One Fullerene Shell

Just as for an isolated atom, the differential in angle cross section for inelastic scattering of fast electrons or other charged particles by endohedrals in first Born approximation is determined by formulas from [LL] and (10.1) for the excitation of the target and (10.2) for its ionization. These formulas relate the cross sections to the so-called generalized oscillator strengths in (10.1) or the density of generalized oscillator strengths in (10.2).

As in Chap. 10, generalized oscillator strengths (or their densities) of the transition $\nu_1 \to \nu_2$ are denoted as $f_{\nu_1\nu_2}(\mathbf{q},\omega)$ (GOS). In the HF approximation for endohedrals, as well as for atoms or molecules, $f_{\nu_1\nu_2}(q,\omega)$ can be presented in two different forms (10.1), named: length $f^{(r)}$ and velocity $f^{(\nabla)}$ forms, respectively:

$$f^{(r)}_{\nu_1\nu_2}(\mathbf{q},\omega) = \frac{2\omega}{q^2}\left|\langle\nu_1|g^{(r)}(\mathbf{q})|\nu_2\rangle\right|^2, \tag{21.1}$$

$$f^{(\nabla)}_{\nu_1\nu_2}(\mathbf{q},\omega) = \frac{1}{2q^2\omega}\left|\langle\nu_1|g^{(\nabla)}(\mathbf{q})|\nu_2\rangle\right|^2, \tag{21.2}$$

M. Ya. Amusia and L. V. Chernysheva, *Computation of Atomic and Molecular Processes*,
Springer Series on Atomic, Optical, and Plasma Physics 117,
https://doi.org/10.1007/978-3-030-85143-9_21

where

$$g^{(r)}(\mathbf{q}) = exp(i\mathbf{q}\mathbf{r}),$$
$$g^{(\nabla)}(\mathbf{q}) = exp(i\mathbf{q}\mathbf{r})(\mathbf{q}\nabla) - (\mathbf{q}\overleftarrow{\nabla})exp(i\mathbf{q}\mathbf{r}) \tag{21.3}$$

and $|\nu_2\rangle$, $\langle\nu_1|$ are the wave functions of final and initial states of an endohedral electron in the HF approximation. The arrow above operator ∇ indicates that it acts to the left; ν_2 and ν_1 are quantum numbers of final and initial endohedral states, respectively, q is the momentum transferred to endohedral atom in the scattering process, $\mathbf{q} = \mathbf{p}_2 - \mathbf{p}_1$, $\omega = E_{\nu_2} - E_{\nu_1}$ is the transferred energy.

The process that corresponds to Eqs. (21.1) or (21.2) for an endohedral in RPAE is depicted by the same diagrams (1.50), (1.51), and (10.6) as for atoms or molecules.

We consider endohedrals as spherical systems. Therefore, it is convenient to represent the interaction operator of fast incident particle with an endohedral using the expansion in terms of Legendre polynomials:

$$\exp(i\mathbf{q}\mathbf{r}) = \sum (2L+1) j_L(qr) P_L(\cos\varsigma). \tag{21.4}$$

Here $j_L(qr)$ is the spherical Bessel function, $P_L(\cos\varsigma)$—Legendre polynomial, and ς is the angle between vectors $\boldsymbol{q}$ and $\boldsymbol{r}$.

In the HF approximation, the endohedral electron wave functions are sought in the same way as described in Chap. 20, i.e., as products of the radial, angular, and spin parts. Separating angular and spin parts of one-electron HF wave function and using (21.4), we analytically perform angular integration in matrix elements of Eq. (21.1) or (21.2). As a result, along with $g_{fi}(\omega, q)$, it is possible to introduce partial one-electron GOS $g^L_{nl,\varepsilon l'}(\mathbf{q}, \omega)$, with l', which values lie within the following limits $|L - l| \le l' \le L + l$.

An endohedral is an atom A of radius r_A, placed in a fullerene shell of radius R and thickness b. We consider only those endohedrals, in which atom A is located at the spherical fullerene shell center. Endohedral radius R is noticeably larger than fullerenes shell thickness b and atom A radius r_A. Thus, as simplifying factors, we use strong inequalities $R \gg r_A$ and $R \gg b$.

We assume that fullerene shell acts upon atom A electrons with its static potential. We also include the fullerene shell polarization under the incident electron action. In the simplest case, we neglect fullerene shell thickness, i.e., assume that $b \to 0$. In this case, the fullerene potential is determined by (20.1), i.e., $V(r) = -V_0\delta(r - R)$. As shown in Chap. 20, accounting for such a potential is reduced to multiplying one-electron functions of the atom A by so-called fullerene shell reflection factor $F_{l'}(p)$, which is presented as (see Sect. 20.2.1):

$$F_l(p) = \cos\Delta\delta_l(p)\left[1 - \mathrm{tg}\Delta\delta_l(p)\frac{\upsilon_{kl}(R)}{u_{pl}(R)}\right]. \tag{21.5}$$

Here $\Delta\delta_l(p)$ corrects the photoelectron elastic scattering phase as a result of potential (20.1) action. It can be expressed via potential V_0 using the following relation:

$$\mathrm{tg}\Delta\delta_l(p) = \frac{u_{pl}^2(R)}{u_{pl}(R)\upsilon_{pl}(R) - p/2V_0} \tag{21.6}$$

Expression for GOS of i- > f transition with multipolarity L in the HF approximation, $f_{fi}^L(\mathbf{q}, \omega_{fi}^{HF})$, follows from formula (21.1) or (21.2)

$$\begin{aligned} f_{fi}^L(q, \omega_{fi}) = \frac{2\omega_{fi}\left|F_{l_f}(p)\right|^2}{q^2}\left|\int \phi_f^*(\mathbf{r}) j_L(qr) P_L(cos\vartheta)\phi_i(\mathbf{r})d\mathbf{r}\right|^2 = \\ \frac{2\omega_{fi}\left|F_{l_f}(p)\right|^2}{q^2}|\langle f|j_L(qr)|i\rangle|^2, \end{aligned} \tag{21.7}$$

where $p^2/2 = \omega_{fi} - I_i$.

To find GOS with account of electron correlations in RPAE, matrix elements of the operator $g^{(r,\nabla)}(\mathbf{q})$ in formula (21.7) should be replaced by solutions of the equation

$$\begin{aligned} \left\langle f\left|\hat{G}(\boldsymbol{q},\omega)\right|i\right\rangle = \langle f|g^{(r,\nabla)}(\boldsymbol{q})|i\rangle \\ + \left(\sum_{n'\le F, p'>F} - \sum_{n'>F, p'\le F}\right) \frac{\left\langle p'\left|\hat{G}(\boldsymbol{q},\omega)\right|n'\right\rangle\left|F_{l_{p'}}(p')\right|^2\left|F_{n'l_{n'}}\right|^2\langle n'f|U|p'i\rangle}{\omega - \varepsilon_{p'} + \varepsilon_{n'} + i\eta(1-2n_{p'})}. \end{aligned} \tag{21.8}$$

This equation is an obvious generalization of Eq. (9.56) for photoionization amplitude to the case of a GOS. For spherically symmetric endohedrals, (21.8) can be reduced to a system of uncoupled equations for the L-components of GOS

$$\begin{aligned} \left\langle f\left|\hat{G}_L(q,\omega)\right|i\right\rangle = \langle f|g^{L(r,\nabla)}(q)|i\rangle \\ + \left(\sum_{n'\le F, p'>F} - \sum_{n'>F, p'\le F} \frac{\left\langle p'\left|\hat{G}_L(q,\omega)\right|n'\right\rangle\left|F_{l_{p'}}(p')\right|^2|F_{n'l'}|^2\langle n'f|U_L|p'i\rangle}{\omega - \varepsilon_{p'} + \varepsilon_{n'} + i\eta(1-2n_{p'})}\right) \end{aligned} \tag{21.9}$$

Using (21.1), one can obtain the following relations for amplitudes $D^{AC(r)}$ and $Q^{AC(r)}$ of an endohedral atom, expressed in terms of corresponding values for an isolated atom, which correspond to $nl \rightarrow \varepsilon l'$ transitions. The length and velocity forms in RPAE give the same results. Therefore, one can omit the superscripts (r, ∇)

$$G_{nl,pl',L}^{AC}(\omega, q) = F_{l'}(p)G_{nl,pl',L}(\omega, q)\exp[i\Delta\delta_{l'}(p)], \tag{21.10}$$

Here, the superscript AC refers to the endohedral, and A to the atom A.

Inclusion of the polarization factor leads, as described in Sect. 20.1.5, to the factor $G_C^L(\omega)$, so we have:

$$G_{nl,pl',L}^{AC(G)}(\omega,q) = F_{l'}(p)G^L(\omega)G_{nl,pl',L}(\omega,q)\exp[i\Delta\delta_{l'}(p)], \tag{21.11}$$

Separating angular and spin variables and performing integration over angles, and summing over spins, we obtain, instead of Eq. (21.9), a system of separate equations for the reduced matrix elements:

$$\langle n_1l_1\|G_L(q,\omega)\|p_2l_2\rangle = \left\langle n_1l_1\left\|g^{L(r)}(q)\right\|p_2l_2\right\rangle + \left(\sum_{\substack{\tilde{\nu}_3=n_3l_3\le F,\\ \tilde{\nu}_4=p_4l_4>F}} - \sum_{\substack{\tilde{\nu}_3=p_3l_3>F,\\ \tilde{\nu}_4=n_4l_4\le F}}\right)\frac{\langle\tilde{\nu}_4\|G_L(q,\omega)\|\tilde{\nu}_3\rangle\left|F_{l_4}(p_4)\right|^2\left|F_{n_3l_3}\right|^2\langle\tilde{\nu}_3\tilde{\nu}_1\|U_L\|\tilde{\nu}_4\tilde{\nu}_2\rangle}{(2L+1)\left[\omega - E_{\tilde{\nu}_4} + E_{\tilde{\nu}_3} + i\delta(1-2n_{\tilde{\nu}_4})\right]} \tag{21.12}$$

The definition of the reduced dipole matrix element is given by formula (9.56).

As a rule, in calculations it is possible to neglect the fullerene potential action upon wave functions of electrons in occupied atomic states. In this case, instead of products of the squares of modulus $\left|F_{l_4}(k_4)\right|^2\left|F_{n_3l_3}\right|^2$, Eqs. (21.9) and (21.12) include only the term $\left|F_{l_4}(k_4)\right|^2$ for excited and continuous spectrum states $\tilde{\nu}_{4(3)} = p_{4(3)}l_{4(3)}$.

Note that the factor $F_l(p)$ as a function of p has strong resonances, which are reflected in $\sigma_{nl,pl'}^{AC}(\omega)$ and are called *confinement resonances*. The dipole polarization factor $G^d(\omega)$ has a maximum in the vicinity of the giant resonance of the fullerene photoabsorption cross section (see Sect. 20.1.5).

If the fullerene shell is described by a potential of a finite rather than zero thickness *V(r)*, one-particle HF states should be found by solving Eq. (20.33) and the characteristics of inelastic scattering should be found using the same formulas as in Chaps. 10 and 13 for an isolated atom.

21.1.2 Angular Distribution of Secondary Electrons

Formulas obtained below differ from analogous formulas in Chap. 13 only by accounting for reflection and polarization factors. Amplitudes $G_{nl,kl',L}^{AC(G)}(q,\omega)$ make it possible to calculate the total and differential in scattering angle of fast incident electron cross sections. In general, the calculation of the differential cross section is extremely cumbersome. Here, it is calculated for small q [ACY, Sect. 3.10].

In this region of transferred momenta, the dipole contribution to the amplitude $G_{nl,kl',L}^{AC}(q,\omega)$ and the GOS itself $G_{nl,kl',L}^{AC}(q,\omega)$ dominates. GOS is simply proportional to the photoionization cross section $\sigma_{nl}(\omega)$ (see Chap. 10 and [AM]).

For the angle-differential GOS density of the endohedral nl-subshell $dF(q,\omega)/d\Omega$, the following relation is valid in RPAE

$$\frac{dF_{nl}^{AC(G)}(q,\omega)}{d\Omega} = \sum_{L'L} \frac{dF_{nl}^{AC\,L',L}(q,\omega)}{d\Omega} = \frac{\omega\pi}{q^2}\sum_{L'L}(2L'+1)(2L''+1)i^{L'-L''}$$
$$\times \sum_{l'=|L'-l|}^{L'+l}\sum_{l=|L''-l|}^{L''+l} G_C^{L'}(\omega)G_C^{L''}(\omega)\tilde{F}_{l'}(k)\tilde{F}_{l''}(k)\tilde{G}_{kl'L'}(q)\tilde{G}_{klL''}(q)i^{l''-l'}$$
$$\times (2l'+1)(2l+1)e^{i(\bar{\tilde{\delta}}_{l'}-\bar{\tilde{\delta}}_{l})}\begin{pmatrix} L' & l & l' \\ 0 & 0 & 0 \end{pmatrix}\begin{pmatrix} l & l & L'' \\ 0 & 0 & 0 \end{pmatrix}$$
$$\times \sum_{L=|l'-l|}^{l'+l} P_L(\cos\theta)(-1)^{L+l}(2L+1)\begin{pmatrix} l' & L & l \\ 0 & 0 & 0 \end{pmatrix}\begin{pmatrix} L & L' & L'' \\ 0 & 0 & 0 \end{pmatrix}\begin{Bmatrix} L & L' & L'' \\ l & l & l' \end{Bmatrix}. \quad (21.13)$$

The partial nl-subshell GOS value $F_{nl}(q,\omega)$ in RPAE is obtained from (21.13) by integration over $d\Omega$, leading to the following expression:

$$F_{nl}^{(G)}(q,\omega) = \sum_L F_{nl}^L(q,\omega) =$$
$$\frac{4\omega\pi^2}{q^2}\sum_L(2L+1)\sum_{l'=|L-l|}^{L+l}[G^L(\omega)\tilde{F}_{l'}(k)\tilde{G}_{kl'L}(q)]^2(2l'+1)\begin{pmatrix} L & l & l' \\ 0 & 0 & 0 \end{pmatrix}^2. \quad (21.14)$$

To compare the results obtained using the well-known formulas for photoionization with allowance for low-order non-dipole corrections, we consider such small values of q for which it is sufficient to include terms with $L \le 2$ only. In this case, the angular distribution of GOS for endohedrals is described by a formula analogous to (13.5), and can be represented, similar to the case of photoionization, as (11.4) (cf. (20.64)):

$$\frac{dF_{nl}^{(G)}(q,\omega)}{d\Omega} = \frac{F_{nl}^{(G)}(q,\omega)}{4\pi}\Bigg\{1 - \frac{\beta_{nl}^{(in)}(\omega,q)}{2}P_2(\cos\theta) +$$
$$q\Big[\gamma_{nl}^{(in)}(\omega,q)P_1(\cos\theta) + \eta_{nl}^{(in)}(\omega,q)P_3(\cos\theta) + \lambda_{nl}^{(in)}(\omega,q)P_4(\cos\theta)\Big]\Bigg\}. \quad (21.15)$$

The obvious difference from the photoionization case is q-dependence of the coefficients and appearance of an additional term $\varsigma_{nl}^{(in)}(\omega,q)P_4(\cos\theta)$. Even so, the expressions for $\beta_{nl}^{(in)}(\omega,q)$, $\gamma_{nl}^{(in)}(\omega,q)$, $\eta_{nl}^{(in)}(\omega,q)$, and $\lambda_{nl}^{(in)}(\omega,q)$ through $g_{kl'L'}(q)$ are too complex compared to the expressions for $\beta_{nl}(\omega)$, $\gamma_{nl}(\omega)$, and $\eta_{nl}(\omega)$ in photoionization. Therefore, it is more convenient to present the results for s-, p-, and d-subshells separately.

Let us start with s-subshells, for which, as follows from (13.5), the differential GOS in the above-mentioned L', $L \le 2$ approximation is given by following relations [ACY, Sect. 3.10].

$$
\begin{aligned}
\frac{\mathrm{d}F_{n0}^{AC(G)}(q,\omega)}{d\Omega} &= \sum_{L',L=0}^{2} \frac{\mathrm{d}F_{n0}^{AC\,L',L}(q,\omega)}{d\Omega} = \frac{F_{n0}^{AC(G)}(q,\omega)}{4\pi} \\
&\times \{1 + \frac{6}{W_0^{AC}} \tilde{G}_C^d \tilde{F}_1 \tilde{G}_{11} \begin{bmatrix} \tilde{G}_C^m \tilde{F}_0 \tilde{G}_{00} \cos(\tilde{\tilde{\delta}}_0 - \tilde{\tilde{\delta}}_1) \\ +2\tilde{G}_C^q \tilde{F}_2 \tilde{G}_{22} \cos(\tilde{\tilde{\delta}}_1 - \tilde{\tilde{\delta}}_2) \end{bmatrix} \\
&\times P_1(\cos\theta) + \frac{2}{7W_0^{AC}} \begin{bmatrix} 21\tilde{G}_C^{d2} \tilde{F}_1^2 \tilde{G}_{11}^2 + 5\tilde{G}_C^q \tilde{F}_2 \tilde{G}_{22} \\ \Big(7\tilde{G}_C^m \tilde{F}_0 \tilde{G}_{00} \cos(\tilde{\tilde{\delta}}_0 - \tilde{\tilde{\delta}}_2) \cos(\tilde{\tilde{\delta}}_0 - \tilde{\tilde{\delta}}_2) \end{bmatrix} \\
&+ \left. 5\tilde{G}_C^q \tilde{F}_2 \tilde{G}_{22}\Big) 5\tilde{G}_C^q \tilde{F}_2 \tilde{G}_{22} \right] P_2(\cos\theta) \\
&+ \frac{18}{7W_0^{AC}} \tilde{G}_C^d \tilde{F}_1 \tilde{G}_C^q \tilde{F}_2 \tilde{G}_{11} \tilde{G}_{22} \cos(\tilde{\tilde{\delta}}_1 - \tilde{\tilde{\delta}}_2) P_3(\cos\theta) \\
&+ \left. \frac{90}{7W_0^{AC}} \tilde{G}_C^{q2} \tilde{F}_2^2 \tilde{G}_{22}^2 P_4(\cos\theta) \right\} \equiv \frac{F_{ns}^{AC(G)}(q,\omega)}{4\pi} \\
&\left\{1 - \frac{1}{2}\beta_{ns}^{AC(in)}(q,\omega) P_2(\cos\theta) + q \begin{bmatrix} \gamma_{ns}^{AC(in)}(q,\omega) P_1(\cos\theta) + \eta_{ns}^{AC(in)}(q,\omega) \\ P_3(\cos\theta) + \lambda_{ns}^{AC(in)}(q,\omega) P_4(\cos\theta) \end{bmatrix} \right\}
\end{aligned}
\tag{21.16}
$$

where

$$
F_{ns}^{AC(G)} = \frac{4\pi^2\omega}{(\omega - I_{n0})q^2} W_0^{AC}, \quad W_0^{AC} = \tilde{G}_C^{m2} \tilde{F}_0^2 \tilde{G}_{00}^2 + 3\tilde{G}_C^{d2} \tilde{F}_1^2 \tilde{G}_{11}^2 + 5\tilde{G}_C^{q2} \tilde{F}_2^2 \tilde{G}_{22}^2,
\tag{21.17}
$$

and $\tilde{\tilde{\delta}}_{lL} = \delta_{l'} + \Delta_{l'} + \Delta\delta_{l'} + \xi_{l'} + \zeta$ (see (20.62)).

To simplify the writing of very cumbersome formulas, we use the following notation. The reflection factor modulus for angular momentum with values from zero to 4 is denoted as $\tilde{F}_{0\div4}(k) \equiv \tilde{F}_{0\div4}$, and the modulus of the monopole m, dipole d, and quadrupole q polarization factors—as $\tilde{G}_C^{m,d,q}(\omega) \equiv \tilde{G}_C^{m,d,q}$, respectively.

Expressions for p- and d-subshells are much more complicated than expressions for s. On the other hand, they are of greater importance and interest, since the corresponding outer p- and d-subshells are much larger than the s-shells. In them, especially in the $4d$-subshell, multielectron effects are very strong, as evidenced by presence of the well-known dipole Giant resonance in the $4d^{10}$-subshell of Xe.

For differential GOS of p-subshells of endohedrals, $l = 1$, expressions similar to (13.9) and (13.10) are valid [ACY, Sect. 3.10]:

$$
\begin{aligned}
&\frac{dF_{np}^{AC(G)}(q,\omega)}{d\Omega} = \sum_{L',L=0}^{2} \frac{dF_{np}^{AC\,L',L}(q,\omega)}{d\Omega} = \frac{F_{np}^{AC(G)}}{4\pi}\{1+ \\
&\frac{1}{5W_1^{AC}}\Big[10\tilde{F}_1\tilde{F}_0\tilde{G}_{01}(2\tilde{G}_C^q\tilde{G}_{12} - \tilde{G}_C^m\tilde{G}_{10})\cos(\tilde{\tilde{\delta}}_0 - \tilde{\tilde{\delta}}_1) + 4\tilde{G}_C^d\tilde{F}_2\tilde{G}_{21}\times \\
&\Big((5\tilde{G}_C^d\tilde{F}_0\tilde{G}_{10} - \tilde{G}_C^d\tilde{F}_2\tilde{G}_{12})\cos(\tilde{\tilde{\delta}}_1 - \tilde{\tilde{\delta}}_2) + 9\tilde{G}_C^q\tilde{F}_3\tilde{G}_{32}\cos(\tilde{\tilde{\delta}}_2 - \tilde{\tilde{\delta}}_3)\Big)\Big]P_1(\cos\theta)+ \\
&\frac{1}{7W_1^{AC}}\Big[7\tilde{G}_C^{d2}\tilde{F}_2\tilde{G}_{21}\Big(\tilde{F}_2\tilde{G}_{21} - 2\tilde{F}_0\tilde{G}_{01}\cos(\tilde{\tilde{\delta}}_0 - \tilde{\tilde{\delta}}_2)\Big) + 7\tilde{G}_C^q\tilde{F}_1^2\tilde{G}_{12}(\tilde{G}_C^q\tilde{G}_{12} \\
&-2\tilde{G}_C^m\tilde{G}_{10}) + 3\tilde{G}_C^q\tilde{F}_3\tilde{G}_{32}\Big((7\tilde{G}_C^m\tilde{F}_1\tilde{G}_{10} - 2\tilde{G}_C^q\tilde{F}_1\tilde{G}_{12})\cos(\tilde{\tilde{\delta}}_1 - \tilde{\tilde{\delta}}_3) + 4\tilde{G}_C^q\tilde{F}_3(k)\tilde{G}_{32}\Big)\Big] \\
&P_2(\cos\theta) - \frac{3}{5W_1^{AC}}\Big[6\tilde{G}_C^d\tilde{G}_C^q\tilde{F}_1\tilde{F}_2\tilde{G}_{21}\tilde{G}_{12}\cos(\tilde{\tilde{\delta}}_1 - \tilde{\tilde{\delta}}_2) + \tilde{G}_C^q\tilde{F}_3\tilde{G}_{32}\times \\
&\Big(5\tilde{G}_C^d\tilde{F}_0\tilde{G}_{01}\cos(\tilde{\tilde{\delta}}_0 - \tilde{\tilde{\delta}}_3) - 4\tilde{G}_C^d\tilde{F}_2\tilde{G}_{21}\cos(\tilde{\tilde{\delta}}_2 - \tilde{\tilde{\delta}}_3)\Big)\Big]P_3(\cos\theta)+ \\
&\frac{18}{7W_1^{AC}}\tilde{G}_C^{q2}\tilde{F}_3\tilde{G}_{32}\Big[\tilde{F}_3\tilde{G}_{32} - 4\tilde{F}_1\tilde{G}_{12}\cos(\tilde{\tilde{\delta}}_1 - \tilde{\tilde{\delta}}_3)\Big]P_4(\cos\theta)\Big\} \equiv \frac{F_{np}^{AC(G)}(q,\omega)}{4\pi}\times \\
&\Big\{1 - \frac{\beta_{np}^{(in)}(\omega,q)}{2}P_2(\cos\theta) + q\Big[\gamma_{np}^{(in)}(\omega,q)P_1(\cos\theta) + \eta_{np}^{(in)}(\omega,q)P_3(\cos\theta) + \\
&\lambda_{np}^{(in)}(\omega,q)P_4(\cos\theta)\Big]\Big\},
\end{aligned}
\tag{21.18}
$$

where

$$F_{np}^{AC(G)} = \frac{4\pi^2\omega}{q^2}W_1^{AC}, \tag{21.19}$$

$$W_1^{AC} = \tilde{G}_C^{m2}\tilde{F}_1^2\tilde{G}_{10}^2 + \tilde{G}_C^{\beta 2}\tilde{F}_0^2\tilde{G}_{01}^2 + 2\Big[\tilde{G}_C^{d2}\tilde{F}_2^2\tilde{G}_{21}^2 + \tilde{G}_C^{q2}\tilde{F}_1^2\tilde{G}_{12}^2\Big] + 3\tilde{G}_C^{q2}\tilde{F}_3^2\tilde{G}_{32}^2. \tag{21.20}$$

For differential GOS of d-subshells of endohedrals, $l = 2$, we have the following expression [ACYa2, Sect. 3.10]

$$
\begin{aligned}
&\frac{dF_{nd}^{AC(G)}(q,\omega)}{d\Omega} = \frac{F_{nd}^{AC(G)}}{4\pi} \\
&\times\Big\{1 + \frac{6}{W_2^{AC}}\Big[14\tilde{G}_C^d\tilde{F}\tilde{G}_{11}\Big(\tilde{G}_C^q\tilde{F}_{22}\tilde{G}_{22} - \tilde{G}_C^m\tilde{F}_2\tilde{G}_{20}\Big)\cos\Big(\tilde{\tilde{\delta}}_1 - \tilde{\tilde{\delta}}_2\Big) \\
&- 14\tilde{G}_C^d\tilde{G}_C^q\tilde{F}_0\tilde{F}_1\tilde{G}_{11}\tilde{G}_{02}\cos\Big(\tilde{\tilde{\delta}}_0 - \tilde{\tilde{\delta}}_1\Big) + 3\tilde{G}_C^d\tilde{F}_3\tilde{G}_{31} \\
&\times\Big(7\tilde{G}_C^m\tilde{F}_2\tilde{G}_{20} - 2\tilde{G}_C^q\tilde{F}_2\tilde{G}_{22}\Big)\cos\Big(\tilde{\tilde{\delta}}_2 - \tilde{\tilde{\delta}}_3\Big) + 12\tilde{G}_C^q\tilde{F}_4\tilde{G}_{42}\cos\Big(\tilde{\tilde{\delta}}_3 - \tilde{\tilde{\delta}}_4\Big)\Big]P_1(\cos\theta) \\
&+ \frac{2}{245W_2^{AC}}\Big[1029\Big(\tilde{G}_C^{d2}\tilde{F}_1^2\tilde{G}_{11}^2 + 6\tilde{G}_C^{d2}\tilde{F}_3^2\tilde{G}_{31}^2\Big) - 18522\tilde{G}_C^{m2}\tilde{F}_1\tilde{F}_3\tilde{G}_{11}\tilde{G}_{31}\cos\Big(\tilde{\tilde{\delta}}_1 - \tilde{\tilde{\delta}}_3\Big) \\
&+ 1225\tilde{G}_C^q\tilde{F}_0\tilde{G}_{02}\Big(7\tilde{G}_C^m\tilde{F}_2\tilde{G}_{20} - 10\tilde{G}_C^q\tilde{F}_2\tilde{G}_{22}\Big)\cos\Big(\tilde{\tilde{\delta}}_0 - \tilde{\tilde{\delta}}_2\Big) \\
&- 125\tilde{G}_C^q\tilde{F}_2\tilde{G}_{22}\Big(98\tilde{G}_C^m\tilde{F}_2\tilde{G}_{20} + 15\tilde{G}_C^q\tilde{F}_2\tilde{G}_{22}\Big) + 450\tilde{G}_C^q\tilde{F}_4\tilde{G}_{42} \\
&\times\Big(\tilde{F}_2(49\tilde{G}_C^m\tilde{G}_{20} - 20\tilde{G}_C^q\tilde{G}_{22}\Big(\tilde{\tilde{\delta}}_2 - \tilde{\tilde{\delta}}_4\Big) + 25\tilde{G}_C^q\tilde{F}_4\tilde{G}_{42}\Big)\Big]P_2(\cos\theta) \\
&+ \frac{18}{W_2^{AC}}\Big[2\tilde{G}_C^d\tilde{F}_1\tilde{G}_{11}\Big(\tilde{G}_C^q\tilde{F}_1\tilde{G}_{22}\Big(\tilde{\tilde{\delta}}_1 - \tilde{\tilde{\delta}}_2\Big) - 6\tilde{G}_C^q\tilde{F}_4\tilde{G}_{42}\Big(\tilde{\tilde{\delta}}_1 - \tilde{\tilde{\delta}}_4\Big)\Big)
\end{aligned}
$$

$$
\begin{aligned}
&+ \tilde{G}_C^d \tilde{F}_3 \tilde{G}_{31}\left(7\tilde{G}_C^q \tilde{F}_0 \tilde{G}_{02}\left(\tilde{\tilde{\delta}}_0 - \tilde{\tilde{\delta}}_3\right) - 8\tilde{G}_C^q \tilde{F}_2 \tilde{G}_{22}\left(\tilde{\tilde{\delta}}_2 - \tilde{\tilde{\delta}}_3\right) + 6\tilde{G}_C^q \tilde{F}_4(k)\tilde{G}_{42}\left(\tilde{\tilde{\delta}}_3 - \tilde{\tilde{\delta}}_4\right)\right]\\
&\times P_3(\cos\theta) + \frac{90}{49W_2^{AC}}\left[20\tilde{G}_C^q \tilde{F}_2(k)\tilde{G}_{22}^2 + \tilde{G}_C^q \tilde{F}_4 \tilde{G}_{42}\left(98\tilde{G}_C^q \tilde{F}_2 \tilde{G}_{02}\left(\tilde{\tilde{\delta}}_0 - \tilde{\tilde{\delta}}_4\right)\right.\right.\\
&- \left.\left. 100\tilde{G}_C^q \tilde{F}_2 \tilde{G}_{22}\left(\tilde{\tilde{\delta}}_2 - \tilde{\tilde{\delta}}_4\right) + 27\tilde{G}_C^q(\omega)\tilde{F}_4 \tilde{G}_{42}\right)\right] P_4(\cos\theta)\}\\
&\equiv \frac{F_{nd}^{AC(G)}(q,\omega)}{4\pi}\left\{1 - \frac{\beta_{nd}^{AC(in)}(q,\omega)}{2} P_2(\cos\theta)\right.\\
&\left.+ q\left[\gamma_{nd}^{AC(in)}(\omega,q)P_1(\cos\theta) + \eta_{nd}^{AC(in)}(\omega,q)P_3(\cos\theta) + \lambda_{nd}^{AC(in)}(\omega,q)P_4(\cos\theta)\right]\right\},
\end{aligned}
\tag{21.21}
$$

where

$$
F_{nd}^{AC(G)} = \frac{4\pi^2\omega}{35q^2} W_2^{AC}, \tag{21.22}
$$

$$
\begin{aligned}
W_2^{AC} &= 35\tilde{G}_C^{m2}\tilde{F}_2^2\tilde{G}_{20}^2 + 42\tilde{G}_C^{d2}\tilde{F}_1^2\tilde{G}_{11}^2 + 63\tilde{G}_C^{d2}\tilde{F}_3^2\tilde{G}_{31}^2 + 35\tilde{G}_C^{m2}\tilde{F}_1^2\tilde{G}_{02}^2 +\\
&50\tilde{G}_C^{q2}\tilde{F}_2^2\tilde{G}_{22}^2 + 90\tilde{G}_C^{q2}\tilde{F}_4^2\tilde{G}_{42}^2.
\end{aligned}
\tag{21.23}
$$

21.2 Numerical Model

21.2.1 Generalized Oscillator Strength

In our programs, for occupied and discrete excited states the reflection factor is assumed, for simplicity, to be equal to 1.

The GOS matrix elements $G_\alpha(\omega)$ in RPAE, including the endohedral atom reflection factor $F_l(\omega)$, are solutions of Eq. (21.9) that are derived in the same way as described in Chap. 10.

Through the GOS matrix elements, the GOS density for an endohedral atom in the RPAE in the length form is expressed via the formula

$$
f_{\nu_1\nu_2}^{(r)}(q,\omega) = 2\frac{\omega}{q^2}\left\langle \nu_1 \middle| \hat{G}(\mathbf{q},\omega) \middle| \nu_2 \right\rangle \tag{21.24}
$$

To include the influence of polarization of the fullerene shell, it is necessary to introduce into (21.24) the polarization factor $G(\omega)$, which for the dipole and quadrupole excitation is determined using (20.36) and (20.38), respectively.

These characteristics are calculated using ***gosful1*** program for an endohedral with one fullerene shell, or, in case of two shells using ***gosful2***. Concrete calculations were carried out for the radii of the fullerene shell $R_1 = 6.72$ ($V_0 = 0.4425$) and $R_2 = 13.5$ ($V_0 = 0.5293$) for endohedrals A@C_{60}, A@C_{240}, or A@C_{60}@C_{240} ($R_1 = 6.72,\ R_2 = 13.5$).

21.2.2 Angular Anisotropy Parameters for s-subshells

Let us start with dipole angular anisotropy parameter for the s-subshell. From (21.15)–(21.16), we obtain the following relation

$$\beta_{ns}^{AC(in)}(\omega, q) = -\frac{4}{7W_0^{AC}} \times \left[21\tilde{G}_C^{d2}\tilde{F}_1^2\tilde{G}_{11}^2 + 5\tilde{G}_C^q\tilde{F}_2\tilde{G}_{22}\left(7\tilde{G}_C^m\tilde{F}_0\tilde{G}_{00}\cos(\tilde{\tilde{\delta}}_0 - \tilde{\tilde{\delta}}_2) + 5\tilde{G}_C^q\tilde{F}_2\tilde{G}_{22}\right)\right], \tag{21.25}$$

where according to (21.17),

$$W_0^{AC} = \tilde{G}_C^{m2}\tilde{F}_0^2\tilde{G}_{00}^2 + 3\tilde{G}_C^{d2}\tilde{F}_1^2\tilde{G}_{11}^2 + 5\tilde{G}_C^{q2}\tilde{F}_2^2\tilde{G}_{22}^2. \tag{21.26}$$

Here $\tilde{G}_{00}, \tilde{G}_{11}, \tilde{G}_{22}$ are the modules of the GOS matrix elements for transitions $s - s(L = 0)$, $s - p(L = 1)$, and $s - d(L = 2)$, respectively, which are expressed via GOS matrix elements by the following relations

$$G_{l'L}(\omega) \equiv \tilde{G}_{l'L}(\omega)\exp[i\Delta_{l'}(\varepsilon)] \tag{21.27}$$

The matrix elements of endohedral atom GOS are determined by the following formula

$$G_{l'L}^{AC(G)}(\omega) = G_C(\omega)F_{l'}(k)G_{l'L}^A(\omega) \tag{21.28}$$

Here $G_{l'L}^A(\omega)$ is partial GOS of an isolated atom, the derivation of which is described in Sect. 10.1.

The non-dipole angular anisotropy parameters are obtained from formulas (21.15) and (21.16):

$$\gamma_{ns}^{AC(in)}(\omega, q) = \frac{6}{q\,W_{ns}^{AC}}\tilde{G}_C^d\tilde{F}_1\tilde{G}_{11} \left[\tilde{G}_C^m\tilde{F}_0\tilde{G}_{00}\cos(\tilde{\tilde{\delta}}_0 - \tilde{\tilde{\delta}}_1) + 2\tilde{G}_C^q\tilde{F}_2\tilde{G}_{22}\cos(\tilde{\tilde{\delta}}_1 - \tilde{\tilde{\delta}}_2)\right], \tag{21.29}$$

$$\eta_{ns}^{(in)}(\omega, q) = \frac{18}{7q\,W_{ns}^{AC}}\tilde{G}_C^d\tilde{F}_1\tilde{G}_C^q\tilde{F}_2\tilde{G}_{11}\tilde{G}_{22}\cos(\tilde{\tilde{\delta}}_1 - \tilde{\tilde{\delta}}_2), \tag{21.30}$$

$$\lambda_{ns}^{AC(in)}(q, \omega) = \frac{90}{7q\,W_0^{AC}}\tilde{G}_C^{q2}\tilde{F}_2^2\tilde{G}_{22}^2. \tag{21.31}$$

It can be seen that all angular anisotropy parameters are expressed in terms of the matrix elements of the GOS for various transitions. Matrix elements are calculated in the ***gos*** program by the formula:

$$\langle kl_2|g_L^r(q)|nl_1\rangle = \sqrt{(2l_1+1)(2l_2+1)}\begin{pmatrix} l_1 & L & l_2 \\ 0 & 0 & 0 \end{pmatrix}\int P_{kl_2}\tilde{j}_L(qr)P_{nl_1}\,\mathrm{d}r$$
$$\equiv c_L\int P_{kl_2}j_L(qr)P_{nl_1}\,\mathrm{d}r. \qquad (21.32)$$

Here $\tilde{j}_{l_\nu}(qr)$ is the spherical Bessel function divided by q^L.

Consider transitions that should be taken into account for s-shell. For a monopole transition $ns-\varepsilon s$ $(L=0)$, the coefficient c_0 is equal to 1. The modulus of the monopole matrix element for each energy ω is expressed in terms of the real and imaginary parts of the GOS $\tilde{G}_{00}=\sqrt{(\mathrm{Re}G_{00})^2+(\mathrm{Im}G_{00})^2}$. The phase of the matrix element is $\bar{\delta}_0=\delta_0^{HF}+\Delta_0$, and the phase correction is calculated by the formula $\Delta_0=\mathrm{arctg}(\mathrm{Im}G_{00}/\mathrm{Re}G_{00})$.

For dipole $ns-\varepsilon p$ $(L=1)$ transition, we have $c_1=(-1)/q$, and the dipole matrix element is found from the expression $\tilde{G}_{11}=\sqrt{(\mathrm{Re}G_{11})^2+(\mathrm{Im}G_{11})^2}$, its phase is $\bar{\delta}_1=\delta_1^{HF}+\Delta_1$, with a correction to the phase equal to $\Delta_1=\mathrm{arctg}(\mathrm{Im}G_{11}/\mathrm{Re}G_{11})$.

For quadrupole $ns-\varepsilon d$ $(L=2)$ transition, we obtain $c_2=1/q^2$ that the quadrupole matrix element is determined by the expression $\tilde{G}_{22}=\sqrt{(\mathrm{Re}G_{22})^2+(\mathrm{Im}G_{22})^2}$, and the phase is equal to $\bar{\delta}_2=\delta_2^{HF}+\Delta_2$, where Δ_2 is determined by relation $\Delta_2=\mathrm{arctg}(\mathrm{Im}G_{22}/\mathrm{Re}G_{22})$. Here δ_0^{HF}, δ_1^{HF}, δ_2^{HF} are the HF phases of corresponding electron wave functions. The wave functions and HF-phases themselves are calculated in the ***fchf*** program, the GOS matrix elements (real and imaginary parts)—in the ***gos*** program, and finally, the dipole and non-dipole angular anisotropy parameters are obtained using ***anglanizful0*** program while the differential in outgoing electron angle GOS density is obtained using the ***difgosful0*** program (21.16, 21.17).

21.2.3 Angular Anisotropy Parameters for p*-subshells*

The dipole angular anisotropy parameters are obtained from formulas (21.18)–(21.20) by the appropriate substitutions

$$\beta_{np}^{(in)} = -\frac{2}{7W_1^{AC}}[7\tilde{G}_C^{d2}\tilde{F}_2\tilde{G}_{21}(\tilde{F}_2\tilde{G}_{21}-2\tilde{F}_0\tilde{G}_{01}\cos(\tilde{\bar{\delta}}_0-\tilde{\bar{\delta}}_2))$$
$$+7\tilde{G}_C^q\tilde{F}_1^2\tilde{G}_{12}(\tilde{G}_C^q\tilde{G}_{12}-2\tilde{G}_C^m\tilde{G}_{10})$$
$$+3\tilde{G}_C^q\tilde{F}_3\tilde{G}_{32}eft((7\tilde{G}_C^m\tilde{F}_1\tilde{G}_{10}-2\tilde{G}_C^q\tilde{F}_1\tilde{G}_{12})\cos(\tilde{\bar{\delta}}_1-\tilde{\bar{\delta}}_3)+4\tilde{G}_C^q\tilde{F}_3(k)\tilde{G}_{32})], \qquad (21.33)$$

where $W_1=\tilde{G}_C^{m2}\tilde{F}_1^2\tilde{G}_{10}^2+\tilde{G}_C^2\tilde{F}_0^2\tilde{G}_{01}^2+2[\tilde{G}_C^{d2}\tilde{F}_2^2\tilde{G}_{21}^2+\tilde{G}_C^{q2}\tilde{F}_1^2\tilde{G}_{12}^2]+3\tilde{G}_C^{q2}\tilde{F}_3^2\tilde{G}_{32}^2$. (see (21.20)).

The non-dipole angular anisotropy parameters are found from the same formula (21.18):

$$\gamma_{np}^{(in)}(\omega, q) = \frac{1}{5qW_1^{AC}}[10\tilde{F}_1\tilde{F}_0\tilde{G}_{01}(2\tilde{G}_C^q\tilde{G}_{12} - \tilde{G}_C^m\tilde{G}_{10})\cos(\tilde{\delta}_0 - \tilde{\delta}_1) + 4\tilde{G}_C^d\tilde{F}_2\tilde{G}_{21}$$
$$\times((5\tilde{G}_C^d\tilde{F}_0\tilde{G}_{10} - \tilde{G}_C^d\tilde{F}_2\tilde{G}_{12})\cos(\tilde{\delta}_1 - \tilde{\delta}_2) + 9\tilde{G}_C^q\tilde{F}_3\tilde{G}_{32}\cos(\tilde{\delta}_2 - \tilde{\delta}_3))], \quad (21.34)$$

$$\eta_{np}^{(in)}(\omega, q) = \frac{3}{5qW_1^{AC}}\Big[6\tilde{G}_C^d\tilde{G}_C^q\tilde{F}_1\tilde{F}_2\tilde{G}_{21}\tilde{G}_{12}\cos(\tilde{\tilde{\delta}}_1 - \tilde{\tilde{\delta}}_2) + \tilde{G}_C^q\tilde{F}_3\tilde{G}_{32}$$
$$\times\left(5\tilde{G}_C^d\tilde{F}_0\tilde{G}_{01}\cos(\tilde{\tilde{\delta}}_0 - \tilde{\tilde{\delta}}_3) - 4\tilde{G}_C^d\tilde{F}_2\tilde{G}_{21}\cos(\tilde{\tilde{\delta}}_2 - \tilde{\tilde{\delta}}_3)\right)\Big], \quad (21.35)$$

$$\lambda_{np}^{(in)}(\omega, q) = \frac{18}{7qW_1^{AC}}\tilde{G}_C^{q2}\tilde{F}_3\tilde{G}_{32}\Big[\tilde{F}_3\tilde{G}_{32} - 4\tilde{F}_1\tilde{G}_{12}\cos(\tilde{\tilde{\delta}}_1 - \tilde{\tilde{\delta}}_3)\Big]. \quad (21.36)$$

To determine the parameters of angular anisotropy, one should account for the GOS in the following transitions:

1. Monopole transition $np - \varepsilon p$. For this transition, one has the coefficient $c_0 = -1.73205$, the monopole matrix element for each energy is determined by real and imaginary parts of GOS, i.e., $\tilde{G}_{10} = \sqrt{(\text{Re}G_{10})^2 + (\text{Im}G_{10})^2}$ and the phase is $\overline{\delta}_{10} = \delta_{10}^{HF} + \Delta_{10}$. Correction to the phase is given by formula $\Delta_{10} = \text{arctg}(ImG_{10}/\text{Re}G_{10})$.
2. Dipole transition $np - \varepsilon s$. For this transition, it is $c_1 = -1/q$. The module of the matrix element is found from the expression $\tilde{G}_{01} = \sqrt{(\text{Re}G_{01})^2 + (\text{Im}G_{01})^2}$. The phase is $\overline{\delta}_{01} = \delta_{01}^{HF} + \Delta_{01}$, and the correction to it is $\Delta_{01} = \text{arctg}(ImG_{01}/\text{Re}G_{01})$.
3. Dipole transition $np - \varepsilon d$. For this transition, one has $c_1 = 1.4142/q$. The modulus of the dipole matrix element is found from relation $\tilde{G}_{21} = \sqrt{(\text{Re}G_{21})^2 + (\text{Im}G_{21})^2}$, and the phase is $\overline{\delta}_{21} = \delta_{21}^{HF} + \Delta_{21}$, with correction Δ_{21} to it equal to $\Delta_{21} = \text{arctg}(ImG_{21}/\text{Re}G_{21})$.
4. Quadrupole transition $np - \varepsilon p$. For this transition, it is $c_2 = 1.0954/q^2$. The quadrupole matrix element modulus is found from the relation $\tilde{G}_{12} = \sqrt{(\text{Re}G_{12})^2 + (\text{Im}G_{12})^2}$, and the phase is $\overline{\delta}_{12} = \delta_{12}^{HF} + \Delta_{12}$, with a correction Δ_{12} to it equal to $\Delta_{12} = \text{arctg}(ImG_{12}/\text{Re}G_{12})$. The phases $\overline{\delta}_{12}$ and $\overline{\delta}_{10}$ are close to each other, so any of them is can be chosen as $\overline{\delta}_1$.
5. Quadrupole transition $np - \varepsilon f$. For this transition, it is obtained $c_2 = (-1.3416)/q^2$. The modulus of the quadrupole matrix element is found from relation $\tilde{G}_{32} = \sqrt{(\text{Re}G_{32})^2 + (\text{Im}G_{32})^2}$, and the phase is determined by formula $\overline{\delta}_{32} = \delta_{32}^{HF} + \Delta_{32}$, with a correction equal to $\Delta_{32} = \text{arctg}(ImG_{32}/\text{Re}G_{32})$.

One-electron wave functions and HF phases are calculated using ***fchf*** program, the matrix elements of the GOS (real and imaginary parts) are calculated using ***gos*** program, and dipole and non-dipole angular anisotropy parameters—using

anglanizfull program, while the differential in outgoing electron angle GOS density is obtained using the ***difgosfull*** program.

21.2.4 Angular Anisotropy Parameters for d-subshells

The dipole angular anisotropy parameter is obtained from formulas (21.21) and (21.23) after appropriate substitutions:

$$\beta_{nd}^{(in)}(\omega, q) = -\frac{4}{245 W_2^{AC}}\Big[1029\Big(\tilde{G}_C^{d2}\tilde{F}_1^2\tilde{G}_{11}^2 + 6\tilde{G}_C^{d2}\tilde{F}_3^2\tilde{G}_{31}^2\Big) - \\ 18522\tilde{G}_C^{m2}\tilde{F}_1\tilde{F}_3\tilde{G}_{11}\tilde{G}_{31}\cos(\tilde{\tilde{\delta}}_1 - \tilde{\tilde{\delta}}_3) + 1225\tilde{G}_C^q\tilde{F}_0\tilde{G}_{02}\Big(7\tilde{G}_C^m\tilde{F}_2\tilde{G}_{20} - 10\tilde{G}_C^q\tilde{F}_2\tilde{G}_{22}\Big) \\ \cos(\tilde{\tilde{\delta}}_0 - \tilde{\tilde{\delta}}_2) - 125\tilde{G}_C^q\tilde{F}_2\tilde{G}_{22}\Big(98\tilde{G}_C^m\tilde{F}_2\tilde{G}_{20} + 15\tilde{G}_C^q\tilde{F}_2\tilde{G}_{22}\Big) + \\ 450\tilde{G}_C^q\tilde{F}_4\tilde{G}_{42}\Big(\tilde{F}_2(49\tilde{G}_C^m\tilde{G}_{20} - 20\tilde{G}_C^q\tilde{G}_{22})\cos(\tilde{\tilde{\delta}}_2 - \tilde{\tilde{\delta}}_4) + 25\tilde{G}_C^q\tilde{F}_4\tilde{G}_{42}+\Big)\Big], \tag{21.37}$$

where

$$W_2^{AC} = 35\tilde{G}_C^{m2}\tilde{F}_2^2\tilde{G}_{20}^2 + 42\tilde{G}_C^{d2}\tilde{F}_1^2\tilde{G}_{11}^2 + 63\tilde{G}_C^{d2}\tilde{F}_3^2\tilde{G}_{31}^2 + 35\tilde{G}_C^{m2}\tilde{F}_1^2\tilde{G}_{02}^2 \\ + 50\tilde{G}_C^{q2}\tilde{F}_2^2\tilde{G}_{22}^2 + 90\tilde{G}_C^{q2}\tilde{F}_4^2\tilde{G}_{42}^2. \tag{21.38}$$

The non-dipole angular anisotropy parameters are obtained in a similar way from the same formulas:

$$\gamma_{nd}^{(in)}(\omega, q) = \frac{6}{q W_2} \times \\ \Big[14\tilde{G}_C^d\tilde{F}_1\tilde{G}_{11}\Big(\tilde{G}_C^q\tilde{F}_2\tilde{G}_{22} - \tilde{G}_C^m\tilde{F}_2\tilde{G}_{20}\Big)\cos(\tilde{\tilde{\delta}}_1 - \tilde{\tilde{\delta}}_2) \\ - 14\tilde{G}_C^d\tilde{G}_C^q\tilde{F}_0\tilde{F}_1\tilde{G}_{11}\tilde{G}_{02}\cos(\tilde{\tilde{\delta}}_0 - \tilde{\tilde{\delta}}_1) + 3\tilde{G}_C^d\tilde{F}_3\tilde{G}_{31} \\ \Big((7\tilde{G}_C^m\tilde{F}_2\tilde{G}_{20} - 2\tilde{G}_C^q\tilde{F}_2\tilde{G}_{22})\cos(\tilde{\tilde{\delta}}_2 - \tilde{\tilde{\delta}}_3) + 12\tilde{G}_C^q\tilde{F}_4\tilde{G}_{42}\cos(\tilde{\tilde{\delta}}_3 - \tilde{\tilde{\delta}}_4)\Big)\Big], \tag{21.39}$$

$$\eta_{nd}^{AC(in)}(q, \omega) = \frac{18}{q W_2^{AC}} \\ \Big[2\tilde{G}_C^d\tilde{F}_1\tilde{G}_{11}\Big(\tilde{G}_C^q\tilde{F}_1\tilde{G}_{22}\cos(\tilde{\tilde{\delta}}_1 - \tilde{\tilde{\delta}}_2) - 6\tilde{G}_C^q\tilde{F}_4\tilde{G}_{42}\cos(\tilde{\tilde{\delta}}_1 - \tilde{\tilde{\delta}}_4)\Big) + \tilde{G}_C^d\tilde{F}_3\tilde{G}_{31} \\ \Big(7\tilde{G}_C^q\tilde{F}_0\tilde{G}_{02}\cos(\tilde{\tilde{\delta}}_0 - \tilde{\tilde{\delta}}_3) - 8\tilde{G}_C^q\tilde{F}_2\tilde{G}_{22}\cos(\tilde{\tilde{\delta}}_2 - \tilde{\tilde{\delta}}_3) + 6\tilde{G}_C^q\tilde{F}_4(k)\tilde{G}_{42}\cos(\tilde{\tilde{\delta}}_3 - \tilde{\tilde{\delta}}_4)\Big)\Big]. \tag{21.40}$$

$$\lambda_{nd}^{AC(in)}(\omega, q) = \frac{90}{49qW_2^{AC}}\Big[20\tilde{G}_C^q \tilde{F}_2(k)\tilde{G}_{22}^2 + \tilde{G}_C^q \tilde{F}_4 \tilde{G}_{42}$$
$$\Big(98\tilde{G}_C^q \tilde{F}_2 \tilde{G}_{02}\cos(\tilde{\bar{\delta}}_0 - \tilde{\bar{\delta}}_4) - 100\tilde{G}_C^q \tilde{F}_2 \tilde{G}_{22}\cos(\tilde{\bar{\delta}}_2 - \tilde{\bar{\delta}}_4) + 27\tilde{G}_C^q(\omega)\tilde{F}_4\tilde{G}_{42}\Big)\Big]. \tag{21.41}$$

To obtain angular anisotropy parameters, the following transitions should be included:

1. Monopole transition $nd - \varepsilon d$. For this transition, one has the coefficient $c_0 = -2.2361$. The modulus of monopole matrix element for each energy ω is expressed in terms of the real and imaginary parts of the GOS, as above, using the relation $\tilde{G}_{20} = \sqrt{(\text{Re}G_{20})^2 + (\text{Im}G_{20})^2}$. There is a phase $\overline{\delta}_{20} = \delta_{20}^{HF} + \Delta_{20}$, and the correction to it is $\Delta_{20} = \text{arctg}(ImG_{20}/\text{Re}G_{20})$.
2. Dipole transition $nd - \varepsilon p$. For this transition, the coefficient is $c_{11} = -1.4142/q$. The modulus of the dipole matrix element for each energy ω is determined by real and imaginary parts of GOS via relation $\tilde{G}_{11} = \sqrt{(\text{Re}G_{11})^2 + (\text{Im}G_{11})^2}$. The phase is $\overline{\delta}_{11} = \delta_{11}^{HF} + \Delta_{11}$, and the correction to it is $\Delta_{11} = \text{arctg}(ImG_{11}/\text{Re}G_{11})$.
3. Dipole transition $nd - \varepsilon f$. For this transition, one has for the coefficient $c_{31} = -1.73205/q$. The matrix element modulus for each energy ω is determined through real and imaginary parts of GOS, using the formula $\tilde{G}_{31} = \sqrt{(\text{Re}G_{31})^2 + (\text{Im}G_{31})^2}$. The phase is equal to $\overline{\delta}_{31} = \delta_{31}^{HF} + \Delta_{31}$, while there is a correction to the phase $\Delta_{31} = \text{arctg}(ImG_{31}/\text{Re}G_{31})$.
4. Quadrupole transition $nd - \varepsilon s$. For this transition, the coefficient is $c_{02} = 1/q^2$. The modulus of quadrupole transition matrix element for each energy ω is determined by real and imaginary parts of the GOS, i.e., $\tilde{G}_{02} = \sqrt{(\text{Re}G_{02})^2 + (\text{Im}G_{02})^2}$. The phase is equal to $\overline{\delta}_{02} = \delta_{02}^{HF} + \Delta_{02}$, while there is a correction to the phase $\Delta_{02} = \text{arctg}(ImG_{02}/\text{Re}G_{02})$.
5. Quadrupole transition $nd - \varepsilon d$. For this transition, the coefficient is $c_{22} = (-1.1952)/q^2$. The modulus of the matrix element for each energy ω is determined by real and imaginary parts of GOS $\tilde{G}_{22} = \sqrt{(\text{Re}G_{22})^2 + (\text{Im}G_{22})^2}$. The phase is equal $\overline{\delta}_{22} = \delta_{22}^{HF} + \Delta_{22}$, while its correction is $\Delta_{22} = \text{arctg}(ImG_{22}/\text{Re}G_{22})$. The phases $\overline{\delta}_{22}$ и $\overline{\delta}_{20}$ are close to each other, so any of them could be selected as $\overline{\delta}_2$.
6. Quadrupole transition $nd - \varepsilon h$. For this transition, one obtains for the coefficient $c_{42} = (-1.6036)/q^2$. The modulus of the matrix element for each energy ω is determined through real and imaginary parts of GOS $\tilde{G}_{42} = \sqrt{(\text{Re}G_{42})^2 + (\text{Im}G_{42})^2}$. The phase is equal $\overline{\delta}_{42} = \delta_{42}^{HF} + \Delta_{42}$, while there is a correction to the phase $\Delta_{42} = \text{arctg}(ImG_{42}/\text{Re}G_{42})$.

Wave functions and HF-phases are calculated using ***fchf*** program, the matrix elements of GOS (real and imaginary parts)—using **gos** program, and finally, dipole and non-dipole angular anisotropy parameters are obtained using ***anglanizful2*** program, while GOS differential in angle of the outgoing electron is obtained using ***difgosful2*** program.

21.2.5 Angular Anisotropy Parameters of Atoms with Semi-Closed Shells

The same parameters—GOS and angular anisotropy parameters—can be calculated for $A@C_N$, where A is an atom with semi-closed subshells. To obtain $\sigma_{nl,kl'\uparrow,\downarrow}^{AC(in)}(\omega)$, $\beta_{nl\uparrow,\downarrow}^{AC(in)}(\omega)$, $\gamma_{nl\uparrow,\downarrow}^{AC(in)}(\omega)$, $\eta_{nl\uparrow,\downarrow}^{AC(in)}(\omega)$, and $\lambda_{nl\uparrow,\downarrow}^{AC(in)}(\omega)$ use (20.63) and (20.66)–(20.69), replacing there $\tilde{D}_{l'}(\omega), \tilde{Q}_{l''}(\omega)$, $\tilde{\tilde{\delta}}_{l'''}$, and $F_{l'}(\omega)$ by respective G-values, their moduli and phases for emitted electrons with spin "up" ($\uparrow$), or spin "down" ($\downarrow$), respectively (see Sects. 3.3 and 9.2).

References

1. M.Ya. Amusia, L.V. Chernysheva, E.Z. Liverts, Phys. Rev. A **80**, 032503 (2009)
2. M.Ya. Amusia, L.V. Chernysheva, E.Z. Liverts, JETP Lett. **90**, 5 (2009)

Chapter 22
Structure of the ATOM-M System

This chapter describes the structure of ATOM-M system software. This system is organized as a set of modules of various types. Executive modules are computational programs that solve certain physical problems using procedures (modules) of other types. Executive modules that solve new, similar to those discussed, physical problems can be easily created using the procedures described in this book. Examples are given of assigning input data for programs solving specific physical problems.

22.1 General Consideration

The ATOM-M system is a complex of interdependent programs and data with a hierarchical structure. At the upper level are executive programs, each of which solves some specific physical problem that are listed in Chap. 2. At the lower levels are applied program modules and a database containing wave functions of atoms and diatomic molecules.

The first programs of ATOM system were written in BESM-ALGOL [1] for the computer BESM-6 [2] and in FORTRAN 77 for EC-1066 [3] in USSR. These programs are currently used on personal computers. The texts of programs in ALGOL and FORTRAN are given in the series of published preprints [4–37]. In [38] and in this book, all programs are given in FORTRAN (both previously published and new programs). Macro-extension of the used language describes the process of assembling modules and allows implementing a modular principle of organization.

Molecular programs are written only in FORTRAN and have not been published before. The structure of these programs largely coincides with the structure of atomic programs. The main difference from the atomic case is that each molecular wave function is represented by a set of radial wave functions (see Chaps. 4 and 6). Having this in mind, one can roughly compare the calculation times for atoms and diatomic molecules. For fairly simple diatomic molecules (N_2,CO), the Hartree–Fock (HF)

M. Ya. Amusia and L. V. Chernysheva, *Computation of Atomic and Molecular Processes*, Springer Series on Atomic, Optical, and Plasma Physics 117,
https://doi.org/10.1007/978-3-030-85143-9_22

wave function of one shell is represented in the form of about ten radial wave functions (see Chaps. 4 and 6). The basic procedures for solving both the HF and RPAE equations require matrix inversion.

As is known, the solution of the problem of inverting a matrix of size $n \times n$ requires proportional computer (processor) time n^3. From this, it is easy to conclude that calculating the HF wave functions for a diatomic molecule with the same number of shells as an atom requires about three orders of magnitude more processing time. To solve problems in RPAE, the number of linear momentum (energy) points is also added, for which the matrix inversion procedure is again necessary. Therefore, it can be roughly estimated that, on average, it takes four orders of magnitude more time and about two orders of magnitude more memory to solve a molecular problem in RPAE than to solve a similar atomic problem.

The system ATOM-M is structured as follows. The algorithms described above are divided into building blocks (modules). Many of them are common to all the studied processes, for example, the calculation of the Coulomb matrix, $3j$-symbol, etc. Each program designed to solve a specific physical problem is built from these modules. This approach allows us to:

- adjust easily, if necessary, the program in accordance with the original task by making changes in some modules,
- most effectively use intermediate results,
- expand the system and supplement it with new programs without spending a lot of time on programming,
- constantly develop the system, improving its characteristics, such as computing time and volume of required memory.

22.2 Description of Modules

All modules included in the system are divided into executive, service, specialized, and general modules Tables 22.1, 22.2, 22.3 and 22.4.

Executive modules are procedures that allow you to solve specific physical problems described in Chaps. 4–21, for example, the calculation of generalized oscillator strengths, photoionization cross sections, etc. (Fig. 22.1).

Generic modules implement standard mathematical calculation methods, the application of which is not limited to the above problems. As such modules, if possible, procedures included in the libraries of standard procedures are taken, for example, matrix inversion, quadratic interpolation according to Newton's method, etc.

Specialized modules solve problems that take into consideration the specifics of the studied physical process or process characteristics, for example, calculating dipole matrix elements.

Service modules provide work with input, output, and intermediate data in a form convenient for the user or program developer, for example, printing tables.

Main modules are the main programs for executive modules.

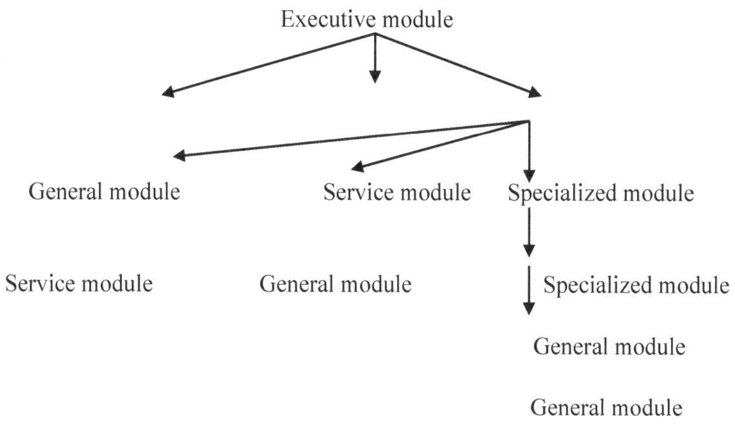

Fig. 22.1 Example of an executive modulus structure

Executive modules enter the system as procedures without specific parameters. Their names are abbreviated names of the corresponding physical processes. Executive modules are assembled from lower-level modules and carry out input, calculation, and printing of results in the form of tables.

22.3 List of Modules

The ***iot31*** module used in atomic programs, intended for calculating $3j$-symbols and including the ***Ifact*** function procedure, has been replaced in molecular programs by a combination of ***defa*** and ***f3js*** modules, since molecular programs have to deal with factorials of large integers many times. Therefore, in molecular programs, an array of factorials is formed in advance (in the ***defa*** module) and then simply used as an input parameter in the ***f3js*** module (Table 22.4).

22.4 Input and Output (I/O) Data

The following types of data are used in the ATOM-M system:

(a) input data—characteristics of an atom (diatomic molecule) and sampling parameters,
(b) wave functions of atoms in their ground and excited states (intermediate data) if they are contained in the database,
(c) output data—characteristics of atomic processes.

Let us explain this in more detail:

Table 22.1 Executive modules

	Executive module	Section, Chapter	Description: what it solves or calculates?	Example
1	***scfhf***	2.1, 4	Self-consistent Hartree–Fock equations for atoms	Ar
2	***rscfhf***	2.2, 5	Ground state wave functions in the Hartree–Fock–Dirac approximation	Ar
3	***fchf***	2.3, 6	Frozen core Hartree–Fock wave functions of electrons, positrons, mesons in an excited state (discrete and continuous spectrum)	Ar
4	***rfchf***	2.4, 7	Wave functions of electrons in an excited state (discrete and continuous spectrum) in the Hartree–Fock–Dirac approximation	–
5	***scfhfm***	2.5, 4	Self-consistent Hartree–Fock equations for simplest molecules	
6	***fchfmd***	2.5, 6	Wave functions of electrons in excited states (discrete spectrum) in a "frozen" field of the simplest molecule	
7	***fchfmc***	2.5, 6	Wave functions of electrons in an excited state (continuous spectrum) in a "frozen" field of the simplest molecule	
8	***scfhfw***	2.6, 4	Hartree–Fock wave functions of electrons in the ground state (with a finite depth well and width for an atom inside a fullerene)	Xe
9	***fchfw***	2.6, 6	Hartree–Fock wave functions of electrons in an excited state (discrete and continuous spectrum) in the "frozen" core of all atomic electrons (with a well of finite depth and width)	Xe

(continued)

Table 22.1 (continued)

	Executive module	Section, Chapter	Description: what it solves or calculates?	Example
10	***phoabs***	2.7, 9	Photoionization cross section, oscillator strengths, and the angular anisotropy parameter in the Hartree–Fock approximation and RPAE in *n*-transitions	Ar (n = 3)
11	***ospp***	2.7, 9	Photoionization cross section for atoms with open shells with allowance for correlations in HF and RPAE in n-transitions	I (n = 5,6,7)
12	***polar***	2.7, 9	Dipole dynamic polarizability of an atom	Ar3p
13	***level***	2.7, 9	Quasi-photoionization cross sections	F^{2+}
14	***pabsout***	2.7, 9	Photoionization cross section in Hartree–Fock and RPAE approximations for the off-mass-shell transitions	–
15	***pabsm***	2.26, 9	Photoionization cross sections for diatomic molecules in the HF and RPAE	
16	***gos***	2.9, 10	Generalized oscillator strengths (GOS) accounting for correlations within the *n*-transitions	Ar (l = 0, 1, 2)
17	***scatlog***	2.10, 10	Fast electron inelastic scattering total cross section with ionization of any atomic electron in the logarithmic approximation	Ar
18	***anglaniz0***	2.11, 2.14, 11	Non-dipole angular anisotropy parameters γ, η, $\gamma^C, \delta^C, (\gamma^C + 3\delta^C)$ of photoelectron, as well as secondary electrons in the scattering of fast electrons, $\gamma^{in}, \eta^{in}, \lambda^{in}$ for *s*—subshells in the HF and RPAE for a linear momentum $q = 0$	Ar3s

(continued)

Table 22.1 (continued)

	Executive module	Section, Chapter	Description: what it solves or calculates?	Example
19	***anglaniz1***	2.11, 2.14, 11	Non-dipole angular anisotropy parameters of photoelectrons γ, η, $\gamma^C, \delta^C, (\gamma^C + 3\delta^C)$, as well as of secondary electrons in the scattering of fast electrons, $\gamma^{in}, \eta^{in}, \lambda^{in}$ for *p*—subshells in the HF and RPAE for a pulse $q = 0$	Ar3p
20	***anglaniz2***	2.11, 2.14, 11	Non-dipole angular anisotropy parameters of photoelectrons γ, η, $\gamma^C, \delta^C, (\gamma^C + 3\delta^C)$, as well as secondary electrons in the scattering of fast electrons $\gamma^{in}, \eta^{in}, \lambda^{in}$ for *d*—subshells in the HF and RPAE for a pulse $q = 0$	Xe4d
21	***difgos0***	2.11, 13	Differential GOS for magic angle $P_2(\cos\theta_m) = 0$ and *s*-subshells of an atom at different *q*	Ar3s
22	***difgos1***	2.11,13	Differential GOS for magic angle $P_2(\cos\theta_m) = 0$ and *p*-subshells of an atom at different *q*	Ar3p
23	***difgos2***	2.11, 13	Differential GOS for magic angle $P_2(\cos\theta_m) = 0$ and *d*-subshells of an atom at different *q*	Xe4d
24	***spinpol***	2.12	Dipole spin polarization parameters	Xe3d
25	***spinpolq***	2.12	Spin polarization parameters for *d*-subshells with allowance for quadrupole terms	Xe3d
26	***Scat***	2.18, 2.20, 14	Scattering phases in HF, SRPAE and with exact consideration of the self-energy part	Ar, Mn
27	***anglscat***	2.18, 14	Differential cross section for elastic scattering of electrons by atoms	Ar, Mn

(continued)

Table 22.1 (continued)

	Executive module	Section, Chapter	Description: what it solves or calculates?	Example
28	***phosep***	2.19, 18	Cross section and angular distribution in photoabsorption and single photoionization of intermediate atomic shells allowing for the photoelectron self-energy part	–
29	***auger***	2.21, 15	Auger decay width of atomic levels allowing for many-electron correlations within the framework of *n*-transitions	Xe
30	***dauger***	2.21, 15	Double-Auger decay probability	
31	***phodec***	2.21,2.192.10, 16	Probability of one-photon decay of two-hole states	–
32	***impact***	2.16, 19	Differential cross section for ionization by electron impact in PWBA	
22	***impact1***	2.16, 19	Differential cross section for ionization by electron impact in DWBA	
	impact2	2.16, 19	Differential cross section for electron impact excitation в DWBA	
24	***autpci***	2.17, 19	Cross section for the excitation of atomic autoionization triplet state allowing for the post-collision interaction	
31	***intripl***	2.23, 19	Probability amplitude of the triplet atomic states excitation	–
32	***twosec***	2.23, 19	Differential excitation cross section of triplet monopole levels	–
33	***brem***	2.24, 17	Bremsstrahlung spectrum of fast electrons colliding with an atom for emitted quanta energies below the ionization potential	–

(continued)

Table 22.1 (continued)

	Executive module	Section, Chapter	Description: what it solves or calculates?	Example
34	***brem1***	2.24, 17	Bremsstrahlung spectrum of fast electrons colliding with an atom for emitted quanta energies exceeding the ionization potential	–
35	***bremco***	2.24,17	Bremsstrahlung spectrum of fast electrons in RPAE	–
36	***bsiea***	2.24, 17	Bremsstrahlung spectrum of intermediate energy particles	–
37	***meson***	2.25, 14	Parameters for muons capture of by atoms	He
39	***pabful1***	2.27, 20	Photoionization cross section in the HF and RPAE within n-transitions for $A@C_{60}$, $A@C_{240}$	Ar ($n = 6$)
41	***pabful2***	2.27, 20	Photoionization cross section in the HF and RPAE approximations within n-transitions for $A@C_{60}C_{240}$	
42	***gosful1***	2.28, 21	Generalized oscillator strengths with allowance for correlations within the n-transitions for $A@C_{60}$ или $A@C_{240}$	Ar ($n = 3$, $l = 1$)
44	***gosful2***	2.28, 21	Generalized oscillator strength with allowance for correlations within n-transitions for $A@C_{60}C_{240}$	Ar (n = 2,l = 0)
45	***anglanizful0***	2.28, 21	Non-dipole angular anisotropy parameters for s-subshells in $A@C_{60}$	Ar3s
46	***anglanizful1***	2.28, 21	Non-dipole angular anisotropy parameters for p-subshells in $A@C_{60}$	Ar3p
47	***anglanizful2***	2.28, 21	Non-dipole angular anisotropy parameters for d-subshells in $A@C_{60}$	Xe4d
48	***difgosful0***	2.28, 21	Angle differential GOS for magic angle $P_2(\cos\theta_m) = 0$ and s-subshells of $A@C_{60}$ at different q	Ar3s

(continued)

Table 22.1 (continued)

	Executive module	Section, Chapter	Description: what it solves or calculates?	Example
49	***difgosfull***	2.28, 21	Angle differential GOS for magic angle $P_2(\cos\theta_m) = 0$ and p-subshells of A@C_{60} at different q	Ar3p
50	***difgosful2***	2.28, 21	Angle differential GOS for magic angle $P_2(\cos\theta_m) = 0$ and d-subshells of A@C_{60} at different q	Xe4d

Table 22.2 Service modules

Module	Description: what does it do?
Disko 1	Writes or reads wave functions from disk (from database with wave functions)
Column	Writing or reading wave functions from RAM
Common	Printing input data in the program ***scfhf***
Comon3	Printing input data in the program ***fchf***
Comonr	Printing input data in the program ***rscfhf***
Comor3	Printing input data in the program ***rfchf***
Intplr	Printing the output in the programs ***rscfhf, rfchf***
Intpol	Printing the output in the programs ***scfhf, fchf***
Veqc	Assigns one specified value to all elements of the array. Used to zero out working arrays
Refb	Reading the ground state molecular wave function from a disk

(a) The volume of input data is small compared to all the information in the system.

For the ground state of an atom, the necessary input data are Z—the charge of its nucleus, configuration (indication of the set of atomic subshells with n, l quantum numbers and the number of electrons Q in each subshell), and a term, i.e., total angular momentum and spin. The object of research is the atomic subshell. Because of its excitation, electron transitions to states with discrete and continuous spectrum become possible. The transition is described by the wave function of the ground state and a set of excited states—m_1 wave functions corresponding to discrete states (with quantum numbers n, l) and m_2 corresponding to the continuous spectrum (with energy E and orbital angular momentum l) for a finite series of linear momenta p. This is the input data for all processes studied using the ATOM-M system.

For the ground state of a diatomic molecule with occupied spin–orbital levels, the required input data are the charges of both nuclei, internuclear distance, the number N of expansion terms used (see Chap. 1), and a set of molecular shells with quantum numbers m.

Table 22.3 Specialized modules

Module	Formulas	Description: what does it do or determines?
accobm	4.40–4.41	Acceleration of convergence in calculating the Hartree–Fock exchange potential
Beta	9.97	Angular anisotropy parameter
Cqd	9.70–9.72	Numerical integration coefficients in the presence and absence of a pole
Coef	4.10—4.13	Direct and exchange coefficients in the Hartree–Fock equations
coef2	5.21, 5.22	Direct and exchange coefficients in the Hartree–Fock–Dirac equations
coef3	4.10—4.13	Direct and exchange coefficients in the Hartree–Fock equations of excited states
Crosst	9.1	The total cross section in ***n***-transitions
Disn	9.93	Energy correction for discrete state
Dfdro	5.3	Derivation of "large" component in HFD
dh1	4.2–4.6	Solution of HF equations
Dhr	5.13–5.21	Solution of HFD equations
direc3	6.6	Direct potential in HF equations (excited state)
Direcr	5.16–5.17	Direct potential in HFD equations (ground state)
Direc	4.5	Direct potential in HF equations (ground state)
direr3	7.5	Direct potential in HFD equations (excited state)
exch3	6.7	Exchange potential in HF equations (excited state)
Exchan	4.5	Exchange potential in HF equations (ground state)
Exchr	5.18	Exchange potential in HFD equations (ground state)
exchr3	7.6	Exchange potential in HFD equations (excited state)
f1	4.10	Direct coefficient in HF equations
f3	4.12	Direct coefficient in HF equations
Fqr	17.2	Atomic form factor
formfact	17.25	Matrix elements for form factor calculations
Fcross	9.6, 9.7	HF and RPAE calculation of photoionization cross section or oscillator strength
g1	4.11	Exchange coefficient in HF equations
g3	4.13	Exchange coefficient in HF equations
g5	4.82	Exchange coefficient in spin-polarized HF equations
Gg	5.20, 5.21	Function $G(r)$ in solving HFD equation
Goscor	10.21, 10.22	GOS for inelastic scattering of fast electrons in HF and RPAE
interG1	20.37	$G_1(\omega)$ for the endohedral $A@C_N$
interG2		$G_{12}(\omega)$ for the fullerene $A@C_{N1}@C_{N2}$
invman	9.81–9.86	Inversion of the Coulomb matrix

(continued)

Table 22.3 (continued)

Module	Formulas	Description: what does it do or determines?
Ksii	5.13–5.14	Function $\xi(r)$
mael1	9.14–9.18	HF dipole matrix elements in "length" and "velocity" form
mael2	9.40, 1.56	Coulomb effective interaction matrix for ***phoabs*** program
maelo2	9.40	Coulomb effective interaction matrix for ***ospp*** program
mael3	10.23–10.24	HF matrix elements of operators $\exp(ipr)$ and $\exp(ipr)\nabla$
mael4	9.34	Matrix of Coulomb matrix elements
mael5	9.34	Vector of Coulomb matrix elements
Maelc	17.32–17.45	Continuous–continuous spectrum matrix elements
main1	4.59–4.61	Solution of one HF equation for the ground state
main13	6.5, 6.12, 6.13	Solution of one HF equation for discrete spectrum
main22	6.5, 6.14, 6.31	Solution of one HF equation for continuous spectrum
Mainr	5.13–5.21	Solution of one HFD equation
Matr	9.32, 9.33	Coulomb matrix elements "two particles–two holes", "three particles–one hole," and "three holes–one particle"
Meco	9.56	Dipole matrix elements in RPAE
Mulma	9.83, 9.84	Matrix multiplication
offdi3	6.44	Non-diagonal parameters for excited states in HF (allowing for exchange before orthogonalization)
offdi31	6.47, 6.51	Non-diagonal parameters for excited states in HF (orthogonalization before allowing for exchange)
Offdia	4.7	Non-diagonal parameters for the ground state in HF
offdr3	5.4, 5.10	Non-diagonal parameters for the ground state in HFD
Phase	6.54	Scattering phase shift in HF approximation
Pot	4.4	Potential in the HF equation
Polar	9.101	Dipole dynamic polarizability in ***n***-transitions
rg1	5.23	Exchange coefficients in HFD equations
rf1	5.24	Direct coefficients in HFD equations
Rpae	9.82–9.86	Solution of RPAE equation
Rpaem		Solution of RPAE equation for diatomic molecules
Yk	4.53–4.61	Coefficients $Y_{ij}^{k}(r)$
var11	4.42–4.48	Change of variables $r \to \rho$ by Newton method
Zero	4.31, 4.32	Initial values of energies and wave functions in HF
Zeror	5.29–5.31	Initial values of energies and wave functions in HFD
Desyg	4.23	Solution of one HF equation of discrete spectrum for molecules, for a fixed λ
Desy0	4.23	Corrections to the discrete energies for molecules

(continued)

Table 22.3 (continued)

Module	Formulas	Description: what does it do or determines?
Cotr0	4.99, 4.100	Calculation of arrays associated with the asymptotic calculation of wave functions near the molecule axis
Dhfsi0	4.23	Formation of the main matrix of the system of equations associated with solution of the molecular HF equation
Dms00	4.23	Solution of the system of HF equations for molecules by the iteration method
Zijk	4.115	Calculation of Coulomb integrals
Suma	4.128, 4.129	Adding Coulomb integrals to the system of equations
Hafel	4.108, 4.109	Computing arrays associated with variable substitution
Bpnm	4.117, 4.118	Direct matrix method
Rdeq	4.110	Solution of one equation in elliptic coordinates
Offdi		Calculation of non-diagonal parameters

Table 22.4 General modules

Module	References	Description
iot31	[38]	$3j$-symbol
iot61	[38]	$6j$-symbol
Ifact	[38]	$n!$
Ic	[38]	$C(n, m)$
Ipari	[38]	Integer part of a number
Simps	[BZ]	Integration by Simpson's method
One	[BZ]	Single integration with a pole
bessel	[39]	Spherical Bessel function of the first kind
invert	[BZ]	The Gauss matrix inversion without choosing the maximum element
kvint	[BZ]	Quadratic interpolation by Newton's method
neyman	[39]	Neumann function
fplg	[40]	$P_L \cos(\vartheta)$
f3js		$3j$-symbol in molecular programs
veqc		Assigning one specified value to all elements of the array. Used to zero out working arrays
defa		Computing an array where the nth element is $n!$
fovi		Computing overlap integrals

Sampling parameters can be divided into two groups. For the case of atoms, the first group includes the properties required to calculate wave functions. These are the range of integration over r for the Hartree–Fock equations; the coefficients c_1 and c_2 in (4.106), which determine the change of variables $r \to \rho$; initial value ρ and step $\Delta\rho$; set of points N at which the wave functions are to be determined; calculation accuracy ε. The second group includes parameters that are determined when calculating characteristics of the atomic process: m_1—the number of considered discrete states with different angular moments nl; m_2 the number of continuous spectrum states, the initial electron linear momentum p, and the step in p. For each process, this data remains the same for all programs.

(b) Atomic and molecular wave functions in the ground and excited states are intermediate data, since, on the one hand, they are calculated by the ATOM-M system, while on the other, they can be used as input data in the study of any process of interaction of an atom (molecule) with other particles. The time required to calculate a wave function, and even more so a set of wave functions, is often comparable to the time required to calculate the characteristics of an atomic (molecular) process. In addition, the wave functions are used multiple times in various processes. Therefore, the ATOM-M system has a database for storing the calculated wave functions. The file name is the criterion for finding the wave function. Depending on the file name, the ***diskol*** executive module requests the required wave functions from the ATOM database, or, on the contrary, writes them to the database for storage.

(c) The output data of the ATOM-M system are the characteristics of atomic and molecular processes: photoionization cross sections, generalized oscillator strengths, angular distribution of photoelectrons, and many others. They are stored and can be used repeatedly by theorists and researchers in their daily work.

22.5 Data Assignment Examples

22.5.1 Ground and Excited States of an Atom in the HF Approximation

Consider the data structure for calculating the ground state of an atom (Fig. 22.2) and excited (Fig. 22.3) in the Hartree–Fock approximation (see Chaps. 4 and 6).

The input data for computing the ground state $Ar\ ^1S\ (1s^22s^22p^63s^2p^6)$ are written to a file named ***ar.dsc*** in the following form:

&CORE

K7 = 10,
R7 = 1.0, 2.0, 3.0, 4.0, 5.0, 6.0, 7.0, 8.0, 9.0, 10.0, 11.0, 12.0,
NU = 13, S = 5, NE = 2000,
Z = 18.0, R = 100.0, BET = 1.0, H = 0.05,
S1 = 1, L1 = 0,
N = 1, 2, 2, 3, 3,
L = 0, 0, 1, 0, 1,
Q = 2, 2, 6, 2, 6,

```
                     | COEF      | G1         |
                     |           | G3         |
                     | COMON     | G5---------| IOT31---IC
                     |           | F1         | IOT61---IC
SCFM---SCFHF---------|-VAR11     | F3         |
                     |           | F4         |
                     | ZERO                    |
                     |          |              | SIMPS2
                     | DH1      | MAIN1        | EXCHAN  POT
                     |          | DISKO1       | DIREC   COLUMN
                     |            SIMPS        | ACCOBM
                     | INTPOL                    OFFDIA
```

Fig. 22.2 *Scfhf* program structure

Fig. 22.3 *Fchf* program structure

```
              | COEF3  | G1  |
              |        | G3  |
              | COEF   | G5  | IOT31   IC
              |        | F1  |
              | COMON  | F3  | IOT61   IC
              |        | F4  |
FCHFM   FCHF  | VAR11
              |
              | ZERO
              |
              | DISKO1
              |
              | DISKO1
              |              | SIMPS2
              | INTPOL       | EXCH3   | POT
              |              | DIREC3  | COLUMN
              | MAIN23 |     | OFFDI3
              |        |     | ACCOMB
              |        |
              | MAIN13 | KVINT
              |        | PHASE
              | INTPOL
```

SPI = 0, 0, 0, 0, 0,
G = 1,
GAM = 0,
KK = 0,
SI1 = 1,
SI2 = 2,
EPS = 0.00001,
KSU1 = 1,
DZ = 0.0, ZK = 1,

&END

For computing as an example $Ar\ ^1P$ in an excited state $(1s^2 2s^2 2p^6 3s^2 3p^5 3d^1)$, the wave functions are written in the *ar3p3d.dfc* file in the following form:

&EXCITE

K7 = 9,
R7 = 1, 2, 3, 4, 5, 6, 7, 8, 9,
S = 6, NU1 = 8, NE = 2000,
Z = 18.0, R = 100.0, BET = 1.0, H = 0.05,
L1 = 1, S1 = 1,
N = 1, 2, 2, 3, 3, 3,
L = 0, 0, 1, 0, 1, 2,
Q = 2, 2, 6, 2, 5, 1,
SPI = 0, 0, 0, 0, 0, 0,
ORT = 0, 0, 0, 0, 0, 1,
G = 1,
GAM = 0,
KK = 1,
SI1 = 5,
AKAP = 0.01, DKAP = 0.1,
K1 = 57, K2 = 3, EPS = 0.00001,
MU = 1, DN = 1, NR = 1,
IT = 5, NU2 = 0,
GAM1 = 0, 0,
GAM2 = 0, 0.

&END

In this data, the following notations are adopted:

K7 a set of radial points at which the wave function values are printed,
R7 an array of radial points, in which the values the wave function values are printed,
Z nuclear charge,
S number of subshells in an atom,
R upper limit of the integration range,
NE number of points of integration over ρ,
H step in ρ,
N array of S-elements: the principal quantum numbers of subshells, starting with the innermost one,
L array of S-elements: the orbital quantum numbers of the subshells, starting with the innermost one,
Q array of S-elements: the number of electrons in the subshell, starting with the innermost one,

ORT an array of S-elements; the element is equal to 1 or 0, depending on whether the selected wave function should be orthogonal to the wave function of the ground state with the same orbital angular momentum l, or not.

The last element of the ORT(S) array determines how the non-diagonal parameters are calculated:

ORT(S) = 1 non-diagonal parameters are calculated by Formula (6.46),
ORT(S) = 0 non-diagonal parameters are calculated iteratively (exchange before orthogonalization),
ORT(S) = 3 non-diagonal parameters are calculated iteratively (orthogonalization before exchange),
SPI auxiliary parameter for setting the spin:
SPI = 0 for all shells in the HF approximation,
SPI = −1 for a shell with spin "down" in SP HF,
SPI = +1 for a shell with spin "up" in SP HF.

The term of an atom is given by two numbers:

S1 is the total spin of the atom ($S1 \equiv 2S + 1$), and L1 is the total orbital angular momentum of the atom.

If $S1 = 100$, then the calculation is performed within the spin-polarized approximation (see Sect. 4.2.6); if $L1 = 111$, then direct $\gamma'_k(ij)$ and exchange $\gamma_k(ij)$ coefficients are entered into the program (arrays GAM, KK, SI1, SI2),

NU1 is the number of coefficients γ'_k and γ_k, if they are not calculated in the program, but are input data,
G is the number of exchange coefficients γ_k, if they are introduced,
EPS calculation accuracy ε,
KSU1 auxiliary parameter:
KSU1 = 1 corresponds to the calculation of the ground state wave function,
KSU1 = 2 corresponds to the calculation of the excited state wave function in a self-consistent field,
KSU1 = 3 corresponds to the calculation of the excited state wave function in the frozen core,
DZ corrections ΔZ to the nuclear charge; usually DZ = 0, but in the case of negative ions, usually DZ = 0.3,
ZK is the number of steps in Z for the case of negative ions, usually ZK = 1.

The following data are required to calculate the wave functions of excited states:

KAP The initial linear momentum p for the wave function in the continuous spectrum,
DKAP linear momentum step size Δp,
K1 is the number of wave functions in the continuous spectrum,
K2 is the number of wave functions in the discrete spectrum,

MU	$(-Q \times M)$, Q—particle charge, M—its mass in at. un. (MU = 1—for an electron, MU = −1for a positron, MU = μ for μ^--meson, MU = $-\mu$ для μ^+-meson),
NR	is the number of the first calculated wave function (usually NR = 1),
DN	is the step in the principal quantum number in the set of wave functions of discrete excited states,
IT	is the number of the core subshells to which the wave functions of excited states should be orthogonal (if ORT (S) = 1),
NU2	the number of coefficients GAM1, GAM2,
GAM1, GAM2	are direct γ' and exchange γ coefficients, which can be additionally introduced for open shell atoms.

Direct γ' and exchange γ coefficients are calculated in the program for the cases described in Chap. 8, i.e.,

(a) for the averaged (with arbitrary configuration, S1 = 1) or a given term (L and S are defined in the LS-coupling or L is defined in the polarized spin method),
(b) for atoms with closed or semi-closed subshells,
(c) for atoms with one electron in one of the subshells; all other subshells are closed,
(d) for atoms with one hole in only one subshell, all other subshells are closed.

In cases not covered in Chap. 8, the coefficients are given as input.

When calculating ground state wave functions of an atom placed inside a fullerene shell using the ***scfhfw*** program and the excited states using the fchfw program, add the parameters of fullerenes well to the above data in the task: RFW1 and RFW2 are the parameters of the well width in atomic units, i.e., RFW1—beginning, RFW2—end of the well in distance r from the center of the system, VW—well depth in the same units, for example, RFW1 = 5.8, RFW2 = 7.7, VW = 0.602.

The hierarchy of programs for solving self-consistent Hartree–Fock field equations is given in Fig. 22.2. After starting the ***scfhf*** program, depending on the total orbital angular momentum L value, the coefficients are either calculated (***coef***) or taken from the input data. The ground state of an atom is determined in LS-coupling approximation or in the polarized spin method by straight ***f1*** and exchange ***g1*** coefficients.

The input data are printed using the ***comon*** procedure. The wave functions and energies initial values are determined in the ***zero*** procedure, the wave functions are calculated according to the algorithm described in Chap. 6 (***dh1***), and the equations are solved with given precision (***main1***) using the modules ***yk, direct, exch, and simps***. The values of the wave functions at points r are printed out (***intpol***) and written to the database (***disko1***). The total energy of the atom (***simps***) is also calculated.

The structure of the program for calculating excited state atomic wave functions in discrete and continuous spectra ***fchf*** is shown in Fig. 22.3.

After starting the program, depending on L value, the coefficients are calculated using procedures ***g1, g3, g5, f1, f3, f5*** (if the total orbital angular momentum L is specified) or are taken from the input data ($L = 111$). All input is printed (***comon3***). As a result of calculation (***main13, main23***), we obtain m_2-discrete spectrum wave functions (for different principal quantum numbers) and m_1 continuous spectrum wave functions with a constant step in linear momentum p. The excited states wave functions are stored at the same points along ρ that are used for the ground state.

The ATOM system uses a database where the ground and excited states wave functions for studied atoms are recorded in sequential files. The atomic ground state wave functions $f_i(\rho),\ i = 1, \ldots, s$, as a result of calculations using the ***scfhf*** program, are recorded in the database file along with additional information characterizing these functions. In front of the wave function array, 15 elements of array A are written:

A1	KSU1 (=1 for the ground state, = 2 for an excited state with a self-consistent core, = 3 for an excited state in a frozen core),
A2	R (upper limit of integration range),
A3	NE (number of integration points),
A4	H (step in ρ),
A5	BET (change of variable parameter),
A6	EPS (calculation accuracy),
A7	Z (core charge),
A8	S (number of subshells in an atom), (S + 1)—for excited state,
A9	ALFA (change of variable parameter),
A10	RO1 (initial value ρ),
A11	ETOT (total energy of an atom),
A12	M2 (for excited states m_2—the number of discrete spectrum wave functions),
A13	M1 (for excited states m_1—the number of continuous spectrum wave functions),
A14	AKAP (for excited states κ_0—initial linear moment),
A15	DKAP (for excited states $\Delta\kappa$—step in linear moment).

After this A array, S arrays of (NE + 15)-elements are written each, which includes

r_1	energy E_i,
r_2	orbital moment l_i,
$r_3 - r_{NE+2}$	wave function $P_i(r)$ calculated at NE points,
r_{NE+5}	principal quantum number,
r_{NE+6}	number of electrons q_i in the subshell,
r_{NE+7}	phase shift δ_i (for continuous spectrum states),
r_{NE+8}	spin projection (=1 for "up spin", = −1 for "down" spin).

The wave functions of the ground state are saved in the ***ar.wsc*** files in unformatted form, used in further calculations and ***ar.rsc*** in the form for printing. Similarly, the wave functions of excited states (e.g., transition $2p^5 \rightarrow 3d^1$ in Ar) are stored in the *ar2p3d.wfc* and *ar2p3d.rfc* files.

22.5.2 *Ground and Excited States of an Atom in the Hartree–Fock–Dirac Approximation*

Consider the data structure for calculating the ground and excited states of an atom in the Hartree–Fock-Dirac approximation (see Chaps. 5 and 7). The input data for calculating the ground state of, for example, $Ar\ ^1S$ are written to a file named *ar.drh* in the following form:

```
&RCORE

    K7 = 6,
    R7 = 0.001, 0.002, 0.003, 0.004, 0.005, 0.006,
    NU = 24, S = 7, NE = 700,
    Z = 18.0, R = 60, BET = 1.0, H = 0.08,
    S1 = 1, L1 = 0,
    N = 1, 2, 2, 2, 3, 3, 3,
    L = 0, 0, 1, 1, 0, 1, 1,
    Q = 2, 2, 2, 4, 2, 2, 4,
    JJ = 0.5, 0.5, 0.5, 1.5, 0.5, 0.5, 1.5,
    SPI = 0, 0, 0, 0, 0, 0, 0,
    G = 22,
    GAM = 0.0,
    KK = 0,
    SI1 = 1,
    SI2 = 2,
    EPS = 0.0001,
    KSU1 = 1,
    DZ = 0.0, ZK = 1, ADD = 40.

&END
```

All parameters are listed above except one: ADD—atomic mass of the nucleus.

To calculate Ar^1P in an excited state $3p_{3/2}^4 - 3p_{3/2}^{-1}nd_{3/2}$ in HFD, the wave functions are written in the *ar3p322d.drf* file in the following form:

```
&REXC

    K7 = 11,
    R7 = 0.1, 1, 2, 3, 4, 5, 6, 7, 8, 9, 10,
    S = 8, NU1-9, NE = 700,
    Z = 18, R = 60, BET = 1.0, H = 0.08,
    S1 = 1, L1 = 1,
    N = 1, 2, 2, 2, 3, 3, 3, 3,
    L = 0, 0, 1, 1, 0, 1, 1, 2,
    Q = 2, 2, 2, 4, 2, 2, 3, 1,
    JJ = 0.5, 0.5, 0.5, 1.5, 0.5, 0.5, 1.5, 1.5,
    SPI = 0, 0, 0, 0, 0, 0, 0, 0,
```

ORT = 0, 0, 0, 0, 0, 0, 0, 0,
G = 9,
GAM = 0.333333, 0.333333, 0.2, 1, 0.2, 0.333333, 0.2, 0.75, 0.15,
KK = 1, 1, 2, 0, 2, 1, 2, 0, 2,
SI1 = 1, 2, 3, 4, 4, 5, 6, 7, 7,
AKAP = 0.1, DKAP = 0.2,
K1 = 3, K2 = 3, EPS = 0.00001,
MU = 1, DN = 1, NR = 1,
IT = 5, NU2 = 0,
GAM1 = 0.0,
GAM2 = 0.0,
ADD = 40,

&END

The ground state wave functions in HFD approximation are saved, unformatted, in the *ar.rrh* files, and used in further calculations and in *ar.rsc* in a suitable for printing form. Similarly, the excited states wave functions (transition $2p^5 \to 3d^1$ in Ar) are stored in the *ar2p322d.wrf* and *ar2p322d.rrf* files.

22.5.3 Wave Functions of Diatomic Molecules in the Hartree–Fock Approximation in the Ground and Excited States

Below are examples of setting data for nitrogen molecule N_2. To calculate the ground state wave functions of N_2, the data file has the following form:

&0i

Mi0 = 1,
Fi0 = 'SC_r2068.Z00',

&0

FIL = 'SC_068a.L00',
FIR = 'SC_068a.Z00',
MI = 0, 1, 0, 1,
Z1 = 7.0,
Z2 = 7.0,
RZ = 2.06800
NSH = 6,
MLN = 0, 0, 8,

0, 1, 8,
0, 0, 8,
0, 1, 8,

0, 0, 8,
1, 1, 8,

AHF = 0.0, 2.0, 2.0, 2.0, 2.0, −4.0,

2.0, 0.0, 2.0, 2.0, 2.0, −4.0,
2.0, 2.0, 0.0, 2.0, 2.0, −4.0,
2.0, 2.0, 2.0, 0.0, 2.0, −4.0,
2.0, 2.0, 2.0, 2.0, 0.0, −4.0,
2.0, 2.0, 2.0, 2.0, 2.0, 0.0,

BHF = −1.0, 1.0, 1.0, 1.0, 1.0, −2.0,

1.0, −1.0, 1.0, 1.0, 1.0, −2.0,
1.0, 1.0, −1.0, 1.0, 1.0, −2.0,
1.0, 1.0, 1.0, −1.0, 1.0, −2.0,
1.0, 1.0, 1.0, 1.0, −1.0, −2.0,
1.0, 1.0, 1.0, 1.0, 1.0, 3.0,

JOR = 0, 0, 0, 0, 0, 0,

0, 0, 0, 0, 0, 0,
0, 0, 0, 0, 0, 0,
0, 0, 1, 0, 0, 0,
0, 0, 0, 0, 0, 0,

NA = 5,
NB = 1000,
RP = 1.020, 60.0,
EPS = 1.D-08,

/END

These data use the following notation:

Mi0 control parameter. For Mi0 = 1, a file with an initial approximation for the desired wave functions is used; for Mi0 = 0, the calculations are performed without any initial approximation,
Fi0 file name with initial approximation for wave functions; this parameter must be set when Mi0 = 1,
FIL name of the file with a short printout of the results,
FIR name of the file in which the molecular wave functions are recorded,
MI control parameter,
Z1, Z2 charges of nuclei in a molecule,
RZ internuclear distance,
NSH number of shells in a molecule,
MLN array of NSH lines, each line contains three elements, which sequentially set the quantum number m of the shell, the initial l in the expansion of the function in spherical harmonics, the number of expansion terms; l should be set equal to m in the case of heteronuclear molecules and in the case

of g-symmetry of the homonuclear molecules shell; for u-symmetry of homonuclear molecules shell $l = m + 1$,

AHF values of direct HF coefficients,

BHF values of HF exchange coefficients,

JOR array that specifies additional enforced orthogonalization of shells of the same symmetry close in energy—if JOR $(i, j) = 1$, then at each iteration the obtained wave function of the j-th shell is orthogonalized to the wave function of the i-th shell to improve the convergence process,

NA number of points of integration near the axis of the molecule at which its asymptotic behavior (representation) is used,

NB total number of points of numerical integration,

RP parameters of integration; from $\xi = 1$ to RP (1), the asymptotic representation of the wave function is used, then from RP(1) to RP(2) difference formulas of numerical integration are used, i.e., RP(2) sets ξ_{max}—the maximum "radius" of integration,

EPS precision of the wave function calculation.

When calculating the continuous spectrum wave functions of the nitrogen molecule, the data file has the form.

```
&S30i
    Mi0 = 1,
    FiG = 'SC_068m.z00',
    Ipc = 1,
    FiC = '1s_068m.z03',
/end
&S30
    Mi = 0, 0, 0,
    Z1 = 0.0,
    Z2 = 7.0,
    RZ = 2.068000,
    NSH = 7,
    MLN = 0, 0, 8,
        0, 1, 8,
        0, 0, 8,
        0, 1, 8,
        0, 0, 8,
        1, 1, 8,
        0, 0, 8,
    IOS = 1,
    AHF = 1.0, 2.0, 2.0, 2.0, 2.0, −4.0, 0.0,
    BHF = --1.0, 1.0, 1.0, 1.0, 1.0, −2.0, 0.0,
    NA = 5,
```

NB = 1000,
RP = 1.020, 60.0,
Mr = 0,
Nl = 4,
Np = 10,
P0 = 0.0100,
Dp = 0.100,
Zf = 1.0,
EPS = 1.D-07,
FiL = 'S0n51_61.L30',
FiR = 'S0n51_61.z30',

/end

In these data, in addition to the parameters already described above, the following notations are used:

Mi0 control parameter. For Mi0 = 1, a file with initial state wave functions is used. Thus, Mi0 = 0 is specified only when calculating one-electron ions, for example H_2^+,
IOS number of the ionized shell,
P0 initial linear momentum, i.e., that for the first continuous spectrum wave function,
Dp step in linear momentum,
Np number of points by momentum,
Nl number of calculated wave functions with different "orbital angular momentum".

When calculating the discrete spectrum wave functions, instead of last four parameters, only one parameter is set:

ND number of discrete spectrum wave functions, which are calculated sequentially in descending order of the eigen energy modulus.

The structure of the program ***scfhfm*** for calculating the molecular ground state wave functions is shown in Fig. 22.4. After addressing the program, if Mi0 = 1, the initial approximation for the calculated wave functions is read from the disk by the ***refb*** procedure, then the shells are sequentially enumerated, starting from the

```
                                  |--ZIJK -------RDEQ -----BPNM
                                  |--VEVR
SCFHFM------DMS00----DHFSI---|---DESYG----DESY0----CPYP0----INVE
            --REFB     --COTR0  |--MASH     --FOVI
                       --H AFA0 |--FOVI
                       --FOVI   |  SUMA
                                |  OFFDI
```

Fig. 22.4 ***Scfhfm*** program structure

```
                                         |--ZIJK --------RDEQ -----BPNM
                                         |--VEVR
FCHFM------DMS30----DHFSI3----|-DESYC3----SYDER----CPYP0----INVE
            --REFB     --COTR0   |--MASH     --FOVI
                       --HAFA0   |--FOVI
                       --FOVI    |--SUMA
                                 |--OFFDI
                                 |--TAIL
```

Fig. 22.5 ***Fchfm*** program structure

innermost to the outer one, and a new approximation is found for each shell in ***dhfsi*** procedure. For this, Coulomb integrals are calculated with the old wave function (procedure ***zijk***), terms with these integrals are added to the HF equation (procedure ***suma***), and with the old energy value, a new wave function is calculated (procedure ***desy0***). Further, depending on the sign and magnitude of the discontinuity of the first derivative, a correction is made to the energy (the ***desyg*** procedure), and then this process is repeated until convergence.

The structure of the program ***fchfm*** for calculating the excited states molecular wave functions is shown in Fig. 22.5. The execution of this program is in many ways similar to the execution of the ***scfhfm*** program. The difference is that there is no enumeration of shells. For continuous spectrum, in addition, the functions are not vanishing at the end point of integration, and there is no iteration in energy.

22.5.4 Photoionization Cross Section and Dipole Angular Anisotropy Parameter

Let us consider the structure of ***phoabs*** executive module for calculating the photoionization cross section of an atom in the HF and RPAE. Consider Ar case as an example. The input data for Ar, allowing for four channels (*3p-es, 3p-ed, 3s-ep*), are prepared in the *ar31.dpa* file in the following form:

```
&PABS
   NES = 3,
   EXPI = 0,
   LAMI1 = 0, 0, 0,
   IZ = 5, 5, 4,
   CHAN = 1, 2, 3,
   RPAE = 1,
   NRES = 0,
   HR = 0.01, 0.01, 0.01, 0.01,
   OMR2 = 2.30, 18.90, 24.40, 226.90,
   OMR3 = 2.55, 19.10, 24.60, 227.20,
```

NE = 2000, S = 5, NW1 = 60,

&END

In these data, the following notations are adopted.

NES	number of taken into consideration transitions,
EXPI	sign; if EXPI = 0, the calculation is carried out with HF values of the vacancy states energies; if EXPI = 1, the experimental energy values are employed specified in the LAML1 array,
LAMI1	array of vacancy states energies,
IZ	an array with shell numbers corresponding to vacancy states,
CHAN	an array of numbers of channel for printing,
RPAE	sign; if RPAE = 0, the calculations are limited by the HF approximation, if RPAE = 1, calculations in RPAE are also performed,
NRES	number of recorded autoionization resonances (NRES = 0—in normal mode),
HR	step in ω a given area of considered resonances (array of length NRES),
OMR2	initial values of photon energies ω, at which the photoionization cross sections in the vicinity of resonances are calculated (an array of length NRES),
OMR3	final photon energy (NRES length array),
NE	maximum number of points in the wave function arrays,
S	maximum number of subshells in the core,
NW1	maximum number of wave functions in excited states.

After the specified file (*ar31.dpa*) has been created and the required sets of wave functions (*ar.wsc, ar3p3d.wfc, ar3p3s.wfc, ar3s3p.wfc*) are calculated in the current database DB1, the *ar31.bat* file is launched on the account of this programs (NDPFOR directory: *phoabs.exe*):

```
copy %DB1%\ar31.dpa %ndpfor%\sys$inpu
copy %DB1%\ar.wsc %ndpfor%\OLD$DB1
copy %DB1%\ar3p3d1.wfc %ndpfor%\OLD$DB2
copy %DB1%\ar3p4s.wfc %ndpfor%\OLD$DB3
copy %DB1%\ar3s4p.wfc %ndpfor%\OLD$DB4
cd %ndpFOR%
phoabs.exe
copy %ndpfor%\SYS$OUTP %DB1%\ar31.rpa
erase %ndpfor%\SYS$INPU
erase %ndpfor%\SYS$outP
erase %ndpfor%\OLD$db1
erase %ndpfor%\OLD$DB2
erase %ndpfor%\OLD$db3
erase %ndpfor%\OLD$DB4
cd %DB1%
```

```
                  |--VAR11
                  |
                  |--DISKO1                          |
                  |                                  |--COLUMN
                  |--MAEL1--|--COLUMN                |--IOT31------IC
                  |         |--SIMPS2                |--IOT61------IC
                  |                                  |--POT
                  |--MAEL02--------------------------|----SIMPS2
PABSM---PHOABS--|
                  |--DISNN--|--CQDN
                  |         |--INVMAN       |----INVERT
                  |--MECO                   |---MULMA1
                  |                         |----MULMA2
                  |--FCROS1
                  |
                  |--CROSST
                  |
                  |--BETA
```

Fig. 22.6 ***Phoabs*** program structure

The result of the photoionization cross section calculation is written to the *ar31.rpa* file. The names of all the listed files can be defined by the user, based on the convenience of working with them.

The hierarchy of modules for solving the RPAE equations is given in Fig. 22.6. After starting the program, the ***var1*** variables are changed, the wave functions are read, and the basic data are printed. Then the dipole matrix elements are calculated in the HF approximation (***mael1***), the Coulomb matrix elements in RPAE (***maelo2***) are determined, finally, RPAE equation (***cqdn, meco***) is solved with matrix inversion (***invman***, ***invert***, ***mulma1***, ***mulma2***), and corrections to discrete excited levels (***disnn***) are obtained. The cross sections and oscillator strengths are calculated at given points of ω by ***fcros***s and dipole angular anisotropy parameter by ***beta***, using a number of auxiliary procedures (***simps2***, ***ipari***, ***iot31***, ***iot61***, ***ic***, ***kvint***, ***pot***).

The program that solves RPAE equations for endohedrals includes in addition the procedure for calculating the F-factor (***ffactor***) and some modifications to a number of procedures (***invmanful***, ***phoabsful***). Data assignment is the same (see above). Only the launcher differs: ***pabsful1*** (for the $A@C_N$ case) or ***pabsful2*** (for the $A@C_{N1}@C_{N2}$ case).

If it is necessary to take into account the polarization of the fullerene, one should first calculate the polarization factor $G(\omega)$ using the programs ***interG1*** (for the case $A@C_{N1}$), ***interG1*** (for the case $A@C_{N2}$), or ***interG2*** (for the case $A@C_{N1}@C_{N2}$) and multiply the obtained cross section by it.

For molecular calculations, an example of setting data looks like this:

```
& D40

   NCH = 1,
```

ID = 0, 3,
IC = 30, 25,
FIR = 'P1n41_65.z40',
FIL = 'P1n41_65.L40',
MiE = 1,
Ech = −0.1506400E + 02

−0.1506000E + 02

FIG = 'SCF_N2.z00',
FIC = 'P1n41_65.Z30',
FID = '1ss_3m.R25',
NEM = 122,
NUR = 312,
irp = 0,

/

The following notations are used here:

NCH	number of channels in the calculation,
ID	number of discrete functions for each channel,
IC	number of continuous spectrum functions for each channel,
FIL	name of the file with the printout of the results for the photoionization cross section,
FIR	name of the file into which the matrix elements are written,
MiE	control parameter; at MiE = 1, the experimental energy thresholds of the shells are used,
Ech	values of the experimental energy thresholds of the shells are set at MiE = 1,
FIG	file with wave functions of the ground state,
FIC	files with wave functions of a continuous spectrum, consistently for all channels,
FID	files with wave functions of a discrete spectrum, consistently for all channels,
IRP	control parameter; when IRP = 1, calculations are performed in RPAE.

22.5.5 Polarizability of an Atom

When calculating the polarizability of an atom or ion, the ***polar*** procedure is used.
The input data for it are set in the *ar022p.dpl* file:

& PLR
Further:

Z	nuclear charge,
NOM	number of points—energy values for which polarizability will be calculated,

OMEGAO	array of photon energies,
NT	number of transitions included in calculation (*3p-ed*, *3p-es*—in *Ar* atom example),
NN	array of length NT of principal quantum numbers values of first discrete excited states,
E	array of hole energy levels with length NT,
M1	number of discrete levels,
M	number of excited states in continuous spectrum,
EEL	array of excited levels energies M1 (discrete) + M (continuous) for the first channel and the same for the next one, length (M1 + M) * NT,
HFL	array ($s1$) of photoionization cross sections in the length form in HF approximation, with length (M1 + M) * NT,
HFV	array (s2) of photoionization cross sections in the velocity form in HF approximation, with length (M1 + M) * NT,
RPAL	array (s3) of photoionization cross sections in the length form in RPAE, with length (M1 + M) * NT,
RPAV	array (s4) of photoionization cross sections in the length form in RPAE, with length (M1 + M) * NT.

All data for this program are taken from the listing of the ***phoabs*** program. Launcher for calculating dynamic polarizability is ***polarAr3p.bat***. The calculation results are written to the *ar022p.rpl* file. For each transition, the input data, dipole dynamic polarizability at the specified points, form the output. After that, the contribution of the oscillator strengths is printed allowing for all up the highest excited states according to formulas (9.101–9.104), i.e., the contribution of the 50 overlying excited states is added, and at the end the sum of contributions of all transitions to the dipole dynamic polarizability is printed.

22.5.6 Fast Electrons Scattering Cross Section and Generalized Oscillator Strengths

Let us consider the data for the executive module ***gos*** for calculating the generalized oscillator strengths in the HF and RPAE. For Ar, allowing for six channels (3*p-es,* 3*p-ed,* 3*s-ep*, 2*p-es,* 2*p-ed,* 2*s-ep*), they are prepared in *ar61.dgs* file in the following form:

```
& OS

  NES = 6,
  HI = 1,
  EXPI = 0,
  LAMI1 = 0, 0, 0, 0, 0, 0,
  IZ = 5, 5, 4, 3, 3, 2,
  CHAN = 1, 2, 3, 4, 5, 6,
```

Y = 0.1, DELY = 0.2, TAU = 1,
LV = 1 RPAE = 1,
NE = 2000, S = 5, NW1 = 58,

& END

Parameters NES, EXPI, LAML1, IZ, CHAN, RPAE, NE, S, and NW1 are described above.

HI number of points along the momentum q,
LV transferred angular momentum (for monopole transitions LV = 0, for dipole transitions LV = 1, for quadrupole LV = 2, etc.),
Y initial value of transferred linear momentum q_0,
DELY initial value of transferred linear momentum step Δq_0,
TAU linear transferred momentum step multiplier (see formula (10.39)).

The results of calculating of the generalized oscillator strengths are written to the *ar61.rgs* file. The names of all the listed files can be defined by the user, based on the convenience of working with them.

The hierarchy of modules for determining the generalized oscillator strengths is shown in Fig. 22.7.

After starting the program, the variables are changed (***var1***), the wave functions are read, and the basic data are printed. Then the matrix elements are calculated in the HF approximation (***bessel, mael3***), the Coulomb matrix elements in RPAE (***mael02***) are determined, finally, RPAE equations (***cqdn, meco***) are solved by matrix inversion (***invman, invert, mulma1, mulma2***), and corrections to discrete excited levels are evaluated (***disnn***). The generalized oscillator strengths are calculated at the given

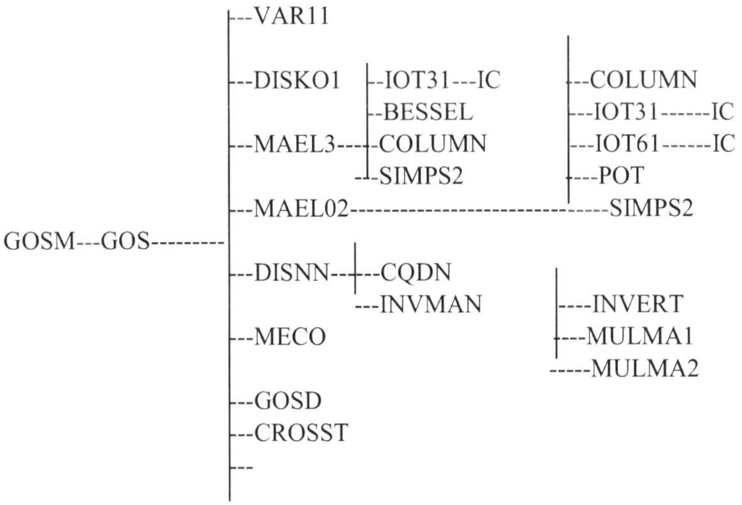

Fig. 22.7 ***Gos*** program structure

points in ω (***gosd***). This uses a number of auxiliary procedures (***simps2, ipari, iot31, iot61, ic, kvint, pot***).

The program for solving RPAE equations for endohedrals differs only by introduction of a procedure for calculating the F-factor (***ffactor***) and by additions to a number of procedures (***invmanful, gosful***). Data assignment is the same (see above). Only the launch program is different: ***gosful1*** (for the case of one fullerene shell) or ***gosful2***—for two fullerene shells.

22.5.7 Non-dipole Angular Anisotropy Parameters of Secondary Electrons at q = 0

To calculate the non-dipole angular anisotropy parameters, three different programs are used depending on the orbital angular momentum of the shell corresponding to the hole state: ***anglaniz0*** ($l = 0$), ***anglaniz1*** ($l = 1$), ***anglaniz2*** ($l = 2$).

1. Let us consider the calculation of the angular anisotropy non-dipole parameters of Ar 3s-subshell using the ***anglaniz0*** executive module. The input data are written in the *ar3sgamRP.dgd* file.

It uses the following notation:

NE	number of points at which non-dipole parameters are determined,
LL	orbital angular momentum of the atomic shell corresponding to the vacancy state,
EELRY1	electron energy at the first point,
RPAE	sign; if RPAE = 0, computations are limited to the HF approximation, if RPAE = 1, computations are performed also in RPAE,
EphRY	array of photon energies (NE length), in which matrix elements were calculated by programs ***phoabs*** and ***gos***.

The dipole matrix elements are taken from the results of the ***phoabs*** program:

RE1	array of matrix elements (real parts of length NE) for the transition $3s$-ep,
IM1	array of matrix elements (imaginary parts of NE length) for the $3s$-ep transition,
DEL1	array of phase shifts (NE length) for $3s$-ep transition.

The GOS matrix elements are taken from the results of ***gos*** program:

RE2	array of matrix elements (NE length) for transition $3s$-ed,
IM2	array of matrix elements (NE length) for transition 3s-ed,
DEL2	array of phase shifts (NE length) for $3s$-ed transition.

The calculation of non-dipole parameters is carried out according to the formulas (11.51–11.66). After executing the program, we have the result in *ar3sgamRP.rgd* file.

2. Let us consider the calculation of non-dipole angular anisotropy parameters for, as an example, Ar 3*p*-subshell using the ***anglaniz1*** executive module. The input data are written in the *Ar3pgamRP.dgd* file.

This file uses the following notations:
Parameters NE, LL, EELRY1, RPAE, and EPHRY are described above.
Dipole matrix elements are selected from the results of calculation using ***phoabs*** program where

RE2	array of matrix elements (real parts of length NE) for transition 3*p*-*ed*,
RE0	array of matrix elements (real parts of length NE) for 3*p*-*es* transition,
IM2	array of matrix elements (imaginary parts of NE length) for 3*p*-*ed* transition,
IM0	array of matrix elements (imaginary parts of NE length) for 3*p*-*es* transition,
DEL2	array of phase shifts (NE length) for *3p-ed* transition,
DEL0	array of phase shifts (NE length) for 3*p*-*es* transition.

The matrix elements of GOS are taken from calculation results using the ***gos*** program:

RE3	array of matrix elements (NE length) for transition 3*p*-*ef*,
RE1	array of matrix elements (NE length) for 3*p*-*ep* transition,
IM3	array of matrix elements (NE length) for transition 3*p*-*ef*,
IM1	array of matrix elements (NE length) for 3*p*-*ep* transition,
DEL3	array of phase shifts (NE length) for 3*p*-*ef* transition,
DEL1	array of phase shifts (NE length) for 3*p*-*ep* transition.

The calculation of non-dipole angular anisotropy parameters is carried out according to formulas (11.67–11.74). After performing the calculations by the program, the results are in *ar3pgamRP.rgd* file.

3. Consider the calculation of non-dipole angular anisotropy parameters for Xe 4*d*-subshell using ***anglaniz2*** executive module. The input data are recorded in the *xe4ddogamRP.dgd* file. This file uses the following notation:

The parameters NE, LL, EELRY1, and EPHRY are described above.
The dipole matrix elements are taken from results of calculations using the ***phoabs*** program:

RE1, IM1, DEL1	arrays of matrix elements (real and imaginary parts of NE length) and phase shifts for 4*d*-*ep* transition,
RE3, IM3, DEL3	arrays of matrix elements (real and imaginary parts of NE length) and phase shifts for 4*d*-*ef* transition.

The matrix elements of GOS are taken from calculation results using the ***gos*** program:

RE0, IM0, DEL0	arrays of matrix elements (real and imaginary parts of NE length) and phase shifts for the 4*d*-*es* transition,
RE2, IM2, DEL2	arrays of matrix elements (real and imaginary parts of NE length) and phase shifts for the 4*d*-*ed* transition,

RE4, IM4, DEL4 arrays of matrix elements (real and imaginary parts of length NE) and phase shifts for the *4d-eg* transition.

The calculation of non-dipole angular anisotropy parameters is carried out according to formulas (11.75–11.84). After performing the calculations according to the program, we have the result in the *xe4dgamRP.rgd* file.

22.5.8 Spin Polarization Parameters of Photoelectrons (A, ξ, α)

The spin polarization parameters are calculated using the ***spinpol*** program. For Xe 3d-"down" level, as an example, the source data for this program are written to the *Xe3ddoRP.dps* file. It uses the following notations:

NE	number of points at which these parameters are calculated,
SJ	spin projection (=1.5 for spin "down", =2.5 for spin "up"),
LL	orbital momentum of the *d*-vacancy subshell,
Eph	array of photon energies, in which the spin polarization parameters are calculated,
RE1, DEL1	array of dipole matrix elements and HF phase shifts for the *3p-es* transition,
RE2, DEL2	array of dipole matrix elements and HF phase shifts for the *3p-ed* transition.

Calculation of the spin polarization parameters is carried out by formulas (12.18–12.22). After performing the calculations according to the program, we have the result in *Xe3ddoRP.rps* file.

The quadrupole corrections to the spin polarization of electrons are included using the ***spinpolq*** program. In this case, the input data, for example, for Xe*3d* are set in the *Xe3dqdoRP.dsp* file in the following form:

NE	number of points in energy,
LL	orbital moment of the core shell 3d LL = 2,
SJ	spin projection (=1.5—"down" spin),
EPH	photon energy array,
EEL	array of electron energies (NE length).

Dipole matrix elements determined by the ***phoabs*** program:

RED1, IMD1, DLT1	dipole matrix elements (real and imaginary parts), determined using the ***phoabs*** program, and HF phase shifts, determined using the ***fchf*** program, for the *3d-ep* "down" transition,
RED3, IMD3, DLT3	dipole matrix elements (real and imaginary parts), determined by the ***phoabs*** program, and HF phase shifts, determined by the ***fchf*** program, for the *3d-ef* "down" transition,

REQ0, IMQ0, DLT0 quadrupole matrix elements determined by the ***gos*** program and HF phase shifts determined by the ***fchf*** program for the $3d$-es "down" transition,

REQ2, IMQ2, DLT2 quadrupole matrix elements determined by the ***gos*** program and HF phase shifts determined by the ***fchf*** program for the $3d$-ed "down" transition,

REQ4, IMQ4, DLT4 quadrupole matrix elements determined by the ***gos*** program and HF phase shifts, determined using the ***fchf*** program for the $3d$-eg "down" transition.

BET is the dipole angular anisotropy parameter for $3d$-ef "down" and $3d$-ep "down" transitions.

The calculation of the spin polarization parameters accounting for the quadrupole terms is carried out by formulas (12.23–12.25). After performing the calculations according to the program, we have the result in *Xe3dqdoRP.rsp* file.

22.5.9 Angular Distribution of Secondary Electrons in Fast Particles Scattering

To calculate the dipole and non-dipole angular anisotropy parameters of secondary electrons, three different programs are used depending on the orbital angular momentum of the shell corresponding to the hole state: ***difgos0*** ($l = 0$), ***difgos1*** ($l = 1$), and ***difgos2*** ($l = 2$).

1. Let us consider the calculation of dipole and non-dipole angular anisotropy parameters for, as an example, Ar $3s$-subshell (transitions $3s$-es ($L = 0$), $3s$-ep ($L = 1$), $3s$-ed ($L = 2$)) using the ***difgos0*** module. The original data re written in the *Ar3sdgRpq01.dag* file, where the first line is written as follows:

& ANGGOS0
Further

Q transferred linear momentum q,
LAM energy of the core subshell (for Ar—3s-subshell),
GRAD magic angle θ_m equal to 54.735621°,
NW number of points at which non-dipole parameters are determined,
RPAE described above,
EEL array of electron energies, in which parameters are calculated.

The GOS matrix elements are selected from the calculation results using the ***gos*** program:

RE00, IM00, DEL0 arrays of matrix elements (real and imaginary parts of length NW) and phase shifts for the $3s$-es transition,

RE11, IM11, DEL1 arrays of matrix elements (real and imaginary parts of NE length) and phase shifts for $3s$-ep transition,

RE22, IM22, DEL2 arrays of matrix elements (real and imaginary parts of NE length) and phase shifts for 3*s*-*ed* transition.

Calculation of dipole and non-dipole parameters is carried out according to formulas (13.16–13.21). After performing calculations using the program, we have the results (GOS, β^{in}, γ^{in}, and ξ^{in}) in the file *Ar3sRPq01.rag*.

2. Consider calculation of the dipole and non-dipole angular anisotropy parameters for Ar 3*p*-subshell (transitions 3*p*-*es*, 3*p*-*ed* ($L = 1$), 3*p*-*ep* ($L = 0$), 3*p*-*ed*, 3*p*-*ef* ($L = 2$)) using the ***difgos1*** module. The input data are recorded in the *Ar3pdgRPq01.dag* file.

In it, the parameters Q, LAM, GRAD, NW, RPAE, and EEL are described above.

RE10, IM10, DEL10 arrays of the GOS matrix elements (real and imaginary parts of length NW), calculated using the ***gos*** program, and HF phase shifts (***fchf*** program) for the 3p-ep transition ($L = 0$),
RE21, IM21, DEL2 arrays of GOS dipole matrix elements (real and imaginary parts of length NW), calculated using the ***gos*** program, and HF phase shifts (***fchf*** program) for the 3*p*-*ed* transition ($L = 1$),
RE01, IM01, DEL0 arrays of GOS dipole matrix elements (real and imaginary parts of length NW), calculated using the ***gos*** program, and HF phase shifts (***fchf*** program) for the 3*p*-*es* transition ($L = 1$),
RE32, IM32, DEL3 arrays of GOS matrix elements (real and imaginary parts of length NW), calculated using ***gos*** program, and HF phase shifts (***fchf*** program), for the 3*p*-*ef* transition ($L = 2$),
RE12, IM12, DEL1 arrays of GOS matrix elements (real and imaginary parts of length NW), calculated using the ***gos*** program, and HF phase shifts (***fchf*** program) for 3*p*-*ep* transition ($L = 2$).

The calculation of the parameters is carried out according to formulas (13.22–13.25). After calculations are performed, we have the results (GOS, β^{in}, γ^{in}, and ξ^{in}) in the file *Ar3pdgRPq01.rag*.

3. Consider calculation of dipole and non-dipole angular anisotropy parameters for, as a concrete example, 4*d*-subshell of Xe (transitions 4*d*-*ep*, 4*d*-*ef* ($L = 1$), 4*d*-*ed* ($L = 0$), 4*d*-*es*, 4*d*-*eg* ($L = 2$)) using the ***difgos2*** module. The input data are recorded in the *Xe4ddgRPq01.dag* file.

Calculations use the Q, LAM, GRAD, NW, RPAE, and EEL parameters described above.

RE20, IM20, DEL2 arrays of matrix elements of GOS (real and imaginary parts of length NW), calculated using the ***gos*** program, and HF phase shifts (***fchf*** program) for 4*d*-*ed* transition ($L = 0$),
RE11, IM11, DEL1 arrays of GOS dipole matrix elements (real and imaginary parts of length NW), calculated using the ***gos*** program, and HF phase shifts (***fchf*** program) for 4*d*-*ep* transition ($L = 1$),

RE31, IM31, DEL3 arrays of GOS dipole matrix elements (real and imaginary parts of length NW), calculated using the ***gos*** program, and HF phase shifts (***fchf*** program) for 4d-ef transition ($L = 1$),

RE02, IM02, DEL0 arrays of the GOS matrix elements (real and imaginary parts of length NW), calculated using the ***gos*** program, and HF phase shifts (***fchf*** program) for the $4d$-es transition ($L = 2$),

RE22, IM22, DEL2 arrays of GOS matrix elements (real and imaginary parts of length NW), calculated using the ***gos*** program, and HF phase shifts (***fchf*** program) for $4d$-ed transition ($L = 2$),

RE42, IM42, DEL4 arrays of GOS matrix elements (real and imaginary parts of length NW), calculated using the ***gos*** program, and HF phase shifts (***fchf*** program) for $4d$-eg transition ($L = 2$).

Calculation of dipole and non-dipole parameters is carried out according to formulas (13.26–13.30). After performing the calculations according to the program, we have the results (GOS, β^{in}, γ^{in}, and ξ^{in}) in the file *Xe4ddgRPq01.rag*.

22.5.10 Scattering of Low and Medium-Energy Electrons by Atoms and Endohedrals

When calculating the scattering phases, cross sections for elastic and inelastic scattering of electrons (positrons) in simplified RPAE or SRPAE and allowing for the polarization corrections, the programs of ATOM-M system are used in the following order:

1. The ***scfhf*** program, the result of which is a set of functions for the ground state of the atom,
2. The ***fchf*** program, which permits to calculate:

   k-wave functions of a scattered electron (positron), $k \geq 3$,
   m-wave functions of the excited state of the atom,
   f-wave functions of the scattered electron intermediate state.

3. The ***scat*** program, which permits to calculate the scattering phases, as well as elastic and inelastic scattering cross sections.

Before using the programs, it is necessary to consider all diagrams that represent the process. Let us do it using the example of s-electron scattering by Ar atom (Tables 22.5 and 22.6).

The input data for the s-wave are in *arwsel.dst* file

& SCTR

ES = 16,
ND = 11,
MU = 1,

Table 22.5 Elastic scattering $e + r \rightarrow e + r(3p)$ (Z = 18, S = 5) for *s*-wave

Diagrams	1	2	3	4	5	6	7
State	L(терм)						
K	$0(^2S)$	$0(^2S)$	$0(^2S)$	$0(^2S)$	$0(^2S)$	$0(^2S)$	$0(^2S)$
F	$0(^2S)$	$1(^2P)$	$1(^2P)$	$2(^2D)$	$2(^2D)$	$3(^2F)$	$3(^2F)$
m	$1(^1S)$	$0(^1P)$	$2(^1P)$	$1(^1D)$	$3(^1D)$	$2(^1F)$	$4(^1F)$
lv	0	1	1	2	2	3	3
Hole	3p	3p	3p	3p	3p	3p	3p

Table 22.6 Elastic scattering $e + r \rightarrow e + r(3s)$ for *s*-wave

Diagrams	1	2	3	4
состояние	L(терм)			
K	$0(^2S)$	$0(^2S)$	$0(^2S)$	$0(^2S)$
F	$0(^2S)$	$1(^2P)$	$2(^2D)$	$3(^2F)$
M	$0(^1S)$	$1(^1P)$	$2(^1D)$	$3(^1F)$
Lv	0	1	2	3
hole	3s	3s	3s	3s

EXPI = 0, IPOS = 0,
NI = 2, IT = 5, 4,
K = 25, KK = 3, KZ = 6,
F = 25, F1 = 0, FZ = 6, 31, 31, 56, 56, 81, 81, 6, 31, 56, 81,
M = 24, M1 = 3, MZ = 131, 155, 179, 203, 227, 251, 275, 299, 322, 347, 371,
LV = 0, 1, 1, 2, 2, 3, 3, 0, 1, 2, 3,
IL = 5, 5, 5, 5, 5, 5, 5, 4, 4, 4, 4,
NE = 700, S = 5, NW1 = 25, EMIN = 0, EMAX = 10.0,

& END

Here

NES	number of wave functions sets,
ND	number of considered diagrams,
MU	sign (=1 for an electron, =−1 for a positron),
EXPI	sign (=0 for an electron, = 1 for a positron,
IPOS	potential for the virtual positronium formation (=0 for an electron, =−0.5 for a positron).
NI	number of holes used in the core,
IT	array of hole numbers in the core,
K	number of scattered electron wave functions,
KK	number *i* of core wave functions, added to *k* functions, with orbital angular momenta $l_k = l_i$,

KZ	number of the beginning of the array, where *k* functions are located,
F	number of scattered electron wave functions in the intermediate state, of it:
F1	number of discrete excitations,
FZ	number of the beginning of the array, where the *f*-functions are located,
M	number of wave functions of the excited state of the atom, among them:
M1	number of discrete excitations,
MZ	number of the beginning of the array, where the *m*-functions are located,
LV	transferred angular momentum,
IL	array of holes,
NE	number of points at which the wave functions were calculated,
S	number of subshells of the atom,
NW1	maximum number of the excited state wave functions,
EMIN, EMAX	energy range in at. un., in which the results are displayed.

In inelastic scattering calculations for positrons, the pole in the discrete spectrum of atomic excitations is not included, since it is believed that the real formation of positronium is possible only in the case when the electron and positron are in unbound states, i.e., in the continuum.

So, results of the ***scat*** program are scattering phases in the HF approximation, SRPAE, and accounting for the polarization corrections. In the same approximations, we have cross sections for elastic and inelastic scattering of electrons (positrons) by atoms and endohedrals.

22.5.11 Differential Cross Section of Elastic Electron Scattering by Atoms and Endohedrals

When calculating the differential scattering cross section using ***anglscat*** program, the phases are used that were previously calculated by ***scat*** program in the HF, SRPAE or including also the polarization corrections. The input data are written, for Mn example, in HF, (only *s*-phase of the projectile is considered) in the file *MndoHFspdfg.dan.*

& ANGL
Further

NTET	length of the GRAD array,
GRAD (1: NTET)	array of degrees at which the cross sections are calculated,
NO	number of points in the EEL array, where the cross sections should be calculated,
NLV	length of the LV array,

LV	an array of transferred angular moments,
NW	number of points at which the phases were calculated,
EEL	array of energies in which the phases were calculated, length NW,
REDEL, IMDEL	phase arrays (real and imaginary parts, respectively) of length (NW1 * NLV).

Calculations are carried out according to formulas (14.9–14.11), results are in the file *MndoHFspdfg.ran*.

22.5.12 Non-radiative (Auger) Vacancy Decay

When calculating Auger matrix elements using the example of Ne (see the diagram in Fig. 15.1), programs of the ATOM system are used in the following order:

1. The ***scfhf*** program, the results of which are a set of atomic ground state wave functions. For example, for Ne ($1s^2\ 2s^2\ 2p^6$) they have numbers from 1 to 3; for hole $1s^2$ it is fz1 = 1; for hole $2s^2$ it is fz2 (1) = 2; for hole $2p^6$ it is fz2 (2) = 2.
2. The***fchf*** program, with the help of which excited states are calculated: *m* excited state wave functions of an atom in the continuous spectrum for 2*s-es*, 2*p-ep*, 2*p-es*, 2*p-ed* transitions, in accordance with Table 22.7,
3. The program ***auger*** that calculates the Auger matrix elements, for which the data, presented in Table 22.7 are set into file *Ne1s_4.dag*.

& AGR
Further

NES	number of included transitions,
FZ1	array of numbers of core functions $1s^2$ (on the diagram ν_1),
FZ2	array of numbers of core functions $2s^2$, $2p^6$ (on the diagram ν_2),
FZ3	array of numbers of core functions $2s^2$, $2p^6$ (on the diagram ν_3),
EXPI	sign (=0 when setting theoretical thresholds, =1 when setting experimental thresholds),
EKIN	eigenvalues of *i*-hole states energies (or experimental values of ionization potentials, if EXPI > 0),
LV1	array of transferred angular moments,

Table 22.7 Diagrams that are taken into consideration for Ne

Ne ^{1}S	ν1	ν2	ν3	ν4	Lv1
1	1s	2s	2s	ε s(^{1}S)	0
2	1s	2s	2p	ε p(^{1}S)	0
3	1s	2p	2p	ε s(^{1}P)	1
4	1s	2p	2p	ε d(^{1}P)	1

RPAE sign (RPAE = 0 for HF, RPAE = 1 for RPAE),
NE number of points at which the wave functions are calculated,
S number of subshells in the core,
NW1 number m of wave functions in an excited state,
EPS an array of calculation accuracy for each diagram.

The calculation results are written to the file *Ne1s_4.rag*.

When calculating the probabilities of double Auger decay, the following programs are used:

1. The ***scfhf*** program, the result of which is a set of functions for the ground state of the atom, for example, for Ne ($1s^2 2s^2 2p^6$),
2. The ***fchf*** program, which calculates the following excited states:
   - wave function of the Auger electron,
   - wave functions of intermediate states with $l = 0$,
   - wave functions of intermediate states with $l = 1$.
3. The ***dauger*** program calculating the necessary characteristics for the double Auger decay $1s^{-2}$ $-2s^{-2}2p^{-1}E$ in Ne, for which the data are set in a file where the first line is

& DAU
Further

ND	number of diagrams,
NES	number of excited states,
AA (11, nd)	two-dimensional array describing the initial diagrams: (AA [1, i]-AA [4, i]—these four numbers determine the numbers of the arrays—free ends of the diagram,
AA [5, i]	number of the array with the first excited intermediate state,
AA [6, i]	total number of discrete and continuous functions in the intermediate state,
AA [7, i]	number of discrete wave functions,
AA [8, i]	transferred angular momentum in the *i*-diagram;
AA [9, i]	diagram sign, defined as follows: If hole states are not included in intermediate states, then AA [9, i] = 0, otherwise AA [9, i] = 1;
AA [10, i]	sign, which specifies the number of the formula by which the angular factor of the *i*-diagram is calculated: AA [10, i] = 1, 2, 3, 4, AA [10, i] = 0, if the angular factors for all diagrams that determine the amplitude of the process under consideration are introduced by the array Y(3, nd).
Y [3, nd]	array of angular and spin factors,
NORM	normalization factor;
EXPI	sign (EXPI = 1, if KK, IND, DLAM, and GAM are input data).
KK	number of elements in arrays IND, DLAM, and GAM,
IND	number of functions with an energy denominator,
DLAM	energy denominator,

GAM	intermediate state width,
NE, S, NW1	described above.

22.5.13 Radiative Decay

When calculating the amplitudes and probabilities of one-photon decay of two-hole states, the following programs are used:

1. The ***scfhf*** program, the result of which is a set of functions for the ground state of the atom, for example, for Ne ($1s^2 2s^2 2p^6$),
2. The ***fchf*** program, with which the excited states are calculated: for example, the wave function of an electron at the discrete level 3p,
3. The ***phodec*** program, which calculates the necessary characteristics of one-photon decay (e.g., of $1s^{-1}$ vacancy in Ne to $2s^{-1}[^2S]$ state. For this program, the data are stored in a file, where the first line is

& PHOD
Further

ND	number of diagrams,
NES	number of excited states,
AA (11, nd)	two-dimensional array (see Subsection 5.13),
ORI	total orbital angular momentum of the initial state,
ORF	total orbital angular momentum of the final state,
SP	total spin,
NORM	normalization factor; if the initial (final) vacancies are equivalent, then NORM = 0.5, otherwise NORM = 1,
EK	energy of the emitted photon in Rydberg,
Y [3, nd]	array of angular and spin factors,
EXPI = 0	(if the arrays KK, IND, DLAM, GAM calculated in the ***dauger*** program are specified, then EXPI = 1),
NE, S, NW1	described above.

22.5.14 Bremsstrahlung of Intermediate and High Energy Incident Particles

When calculating the characteristics of electron bremsstrahlung, the programs are used in the following order:

- the ***scfhf*** program for calculating the ground state wave functions of an atom in the HF approximation,
- program ***fchf*** for calculating discrete excited state $m1$-wave functions in the HF approximation and (m-$m1$) continuous spectrum wave functions of a virtually excited atomic electron with a constant step $p = \sqrt{E}$ in linear momentum,

- programs for calculating the bremsstrahlung characteristics,
- program ***brem*** for calculating the bremsstrahlung amplitude and cross section of an electron colliding with an atom, accounting for the polarizability at photon energy higher than the ionization potential of a given subshell of an atom, or
- the ***brem1*** program for calculating the bremsstrahlung amplitude and cross section of an electron colliding with an atom, accounting for the polarizability at photon energy lower than the ionization potential of a given subshell of an atom.

Input data for an example for the ***brem*** program calculating the electron bremsstrahlung cross section (MU = 1) with momentum p = 27 at.un. on Xe atom (*4d-ef*) including the target polarizability (nu1 = 1, nu2 = 1) are written in the file *xe.dbr* as follows:

& BRM
Further

NES	number of transitions,
HI	number of points in the linear momentum q transferred to the atom,
IT	number of the subshell, which is exited,
MU	incident particle mass, in at. un.,
NU1, NU2	constants that determine the process:
NU 1 = 1, NU 2 = 1	electron bremsstrahlung allowing for polarizability,
NU 1 = 1, NU 2 = −1	positron bremsstrahlung,
NU 1 = 0, NU 2 = 1	proton bremsstrahlung,
EXPI	sign (when EXPI = 0, calculations are performed with theoretical eigenvalue of the hole state energy, otherwise with the experimental one),
LAMI1	eigenvalue of the hole state energy in Rydberg, (if EXPI = 0, then LAML1 = 0),
P	linear momentum of the incident particle,
NE	number of points at which the wave functions are calculated,
S	is the number of subshells of the atomic core, including that from which the excitation occurs,
NW1	number of wave function sets.

The result is written to the *xe.rbr* file.

As an example, below is the input data for the ***brem1*** program when calculating the bremsstrahlung of an electron (mu = 1) with linear momentum p = 27 on Xe atom (4*d-ef*) accounting for the polarizability (nu1 = 1, nu2 = 1), which are written in the file *xe.db1* as follows:

& BEFORE

NES = 1,
HI = 9,
IT = 9,

MU = 1,
NU1 = 1, NU2 = 1,
EXPI = 0, LAMI1 = 0,
P = 27,
OM0 = 1, DOM = 0.5, NOM = 10,
NE = 1000, S = 11, NW1 = 25

& End

The parameters notations are the same as in the ***brem*** program. The result is placed in the *xe.rb1* file.

22.5.15 Photoabsorption Cross Section

When calculating the photoabsorption cross section (Chap. 18), the following programs are used:

1. The ***scfhf*** program for calculating the wave functions of the ground state of an atom in the HF approximation,
2. Program ***fchf*** for calculating *m1*-wave functions in a discrete excited state in the HF approximation and *(m-m1)* functions in the continuous spectrum of a virtually excited atomic electron with a constant linear momentum step $p = \sqrt{E}$,
3. The program ***phosep*** for calculating the photoionization cross section accounting for the self-energy part, for which the following input data are specified.

&PHO

NES	number of variants,
NC	is the number of sets of wave functions,
K, KZ	is the number of photoelectron functions and the number of the first array with functions,
F, F1, FZ	the number of functions (full and ν_3-discrete) describing the photoelectron in the intermediate state and the number of the first array,
M, M1, MZ	similarly for functions describing excitation from the ν_4-shell,
IT, IL	number of holes ν_2 and ν_4 in the core,
LV	transferred angular moments in SEP (see Chap. 18),
EXPI	sign (EXPI = 0 for the theoretical threshold of ionization LAML1 and EXPI = 1 for the experimental threshold),
LAML1	experimental subshell ionization threshold,
NE, S, NW1	these parameters are described in the previous section.

22.5.16 Capture of μ-Mesons by Atoms

When calculating the characteristics of muon capture by atoms, the programs are used in the following order:

1. The ***scfhf*** program for calculating the atomic ground state wave functions in the HF approximation,
2. The ***fchf*** program for calculating the wave functions of a meson in initial and final states in the Hartree approximation in discrete (or continuous) spectrum with a constant step in the principal quantum numbers (or in linear momentum $p = \sqrt{E}$),
3. The ***meson*** program for calculating the characteristics of the capture of a μ-meson by atoms. In this program, the wave functions of the knocked-out Auger electron in the field of the core and meson are determined. Their number is equal to the number of wave functions of the meson in the final state. Then, the cross sections for μ-meson capture ($E > 0$) or probability of the Auger transition ($E < 0$) are found. The data for this program are written, as an example, for He in the *he.dms* file in the following form:

```
& MSN
K7 = 3,
R7 = 1, 2, 3,
S = 3, NU1 = 1, NE = 2000,
Z = 2, R = 25, BET = 1, H = 0.1,
L1 = 111, S1 = 1,
N = 1, 2, 2,
L = 0, 0, 1,
Q = 1, 1, 1,
SPI = 0, 0, 0,
ORT = 0, 0, 0,
G = 1,
GAM = −0.33333333,
KK = 1,
SI1 = 1,
AKAP = 1, DKAP = 1,
K1 = 5, K2 = 0, EPS = 0.00001,
MU = 201.1, DN = 2, NR = 1,
IT = 1, NU2 = 0,
GAM1 = 0, 0, GAM2 = 0, 0
& END
```

The data notations for the ***meson*** program are the same as for the ***fchf*** program. The calculation is performed according to the formulas of Section 14.2.4, and the calculation results are stored in the *He.rms* file.

22.5.17 Photoioniz*ation of Endohedrals* $A@C_N$ *and* $A@C_{N1}@C_{N2}$

The ***pabsful1 and pabsful2*** programs are used to calculate the photoionization cross section and the dipole angular anisotropy parameter for atoms placed inside the fullerene, $A@C_N$, and $A@C_{N1}@C_{N2}$. The input data for them are set in the same way as for the program calculating these characteristics for electrons in an atom (the ***phoabs*** program), and at the end two parameters are added:

RF fullerene radius,
V0 potential of the fullerene shell.

When calculating the GOS for electrons in $A@C_N$, the ***gosful1*** and ***gosful2*** programs are used, and the input data for them are set in the same way as for the program calculating these characteristics for electrons in an atom (the ***gos*** program), and at the end, the fullerene radius and the fullerene shell potential are set.

When calculating the angular distribution of secondary electrons in the scattering of fast particles by endohedrals, the ***difgosful0*** (for *s*-electrons), ***difgosful1*** (for *p*-electrons), and ***difgosful2*** (for *d*-electrons) programs are used, and the input data for them are written in almost the same way as for the programs ***difgos0***, ***difgos1***, and ***difgos2***. The differences are in the addition of the polarization factor amplitude and phase.

If the calculation is carried out in GFRPAE approximation (i.e., accounting for polarization), then the input data are specified, for example, for Ar *s*-electrons at $q = 0.1$ in the file *Ar3sdgGFRPq01.dag*, where the first line is written as follows:

& DGOSFUL0

Further go the same data as for similar programs ***gos*** (see Subsection 5.6), ***difgos*** (see Subsection 5.9), and at the end of the data two arrays are added containing the absolute values of the polarization factor amplitude (GFUNC) and phase (GPH):

GFUNC array of length NW (GFUNC (I) = 1, if the calculation is carried out in FRPAE),
GPH array of length NW (GPH (I) = 0, if the calculation is carried out in FRPAE).

For *p*-electrons and $q = 0.1$, the data are set similarly in the *Ar3pdgGFRPq01.dag* file with the first line of the form and DGOSFUL1.

The polarization factor (GFUNC and GPH arrays) is calculated preliminarily at specified points in photon energy using the ***interG1*** programs (for the case of one fullerene shell C_{60} or C_{240}) and ***interG2*** (for the case of two fullerene shells C_{60} and C_{240}).

The data for the ***interG1*** program are contained in the *Ar3pGC60.dit* file (or in the *Ar3pGC240.dit* file), and the result is in the *Ar3pGC60.rit* file (or in the *Ar3pGC240.rit* file).

Ar3sGC60.dit file contains input data:

& QPOL

Further

RF	fullerene radius (RF = 6.72 for C_{60}, R = 13.5 for C_{240}),
NE1	length of arrays EPHEVP, REALFA, IMALFA,
NE2	length of EPHRY1 array,
EPHEVP, REALFA, IMALFA	arrays of energies, as well as real and imaginary parts of the experimental values of the fullerene shell dipole polarizability,
EPHRY1	array of photon energies at which to calculate the arrays GFUNC and GPH.

The data for the ***interG2*** program are contained in the *Ar3pGC60C240.dit* file, and the result is in *Ar3pGC60C240.rit* file, while the first line is

QPOL2

Further

NE1, NE2, EPHEVP, REALFA, IMALFA, and EPHRY1 described for ***interG1***.

22.6 The Limits of Applicability of ATOM-M Program System

The capabilities of the software are determined, on the one hand, by the area of applicability of the method, and, on the other, by the computer configuration.

Consider the constraints associated with the formulation of the problem and its numerical solution.

1. In the HF approximation, the ground state of an atom or molecule should be non-degenerate, so it is possible to carry out calculations of atoms and ions containing
    - Only closed subshells,
    - One electron (vacancy) in an open subshell, the rest are closed.
2. In the spin-polarized Hartree–Fock method, atoms and ions with semi-closed subshells can be calculated.
3. For atoms and ions containing open subshells, the method of the averaged term is used. In molecules, this problem usually does not arise, since the subshells of most molecules of interest are closed.
4. Based on theoretical estimates, it can be concluded that the fundamental error of the RPAE method is on average about 5–10%, and the error of experiments, as a rule, is not less than 5%. This was accounted for when choosing numerical methods for calculations. So, the accuracy of the calculated matrix elements is 4–5 significant digits.
5. Allowance for correlations in atomic processes is carried out in the range of up to $n = 10$ transitions, but their number can be increased significantly if the computer memory allows it. Up to 12 transitions can currently be included.

6. Calculation of the correction to the electron (hole) energy when accounting for discrete excitations is carried out within the framework of perturbation theory. Therefore, it is impossible to consider such two transitions, the discrete excitation energies of which are close to each other, since this means their strong interaction, where the perturbation theory is inapplicable.
7. Wave functions cannot be calculated with linear momentum $p = 0$ $(p^2 = E = 0)$. Usually, the minimum starting linear momentum value is $p_0 = 0.1$ $(E_0 = 0.01Ry)$.
8. When solving RPAE equation, integration is carried out with a constant linear momentum step $p = \sqrt{E}$. The number of points in the continuous spectrum is determined based on the decrease rate of the Coulomb matrix elements as functions of electron energies.
9. When calculating wave functions, the energy is given in Rydberg. In programs ***mael1, mael4, and mael5,*** matrix elements of the Coulomb interaction have the dimension $(Ry)^{(1-n/2)}$, where n is the number of wave functions in the continuous spectrum. To convert matrix elements to atomic units, a factor $2^{(n/2-1)}$ should be introduced. In the ***mael2*** program, dimensionless matrix elements are printed without accounting for the spin factor equal to 2. In the ***mael3*** program, the dimension of the matrix elements is $(Ry)^{n/2}$.

Computer configuration imposes upper restrictions on the following program parameters:

1. The number of points N at which the wave function is calculated,
2. Integration step h,
3. Distance r from the nucleus of an atom, or the axis of a diatomic molecule, in at. un.,
4. The upper electron energy E in a continuous spectrum, since the calculation of strongly oscillating functions requires a significant increase in the number of points N at which it is calculated,
5. The number of excited states m, participating in the calculation of the wave functions, including the number of discrete functions $m1$.
6. Momentum q transferred to an atom or molecule.

Usually, these parameters for atoms and molecules take the following values: $N = 700, 1000, 2000, 20{,}000$; $h = 0.1, 0.08, 0.05$; $r = 60, 100, 150$; $E_{\max} = 100Ry$ $(\approx 1300eV)$; $m = 20$–30, $m1 = 1$–4. For diatomic molecules with more than ten shells, it seems that the use of multiprocessor computers is required, since ordinary calculations in this case require too much space in the RAM and take a lot of time (a day or more).

22.7 Execution Levels of ATOM-M System

The ATOM-M system consists of several levels. The first level is the calculation in initial approximation for the problem under consideration. When calculating various physical characteristics of atoms, diatomic molecules, endohedrals, and processes in them, accounting for multielectron correlations, as an initial serves the Hartree–Fock approximation. The Hartree–Fock approximation is used to find the wave functions of the ground and excited states of atoms, diatomic molecules, and endohedrals. Wave functions and related characteristics (energy, orbital angular momentum, principal quantum number of the subshell) are stored in a database.

The second level is the calculation of the properties of atoms, diatomic molecules, and endohedrals and the probabilities of their interaction with external fields in RPAE, for example, photoionization cross sections, generalized oscillator strengths, scattering phases, etc. In this case, the necessary wave functions are taken from the database.

The system is currently limited to these two levels. It is possible to move to the next level, where RPAE would be used as an initial approximation.

References

1. V.M. Kurochkin *BESM-ALGOL* (Moscow State University, Moscow, 1969)
2. I.D. Bokova, S.A. Zeldinova, *Operating system for BESM-6* (User's Manual, Institute of Applied Mathematics, Acad. Sci. USSR, Moscow, 1973)
3. V.J. Calserbank, *A Course on Programming in FORTRAN* (Chapman and Hall, London, 1983)
4. L.V. Chernysheva, N.A. Cherepkov, *Numerical calculation of Hartree-Fock wave functions* (Ioffe Institute, No 337, Leningrad, 1971) (in Russian)
5. M.Ya. Amusia, N.A. Cherepkov, L.V. Chernysheva, *Numerical calculation of atomic photoionization cross sections with account of multielectron correlations* (Ioffe Institute, No 375, Leningrad, 1971) (in Russian)
6. L.V. Chernysheva, M.Ya. Amusia, N.A. Cherepkov, *Numerical calculation of photoionization cross sections of atoms with allowance for correlations in one and two transitions* (Ioffe Institute, No. 459, Leningrad, 1974) (in Russian)
7. L.V. Chernysheva, N.A. Cherepkov, V. Radojevič, *Numerical solution of the Hartree-Fock self-consistent field equations for Atoms* (Ioffe Institute, No. 486, Leningrad, 1975) (in Russian)
8. L.V. Chernysheva, N.A. Cherepkov, V. Radojevič, *Numerical calculation of Hartree-Fock wave functions of discrete and continuous states in the frozen core field* (Ioffe Institute, No. 487, Leningrad, 1975) (in Russian)
9. L.V. Chernysheva, M.Ya. Amusia, S.I. Sheftel, *Numerical calculation of atomic generalized oscillator strength with wave functions in the Hartree-Fock approximation and with account of many-electron correlations in one transition* (Ioffe Institute, No. 493, Leningrad, 1973) (In Russian)
10. L.V. Chernysheva, M.Ya. Amusia, S.I. Sheftel, *Numerical calculation of atomic generalized oscillator strength allowing for many-electron correlations within two transitions* (Ioffe Institute, No. 495, Leningrad, 1975) (In Russian)
11. L.V. Chernysheva, M.Ya. Amusia, S.G. Shapiro, *Numerical calculation of the diagonal matrix element of the one-particle Green's function self-energy part in simplified random phase approximation with exchange* (Ioffe Institute, No. 504, Leningrad, 1976) (In Russian)

12. L.V. Chernysheva, N.A. Cherepkov, V. Radojevič, Self-consistent field Hartree-Fock program for atom. Comp. Phys. Comm. **11**, 57 (1976)
13. L.V. Chernysheva, M.Ya. Amusia, V.K. Ivanov, S.G. Shapiro, *Numerical calculations of the Auger decay width of atomic levels allowing for multielectron correlations in two transitions* (Ioffe Institute, No. 536, Leningrad, 1977) (In Russian)
14. L.V. Chernysheva, N.A. Cherepkov, V. Radojevič, *Numerical calculations of photoionization cross sections and optical oscillators strengths of atoms in Hartree-Fock approximation and allowing for multielectron correlations in one transition* (Ioffe Institute, No. 589, Leningrad, 1978) (In Russian)
15. L.V. Chernysheva, N.A. Cherepkov, V. Radojevič, Frozen-core Hartree-Fock program for atomic discrete and continuum states. Comp. Phys. Comm **18**, 87 (1979)
16. L.V. Chernysheva, M.Ya. Amusia, N.A. Cherepkov, *Numerical calculation of scattering phases in the simplified random phase approximation with exchange* (Ioffe Institute, No. 663, Leningrad, 1980) (In Russian)
17. L.V. Chernysheva, N.A. Cherepkov, S.I. Sheftel, *Numerical calculations of the coefficients in the Hartree-Fock equations* (Ioffe Institute, No. 760, Leningrad, 1982) (In Russian)
18. L.V. Chernysheva, M.Ya. Amusia, S.A. Sheinerman, *Numerical calculation of the electron-induced triplet excitation cross section in the second order of the distorted-wave approximation* (Ioffe Institute, No. 776, Leningrad, 1982) (In Russian)
19. S.A. Sheinerman, M.Ya. Amusia, M.Yu. Kuchiev, *Numerical calculation of the cross section of atomic ionization near thresholds of inelastic processes* (*Post-collision interaction*) (Ioffe Institute, No. 777, Leningrad, 1982) (In Russian)
20. S.K. Semenov, N.A. Cherepkov, *Numerical calculation of the dipole dynamic polarizability of atoms* (Ioffe Institute, No. 835, Leningrad, 1983) (In Russian)
21. L.V. Chernysheva, N.B. Avdonina, M.Ya. Amusia, M.Yu. Kuchiev, *Numerical calculation of the "atomic" bremsstrahlung spectrum of fast electrons* (Ioffe Institute, No. 865, Leningrad, 1983) (In Russian)
22. L.V. Chernysheva, N.A. Cherepkov, *Numerical calculation of the wave functions of a meson in the field of a "frozen" core and the cross section for meson capture of an atom* (Ioffe Institute, No. 943, Leningrad, 1985) (In Russian)
23. L.V. Chernysheva, M.Ya. Amusia, V.A. Kilin, I.S. Li, *Numerical calculations of the double Auger decay probability* (Ioffe Institute, No. 944, Leningrad, 1985) (In Russian)
24. S.A. Sheinerman, M.Yu. Kuchiev, *Numerical calculations of the photoionization cross section of inner shells of atoms with account of post-collision interaction* (Ioffe Institute, No. 977, Leningrad, 1985) (In Russian)
25. L.V. Chernysheva, M.Yu. Kuchiev, V.L. Yakhontov, *Numerical calculations of ground-state atomic wave functions in Hartree-Fock-Dirac approximation* (Ioffe Institute, No 1015, Leningrad, 1986) (in Russian)
26. L.V. Chernysheva, M.Ya. Amusia, V.A Kilin., A.N. Kolesnikova, I.S. Li, *Numerical calculation of the probability of one-photon decay of two-hole states* (Ioffe Institute, No. 1054, Leningrad, 1986) (In Russian)
27. L.V. Chernysheva, N.A. Cherepkov, *Numerical calculation of the photoionization cross section of an open shell atom with account of electron correlations in two transitions* (Ioffe Institute, No. 1079, Leningrad, 1986) (In Russian)
28. L.V. Chernysheva, B.V. Gul'tsev, M.M. Dovganich, *Numerical calculation of atomic wave functions in the Hartree-Fock approximation on an ES computer* (Ioffe Institute, Leningrad, 1987). (In Russian)
29. L.V. Chernysheva, G.F. Gribakin, *Numerical Calculation of the Wave Functions of an Electron in the Field of a "Frozen" Core for States with LS Term* (Ioffe Institute, Leningrad, 1987). ((In Russian))
30. G.F. Gribakin, V.K. Ivanov, M.Yu. Kuchiev, L.V. Chernysheva. *Many-body theory calculation of negatively charged ions* (Ioffe Institute, No. 1096, Leningrad, 1987) (In Russian)
31. L.V. Chernysheva, N.B. Avdonina, M.Ya. Amusia, A.V. Korol, *Numerical calculation of the intermediate energies particles bremsstrahlung spectrum* (Ioffe Institute, No. 1314, Leningrad, 1989) (In Russian)

32. L.V. Chernysheva, M.Ya. Amusia, V.F. Orlov, S.K. Semenov, N.A. Cherepkov. *Numerical calculations of electron wavefunctions in the frozen-core field with input of various off-diagonal energy parameters* (Ioffe Institute, No 1319, Leningrad, 1989) (in Russian)
33. L.V. Chernysheva, M.Ya. Amusia, K.L Tsemekhman, *Numerical calculation of the cross sections for photoabsorption and single photoionization of intermediate atomic shells with account of the photoelectron self-energy part* (Ioffe Institute, No. 1492, Leningrad, 1990) (In Russian)
34. L.V. Chernysheva, M.Ya. Amusia, A.S. Kheifets, *Numerical calculation of the differential cross sections of excitation and ionization of an atom by electron impact* (Ioffe Institute, No. 1531, Leningrad, 1990) (In Russian)
35. L.V. Chernysheva, S.K. Semenov, N.A. Cherepkov, *Numerical calculation of the Hartree-Fock wave functions of an electron in the field of a frozen core on an ES computer* (Ioffe Institute, No 1574, Leningrad, 1991) (in Russian)
36. L.V. Chernysheva, S.K. Semenov, N.A. Cherepkov, *Numerical calculation of the photoionization partial cross sections for atoms with allowance for correlations in several transitions* (Ioffe Institute, No 1620, Leningrad, 1993) (In Russian)
37. L.V. Chernysheva, *Complex of Programs for Automation of Atomic Calculations* (Ioffe Institute, Leningrad, 1981). ((In Russian))
38. M.Ya. Amusia, L.V. Chernysheva, *Computation of Atomic Processes* (Institute of physics Publishing, Bristol and Philadelphia, 1997)
39. D.A. Varshalovich, A.N. Moskalev, V.K. Khersonskii, *Quantum Theory of Angular Momentum* (World Science, Singapore, 1988)
40. M. Abramowitz, I.A. Stegun (eds.) *Handbook of Mathematical Functions with Formulas, Graphs and Mathematical Tables* (NBS, AMS-55, 1964)

Chapter 23
Examples of Calculation Results Using ATOM-M

This chapter provides examples of results obtained using the ATOM-M system. Examples of both intermediate and final calculation results are given—wave functions and matrix elements, as well as cross sections of processes and excitation probabilities for absorbing photons, scattering of electrons, positrons, and muons. Atoms, endohedrals, and diatomic molecules are considered as targets.

23.1 General Remarks

In this chapter, the calculation methods described above are illustrated by some concrete examples. The automated system for studying the structure of atoms ATOM-M has been successfully and widely used in various institutes in the USSR, USA, Russian Federation, Israel, Germany, and a number of other countries. Numerous interesting results were obtained, which were published in hundreds of articles and a number of monographs (see, e.g., [ACY], in which concrete data are limited to 2008. The monograph [ACY] is not readily available due to its high price. Therefore, later results, mainly on endohedral atoms, are illustrated by the original papers [1–10].

Relatively recently, the ATOM complex was modified so that it became possible to study the properties of endohedral atoms and find the cross sections of processes with their participation (see, e.g., [8]). In recent years, important and recognized results on the study of diatomic molecules have been added here. This became possible only due to generalization and development of a number of programs of the ATOM system for the case of diatomic molecules of cylindrical symmetry, as opposed to centrally symmetric atoms (see, e.g., [11–14]). As a result, the ATOM-M system appears, which makes it possible to uniformly consider atoms, endohedrals and diatomic molecules. This system can develop further, including previously unconsidered processes and new objects, for example, atoms in strong fields and triatomic molecules.

M. Ya. Amusia and L. V. Chernysheva, *Computation of Atomic and Molecular Processes*, Springer Series on Atomic, Optical, and Plasma Physics 117,
https://doi.org/10.1007/978-3-030-85143-9_23

Before presenting a brief account of the results obtained, we would like to mention several general problems associated with the design and organization of systems of this kind.

In our opinion, when developing such systems, the following goals should be pursued:

1. Creation of some new research tool for a theoretical physicist,
2. Performing a complex set of studies, which can be called a theoretical or numerical experiment,
3. Carrying out mass calculations of various characteristics that are of interdisciplinary importance, and can and should become a tool of an experimental physicist.

The first, and most important, goal involves the development of numerical models for a fairly wide range of problems within the framework of the main chosen theoretical approach, if this approach cannot be realized without using a computer. At this stage, a software package should be developed to ensure that the desired results are achievable within a reasonable time frame with the specified accuracy. The software should also make it possible to use the results in further research. On the other hand, it must be borne in mind that the theoretical physical models underlying mathematical programs must be brought to a level that permits its implementation on a computer. The computer itself makes special demands on the preparatory work of a theoretical physicist. We think that the development of the system is possible only by combining the efforts of theorists and specialists in the field of software development, whose close interaction allows them to agree on common requirements. Such cooperation, as our experience shows, dramatically increases the effectiveness of each specialist, freeing him from the need to deal with problems outside the area of his/her competence.

After completing the first task, it is natural to take care of further improving the theory, i.e., going beyond the theoretical approach used initially. This second goal requires conducting exploratory calculations, testing various hypotheses and assumptions, etc. At this stage, the absence of a rigorous theoretical model is compensated by the possibility of testing a large number of hypotheses within the framework of the chosen theoretical approach. In fact, the work of the theoretician becomes similar to the work of the experimentalist, but here the object of research is a numerical model. In other words, the implementation of the second goal is a kind of theoretical or numerical experiment. A special requirement for the software package to fulfill this goal is the maximum possibility of changing data at intermediate stages of calculations using additional conditions, as well as the possibility of direct use of intermediate calculations for further research.

The natural development in achieving of the two described goals, if they provide good agreement between the calculated and experimental data, is to carry out mass calculations. In this case, the developed software package should provide the final results in the required form—as curves, tables, databases, etc. In contrast to the case of first two goals, where it is most important to reduce the time for creating new programs, the third goal requires the fastest execution of calculations.

Almost all physical problems are now solved with the help or participation of computers. However, these problems are usually posed to the computer by the theorists themselves, who solve only their own question, and therefore the created by them program turns out to be adapted only for one or another specific problem. An integrated approach to solving a wide class of problems, combining the joint efforts of experts in various fields and a special organization of software as a flexible system, user-friendly, implementing new approaches and methods in the studied area, as well as easily expandable, opens up new horizons in theoretical research and increases its effectiveness.

The ATOM-M system developed by us implements a technique for studying the structure of the electron shells of atoms, endohedrals, and diatomic molecules based on the many-body theory. Previous approaches in this area gave results that were in poor agreement with experimental data. In contrast to the previously widely used Hartree–Fock method, the approach that implements the random phase approximation with exchange (RPAE) is the next step in the study of atoms, endohedrals, and diatomic molecules, since this makes it possible to account, as noted above, multi-electron correlations. The Hartree–Fock method is used in the ATOM-M system as an initial approximation. As a result, the ATOM-M system enables a more complete understanding of the electronic structure of both atoms and various polyatomic formations—molecules and solids, contributing to the improvement of their theoretical description and stimulating the development of experimental methods.

Some calculation results are obtained faster than that in experiments and with a sufficiently high accuracy, which makes it possible to obtain the necessary data without performing experimental research, and sometimes predict new physical effects. These considerations made it possible to use the ATOM-M system for performing of mass calculations. In some cases, when experimental research is technically very difficult or simply impracticable, ATOM-M turns out to be the only reliable source of information.

At present, the ATOM-M system makes it possible to go beyond the RPAE framework, using it as an initial estimate for further investigation of the structure of atoms, endohedrals, and diatomic molecules and their interaction with external fields and different projectiles—electrons, positrons, mesons, etc. Calculations can be performed in various modifications of the many-particle perturbation theory.

The results of calculations performed can be divided into two groups. The first includes those obtained directly using the ATOM-M system. The second group is formed by the final results obtained by combining intermediate data from the ATOM-M system with analytical methods.

23.2 Results of the First Group of Calculations

The most complete results of the calculations of physical processes probabilities of interaction of photons, electrons, positrons, and mesons with atoms and endohedrals within the HF and RPAE can be found in the monograph [ACY], which included the data, as mentioned above, up to 2010. The data presented in this monograph are not a repetition of what was given in previously published articles, but represent the results of newly performed calculations, using the same formulas, but employing significantly greater computational capabilities that are a consequence of improvement of computer technology.

Much less extensive data on the study of similar processes in the same approximation, but for diatomic molecules, are so far mainly contained only in original articles (see, e.g., [11–14]). Selected results of these studies are given in [AICC].

In this book, unlike in [ACY], we present only a small number of figures, just to give examples of the results obtained Figs. 23.1, 23.2, 23.3, 23.4, 23.5, 23.6, 23.7, 23.8, 23.9, 23.10, 23.11, 23.12, 23.13, 23.14, 23.15, 23.16, 23.17, 23.18, 23.19, 23.20, 23.21, 23.22, 23.23, 23.24, 23.25, 23.26, 23.27, 23.28, 23.29, 23.30, 23.31, 23.32, 23.33, 23.34, 23.35, 23.36, 23.37, 23.38, 23.39, 23.40, 23.41, 23.42, 23.43, 23.44, 23.45, 23.46, 23.47, 23.48, 23.49, 23.50, 23.51, 23.52 and 23.53 . For simplifying the perception, the figures for atoms follow first, and then goes that for molecules.

The results of the first group include the following:

1. Hartree–Fock ground and various excited states wave functions $\phi^{N+1}, \phi^{N}, \phi^{N(LS)}$ and that with account of the rearrangement for a number of atoms (see examples in Figs. 23.1, 23.2, 23.3). Similar are results for light diatomic molecules (Figs. 23.4, 23.5, 23.6, 23.7, 23.8, 23.9, 23.10, 23.11, 23.12, 23.13, 23.14, 23.15, 23.16, 23.17, 23.18, 23.19, 23.20, 23.21, 23.22, 23.23, 23.24, 23.25, 23.26, 23.27, 23.28, 23.29, 23.30, 23.31, 23.32, 23.33, 23.34, 23.35, 23.36, 23.37, 23.38, 23.39, 23.40, 23.41, 23.42, and 23.43.).

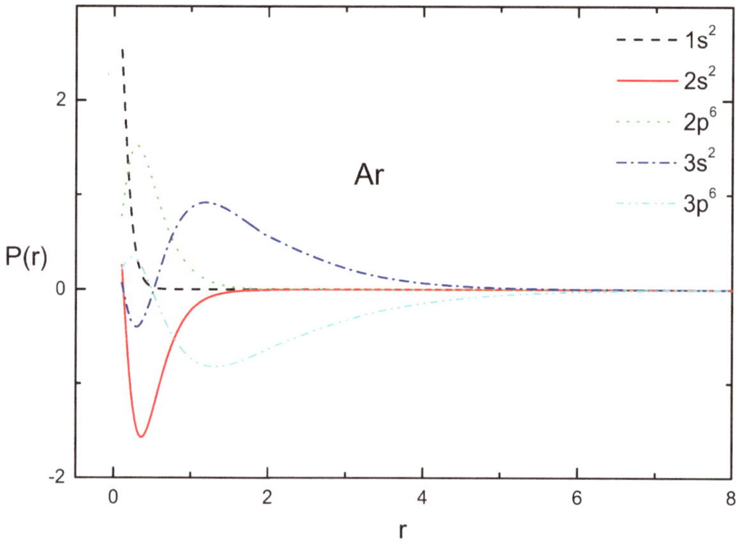

Fig. 23.1 Wave functions of Ar ground state, in HF (program ***scfhf***)

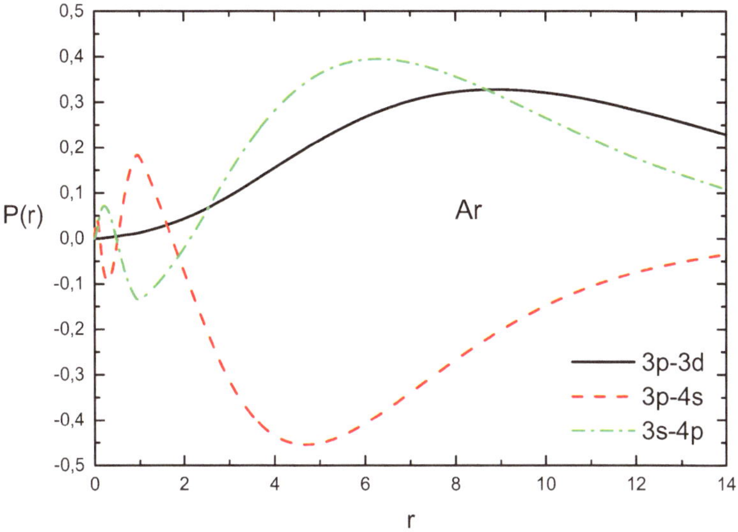

Fig. 23.2 Wave functions of Ar excited discrete states, in HF (program a ***fchf***)

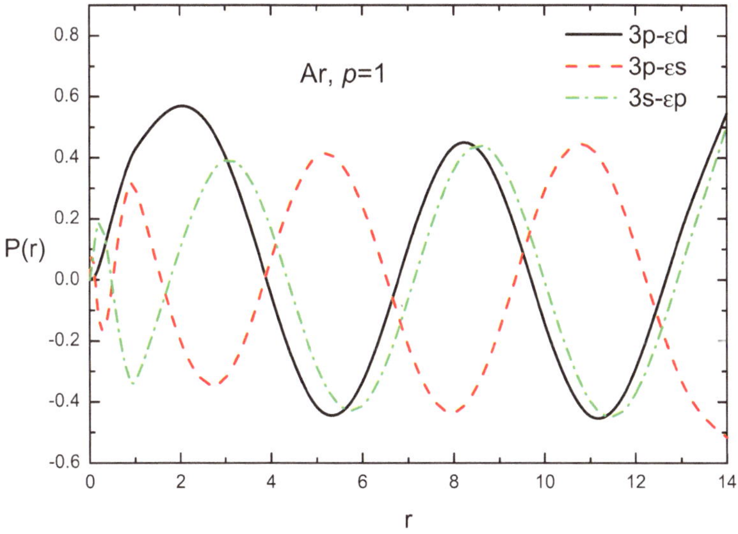

Fig. 23.3 HF wave functions of Ar in continuous spectrum for linear momentum equal $p = 1$ (program ***fchf***)

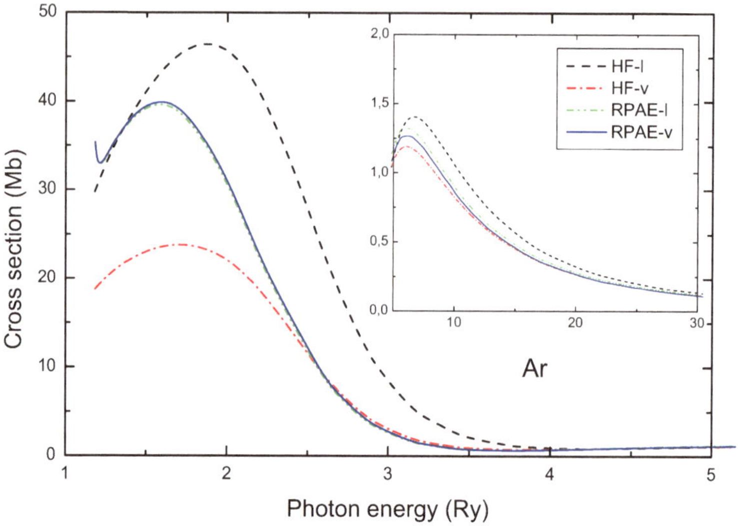

Fig. 23.4 Photoionization cross section of the Ar atom in HF and RPAE (program ***phoabs***), [ACY]

Fig. 23.5 Photoionization cross section of the Xe atom beyond the threshold of the 4d-subshell—giant resonance (program ***phoabs***), [ACY]

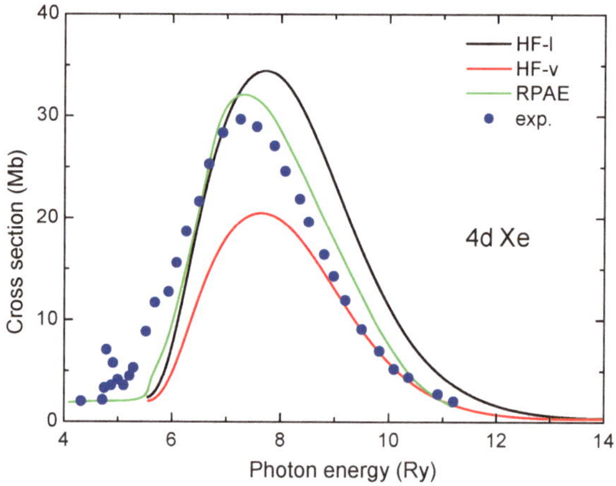

2. Total energies of all considered atoms and molecules in the ground and some excited states; ionization potentials and energies of single-particle excitation levels.
3. Total and partial photoionization cross sections of two or three outer, some intermediate and inner subshells in the HF and RPAE approximations, accounting for electronic correlations in one or several (up to 12) transitions:

   (a) Noble gas atoms He, Ne, Ar, Kr, and Xe,
   (b) Alkaline atoms Li, Na, K, Rb, and Cs,
   (c) Alkaline earth atoms Be, Mg, Ca, Zn, Ba, and Hg,

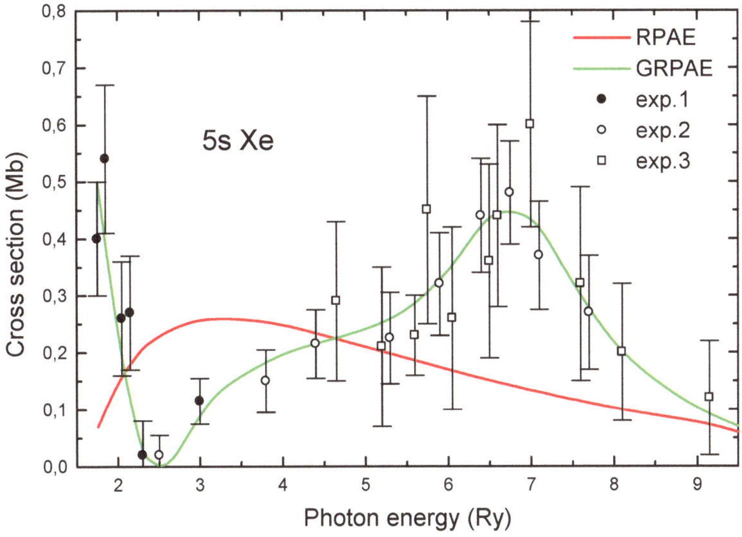

Fig. 23.6 Photoionization cross section—interference resonance (program ***phoabs***) [ACY]

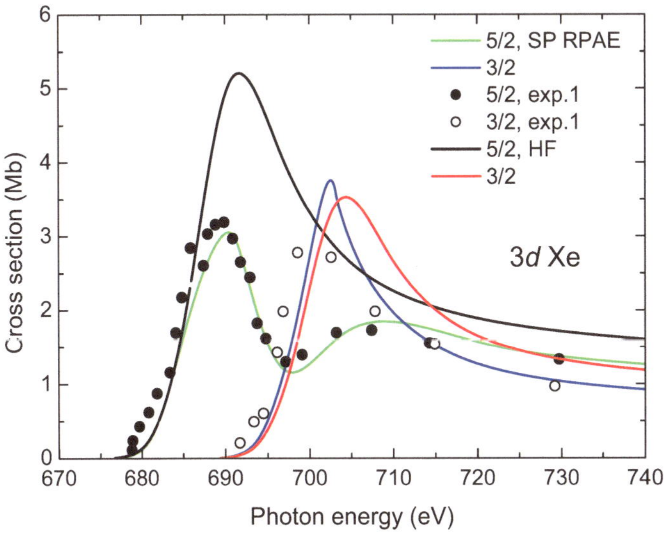

Fig. 23.7 Photoionization cross section of the $3d^{3/2}$ и $3d^{5/2}$ levels of the Xe atom—spin-doublet resonance (program ***phoabs***) [ACY]

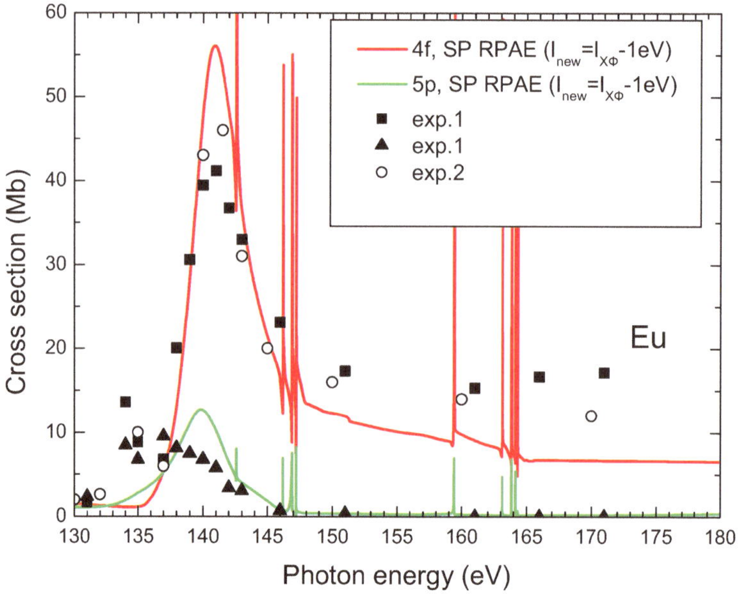

Fig. 23.8 Photoionization cross section of $4f\uparrow$ near the discrete excitation level $5p\downarrow -4f\downarrow$ of the Eu atom—giant autoionization resonance (program ***phoabs***) [ACY]

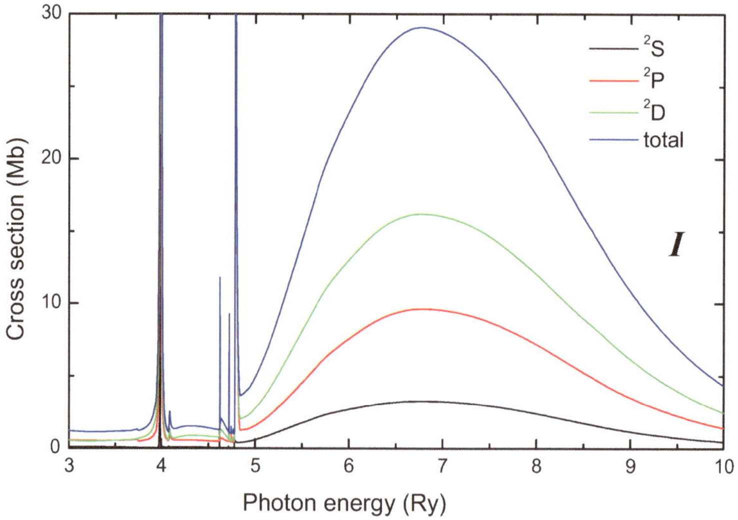

Fig. 23.9 Total photoionization cross section and contributions of individual terms for atom I in RPAE (program ***ospp***)

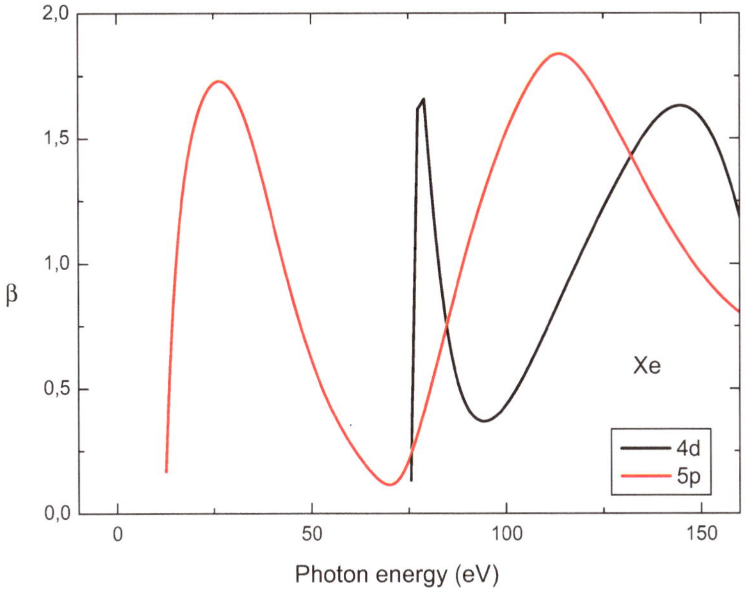

Fig. 23.10 Dipole angular anisotropy parameter for 5p and 4d Xe subshells. (***betta*** procedure in the ***phoabs*** program)

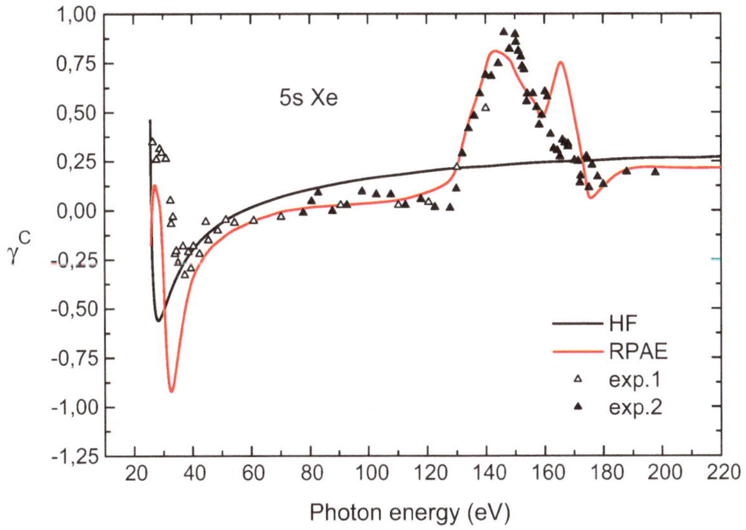

Fig. 23.11 Nondipole angular anisotropy parameter γ_{5s} for the Xe 5p-subshell. (program ***anglaniz0***) [ACY]

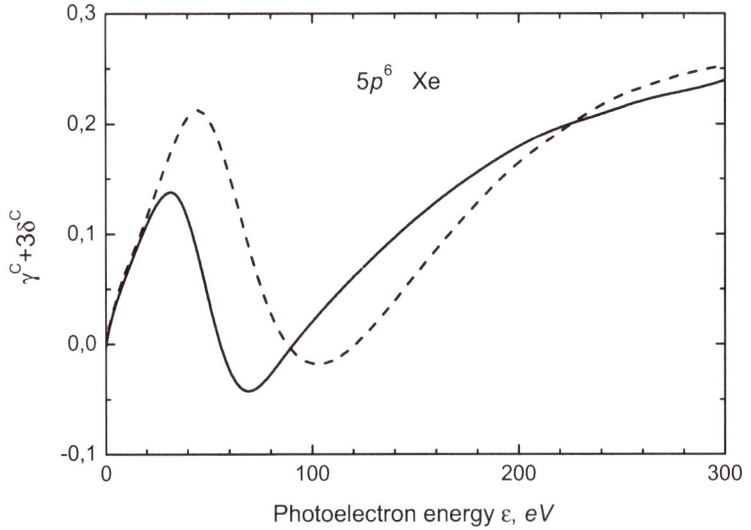

Fig. 23.12 Nondipole angular anisotropy parameter δ_{5p} for Xe 5p-subshell. (***anglaniz1*** program)

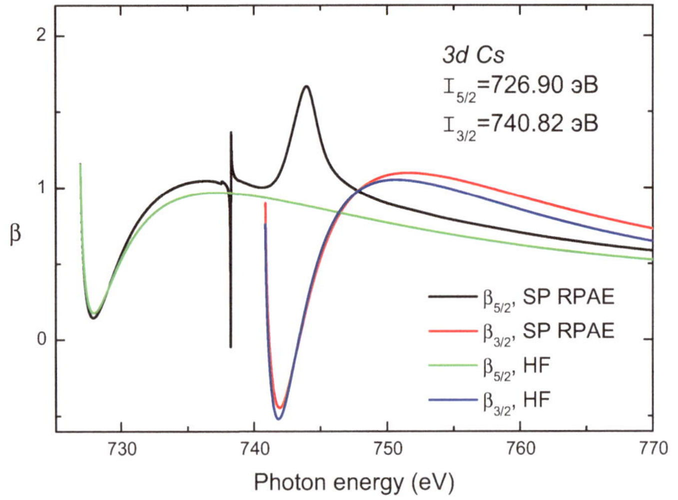

Fig. 23.13 Dipole angular anisotropy parameters $\beta(\omega)$ of Cs $3d_{5/2}$ and $3d_{3/2}$ levels Cs. Calculation in SPRPAE (***betta*** procedure in ***phoabs*** program)

(d) Atoms with semi-closed subshells N, Cr, Mn, Tc, Eu, Pd, In, and La,
(e) Atoms with one vacancy in the closed subshell, or an electron outside the closed subshell,
(f) A whole series of positive and negative ions of some of the said atoms,

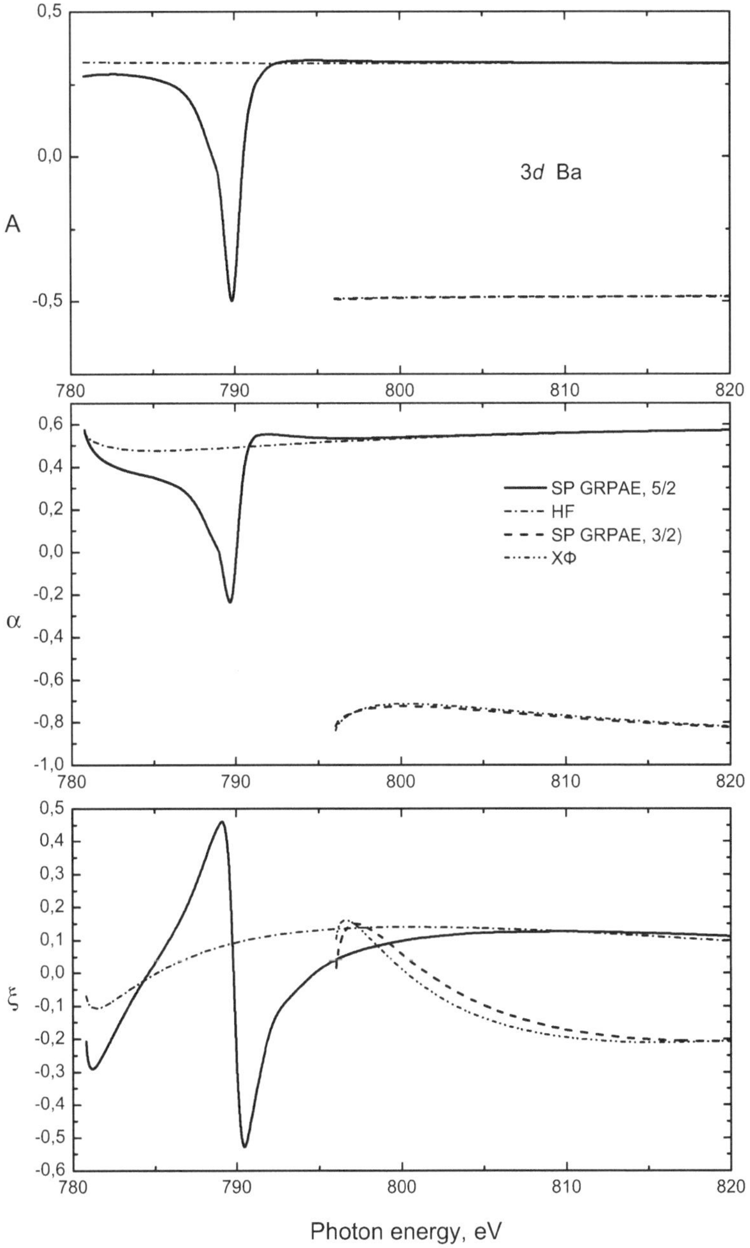

Fig. 23.14 Spin polarization parameters $A^j(\omega)$, $\alpha^j(\omega)$ and $\xi^j(\omega)$ for $3d_{5/2}$ and $3d_{3/2}$ levels of Ba in the SP GRPAE and HF. (***spinpol*** program)

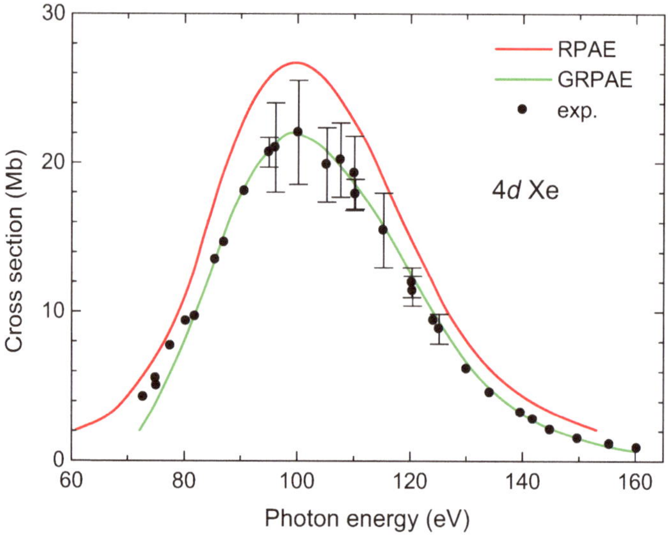

Fig. 23.15 Single-electron photoionization cross section of Xe $4d^{10}$ subshell GRPAE with the inclusion of the effect of "knocking out on the way out" [ACY]

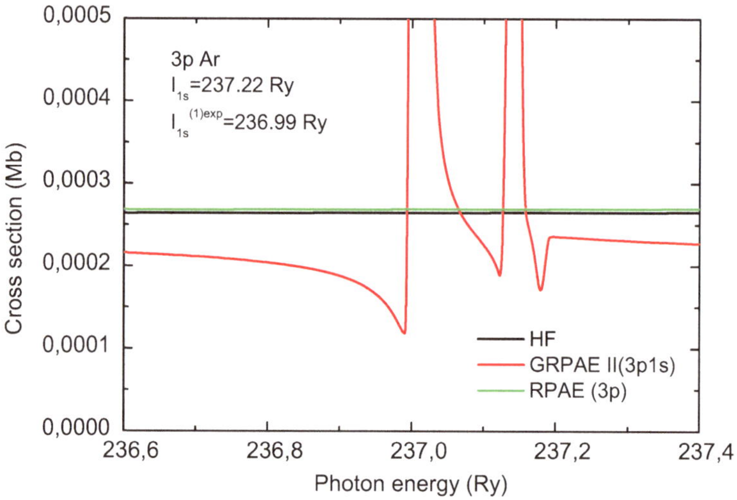

Fig. 23.16 Photoionization cross section of $3p$ electrons near the 1 s-threshold in Ar

(g) Light diatomic molecules.

Examples of calculating the photoionization cross sections for a part of the subshells for atoms are shown in Figs. 23.4, 23.5, 23.6, 23.7, 23.8, and 23.9, and for light diatomic molecules, in Figs. 23.44, 23.45, 23.46,

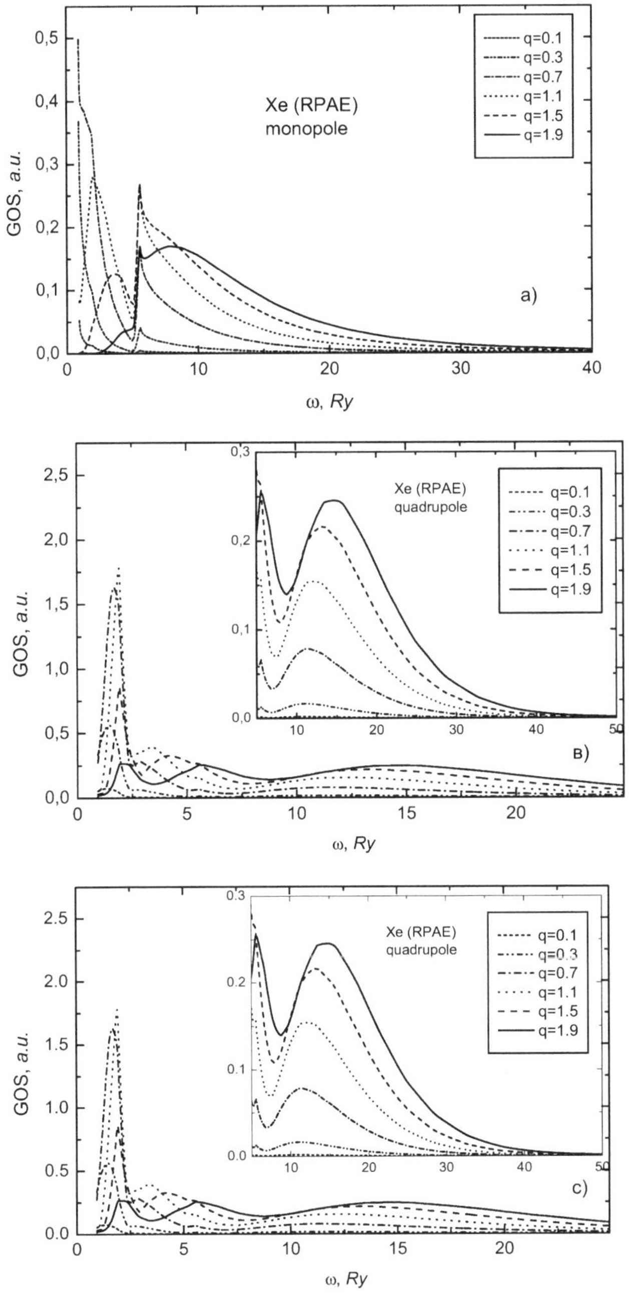

Fig. 23.17 Generalized oscillator strengths of the Xe atom as a function of the transferred energy ω (Ry) and momentum q (at. un.). The total GOS density in RPAE is **a** monopole, **b** dipole, **c** quadrupole. (***gos*** program)

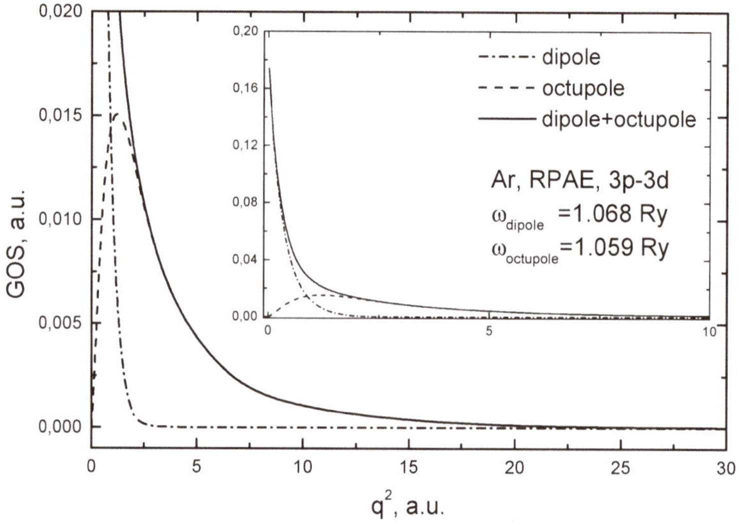

Fig. 23.18 Generalized oscillator strengths of Ar 3*p*-3*d* transition (***gos*** program)

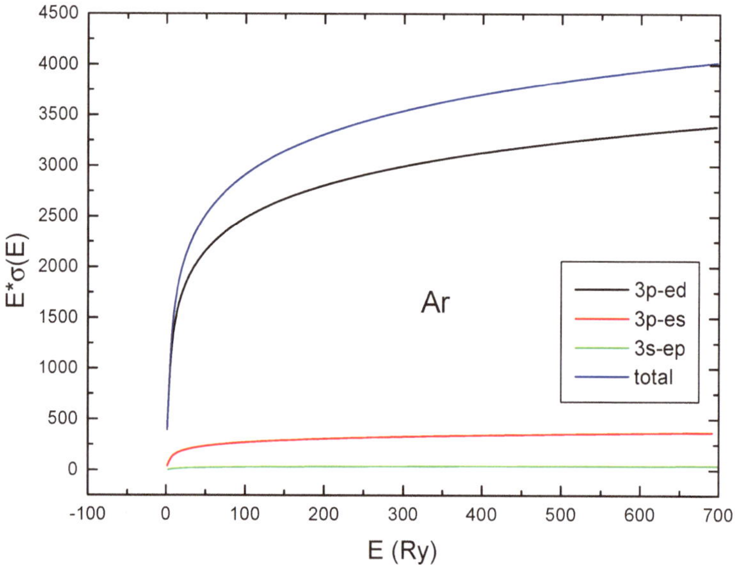

Fig. 23.19 Non-logarithmic correction to the total cross section of fast electrons inelastic scattering on Ar atom (***scatlog*** program)

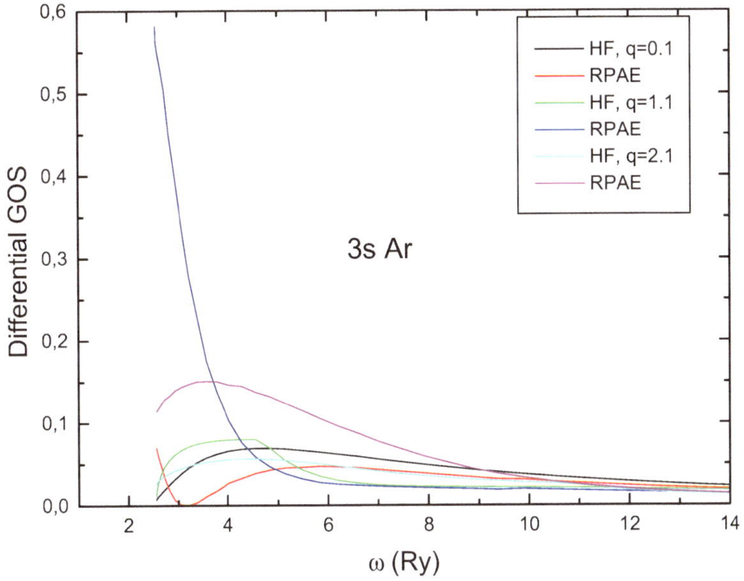

Fig. 23.20 Differential GOS of Ar 3 s-subshells in the HF and RPAE at q = 0.1, 1.1, and 2.1. (***difgos*** program)

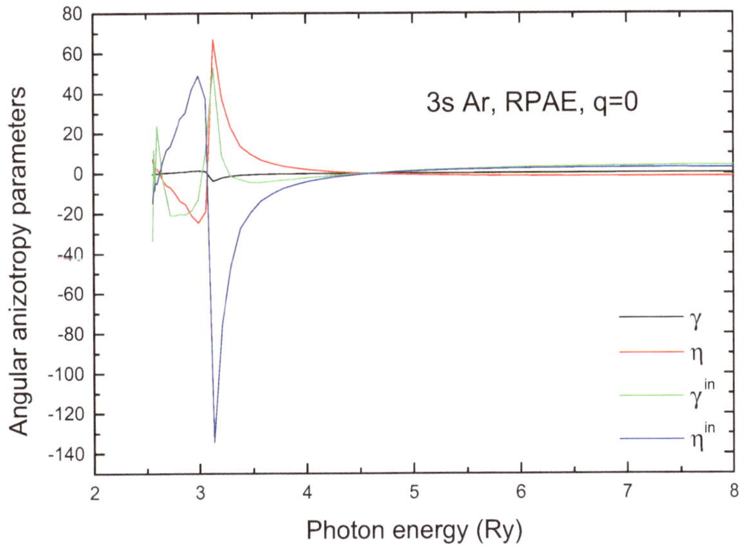

Fig. 23.21 Comparison of the angular anisotropy parameters for photoionization and inelastic scattering of fast electrons at linear momentum transfer $q = 0$ (index 'in') of Ar 3 *s*-subshell (***anglaniz0*** program)

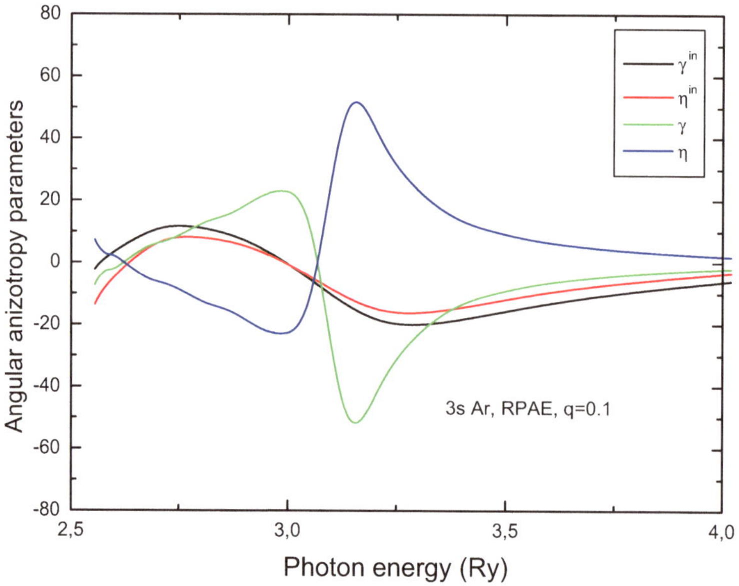

Fig. 23.22 Same, but for $q = 0.1$

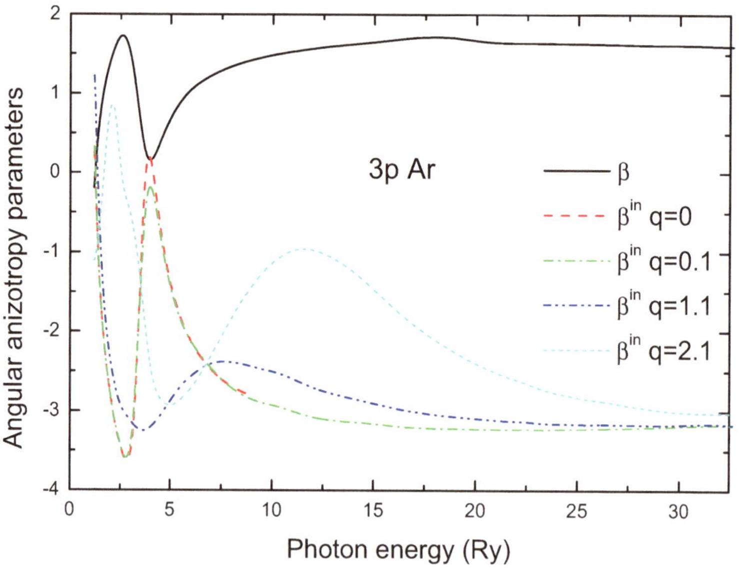

Fig. 23.23 Dipole angular anisotropy parameter of photoelectrons and secondary electrons emitted during inelastic scattering, Ar $3p$-subshell, $q = 0, 0.1, 1.1, 2.1$

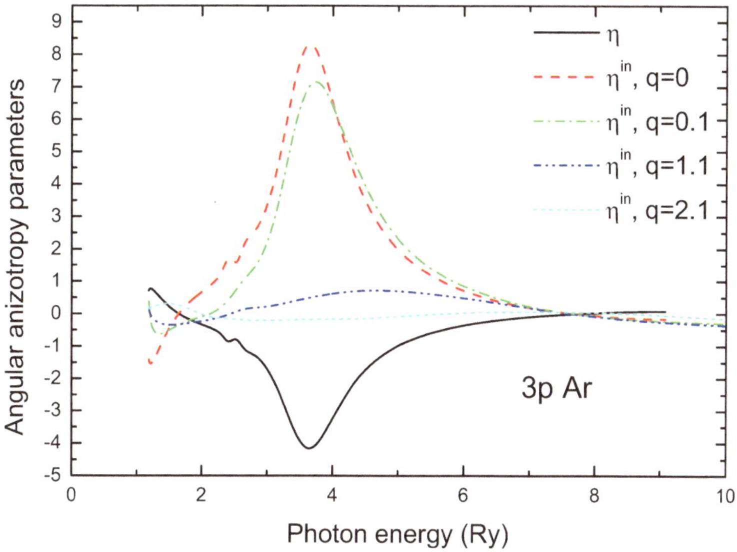

Fig. 23.24 Nondipole angular anisotropy parameter of photoelectrons and secondary electrons emitted in inelastic scattering, Ar *3p*-subshell, $q = 0, 0.1, 1.1$, and 2.1

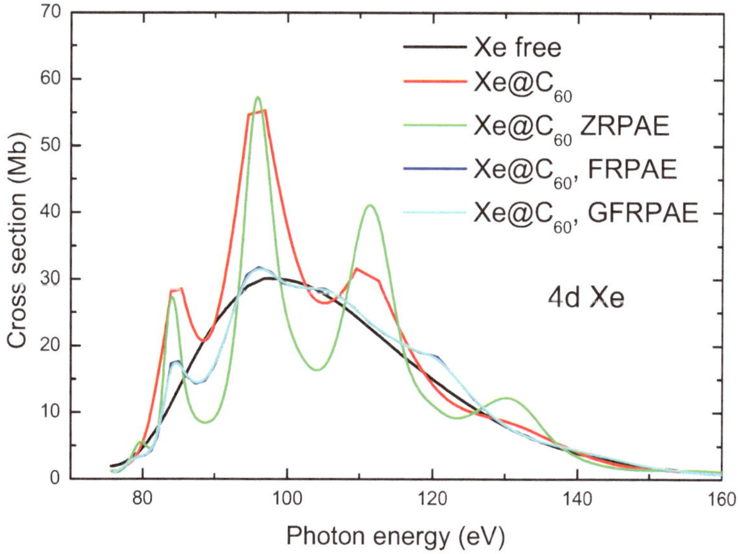

Fig. 23.25 Destruction of the atomic giant resonance under the influence of the fullerene shell. The endohedral Xe@C_{60} is considered and compared with atomic Xe. The results of the model of zero (ZRPAE) and nonzero fullerene potential well thicknesses (FRPAE) and also, with account of its polarizability (GFRPAE), are compared with the results of RPAE for Xe

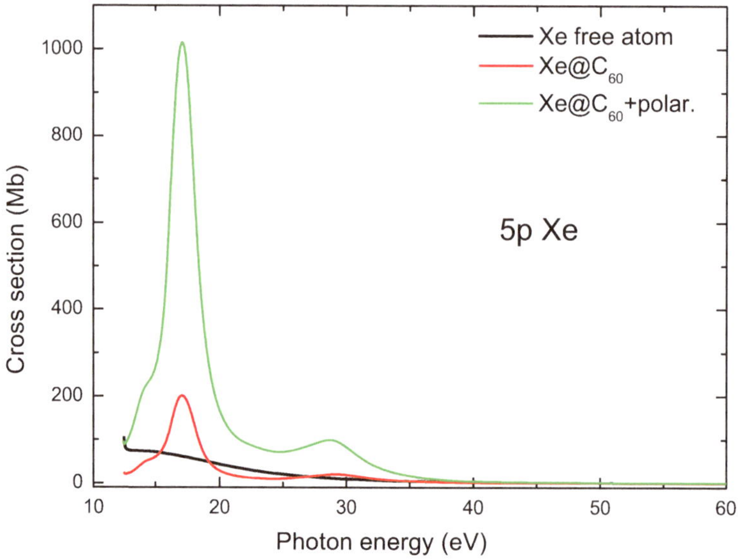

Fig. 23.26 Giant endohedral resonance of the 5p-subshell of Xe@C_{60} in comparison with the Xe atom. The calculations were carried out in RPAE for Xe and in FRPAE and GFRPAE, i.e., taking into account the finite thickness of the fullerene shell and its polarization, for Xe@C_{60} (***pabsfull*** program)

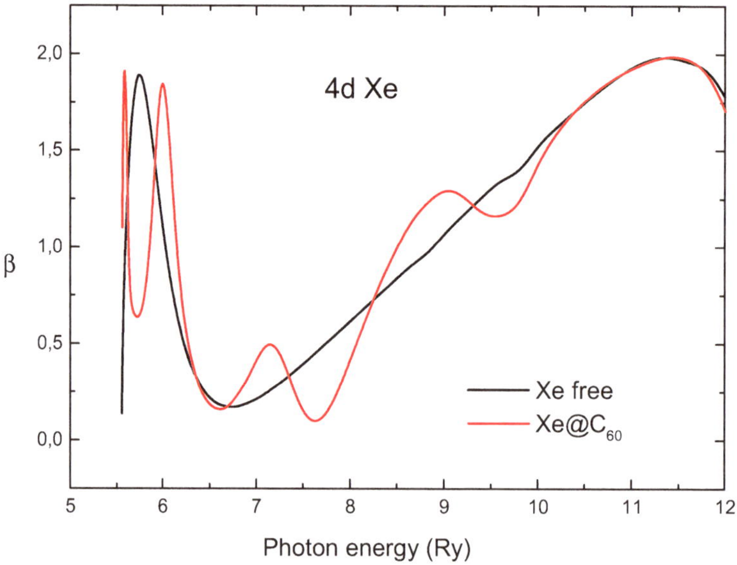

Fig. 23.27 Dipole angular anisotropy parameter of the 4d electrons of the endohedral Xe@C_{60} and the Xe atom in the framework of RPAE and FRPAE. (***betta*** procedure in ***pabsfull*** program)

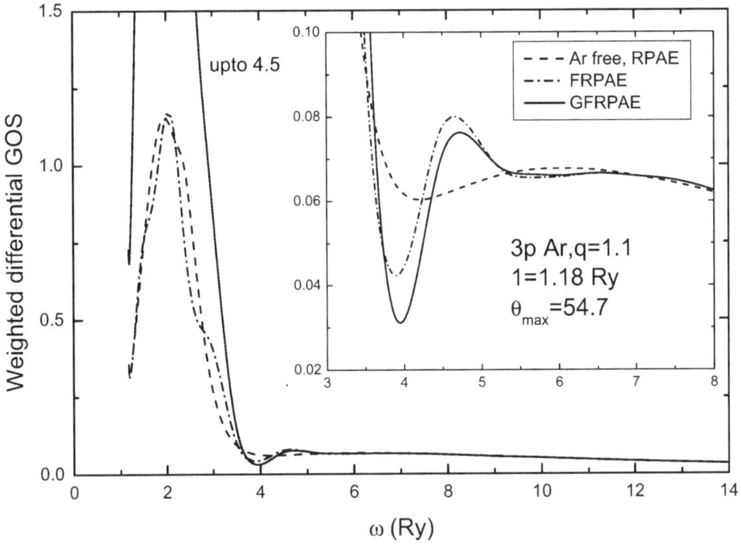

Fig. 23.28 Differential in the emission angle of atomic electrons GOS at the magic angle $P_2(\cos\theta_m) = 0$ for $3p$-subshell of Ar@C_{60} at q = 1.1 in WRPAE, GWRPAE and in RPAE for free Ar. (***difgos*** program)

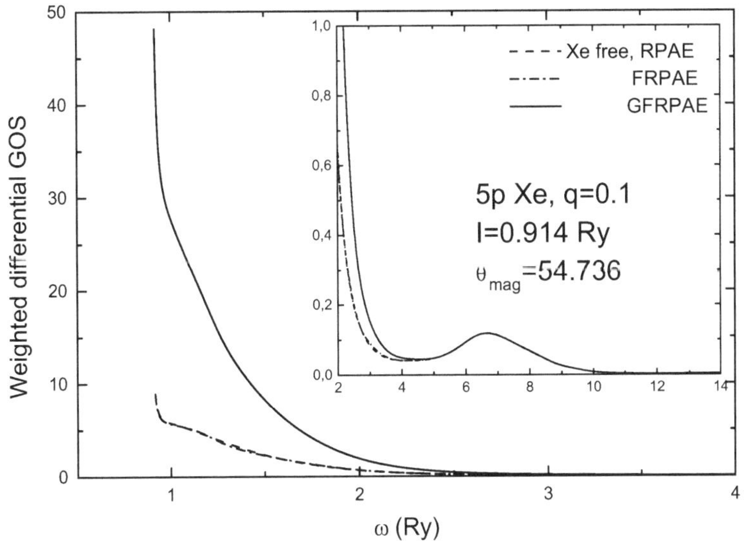

Fig. 23.29 Differential in the emission angle of atomic electrons GOS at the magic angle for $5p$-subshell of Xe@C_{60} at $q = 0.1$ in FRPAE, GFRPAE and for free Xe in RPAE

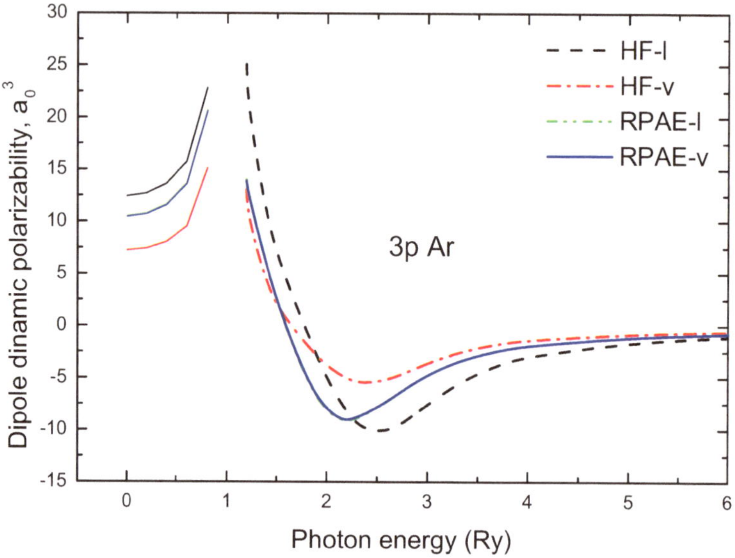

Fig. 23.30 Dipole dynamic polarizability Ar. 3*p-ed* transition contribution (***polar*** program)

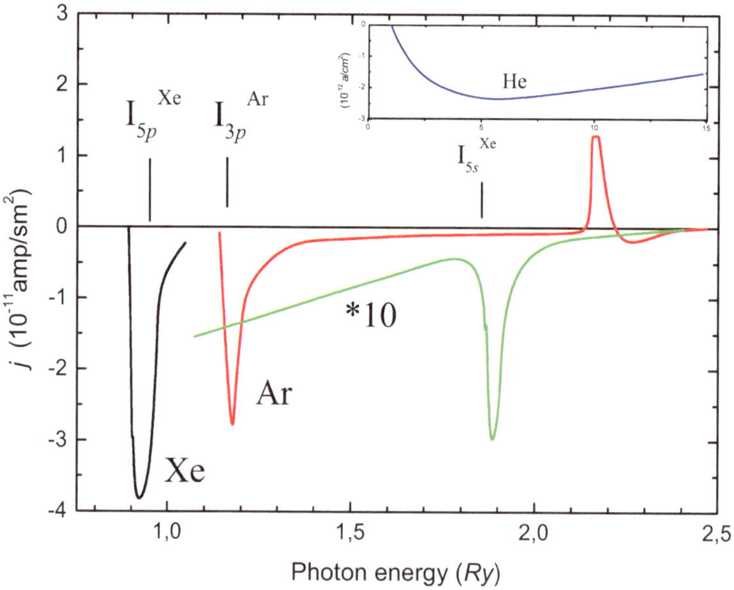

Fig. 23.31 "Drag currents" for Xe in comparison with the same characteristic for Ar and He. The green line is the Xe values multiplied by 10

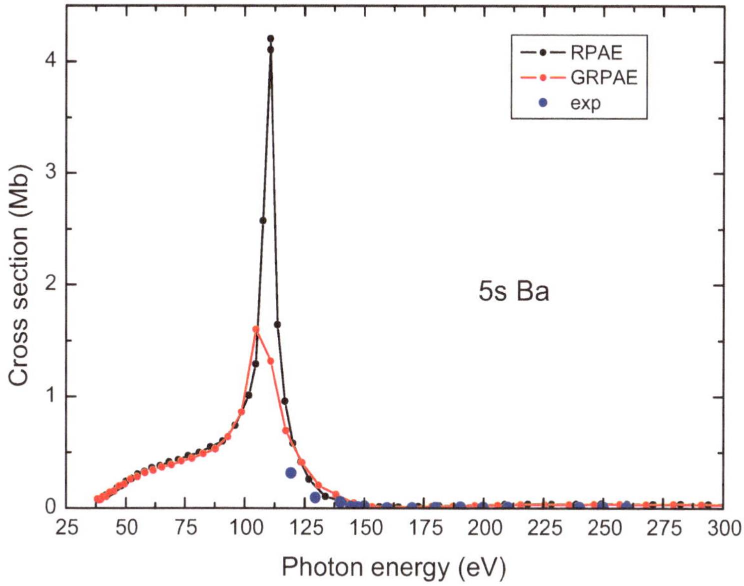

Fig. 23.32 Photoionization cross section of atomic Ba 5 *s*-subshell (interference resonance) [ACY]

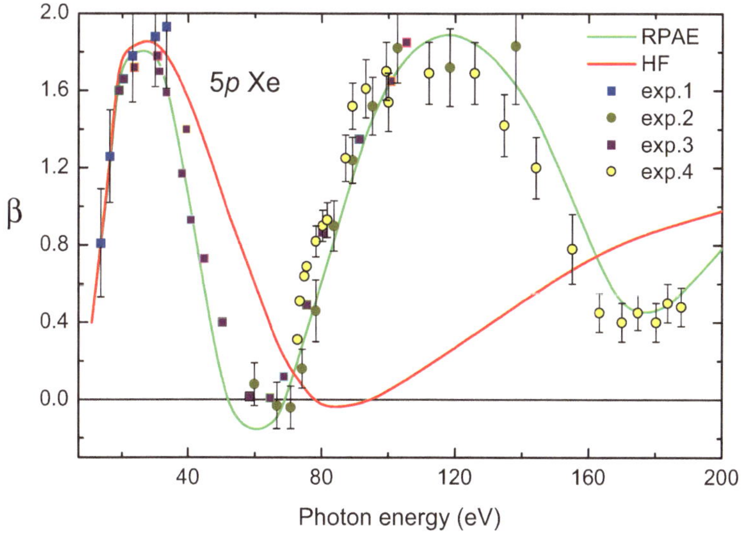

Fig. 23.33 Dipole angular anisotropy parameter of 5p-subshell in the vicinity of Xe 4d-subshell [ACY]

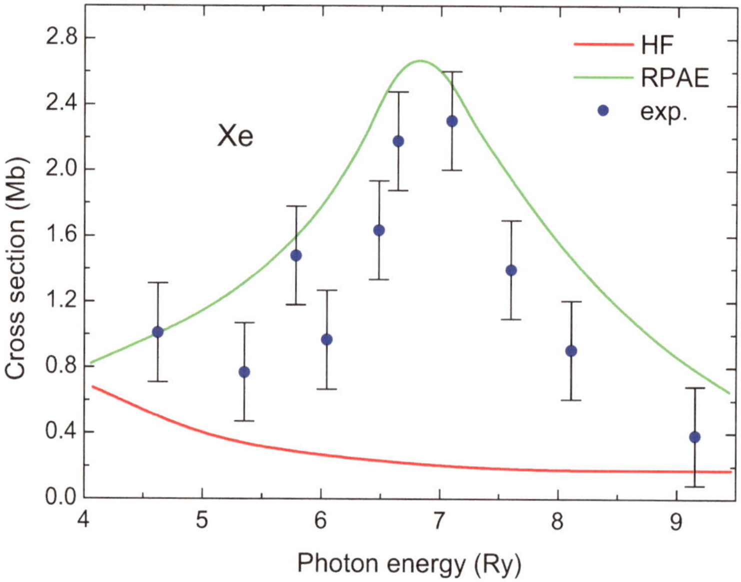

Fig. 23.34 Cross section for the production of singly-charged ions after absorbing photons by Xe atom in the region of 4*d*-subshell [ACY]

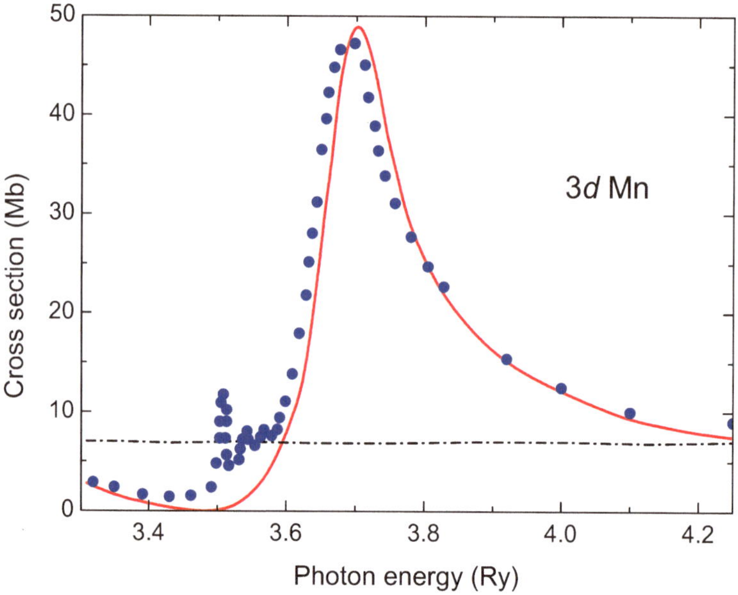

Fig. 23.35 Photoionization cross section of 3d-electrons of Mn atom with allowance for the 3p-transition (autoionization giant resonance) [ACY]

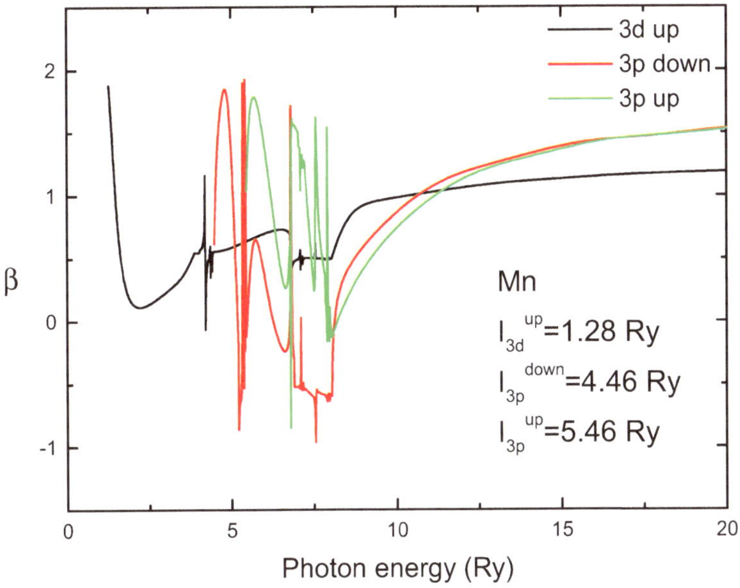

Fig. 23.36 Dipole angular anisotropy parameter of the $3d \uparrow$ and $3p \uparrow\downarrow$ levels of Mn atom in the SP RPAE

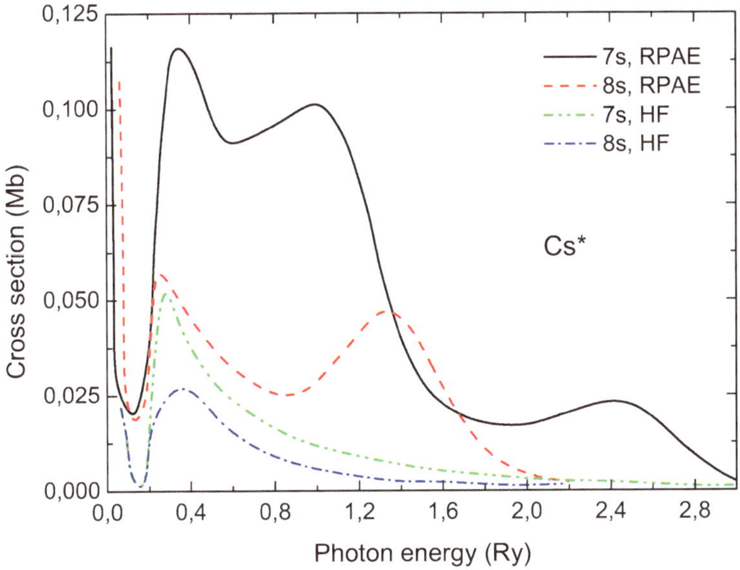

Fig. 23.37 Photoionization cross section for highly excited states of Cs* atom

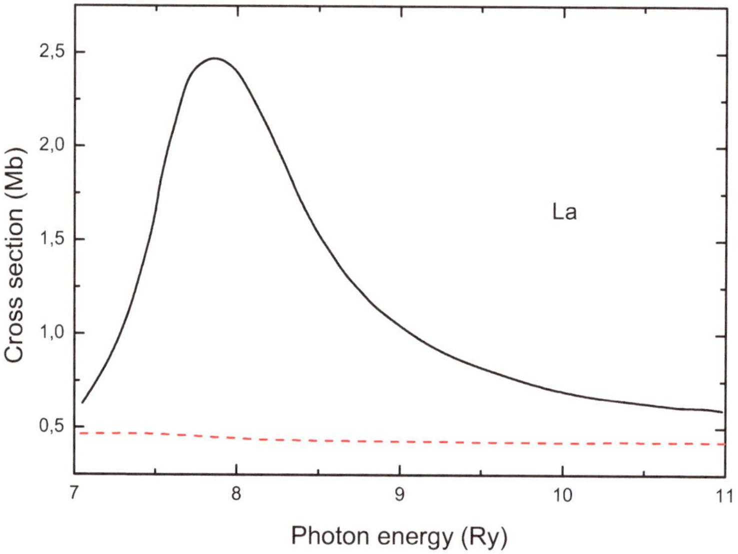

Fig. 23.38 Bremsstrahlung spectrum of fast electrons scattered by the La atom (***brem*** program)

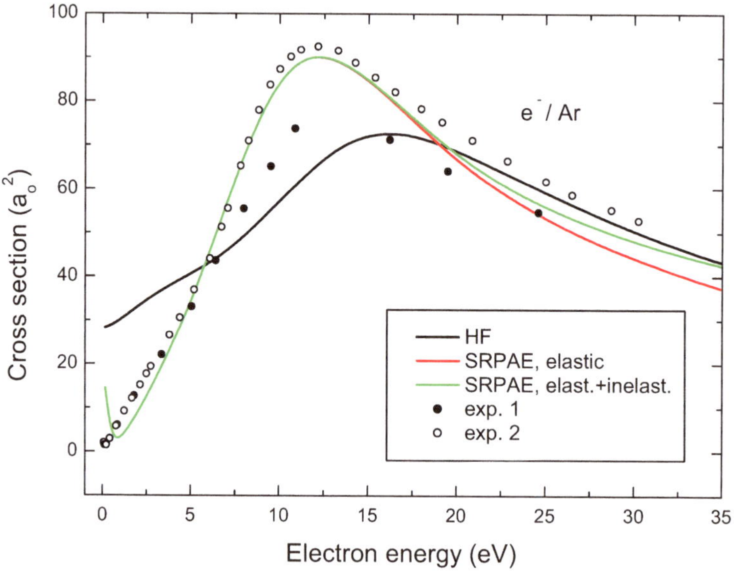

Fig. 23.39 Cross sections for elastic (in HF and SRPAE) and inelastic (in SRPAE) scattering of electrons by an Ar atom. (***scat*** program) [ACY]

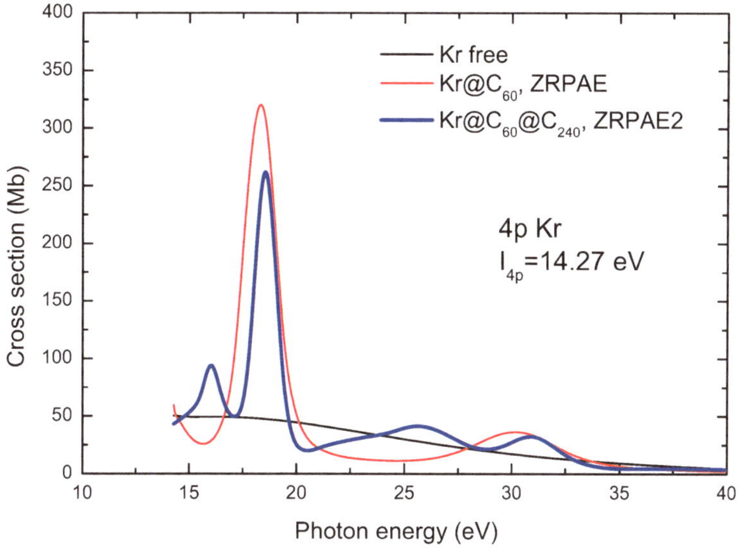

Fig. 23.40 Photoionization cross section of 4p-subshells of the "onion-type" endohedral $Kr@C_{60}@C_{240}$ in ZRPAE

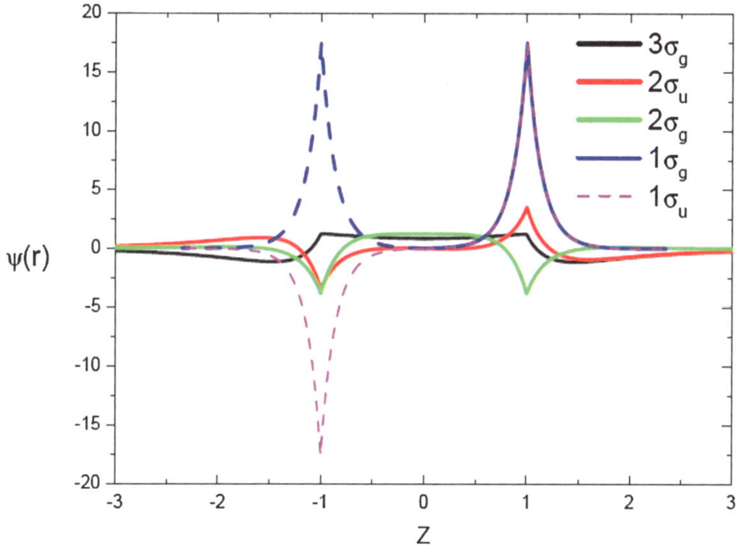

Fig. 23.41 Wave functions of a homonuclear nitrogen molecule along the axis of the molecule

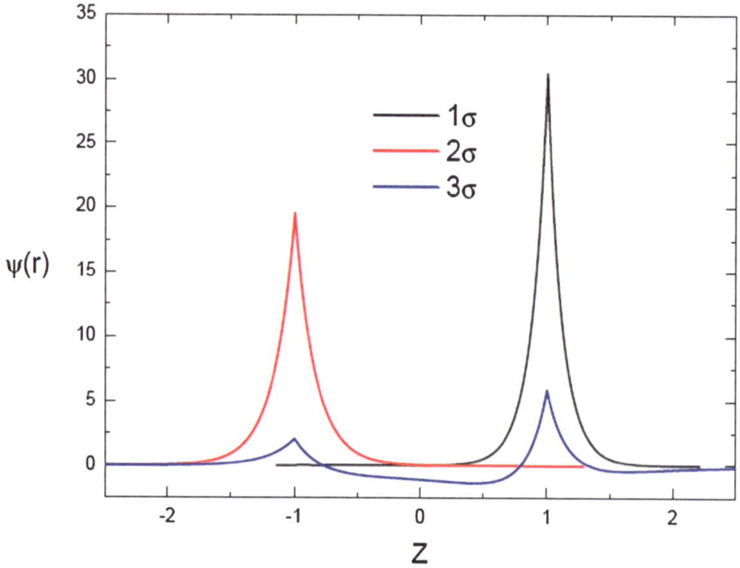

Fig. 23.42 Wave functions of a heteronuclear CO molecule along the molecule axis

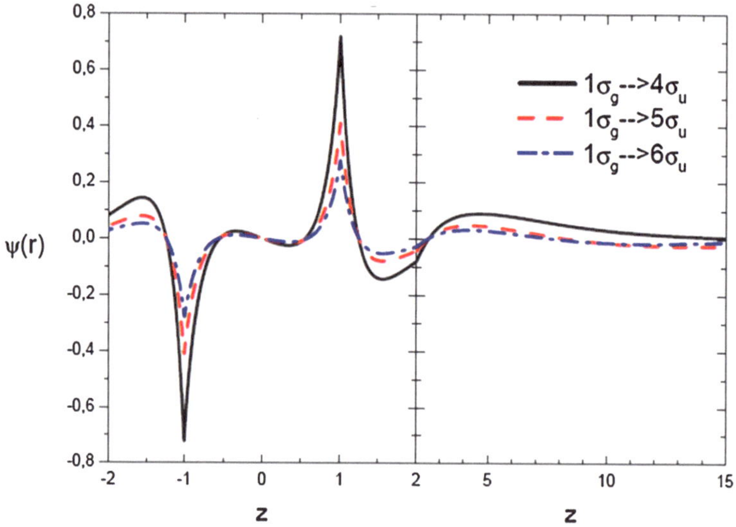

Fig. 23.43 Wave functions of discrete spectrum excited states for the N_2 molecule along its z-axis

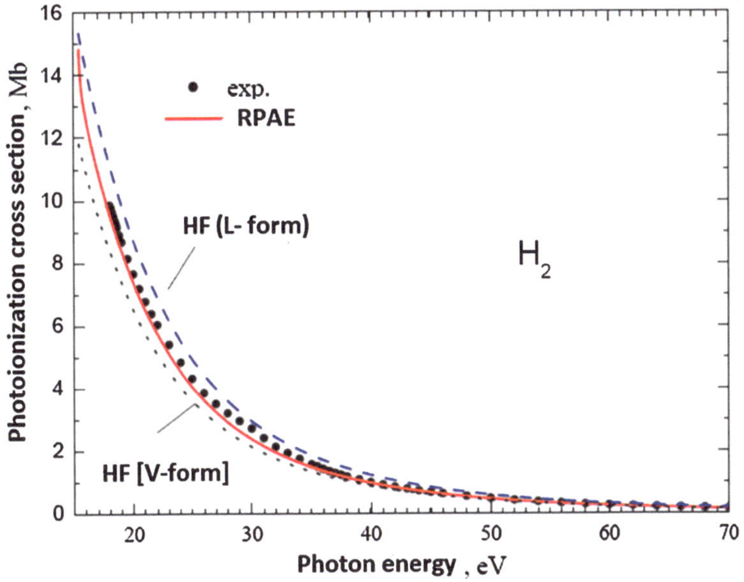

Fig. 23.44 Total cross section of photoionization of a hydrogen molecule, exp.—Samson and Haddad (1994)

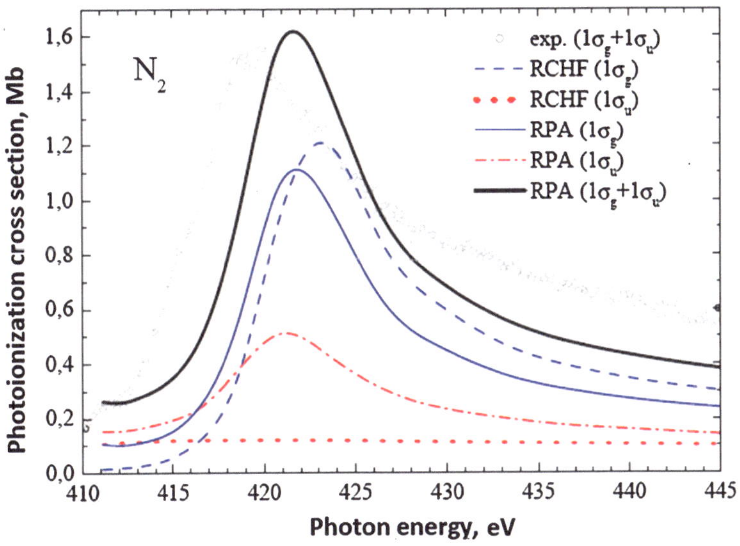

Fig. 23.45 Photoionization cross section of the nitrogen molecule K shell

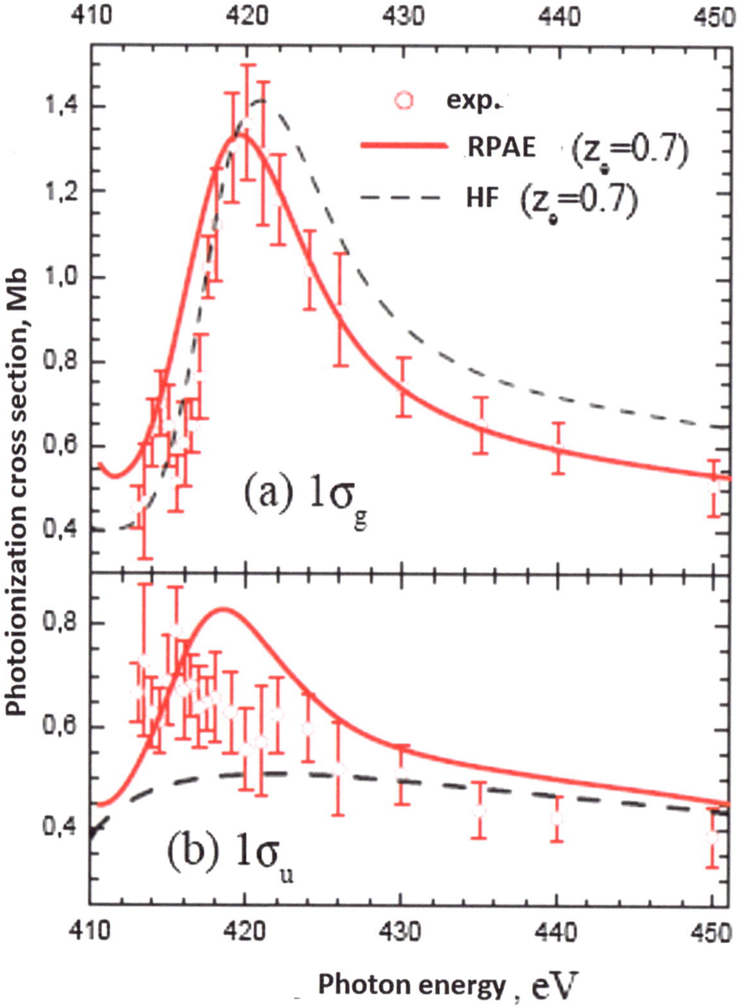

Fig. 23.46 Partial cross sections of the K shell subshells of the N_2 molecule

and 23.47.

4. A number of subshells in medium and heavy atoms were calculated accounting for the rearrangement of electron shells due to formation of a vacancy in the intermediate shells (within the framework of GRPAE) or accounting for the Auger decay of an inner shell vacancy—within the framework of GRPAEII.
5. Dipole and nondipole angular anisotropy parameters of photoelectrons in the HF and RPAE for the same subshells and atoms as mentioned in the previous paragraphs (examples in Figs. 23.10, 23.11, 23.12 and 23.13). Examples of the calculations of dipole parameter β for CO molecule are shown in Fig. 23.48.

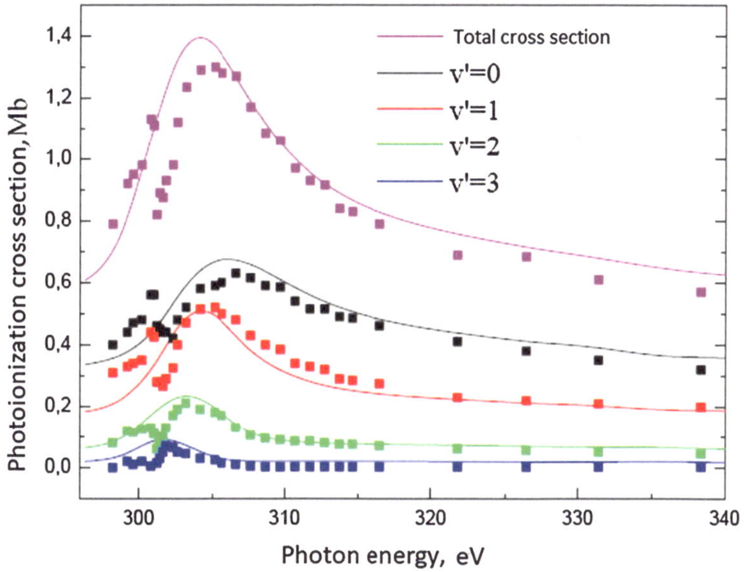

Fig. 23.47 Total and partial cross sections for photoionization of the C K- shell of the CO molecule for various vibrational transitions, calculated using the relaxed core model with a fractional parameter $z_e = 0.5$ (Chap. 4)

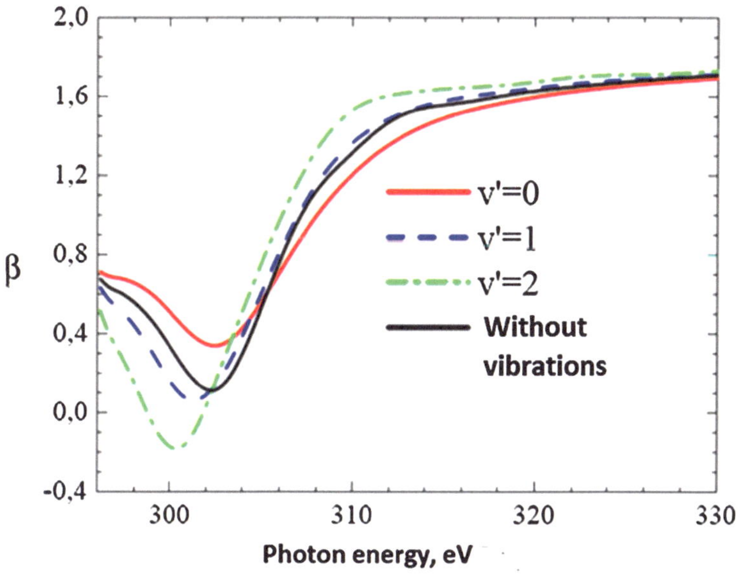

Fig. 23.48 Angular anisotropy parameter for the C K shell of the CO molecule for various vibrational transitions, calculated using the relaxed core model with a fractional parameter $z_e = 0.5$

6. Oscillator strengths of the first few discrete levels in all considered subshells of the above-mentioned atoms,
7. Photoionization cross sections of one-, two-, and higher multiplicity ions of alkali, alkaline-earth, and some other atoms. Usually, those atoms are chosen whose ions have completely closed or semi-closed subshells. The calculations are carried out in the Hartree–Fock, RPAE, and GRPAE. Sequences of ions with the same electronic configuration are also considered, for example, Cs^+, Ba^{++}, La^{+++}, whose configuration coincides with Xe.
8. Photoionization cross sections of negative ions of alkali and some other atoms, which, after formation of negative ions, have only closed or semi-closed subshell. The calculation was carried out in the Hartree–Fock and RPAE.
9. Photoelectron spin polarization parameters (example—$3d_{5/2}$ and $3d_{3/2}$ levels of Ba, Fig. 23.14).
10. The yield of single ions Ar^+, Kr^+, Xe^+, Ba^+ and a number of others in the process of photoionization near the inner shell thresholds in the Hartree–Fock and RPAE, as well as including the rearrangement in GRPAE (the cross section for the Xe^+ yield in photoionization in the region of Xe $4d^{10}$-subshell is given in Fig. 23.15).
11. Photoionization cross sections of excited states of many atoms and ions in the Hartree–Fock, RPAE and GRPAE.
12. Photoionization cross sections with account of vacancy decay in inner shells of atoms. Atomic electrons are calculated within the framework of GRPAEII, i.e., in the field of an ion that appears as a result of the vacancy Auger decay. A concrete example of photoionization is $1s$-shells of Ar and Kr.
13. Cross sections of Compton scattering accompanied by excitation or ionization of the target atom.
14. Cross sections of photoionization near the inner subshells threshold, accounting for a sudden change of the field in which the photoelectron moves—from the field of an internal vacancy to the field of an ion formed after its Auger decay, i.e., with account for the so-called post-collision interaction (PCI).
15. PCI influence on the spectral shape of the Auger electron.
16. PCI influence on the single ions yield near the intermediate shell ionization threshold. A specific example is the Ar $2p$-subshell.
17. Peculiarities of the outer subshell photoionization cross section near the inner shell threshold [15] (example in Fig. 23.16).
18. Generalized oscillator strengths of intermediate and outer shells of a number of atoms, in particular, He, Be, Ne, Ar, and Xe in the Hartree–Fock approximation, RPAE, and GRPAE (examples in Figs. 23.17 and 23.18).
19. Total cross section of fast electron inelastic scattering by atoms (example Ar in Fig. 23.19).
20. Angular distribution of secondary electrons arising from inelastic scattering of fast electrons by atoms (examples in Figs. 23.20, 23.21, 23.22, 23.23, and 23.24).
21. Total and angle-differential cross sections of slow electrons elastic scattering by some noble gas, alkali, and alkaline-earth atoms, as well as by a number

of atoms with semi-closed subshells. The calculations were performed in the Hartree–Fock approximation and allowing for polarization of the target atom by the incident particle. For examples of calculations for Mn and Cr see in [16–18].

22. The yield of singly charged ions in the vicinity of the $4d$-subshells giant resonance for a number of atoms and ions, taking into account inelastic scattering of fast electrons in the Hartree–Fock approximation and SRPAE, taking into account the intra- and intershell correlations.
23. Corrections to the Hartree–Fock ionization potentials due to the rearrangement of atomic subshells for a number of atomic subshells, caused by formation of a vacancy, obtained in the frame of SRPAE [19].
24. Electron affinity of negative alkaline-earth ions [20].
25. Electron affinity of negative ions of He in an excited state [21].
26. Probability of Auger and radiative decay in the Hartree–Fock approximation and in RPAE for a number of transitions in atoms with both closed and semi-closed subshells.
27. Probability of double Auger and radiative decay of deep two-hole states in a number of atoms [21].
28. Excitation energies and probabilities of low-energy correlation satellites and hyper-satellite transitions in states with two arbitrary vacancies in noble gas atoms (see [21]).
29. The probability of low-energy photon emission that follows simultaneous decay of two vacancies in a number of atoms (see [21]).
30. Probabilities of triplet excitations of an atom by electrons in the second Born approximation and in the distorted wave Born approximation.
31. Double differential in angle and energy and differential only in energy cross section of fast electron bremsstrahlung on Ar, Xe, La in the Hartree–Fock and RPAE approximations accounting for the virtual excitation of the target [22].
32. The bremsstrahlung spectrum of intermediate energy electrons (as an example, see the results for Xe [22]).
33. Single- and double-photoionization cross sections of atoms with account for the dynamic polarization of the residual ion by the knocked-out photoelectron. Examples of Xe and Ba are given in [23].
34. Cross section of elastic and inelastic scattering of muons by atoms.
35. Cross sections of positron elastic and inelastic scattering by atoms with account for the formation of positronium in intermediate states [24].
36. Total and differential in angle of electron emission photoionization cross sections of endohedrals (examples in Figs. 23.25, 23.26, and 23.27).
37. GOS of endohedral. Total and differential in angle of secondary electrons emission cross sections for inelastic scattering of fast electrons by endohedrals (examples in Figs. 23.28 and 23.29).

All photoionization cross sections and generalized oscillator strengths are calculated in "length" and "velocity" forms, and the corresponding results closeness serve as an important criterion for assessing the numerical accuracy.

23.3 Results of the Second Group of Calculations

An approach to calculating many of these characteristics is given already in the relatively old review [25]. Most recent is the book [ACY].

1. The photoionization cross section in the vicinity of autoionization levels in atoms with closed and semi-closed subshells, as well as with one electron outside the closed shell or with one vacancy in it. It is possible, very approximately, to take into account the interaction with excited states of the "two particles—two holes" type;
2. Angular distribution anisotropy parameters of photoelectrons in the vicinity of the autoionization level for all the listed atoms. Note that with the development of computer technology, it becomes possible to find the cross sections near discrete excitations numerically;
3. Dipole dynamic polarizabilities and some other properties expressed in terms of photoabsorption cross sections and oscillator strengths. An example is the polarizability of the Ar atom (see Fig. 23.30);
4. The constant of long-range attraction $\sim 1/r^6$of van der Waals forces C_6, which is expressed via integral of photoionization cross sections' product with a known weight factor for any pair of interacting atoms;
5. "Drag currents" appearing in photoionization of gases of considered atoms in the HF and RPAE (an example for atomic Xe in comparison with Ar and He is shown in Fig. 23.31);
6. The parameter of interatomic interaction at intermediate distance in the Hartree–Fock and RPAE. It is expressed in terms of the quadrupole amplitudes of the interaction of atomic electrons with electromagnetic field.

There are many other characteristics of atoms, simple molecules, and endohedrals and processes with their participation, which can be obtained using those matrix elements that are calculated using the programs included in the ATOM-M complex.

23.4 Main Physical Results

With the ATOM-M program system, for the first time, it became possible to calculate various atomic properties with an accuracy that, in some cases, exceeds the accuracy of experimental data.

The most significant achievement is the demonstration of important, sometimes decisive role played by the residual interaction and the electron correlations determined by them in the quantitative and often qualitative description of physical processes in atoms and simple molecules.

Below are the main scientific results obtained using the ATOM-M system.

1. It was found that the photoionization of the outer and intermediate subshells of almost all atoms is of collective nature, so that at least all electrons of the ionized subshell take part in this process, as well as in the process of inelastic scattering of fast electrons by an atom (see, for example, Fig. 23.4).
2. It is shown that the giant resonance in the photoabsorption cross section is a manifestation of the coherent reaction of all electrons of the intermediate multielectron subshell (example—$4d^{10}$-subshell of Xe, Fig. 23.5).
3. Very strong intershell correlations have been found, which lead to the collectivization effect of few-electron subshells. This effect manifests itself in a qualitative change in the dependence of the photoionization cross section and generalized oscillator strengths upon the energy obtained by the atom, in comparison with the one-electron approximation. As an example, we present the strong interaction of three subshells in RPAE, which completely changes the photoionization cross section of the 5 s-subshell in Xe and Ba as compared to the Hartree–Fock approximation, which is illustrated in Figs. 23.6 and 23.32.
4. Very strong manifestations of intersubshell and intertransition correlations in the angular distribution were found, where they significantly change the anisotropy parameters and the degree of photoelectron polarization (see, e.g., Fig. 23.33).
5. It is shown that a photoelectron removed from the intermediate or even inner subshells of atoms excites or ionizes the outer electrons of the residual ion with a high probability, which is illustrated in Fig. 23.15.
6. It is demonstrated that virtual excitation (or polarization) of the core leads to a rapid change in the photodetachment cross section at the threshold of the negative ion due to the polarization of the residual atom by the removed photoelectron.
7. It is shown that the broad maximum in the cross section for the formation of a single charged ion in photoabsorption or inelastic scattering of fast electrons near the $4d$-subshell threshold of a number of atoms has a collective nature. Namely, all electrons of the subshells outside $4d$, and this subshell itself, are responsible for its formation (see, e.g., Fig. 23.34).
8. It is shown that intersubshell correlations significantly affect the optical and generalized oscillator strengths in a wide range of linear momentum transfers.
9. It is shown that the broad maximum in the photoabsorption cross section of atoms with semi-closed subshells is an autoionization resonance arising from the decay of a discrete transition of an atomic electron to free levels of a semi-closed subshell (an example is Mn atom in Fig. 23.35). This maximum also affects other partial cross sections and all angular distributions of a given atom (see example in Fig. 23.36).
10. It was found that the photoionization cross section of a highly excited electron depends sharply on the photon energy and changes significantly under the influence of other atomic electrons (example in Fig. 23.37).

11. It is shown that the interaction between different transitions of the continuous spectrum in some negative ions, for example, Si^- leads to the formation of autoionization resonances due to interaction of two continuous spectra.
12. The effect of suppression of radiative and Auger decays was discovered: a sharp decrease in the probability of decay under the influence of the interaction between electrons of a subshell with a vacancy, as well as those adjacent to it. For example, in the case of decay 4 *s*-vacancy in Kr, correlations decrease the radiative width by almost six times—from 1.034×10^{-5} eV in HF to 0.16×10^{-5} eV in RPAE.
13. It has been shown, mainly by the example of Ar and Xe, that correlation corrections due to complex excitations "two electrons—two holes", i.e., those that are not included in the framework of RPAE, play an important role in determining the shape and position of autoionization resonances in the outer shells of noble gas atoms.
14. It is shown that the photoabsorption cross section of noble gas atoms outer *s*-subshells due to mixing of the *s*-vacancy with more complex excitations of *ns* and outer *np*-subshells decreases by two to three times as compared to the results of calculations within the HF and RPAE.
15. The macroscopic "drag" currents of photoelectrons and free electrons in a weakly ionized plasma under the influence of a flux of electromagnetic radiation are calculated. The current arises when, in addition to dipole transitions, also quadrupole electromagnetic transitions are included. Accounting for electron correlations significantly changes the value and can even change the direction of the current. (see, e.g., Fig. 23.31). The current is especially strong near autoionization resonances.
16. It is shown that the bremsstrahlung of photons with an energy comparable to the ionization threshold of a target atom is produced not by an incident electron, but mainly by a virtually excited target atom (see, e.g., Fig. 23.31).
17. The probability of radiative and Auger decay is calculated for a state with two vacancies, where all the excitation energy is carried away by one photon or electron.
18. A technique has been developed for calculating the scattering cross section of low-energy $\mu^{\pm}$ and $\pi^{\pm}$-mesons by atoms.
19. It is shown that elastic scattering of slow electrons by atoms can be described in good agreement with experimental data in the absence of any adjustable parameters only when one accounts for the dynamic polarization of the target atom during the collision (see, e.g., Fig. 23.39).
20. Using the ATOM-M complex, the electron affinity of alkaline-earth atoms was calculated and the Ramsauer minima were found in elastic scattering cross section of these atoms at very low energies of incident electrons.
21. It is shown that the ionization potential of atoms or molecules change significantly when taking into account the correlation corrections, as compared to the corresponding HF values.

22. It is shown that account of the fullerene shell potential leads to additional oscillations in the total photoionization cross sections and angular distributions of endohedrals.
23. It is shown that the fullerene potential actually destroys the 4*d*-giant resonance (see, e.g., Fig. 23.25).
24. It was found that the combined action of the fullerene shell static potential and its dipole polarization leads to appearance of the giant endohedral resonance (see, e.g., Fig. 23.26) [1]. It is shown that at the same time autoionization giant resonances are preserved in endohedrals almost unchanged [2].
25. Interference effects in the photoionization of endohedral subvalent shells that appear under the joint action of atomic and fullerene shells have been discovered [26].
26. It is shown that accounting for the fullerene static potential action and its dynamic polarization radically affects the probabilities of radiative and Auger decay of an endohedral (see [27] and references therein).
27. Strong interference effects were found in endohedrals with a two-layer fullerene shell (see, e.g., Fig. 23.40, 4p-subshell of $Kr@C_{60}@C_{240}$) [3].
28. It is shown that the elastic scattering phase of an electron on an endohedral in a given partial wave can differ significantly from the analogous phase of an empty endohedral and is close to the sum of the fullerene and inner atom phases (see, e.g., e + $Ar@C_{60}$, p-phase, Fig. 23.49) [28].
29. It is demonstrated that the fullerene shell can essentially modify the polarization potential of the inner atom of the endohedral leading to violation of scattering phase additivity (see, e.g., e+ $Ar@C_{60}$, s-phase, Fig. 23.50) [29].
30. Strong influence of target polarization in elastic scattering of electrons by endohedrals is demonstrated [30].

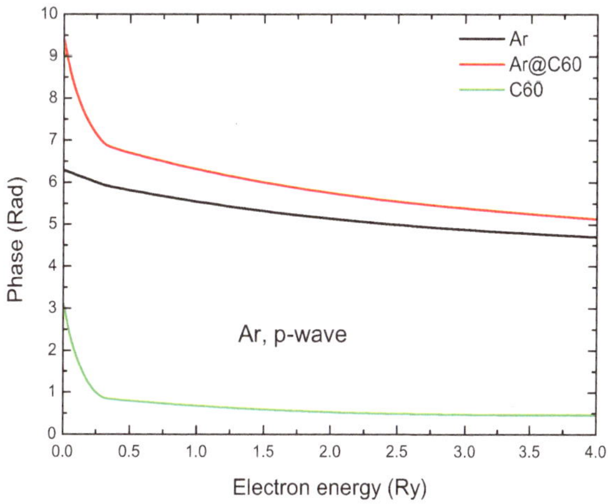

Fig. 23.49 Elastic scattering *p*—phase of electrons upon atom Ar, fullerene C_{60}, and endohedral $Ar@C_{60}$

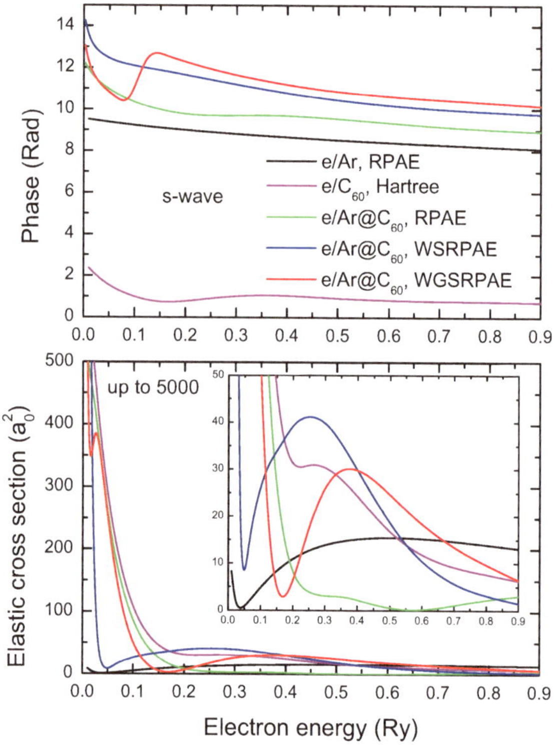

Fig. 23.50 Phases and cross sections of the s-wave contribution in electron collisions with Ar@C_{60}

31. It is demonstrated that the inner atom impressively modifies the positron-endohedral elastic scattering cross section, as compared to that for an empty fullerene (see, e.g., $e^+ + He$, s-phase, Fig. 23.51) [31].
32. The partial photoionization cross section essentially modifies the partial sum role for an atom or endohedral due to intersubshell interaction (see, e.g., Table. 23.1, Xe) [32].
33. It appeared that fullerenes' polarization prominently increases the endohedral photoionization cross section mainly close to the thresholds (see, e.g., photoionization of Ar@C_{60}, Fig. 23.52) [33].
34. The investigation of H_2@C_{60} in comparison with He@C_{60} photoionization demonstrates that confinement resonances are much stronger in molecular than in atomic case (see, e.g., Fig. 23.53) [34].

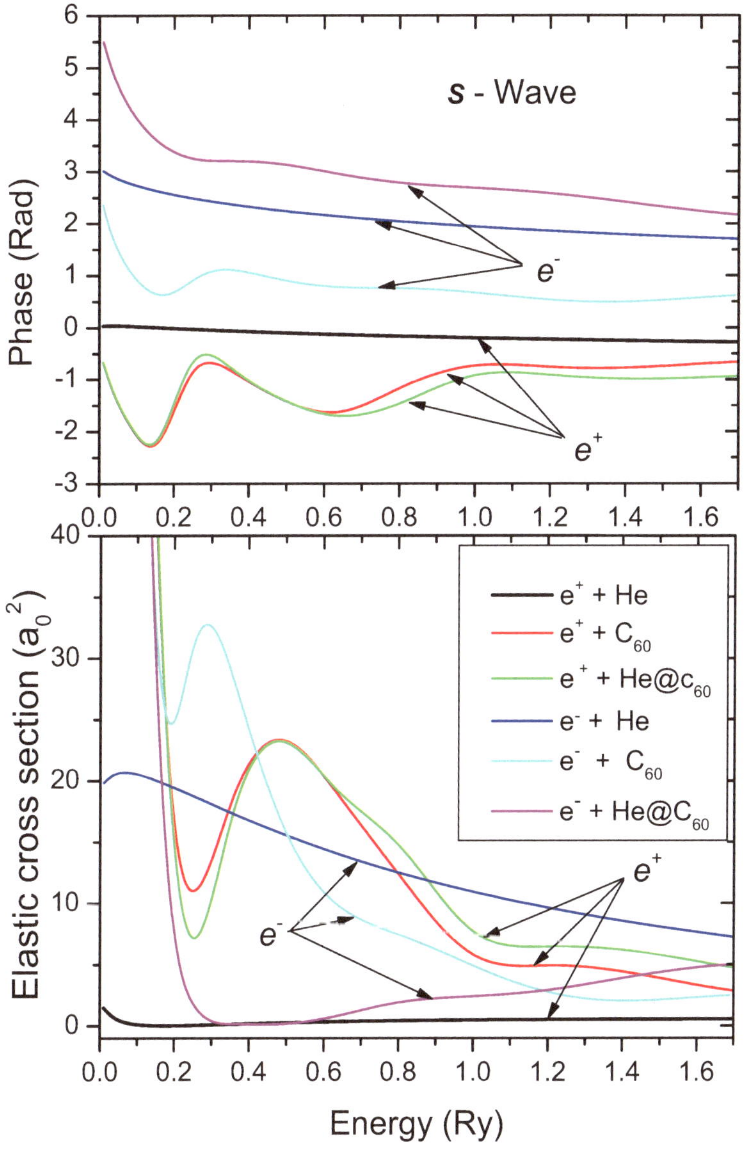

Fig. 23.51 Role of e^+e^- interaction upon phases and cross sections of e^+ scattering on He, C_{60}, and He@C_{60}, **a** *s*-phase with and without I_{Ps} shift, **b** partial *s* cross sections of e^+ with and without I_{Ps} shift

Table 23.1 Partial and total sums $S^{L}_{i,HF}$, $S^{\nabla}_{i,HF}$, $S_{i,RPAE}$, $\Delta_i \equiv S_{i,RPAE} - N_i$, and $\sum_{\leq i} S_{i,RPAE}$ of Xe atom

Xe, $N = Z = \mathbf{54}$	Subshell i	Ni	$S^{L}_{i,HF}$	$S^{\nabla}_{i,HF}$	$S_{i,RPAE}$	Δ_i	$\sum_{\leq i} S_{i,RPAE}$
1	1 s	2	0.50	0.50	0.50	−1.50	0.50
2	2 s	2	0.98	0.97	0.93	−1.07	1.43
3	2p	6	3.59	3.45	3.57	−2.43	5.00
4	3 s	2	1.05	1.01	0.92	−1.08	5.92
5	3p	6	4.34	4.15	4.17	−1.83	10.09
6	3d	10	13.30	11.30	12.31	+2.31	22.40
7	4 s	2	0.75	0.66	0.68	−1.32	23.08
8	4p	6	2.35	2.07	2.28	−3.72	25.36
79	4d	10	18.56	11.93	14.82	+4.82	40.18
10	5 s	2	0.33	0.26	0.53	−1.47	40.71
11	5p	6	12.65	6.18	8.73	+2.73	$S_{\mathrm{RPAE}} = \mathbf{49.44}$

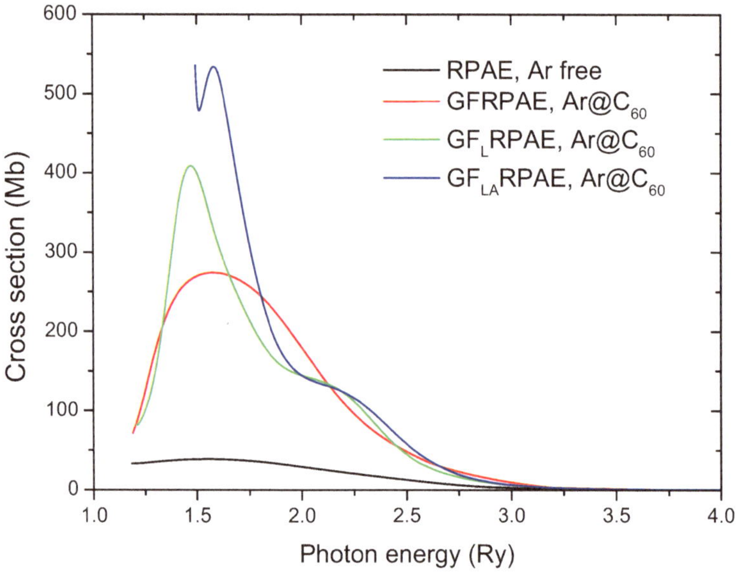

Fig. 23.52 RPAE photoionization cross section of $3p^6$ and $3s^2$-subshells of Ar and Ar@C_{60}. The lower index L denotes the Lorentz shape of the fullerene potential, index A denotes that the atomic polarization potential is also taken into account

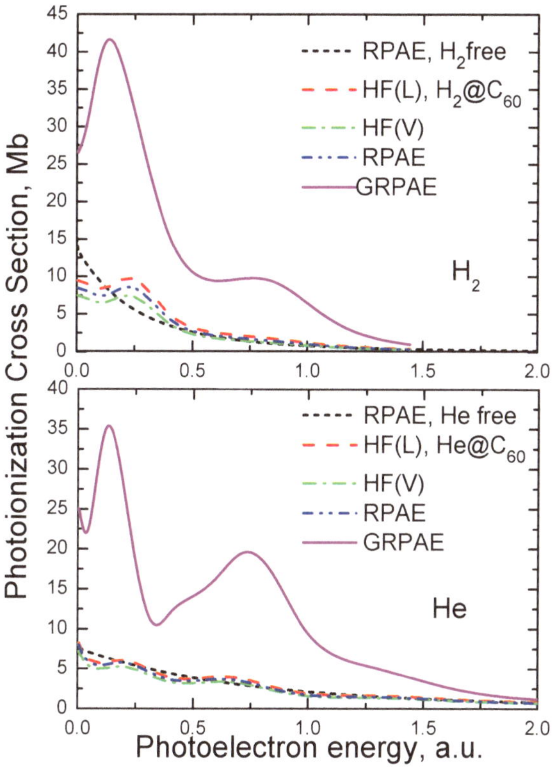

Fig. 23.53 Photoionization cross section of H_2@C_{60} and He@C_{60} in RPAE with account of fullerenes polarization in GRPAE

References

1. M.Ya. Amusia, A.S. Baltenkov, L.V. Chernysheva, JETP **134**, 2 (2008)
2. M.Ya. Amusia, A.S. Baltenkov, L.V. Chernysheva, JETP Lett. **89**, 6 (2009)
3. M.Ya. Amusia, E.Z. Liverts, L.V. Chernysheva, Phys. Rev. A **80**, 032503 (2009)
4. M.Ya. Amusia, Fullerene, nanotubes, and carbon nanostructures. **18**, 1 (2010)
5. M.Ya. Amusia, L.V. Chernysheva, JETP Lett. **95**, 2 (2012)
6. M.Ya. Amusia, E.Z. Liverts, L.V. Chernysheva, Phys. Rev. A**85,** 042722 (2012)
7. M.Ya. Amusia, E.Z. Liverts, L.V. Chernysheva, Int. J. Quantum Chem. **112**, 18 (2012)
8. M.Ya. Amusia, L.V. Chernysheva, V.G. Yarzhemsky, JETP Lett. **97** (2013)
9. V.K. Dolmatov, M.Ya. Amusia, L.V. Chernysheva, Phys. Rev. A**90**, 032717 (2014).
10. M.Ya. Amusia, L.V. Chernysheva, E.G. Drukarev, Nuclear instruments and methods B, proceedings **368** (2016)
11. S.K. Semenov, N.A. Cherepkov, Chem. Phys. Lett. **291** (1998)
12. N.A. Cherepkov, S.K. Semenov, Y. Hikosaka, K. Ito, S. Motoki, A. Yagishita, Phys. Rev. Lett. **84** (2000)
13. S.K. Semenov, N.A. Cherepkov, G. Fecher, G. Schönhense, Phys. Rev. A**61**, 032704 (2000)
14. S.K. Semenov, K. Kreidi, T. Jahnke, Th. Weber, T. Havermeier, R.E. Grisenti, X. Liu, Y. Morisita, L.P.H. Schmidt, M.S. Schöffler, M. Odenweller, N. Neumann, L. Foucar, J. Titze, B. Ulrich, F. Sturm, H. K. Kim, K. Ueda, A. Czasch, O. Jagutzki, N.A. Cherepkov, R. Dörner, Phys. Rev A**85**, 043421 (2012)
15. M.Ya. Amusia, L.V. Chernysheva, E.G. Drukarev, Eur. Phys. J. D**69**, 11, Art. 245 (2015)

16. V.K. Dolmatov, M.Ya. Amusia, L.V. Chernysheva, Phys. Rev. A**88**, 042706 (2013)
17. V.K. Dolmatov, M.Ya. Amusia, L.V. Chernysheva, Phys. Rev. A**90**, 032717 (2014)
18. M.Ya. Amusia, V.K. Ivanov, Sov. Phys. Usp. **30**, 449 (1987)
19. V.K. Ivanov, Many-body effects in negative ion photodetachment (Review). J. Phys. B: At. Mol. Opt. Phys. **32**, R67 (1999)
20. L.V. Chernysheva, G.F. Gribakin, V.K. Ivanov, M.Y. Kuchiev, J. Phys. B: At. Mol. Opt. Phys. **21**, L419 (1988)
21. M.Ya. Amusia, I.S. Lee, Correlative Decay of Highly Excited Atomic States, Atomic Physics 13 ed by H. Walters, T.W. Hansch, B. Neizerk, AIP Conference Proceedings 275 (American Institute of Physics, New York, 1993), p. 375
22. M.Ya. Amusia, *Bremsstrahlung* (Energoatomizdat, Moscow, 1990). ((In Russian))
23. M.Ya. Amusia, L.V. Chernysheva, G.F. Gribakin, K.L. Tsemekhman, J. Phys. B: At. Mol. Opt. Phys. **23**, 393 (1990)
24. M.Ya. Amusia, N.A. Cherepkov, L.V. Chernysheva, S.G. Shapiro, J. Phys. B: At. Mol. Phys. **9**(17) L531 (1976)
25. M.Ya. Amusia, N.A. Cherepkov, *Many-Electron Correlations in the Scattering Processes*, Case Studies in Atomic Physics, vol. 5 (North-Holland Publishing Company, 1975), p. 2
26. M.Ya. Amusia, A.S. Baltenkov, L.V. Chernysheva, J. Phys. B: At. Mol. Opt. Phys. **41,** 165201 (2008)
27. M.Ya. Amusia, J. Electron Spectrosc. Relat. Phenom. **161**, 112 (2007)
28. M.Ya. Amusia, L.V. Chernysheva, JETP Lett. **101**, 556 (2015)
29. M.Ya. Amusia, L.V. Chernysheva, JETP Lett. **103**(4), 260–264 (2016)
30. V.K. Dolmatov, M.Ya. Amusia, L.V. Chernysheva, Phys. Rev. A **95**, 012709 (2017)
31. M.Ya. Amusia, L.V. Chernysheva, JETP Lett. **106**(1), 1–6 (2017)
32. M.Ya. Amusia, L.V. Chernysheva, JETP Lett. **108**(7), 435 (2018)
33. M.Ya. Amusia, L.V. Chernysheva, JETP Lett. **109**(6), 345 (2019)
34. M.Ya. Amusia, L.V. Chernysheva, S.K. Semenov, JETP Lett. **110**(2), 102 (2019)

Appendix

Here, one can find the link to the texts of the mathematical programs included in the ATOM-M system that are stored under the name “ATOM-M system programs for calculation of atomic and molecular structure and processes and examples for calculating various physical characteristics” (Database of Springer). Here, it is:.

There are included a number of file directories for use on VAX, VMS, UNIX, and IBM PCs or compatible with the mentioned personal computers. It includes programs of the ATOM-M system and examples for calculating various physical properties:

- Atoms using the operating systems WINDOWS, UNIX, VMS, MS DOS in the directories COM, DBASE, MSDOS, DB1, NDPFOR FOR;
- Molecules in the WINDOWS operating system.

ATOM catalog includes atomic programs and calculation examples in the following directories:

FOR: This directory contains the Fortran-77 programs described in the book. They can be compiled in WINDOWS OS (e.g., using Fortran PowerStation 4.0 and UNIX right away, since they have the extension <*. F >. For VMS-OS , it is necessary to change this extension to <.for> in all programs. The fr00, fr01, … files are for compiling and linking programs on UNIX OS. This can be done with the command:

make -f fr00.

make -f fr01 ……. etc.

COM: This directory contains the WINDOWS, UNIX, and VMS folders, since the corresponding command files are different for these operating systems. The WINDOWS subdirectory contains command files with extension <*. Bat> . The COM/VMS subdirectory contains both batch files for examples and files fr00.com, fr01.com ……. for compiling and linking programs in the VMS-OS. The COM/UNIX folder contains batch files with the extension <*. Sh> .

DBASE: This directory contains sample files with input (extension <*. d>) and output (extension <*. r>) for all programs.

M. Ya. Amusia and L. V. Chernysheva, *Computation of Atomic and Molecular Processes*, Springer Series on Atomic, Optical, and Plasma Physics 117,
https://doi.org/10.1007/978-3-030-85143-9

To use the above programs on a PC with WINDOWS OS, one has to create the following folders on one of the PC disk, for example, disk D: **MDOS** (the folder includes batch files with the extension <* bat>), **DB1** (the folder contains the input and output data for the selected example), **NDPFOR** (folder contains exe files of compiled programs). In addition, in the information about the OS, one should change the environment variables: the variables DB1 and NDPFOR should be substituted by D: \ DB1, D: \ NDPFOR, respectively.

MOLECULE catalog includes programs and calculation examples in the following folders:

For_s_M—folder with Fortran files (programs and procedures);

Exe_s_M—folder with exe-applications of molecular programs, which can be used if it is unnecessary to make changes in the programs;

Data_M—folder with examples of setting data for molecular calculations;

Results_M—folder with examples of calculations, including calculated wave functions.

Separately goes the reference to "List of publications" in which the ATOM-M system was used that includes about 700 items..

Conclusions

This book summarizes the original half-century long work of researchers in the field of atomic and molecular processes theory. Earlier, a book was published that outlines the basic theoretical approaches developed and used by us and our colleagues [AICC]. This was followed by two books [ACYr, ACY], in which numerous physical results obtained by us were presented.

Of course, no matter how extensive the range of results given in them is, it can and should be expanded by introducing new objects, processes, and their characteristics into consideration—unconsidered atoms and diatomic molecules as well more complex targets. This determined the need to create a book that would be devoted primarily to the mathematical programs used in calculations. Such an attempt is presented in this book.

The description of the programs is all the more important because, as we believe, they can not only be used in the form shown, but also become the basis for expanding the proposed system so that it could allow the study of new objects, like, e.g., an atom in external static fields and more complex than diatomic molecules. This is all the more important because computing capabilities develop very rapidly. And if yesterday the objects of the solutions were mainly one-dimensional equations, then similar methods tomorrow or the day after tomorrow will make it possible to numerically solve two- and even three-dimensional equations with the same computational accuracy that is achievable today in the case of one dimension.

An interesting recent example of a new direction in investigation of atomic or molecular processes is to develop their temporal picture that is characterized by time delay of electron emission in the photon absorption, electron scattering, and vacancy decay. Using the existing formulae that connect the time delays with derivatives in energy of the phases of the investigated amplitudes of physical process, the approach based on ATOM-M system permitted to calculate the required derivatives that demanded employing several already developed programs and procedures. This led to obtaining very fast calculation results [1] to compare with experimental data [2] and to predict a number of targets for experiments in future [3, 4].

The approach developed by us did not set the task of achieving the best possible agreement of the theory with the experimental results at a given level of theory. This, as a rule, requires very cumbersome calculations and is by no means universal, but rather an individual approach—for each process, its characteristic, atom, or molecule.

As to us, we aimed to develop a universal approach that would allow at the same time to dramatically improve the accuracy of the usually applicable initial approximation, which, before our work, was usually a one-electron model with a very crudely chosen local potential.

The results obtained by us are also needed by other theorists. Now, the difference between the results of calculations and reliable experiment is an evidence of a noticeable role played by complex many-electron excitations.

The ATOM-M system proved to be very effective in solving a broad class of problems in the study of atomic processes. To a certain extent, this also applies to processes involving diatomic molecules. This system also turned out to be useful in a number of problems in molecular and solid-state physics.

The development of the proposed approach will raise the level of research in the field of molecules consisting of a small number of atoms. Such programs will make it possible to carry out mass molecular calculations, as well as calculations of atoms placed in strong electric or magnetic fields. The results obtained will become the new theoretical basis, the starting point for comparing theory and experiment at a much higher than previously level. Thus, the region of applicability of the approach, which we have developed with obvious benefit to the theory of atom, will expand covering new areas of research in atomic and molecular physics.

References

1. A.S. Kheifets, Phys. Rev. A**87**, 063404 (2013)
2. M. Schultze, M. Fieß, N. Karpowicz, J. Gagnon, M. Korbman, M. Hofstetter, Science **328**, 1658 (2010)
3. M.Ya. Amusia, L.V. Chernysheva, JETP Lett. **112**(4), 233 (2020)
4. M.Ya. Amusia, L.V. Chernysheva, JETP Lett. **112**(10), 673 (2020)

Index

M. Ya. Amusia and L. V. Chernysheva, *Computation of Atomic and Molecular Processes*, Springer Series on Atomic, Optical, and Plasma Physics 117,
https://doi.org/10.1007/978-3-030-85143-9

Printed by Printforce, the Netherlands